THE TESTING AND INSPECTION OF ENGINEERING MATERIALS

McGRAW-HILL CIVIL ENGINEERING SERIES

Harmer E. Davis, *Consulting Editor*

Babbitt · Engineering in Public Health
Benjamin · Statically Indeterminate Structures
Chow · Open-channel Hydraulics
Davis, Troxell, and Wiskocil · The Testing and Inspection of Engineering Materials
Dunham · Advanced Reinforced Concrete
Dunham · Foundations of Structures
Dunham · The Theory and Practice of Reinforced Concrete
Dunham and Young · Contracts, Specifications, and Law for Engineers
Gaylord and Gaylord · Structural Design
Hallert · Photogrammetry
Hennes and Ekse · Fundamentals of Transportation Engineering
Krynine and Judd · Principles of Engineering Geology and Geotechnics
Leonards · Foundation Engineering
Linsley and Franzini · Elements of Hydraulic Engineering
Linsley, Kohler, and Paulhus · Applied Hydrology
Linsley, Kohler, and Paulhus · Hydrology for Engineers
Lueder · Aerial Photographic Interpretation
Matson, Smith, and Hurd · Traffic Engineering
Mead, Mead, and Akerman · Contracts, Specifications, and Engineering Relations
Norris, Hansen, Holley, Biggs, Namyet, and Minami · Structural Design for Dynamic Loads
Peurifoy · Construction Planning, Equipment, and Methods
Peurifoy · Estimating Construction Costs
Troxell and Davis · Composition and Properties of Concrete
Tschebotarioff · Soil Mechanics, Foundations, and Earth Structures
Urquhart, O'Rourke, and Winter · Design of Concrete Structures
Wang and Eckel · Elementary Theory of Structures

THE TESTING AND INSPECTION OF ENGINEERING MATERIALS

Harmer E. Davis

Professor of Civil Engineering
University of California

George Earl Troxell

Professor of Civil Engineering
University of California

Clement T. Wiskocil

Late Professor of Civil Engineering
University of California

Third Edition

McGraw-Hill Book Company

New York San Francisco Toronto London

The Testing and Inspection of Engineering Materials

Library of Congress Catalog Card Number 63-21885

15655

PREFACE

An essential factor in the development of new materials and commodities is properly organized experimentation. One of the necessary steps in the production of materials for engineering uses or in the manufacture of finished commodities for industrial purposes is control of the quality of the product; this control is applied through appropriate inspection processes. Both experimentation and inspection imply some method of obtaining pertinent physical facts.

The industrial and technical world is becoming increasingly test-minded. Generally, architects, builders, engineers, and industrial designers and managers are familiar with the idea of testing and are coming to rely more and more on tests as a basis for making many important decisions.

In view of the important place that the making of tests has attained in the technical world, it seems appropriate to devote some time to the study of testing as a subject by itself. It was a natural development that methods of testing should have been first considered in relation to particular materials or commodities. In the technical schools the testing laboratory has evolved largely as an adjunct to the study of properties of materials or of principles of theoretical mechanics. But it is possible through such study to develop a basis for the preparation of adequate, enforceable specifications and a background for intelligent inspection, as well as to develop an appreciation of the relation between properties of materials and the means to obtain measures of those properties. It is the aim of the authors to provide a general treatment of the principles and problems of testing with specific reference to the mechanical properties of engineering materials and to establish the basis for the inspection of these materials.

A course involving the testing of materials has an accepted place in every collegiate curriculum of engineering. Some or all of the following benefits are often claimed for such a course: training in methods of observation, knowledge of accepted methods of testing, appreciation of the significance of data derived from tests, familiarity with the properties of common materials, exemplification of principles studied in strength of materials, knowledge of standards and specifications, acquaintance with the technical periodical literature and other sources of recorded data on properties of materials, and practice in preparing reports giving the results of investigations. The authors have observed, however, that the ends claimed are not always fully attained and that students are not infrequently dissatisfied because of insufficient return for effort expended. Perhaps these shortcomings are due in part to the brief time actually allotted to the testing course, whether it is given as an adjunct to courses in strength of materials or materials of construction, or as a course by itself. On the other hand, it may be possible that part of the difficulty lies in the fact that too much is expected and that sufficient background information is not readily available. Accordingly, this text is divided into two sections, one devoted to building up the general concepts of testing, and the other to providing, in such form as to be applicable to most ordinary apparatus, methods of conducting common tests. It is suggested that instructors arrange or regroup the tests to suit their program.

The difficulty of obtaining access to the vast reference material, or lack of time available to consult it, has been a serious handicap to the student. The authors have attempted to gather together in condensed form the more important and pertinent information on testing that is contained in the now voluminous periodical literature and in the specifications published by a variety of organizations. A classified list of references is appended. A brief summary of properties of materials for use in estimating load schedules, for comparison with the results of student tests, etc., is given in other appendices.

It appears undesirable to dispense with the use of specific directions to students for the performance of tests. However, a manual of instructions set up for a particular laboratory is of limited value for general use because of differences in available testing equipment as well as differences in methods of conducting testing laboratory courses at various institutions. It is possible to reduce the length of detailed instructions greatly in a particular laboratory by the use of a text that covers the general principles and common methods of testing.

In the period since publication of the earlier editions there have been many changes in testing equipment, and much additional information has appeared in the technical literature concerning the testing and the properties of engineering materials. As much as possible of this new material has been included in this third edition.

Various new testing machines and other testing devices are described. The increasing use of nondestructive tests has led to the inclusion of additional material on a variety of ultrasonic tests and the use of penetrants for detecting defects. Experimental stress analysis by use of photoelasticity and photoelastic strainometers is included. Among several new subjects are modes of failure, plasticity and strain hardening, the Ramberg-Osgood stress-strain diagram, stress relaxation, tests of metals at low temperatures and short-time tests at high temperatures, surface roughness of metals, splitting tension tests of concrete, and the concrete test hammer. Two new test problems on concrete have been introduced in Part 2 to round out this phase of materials testing.

In recent years, a renewed interest in the physics of the solid state and advancement in knowledge of the microstructure of materials, particularly metals, are bringing about a more basic understanding of the behavior of materials. This new knowledge will provide better control of the properties of materials and will lead to more pointed interpretation of phenomena exhibited by materials tested on a macroscale. Thus, it is believed that testing for macroproperties will continue to play an important part in the development and selection of materials, and the control of their production, although particular procedures of present types of tests may be expected to undergo gradual modification in the years to come.

The authors are indebted to many organizations and individuals for their permission to make use of published material and their cooperation in supplying information on testing methods and on engineering materials and for photographs of materials-testing apparatus. It would be impossible to list all acknowledgments here, but throughout the text an attempt has been made to indicate the sources of all material used. Nevertheless special acknowledgments are due the following organizations for their contributions: American Society for Testing and Materials, American Society for Metals, American Concrete Institute, National Bureau of Standards, U.S. Navy, U.S. Ordnance Department, American Instrument Co., Baldwin-Lima-Hamilton Corp., Federal Products Corp., General Electric X-Ray Corp., Magnaflux Corp., Wilson Mechanical Instrument Co., Shore Instrument Co., Sperry Products, Inc., Tinius Olsen Testing Machine Co., Wiedemann Machine Co., and Riehle Testing

Machine Division of American Machine and Metals, Inc. The authors also desire to acknowledge the helpful association of the late Prof. Clement T. Wiskocil who collaborated in the preparation of the first edition, and the advice and helpful criticism of Profs. H. D. Eberhart and J. W. Kelly of the University of California who reviewed parts of the manuscript.

Harmer E. Davis
George Earl Troxell

CONTENTS

Problems 343

Part 1.

PRINCIPLES OF TESTING AND INSPECTION

Chapter 1.

THE NATURE OF THE PROBLEM

1.1. Introduction. The extensive use of experimental studies preliminary to the design and construction of new mechanical or structural elements and the use of testing procedures for control of established processes of manufacture and construction are significant and well-recognized features of our technical development. Practically all branches of engineering, especially those dealing with structures and machines, are intimately concerned with materials, the properties of which must be determined by tests. Successful mass production depends upon inspection and control of the quality of manufactured products, which implies a system of sampling and testing. The preparation of adequate specifications and the acceptance of material purchased under specifications involve an understanding of methods of testing and of inspection. Settlement of disputes regarding failures and substandard quality almost invariably calls for investigations involving physical tests. Engineering research and development function in large measure on an experimental basis and call for carefully planned, well-devised tests.

For the intelligent appraisal and use of the results of tests, it is important for engineers, even for those not engaged in actual testing work, to have a general understanding of the common methods of testing for the properties of materials, and of what constitutes a valid test. Further, in dealing with the specification and acceptance of materials, an understanding of the limitations imposed by methods of testing and inspection is important.

The following subjects are basic in the study of materials testing:

1. *Technique of Testing.* How do commonly used types of equipment operate? Is the apparatus in widespread use? What are common variants of ordinary apparatus? What limitations are imposed by the apparatus upon the accuracy that can be obtained? Lacking first-class equipment, how can one get rough but significant results from crude tests in the field? Has the theory of models any application in the given test?

2. *Physical and Mechanical Principles Involved in the Testing Apparatus and Procedure.* Are assumed conditions satisfied? What is most likely to go wrong or produce incorrect results? What apparent crudities can be overlooked? What refinements can be made to obtain greater accuracy?

3. *Theory of Measurements.* What is the precision of the results? Which of the measurements involved controls the precision of the final results? Are time and effort wasted securing needless precision in some of the measurements?

4. *Variability of Materials.* How many tests are necessary to give a significant average? What variation from the average is cause for rejection of individual values? What range in strength (or other property) may be expected from a given material as it is used under job conditions?

5. *Interpretation of Results.* What is the significance of the test results? Can the numerical results be applied directly to design and similar uses, or are they of value only for comparison with other results? How can the results of arbitrary tests be interpreted? Do the results of arbitrary tests have meaning if the conditions of test are outside the range of those for which correlations have been set up? Considering the methods of testing and the kind of material, what are the limitations of the test results, or how reliable are the test data? How are limitations of test method and variability of materials reflected in specifications? How should a satisfactory specification for a given material be written?

With advance in our technologic development have come notable improvements in the older type of materials, many discoveries of new materials, and a variety of new uses for all materials. These have greatly extended the scope of materials testing and have complicated its practice.

However, the fundamental principles involved in the conduct of valid and reliable tests are common to all testing, and it is the purpose of this book to treat these principles by discussion of the ordinary methods of testing of the common materials of construction.

1.2. Engineering Materials. The principal materials used in the building of structures and machines include metals and alloys, wood, portland-cement concrete, bituminous mixtures, clay products, masonry materials, and plastics. The principal function of constructional materials is to develop strength, rigidity, and durability adequate to the service for which they are intended. These requirements largely define the properties that the materials should possess and hence broadly determine the nature of the tests made on such materials. Full appreciation of the significance of tests of constructional materials therefore requires some background of knowledge of the general properties of these materials and of the subject of structural mechanics. Reference is made to the many available treatises on materials and mechanics for detailed review of such matters. For convenience in this study of testing, the principal properties of representative materials are summarized in Appendixes A to G. A list of references is given in Appendix H.

A distinguishing feature of auxiliary constructional materials, such as preservative coatings, and perhaps of the majority of specific-use materials is the requirement of some particular qualities or properties the determination of which is often made by special and highly arbitrary tests. The particular problems involved in specialized fields are not within the scope of this treatment; but it is worthy of emphasis that the basic considerations given in Art. 1.1 apply with as much force to specialized tests as to the common ones.

Although in a general sense fluids as well as solids are engineering materials whose properties must be found by test, and although the performance of machines and structures as well as the characteristics of materials are subjects of engineering testing, unless otherwise qualified, the term *materials testing* refers to tests of solid constructional materials or to assemblages of materials for constructional purposes.

1.3. Selection of Materials. Serviceability, in a broad sense, is the ultimate criterion in a choice of materials. One important object of materials testing is to aid in predicting or assuring the desired performance of materials under service conditions. However, in the selection of materials for the building of structures and machines, the problems of quality of material, of design, and of use are interwoven. It should be noted in passing that sound material and correct design can do no more than give assurance that a particular construction will be satisfactory within the limits of its intended use, although the material which can withstand the greatest abuse most certainly has an advantage over its competitors [164].*

In the selection of materials, the designer has two sources from which he can obtain information: (1) knowledge or records of performance of materials in actual service and (2) the results of tests made to supply data on performance. On the basis of such information a specification is prepared.

To transform a design into actual construction, it is necessary for the constructor (or manufacturer) to select, from a variety of available grades of material, the one that the designer had in mind and tried to specify. Tests are then required to "identify" the desired material.

Following is a summary of the considerations involved in the selection of materials as regards the problems of design and manufacture.

1. Kinds of materials available
2. Properties of various materials
3. Service requirements for materials
4. Relative economy of various materials and of various forms of a particular material
5. Methods of preparation or manufacture of various materials or products and the influence of processing on their properties
6. Methods of specification and relation thereof to uniformity and dependability of product secured
7. Methods of testing and inspection and significance thereof with respect to measures of desired properties

1.4. Properties of Materials. A partial classification of the properties of engineering materials is given in Table 1.1. In general, the determination of any or all of these properties may be the subject of engineering testing. However, the major work of the ordinary materials-testing laboratory has to do with mechanical properties. This work is often referred to as "mechanical testing." Because the major factor in the life and performance of structures and machines is applied force, strength is of utmost importance; a first requirement of any engineering material is adequate strength. In its most general sense the term *strength* may be taken to refer to the resistance to failure of an entire piece of a material, a small part of it, or even the surface. The criterion of failure may be either rupture or excessive deformation. From a historical point of view, the earliest tests were concerned with the strength of materials; therefore today the term *testing machine,* used without qualification, refers to a machine for applying known loads.

A complete knowledge of the behavior of a given material would involve study of all its properties under a very wide range of conditions,

* Numbers in brackets designate references listed by number in Appendix H.

Table **1.1. A classification of the properties of engineering materials**

Class	Property	Class	Property
Physical	Dimensions, shape Density or specific gravity Porosity Moisture content Macrostructure Microstructure	Mechanical	Strength: Tension, compression, shear, and flexure Static, impact, and endurance Stiffness Elasticity, plasticity Ductility, brittleness Hardness, wear resistance
Chemical	Oxide or compound composition Acidity or alkalinity Resistance to corrosion or weathering, etc.	Thermal	Specific heat Expansion Conductivity
		Electrical and magnetic	Conductivity Magnetic permeability Galvanic action
Physico-chemical	Water-absorptive or water-repellent action Shrinkage and swell due to moisture changes	Acoustical	Sound transmission Sound reflection
		Optical	Color Light transmission Light reflection

but the conduct of the exhaustive tests necessary to obtain complete information usually would not be necessary or economically feasible. The problem, then, is to secure data on such properties as have a bearing on the economic value and serviceability of a material, or a product made from the given material, for a given purpose. The relative efficiency of a material for a specific use depends on the extent to which pertinent properties are present. For some uses, a property may be highly desirable, whereas for other uses it may be undesirable or even dangerous.

1.5. Materials Testing. The testing of materials may be performed with one of three objects in view: (1) to supply routine information on the quality of a product—commercial or control testing; (2) to develop new or better information on known materials or to develop new materials—materials research and developmental work; or (3) to obtain accurate measures of fundamental properties or physical constants—scientific measurement. These objectives should be clearly discerned at the outset, since they generally affect the type of testing and measuring equipment to be used, the desired precision of the work, the character of the personnel to be employed, and the costs involved.

Commercial testing is concerned principally either with checking the acceptability of materials under purchase specifications or with the control of production or manufacture. Generally, the type of test has been specified, although as a guide to measuring quality it may be entirely arbitrary; standard procedures are used, and the object is simply to determine whether the properties of a material or of a part fall within required limits. A high degree of refinement is not required, although limits of accuracy are often specified.

Common purposes of materials research are (1) to obtain new understanding of known materials, (2) to discover the properties of new materials, and (3) to develop meaningful standards of quality or test procedures. In addition, there may be the specific objective of choosing a material for a particular use, of determining principles to improve design with a chosen material, or of studying the behavior of the part or structure after it has been made [103]. Although many investigations are more or less routine in nature, there are also many that call for a very wide variety of tests and measurements, require an appreciation of all phases of the general problem, and make extreme demands on the skill, ingenuity, and resourcefulness of the experimenter if success is to be attained.

The aim of what is here called *scientific testing* is the accumulation of an orderly and reliable fund of information on the fundamental and useful properties of materials, with an ultimate view to supplying data for accurate analyses of structural behavior and for efficient design. Work of this kind, above all, calls for care, patience, and precision. In a student laboratory, experiments may serve to give insight into principles developed in texts on mechanics of materials.

Distinction may be made between *experiment* and *test*, although usage as regards these ideas is often loose. Experimentation involves the idea that the outcome may be uncertain, that hitherto unknown results may be forthcoming. Testing involves the idea of a more or less established procedure and that the limits of the results are generally defined. Experimentation, especially on a planned or large scale, ordinarily involves many routine tests. Many of the large materials-testing laboratories serve the dual purpose of experimental research bodies and routine-control testing agencies. Although the purpose, the point of view, and the method of attack may differ widely as between research and routine testing, many of the detailed procedures may be exactly the same in both kinds of testing. For example, in research on heat-treatment of steel plates, the tension-test specimens and the test method would very likely be the same as those used the country over for acceptance tests of boiler plate.

For convenience, one may differentiate between *field tests* and *laboratory tests.* Because of difficult or hazardous working conditions, interference, time limitations, and variable weather conditions, tests conducted in the field usually lack the precision of similar tests conducted in the laboratory; however, performance of work in the laboratory does not necessarily ensure precision. Certain types of tests, as, for example, a sieve analysis of gravel, may be made just as accurately by an inspector on the job as by a technician in the laboratory. On the other hand, some tests cannot be made in the laboratory so that the question of field vs. laboratory does not pertain.

With respect to general method of attack and to interpretation of results, it is desirable to distinguish between

1. Tests on full-size structures, members, or parts
2. Tests on models of structures, members, or parts
3. Tests on specimens cut from finished parts
4. Tests on samples of raw or processed materials

In passing, it should be noted that model testing, interest in which has grown markedly in recent years, often demands satisfaction of a number of exacting requirements in order that valid results be obtained.

With respect to usability of material or a part after test, tests may be classified as *destructive* or *nondestructive*. Tests to determine ultimate strength naturally mean destruction of the sample. Since an entire lot cannot be thus tested, there enter the problems of obtaining a reliable indication of the strength of the lot by use of a sufficient number of samples as well as of keeping the expense of sample material within reasonable bounds. For finished products it is desirable to use nondestructive tests if possible. Some hardness tests are of this type, e.g., the scleroscope tests may be used to determine the surface hardness of ground-surfaced heat-treated steel shafting. Proof tests, applied to fabricated parts or structural elements, are of the nondestructive type; e.g., a proof test of a crane hook involves the application of a load somewhat in excess of the working load but less than any damaging load, in order to give assurance that no harmful defects, which might cause failure in service, are present. Nondestructive tests are of particular interest to the inspector on the job. Radiographic and magnetic methods of nondestructive examination of metals are discussed in Chap. 10.

1.6. Inspection of Materials. Although their functions overlap, it is desirable to distinguish between *testing*, as such, and *inspection*. Specifically, testing refers to the physical performance of operations (tests) to determine quantitative measures of certain properties. Inspection has to do with the observation of the processes and products of manufacture or construction for the purpose of ensuring the presence of desired qualities. In many instances inspection may be entirely qualitative and involves only visual observation of correctness of operations or dimensions, examination for surface defects, or possibly the indication of presence or absence of undesirable conditions such as excessive moisture or temperature. On the other hand, inspection may involve the performance of complicated tests to determine whether specification requirements are satisfied. Inspection aims at the *control* of quality through the application of established criteria and involves the idea of rejection of substandard material. In testing, the aim is to *determine* quality, i.e., to discover facts regardless of the implications of the results.

In some organizations, the inspection forces may be thought of as an administrative group, making only simple examinations and sending selected samples to the testing department. In other organizations, the testing engineer is also the chief of inspectors as well as being charged with routine testing and with research and development work.

Inspection involves human relationships as well as technical duties. Not all men can be good inspectors. Some of the principles of inspection are discussed in Chap. 12.

1.7. Significance of Tests. Our concepts of the properties of materials are usually idealized and oversimplified. Actually, we do not *determine properties*, in the sense that some unchanging values are obtained which once and for all describe the behavior of some material. Rather, we obtain only *measures* or *indications* or *manifestations* of properties found from samples of materials tested under certain sets of circumstances. For example, in a tension test of steel the percentage elongation in a given gage length is used as a measure of ductility. At the same time, to the extent possible, we attempt to avoid conditions of test which arbitrarily restrict the general significance of the quantitative measure obtained from a test; for example, we believe that the results of the tension test give a reliable indication of the macroscopic property of tensile strength in uniaxial tension. The measures we obtain depend upon the conditions of test, which include the way the sample is taken and prepared, as well as upon the particular procedures involved in making the test. Thus one implication of "significance of tests" has to do with the *reliability* of tests in yielding measures of the properties they are supposed to determine. For example, the use of the proportional limit as a practical measure of the elastic strength of a material has been questioned because the test results are affected by a number of factors which cannot be properly controlled in ordinary testing and which make the determination of the true magnitude of the proportional limit uncertain.

The real significance of any test lies in the extent to which it enables us to *predict the performance* of a material in service. A test may have significance in one of two ways: (1) it may measure adequately a property that is sufficiently basic and representative that the test results can be used directly in design, or (2) the test, even though highly arbitrary, serves to identify materials that have been proved by experience to give satisfactory performance. For example, in connection with the design of a tension eyebar for a bridge structure, a tension test on a properly selected sample of the steel will give a value that, when modified by a suitable factor of safety, can be taken as the allowable working stress. On the other hand, for example, the Charpy impact test of metals is one that has significance in relation to use of a material only by correlation of test results with the performance of such material in service. This test gives values that are stated in terms of foot-pounds of energy absorbed during the rupture of small standardized, arbitrarily shaped specimens of the metal. The results cannot be utilized directly in any design as can strength values. Yet it has been found that, for certain types of service, failures may be expected if the Charpy values fall much below a given value. This test, then, has significance in that it aids in the elimination of steels unsuitable for a particular use. The test that can be made to give a direct indication of expected performance depends to a large extent upon the state of development of the arts of testing and stress analysis.

One striking fact to be noted in a study of detailed test data, and in the results of investigations in general, is the variation in the quantitative measures of given properties. This may be due in part to lack of absolute precision in the operations of testing, but also it may be due to actual variation in a given property between samples. Our materials are not homogeneous; within limits, their composition may be governed entirely by chance, so that a description of their behavior may rest to a great extent upon a statistical basis. For an intelligent interpretation of results, both with respect to the meaning of variation between samples or parts of a sample and with respect to the relation of samples to an entire lot or the full-sized piece of material, due regard should be given to the requirements of sampling theory and to the statistical nature of the data (see Chap. 11). Further, in the interest of efficient testing and reliable results, the test should be "designed" so that the precision of the various measurements or operations involved is consistent throughout.

1.8. Design of Tests. In the design of a test, the following are suggested as fundamental questions to be considered:

What is the nature of the answer sought?
What test can be made to provide an answer?
How will the test results be related to performance?
What are the limitations of the type of test selected?
How should the precision of the work be adjusted to be in accord with the limitations so as to achieve economy of effort and consistent reliability of results?
What type of specimen is best suited for the test?
How many samples are necessary to obtain representative results?

The ideal test should be meaningful, reliable, reproducible, of known precision, and economical. The selection of a procedure should be controlled by the significance of the test, guided by economy of effort, and influenced by a sense of proportion.

The following observations bearing upon the design of tests are abstracted from an early U.S. Bureau of Standards handbook on materials testing [118]: An adequate measure of a given property is possible when (1) the property can be defined with sufficient exactness, (2) the material is of known composition or purity, (3) the attending conditions are standard or are known, (4) the experimental methods are theoretically correct, (5) the observations and their reductions are made with due care, and (6) the order of accuracy of the results is known. This ideal is rarely if ever reached, but as it is striven for the results pass from the qualitative to the quantitative stage and are called *constants* because redeterminations will not yield sensibly different results. Approximate results are improved upon steadily as more precise instruments and methods are devised. The degree of accuracy to be sought becomes a very practical matter in a testing laboratory. The time and labor involved in tests may well increase out of proportion as the limits of attainable accuracy are approached. For the determination of physical constants or fundamental properties of materials the degree of accuracy sought may be the maximum possible. In general the degree of accuracy striven for should be that which is strictly good enough for the purpose in hand.

1.9. Specification of Materials. A specification is the attempt on the part of the consumer to tell the producer what he wants [182]. Obviously, the skill and accuracy with which a thing *can* be specified depends on the state of knowledge concerning it and on the precision with which its qualities can be determined. As the art and science of testing are advanced, so is the basis for preparation of adequate specifications improved. However, the effectiveness with which a thing *is* specified depends also on how well the specification is written and how enforceable the provisions are. This problem, although it often concerns the testing engineer in particular, may confront any engineer.

Time was when it was customary to specify merely a given brand, "or equal"; past performance and the integrity of the producer were the only guaranty of potential quality. Early specifications were often necessarily crude because the consumer knew little about the material he tried to specify; many present-day specifications are just as crude, and for the same reason. With increased complexity in our industrial system, more adequate specifications have become necessary, and, with advance in our scientific knowledge of materials, more adequate specifications have become possible.

A specification is intended to be a statement of a standard of quality. The ideal specification would uniquely define the qualities of a material necessary to serve most efficiently for a given use, and it can be approached if truly significant tests can be made to determine the presence of the required qualities. A specification often falls short of the ideal for a number of reasons, some of which are the following: (1) it may be so loose as to admit material of inferior quality, (2) it may be overly restrictive and so exclude an equally or more efficient material, (3) it may be based on inadequate or improper criteria with respect to type of service required, and (4) it may make no provision or inadequate provision for proper enforcement. Defects such as these lead not only to the procurement of unsatisfactory materials but often to disproportionate costs and endless disputes. It is important to note also that a specification may admittedly and necessarily be imperfect because it would be impracticable to produce the ideal material. All things considered, it may be just as inefficient to require too high a quality as to accept too low a quality. Practically, specifications are drawn up not for an ideal material, but for a material that it is possible to obtain at reasonable cost under existing conditions of manufacture.

Several considerations fix the limits within which a specified property may be allowed to vary. The maximum and minimum to be set may be based upon experiment but should recognize the limitations of the manufacturing process. These limits correspond to the size limits allowed in making machine parts, where such variation in size of each part is allowed as leads to economy in manufacturing the parts without unduly impairing the efficiency of the assembled machine. In fixing these limits of tolerance for material, care must be exercised to avoid too narrow ranges on the one hand and too wide variations or poor quality on the other. These limits often involve safety and generally involve durability and efficiency [118].

Specifications for materials of construction may define the requirements for acceptability of the material in one or all of the following ways:

1. By specifying the method of manufacture
2. By specifying form, dimensions, and finish
3. By specifying desired chemical, physical, or mechanical properties

Another type of requirement, although rarely used in the materials field, is that a product shall not exhibit stated defects within a certain period after purchase. Performance specifications are commonly used for machines. Often included in materials specifications are requirements with regard to methods of sampling, testing, and inspection.

1.10. Standard Specifications. A notable development of the last several decades, particularly as regards materials, has been the preparation and use of "standard" specifications. A standard specification for a material is usually the result of agreement between those concerned in a particular field and involves acceptance for use by participating agencies.

It does not necessarily imply, however, the degree of permanence usually accorded to dimensional standards, because technical advance in a given field usually calls for periodic revision of the requirements. Some of the various types of standardizing agencies are independent companies, trade associations, technical and professional societies, and bureaus and departments of municipal, state, and national governments. The breadth of acceptance depends to an extent on the scope of influence and of authority of the standardizing agency. Under the standardizing procedure followed by important agencies in this country, a period of negotiation, formulation, and trial usually precedes the use of a specification as a standard, so that it has assurance of being workable [182–184].

A standard specification implies standard methods of testing and sometimes also standard definitions. In some instances, the methods of testing are incorporated within a materials specification. On the other hand, some standardizing agencies set up standard methods of test separately from the materials specifications and make mandatory reference to the test methods.

Properly devised and enforceable standard specifications can be of immense value to industry. Some of the advantages that may be cited for standard specifications for materials are

1. They usually represent the combined knowledge of the producer and consumer and reduce the possibility of misunderstanding to a minimum.
2. They give the manufacturer a standard of production, tend to result in a more uniform product and to reduce the number of required varieties of stock, lowering the attendant waste and therefore lowering the cost.
3. They lower unit costs by making possible the mass production of standardized commodities.
4. They permit the consumer to use a specification that has been tried and is enforceable.
5. They permit the designer to select a material which there is reasonable assurance of getting.
6. They simplify the preparation of special-use specifications because published standard specifications can be incorporated by reference.
7. They aid the purchasing agent in securing truly competitive bids and in comparing bids.
8. They set standards of testing procedure in commercial testing and hence permit comparison of test results obtained from different laboratories.

In the initial development of a standard test procedure, considerable research is often conducted by cooperating organizations to develop a procedure which will yield reproducible and meaningful test results.

The disadvantage of standard specifications is that they tend to "freeze" practices that may be only in the developmental stage and thus hinder progress where most needed. For this reason, standard specifications should be under the jurisdiction of a well-informed and thoroughly open-minded agency.

Specifications for both materials and methods of testing should be subject to continuous review to determine their suitability under changing conditions. Also, various codes based on these standards should be reviewed frequently.

1.11. Standardizing Agencies. Because standardization has such an important influence upon ordinary methods of testing, it is desirable for the engineer to have some familiarity with the nature and the publications of the agencies* that have promulgated some of the widely used materials specifications and methods of testing.

The standards promulgated by the American Society for Testing and Materials are of particular interest and importance to those concerned with materials testing and inspection [1]. This national technical society, formed in 1902, had a membership in 1962 of about 11,000, which may be roughly classified into three groups: consumers, producers, and a general-interest group comprising engineers, scientists, educators, testing experts, etc.

The ASTM performs the dual function of (1) standardization of materials specifications and methods of testing, which is carried out by standing committees, each of which has under its jurisdiction engineering materials in a definitely prescribed field or some specific phase of materials testing; and (2) improvement of engineering materials, which is effected through investigations and research by committees and individual members, the results of which are made public through the society publications. The specifications are published in a separate series of volumes called the *ASTM Standards*, which are issued in new edition every three years.

The standardization work comprises in general (1) the development of methods of test for materials, (2) the setting up of standard definitions, (3) the formulation of materials specifications, and (4) the formulation of recommended practices having bearing on various processes in the utilization of materials. Committees concerned with the development of specifications first study the materials in their respective fields and foster the necessary research upon which the standardization work must be based. On committees dealing with materials having a commercial bearing, the policy is generally to maintain a balance between representatives of producer and consumer interests.

After completion of studies involving methods of testing, nomen-

* For a list of such agencies and addresses of their headquarters, see the first section of Appendix H.

clature, and requirements, a proposed standard is evolved and submitted at a meeting of the committee having jurisdiction over materials in the particular field concerned. If approval is secured at this meeting and also later by letter ballot of the entire committee, a proposed standard is published for information in a committee report at the next annual meeting of the society. If accepted by the membership of the society at this meeting, the specification or test method is published as tentative for at least one year to elicit criticism. After due consideration of comments received, the committee may recommend that the tentative specification be adopted as standard. Each standard, before adoption, must receive proper approval in a letter ballot submitted to the entire membership of the society. Revisions of standards may be considered at any time by the standing committee concerned. Revisions must be published as tentative before they can be incorporated in a standard. Standards may be withdrawn at any time by appropriate action.

In the development of specifications,

> full account must be taken of the influence of manufacturing processes, the nature of the stresses and other conditions to which the materials will be subjected in service, and the particular properties of the material that enable it to give satisfactory service. Painstaking investigation and study of experience over years of service are often required before an adequate specification can be prepared. The committee must come to an agreement upon the properties of the material to be specified, the methods of test, such details of manufacture as may be necessary, methods of inspection and of marking, etc. In all these things it seeks to follow the best commercial practice that has been developed in supplying the particular material or commodity to the trade. Specifications for materials upon whose strength and reliability the safety of human life may depend must be especially carefully drawn and provided with adequate safeguards in testing and inspection. At times, a compromise between the somewhat extreme views that may be held by the producer and consumer is necessary in reaching at least a tentative agreement on certain details, although the more clearly the problems involved are understood and the more complete are the technical data that can be presented on the subject, the more easily can a logical rational agreement be reached [183].

For complete details, the *ASTM Yearbook* and other publications of the society, from which many of the above statements have been drawn, may be consulted.

The American Standards Association was organized in 1918 by the ASCE, ASME, AIEE, AIMME, and ASTM to provide a means for industry, technical organizations, and governmental departments to work together in developing national industrial standards acceptable to all groups and to provide a means whereby standardizing organizations might coordinate their work and prevent duplication of effort [2]. By one method, ASA Standards are developed and approved in a manner very similar to that of the ASTM except that the guiding group is a sectional committee made up of representatives of all groups concerned. By another method, the ASA may approve existing "proprietary" standards as American Standards; many of the ASTM Standards have been accepted by the ASA in this manner.

Although many of the large bureaus of the Federal government have issued their own "standard" specifications, perhaps those of most general interest are the ones developed by the Department of Commerce acting particularly through the National Bureau of Standards; those developed by the U.S. Bureau of Reclamation and the U.S. Engineer Department; and those issued by the Ordnance and Material Departments of the U.S. Army and Navy.

For specialized uses, many specifications of national scope have been sponsored by particular technical societies. The Society of Automotive Engineers has developed a comprehensive series of specifications for steels, including the alloy steels. The useful SAE method of designating steels is widely used in industry. The American Petroleum Institute has developed widely accepted specifications for wire rope. The American Concrete Institute has developed a useful set of specifications relating to concrete construction. The American Bureau of Shipping has adopted "Rules for Building and Classing Steel Vessels." The American Association of State Highway Officials publishes "Standard Specifications for Highway Materials and Methods of Sampling and Testing." A large number of these standards are identical with corresponding ASTM Standards.

Although the American Society for Metals is not a standardizing agency, it should be mentioned as one of the important technical societies concerned with the development of metals and the determination of their properties and characteristics. The *Metals Handbook* [143], issued by this society, is a most useful compilation of information covering all phases of the subject of metals.

1.12. Testing and Common Sense. Scientific experimentation and testing, as well as mathematics, have come to be an important tool of the engineer. Testing should not be used as a substitute for thought, although it may be found that an appropriate experiment may aid in an analysis.

Before it is undertaken, the purpose of a test should be well understood, and the general character of the results should be envisioned. The magic of tests lies not in turning them on and hoping for the best, but results from careful, intelligent planning and the slow, painful process of overcoming difficulties.

It is important for the engineer concerned with the performance of tests to have developed the ability to visualize what goes on behind the physical operations of tests—the paths of stress and deformation, reactions, the movements of component parts, circuits of flow, etc. He must be aware of opportunities for error and be quick to see where mistakes might occur. He should be alert to note the unusual, for herein lies the embryo of new discovery. He should be the *first* to test his results by the "seem-reasonable" criterion and be ready to check them if they do not so appear.

An experiment or test is unfinished until it is summarized, checked, and interpreted. It should be the pride, as it is the duty, of the engineer to present the results of his findings in a clear, forceful, understandable, and pleasing way. The nature of a report should be adjusted to fit the needs of the audience. Nontechnical persons and uninformed users of materials have a tendency to think of tests, especially acceptance tests, as precise, infallible, and of general application. Tests are always made subject to limiting conditions, and the results are not properly reported until a practical interpretation has been placed upon them.

The following remarks, paraphrased from the closing article of the late George Fillmore Swain's book *Structural Engineering—Strength of Materials* [137], are pertinent: The point to remember is that in testing, as in the use of mathematics, common sense should always be in command. The young engineer will do well to cultivate the habit of distrusting what does not seem to him reasonable. At times his "seeming" may be wrong—the result may be correct; in such case his senses need cultivating. Finally, it may be appropriate to refer to a remark of Aristotle who observed that it is the mark of an instructed mind to rest satisfied with that degree of precision which the nature of the subject admits, and not to seek exactness where only an approximation of the truth is possible.

Chapter 2.

GENERAL FEATURES OF MECHANICAL BEHAVIOR

2.1. Mechanical Properties. In a broad sense, strength refers to the ability of a structure or machine to resist loads without failure, which may occur by rupture due to excessive stress or may take place owing to excessive deformation. The latter cause of failure, in turn, may be the result of a limiting stress having been exceeded or the result of inadequate stiffness. The properties of materials that are of significance in relation to this general problem are the *mechanical properties.*

Mechanical properties may be specifically defined as those having to do with the behavior (either elastic or inelastic) of a material under applied forces. Mechanical properties are expressed in terms of quantities that are functions of stress or strain or both stress and strain.

Mechanical testing is concerned with the determination of measures of mechanical properties. The primary measurements involved are determination of load and of change in length. These are translated into terms of stress and strain through consideration of dimensions of the test piece.

Fundamental mechanical properties are strength, stiffness, elasticity, plasticity, and energy capacity. The strength of a material is measured by the stress at which some specified limiting condition is developed. The principal limiting conditions or criteria of failure are termination of elastic action and rupture. Hardness, which is usually indicated by resistance to indentation or abrasion at the surface of a material, may be considered as a particular type or measure of strength. Stiffness has to do with the magnitude of deformation which occurs under load; within the range of elastic behavior, stiffness is measured by the "modulus of elasticity." Elasticity (but not the "modulus of elasticity") refers to the ability of a material to deform without permanent set upon release of stress. The term *plasticity* is here used as a general term to indicate the ability of a material to deform in the inelastic or plastic range without rupture; plasticity may be expressed in a number of ways; e.g., in connection with tension tests of ductile metals, it is referred to as "ductility." The capacity of a material to absorb elastic energy depends upon both strength and stiffness; energy capacity in the range of elastic action is termed resilience; the energy required to rupture a material is taken as a measure of its toughness.

In this chapter are discussed the general features of mechanical testing associated with the above-mentioned mechanical properties of materials. The definitions given are based upon the ASTM definitions of terms relating to testing, although they differ in some respects (ASTM E 6 and E 24).* For this introductory discussion, only the general procedures of mechanical testing are outlined; details of particular types of tests are given in separate chapters under pertinent headings. In order to simplify the developments of the various concepts, the discussion in this chapter refers primarily to tests of specimens of material in which the effects of dimension, shape, stress concentration, etc., are minimized.

It is not to be inferred that in particular cases all the mechanical properties are determined. For reasons of economy, the number and difficulty of tests are kept at a minimum. For a particular problem only a few pertinent tests need ordinarily be made, and for control work a single type of test of the simplest sort, selected because it appears to be a significant indicator of required quality, is often sufficient. For example, in the commercial production of some steel products, simple hardness tests made at suitable intervals are often sufficient to indicate whether or not the quality of the steel is being maintained within specification limits.

2.2. Types of Mechanical Tests. In order to approximate the conditions under which a material must perform in service, a number of test procedures are necessary. The relationship between various test pro-

* Numbers in parentheses preceded by letters indicate the serial designation of ASTM Standards. For explanatory list of letters, see Appendix H, Ref. 1.

cedures can be made evident by an orderly classification of test conditions, the principal types of which are (1) those having to do with the manner in which load is applied, (2) those having to do with the condition of the material or specimen itself at the time of the test, and (3) those having to do with the condition of the surroundings (ambient condition) during the progress of the test, as in atmospheric exposure studies.

The method of loading is the most common basis for designating or classifying mechanical tests. There are three factors involved in defining the manner in which load is applied: the kind of stress induced, the rate at which the load is applied, and the number of times the load is applied.

In the mechanical testing of prepared specimens there are five primary types of loading, as governed by the stress condition to be induced: tension, compression, direct shear, torsion, and flexure. In tension and compression tests, an attempt is made to apply an axial load to a test specimen so as to obtain uniform distribution of stress over the critical cross section. In direct shear tests, an attempt is made to obtain uniform distribution of stress, but this ideal condition is never satisfied in practice because of the way in which the shear stresses are developed within the body under direct shear loads, and because of incidental stresses set up by the holding devices. Pure shear can be developed in cylindrical bars subjected to torsion, although the intensity of shearing stress varies from zero at the center to a maximum at the periphery of the cross section. Torsion tests have an advantage over direct shear tests in that strains can be determined by measurement of the angle of twist. In flexure tests, both tension and compression are involved (and also shear, if other than pure bending is induced), and composite effects are studied; for example, deflections are measured directly, and the modulus of rupture is determined.

In certain special tests, a complex stress condition may be induced by superposition of the primary types of loading; for example, a triaxial compression test involves compression in three directions, or a test may be made in which, say, torsion, is combined with tension. In some cases, a direct primary stress may be combined with secondary effects of bending such as when buckling occurs in a column.

A complex stress condition also occurs when the intensity of stress varies from point to point in a piece of material as the result of localized load application or abrupt changes in shape of the piece. This stress condition is an inherent and significant condition in such tests as indentation hardness tests and notched-bar impact tests. It may be noted in passing, however, that in so far as routine testing is concerned tests of this sort involve only the observation of simple phenomena (e.g., resultant deformation) and do not involve determination of stress distribution.

With respect to the rate at which load is applied, tests may be classified into three groups. If the load is applied over a relatively short time and yet slowly enough so that the speed of testing can be considered to have a practically negligible effect on the results, the test is called a *static* test. Such tests may be conducted over periods ranging from several minutes to several hours. By far the majority of tests fall in this category. If the load is applied very rapidly so that the effect of inertia and the time element are involved, the tests are called *dynamic* tests; in the special case where the load is applied suddenly as by striking a blow, the test is called an *impact* test. If the load is sustained over a long period, say months or even years, the test is a *long-time* test, of which *creep* tests are a special type.

With respect to the number of times load is applied, tests may be classified into two groups. In the first group, which includes the greatest number of all tests made, a single application of load constitutes the test. In the second group, the test load is repeated many times, millions if necessary; the most important category of tests in this group is the *endurance* or *fatigue* tests, whose purpose is to determine the endurance or fatigue limit of a material (of which specimens are made) or of an actual part.

Obviously, combinations of loading conditions defined by the several above-mentioned classifications give rise to a number of particular kinds of tests. The type of test may also be defined by the test conditions other than the type of loading, such as those described in the next article.

2.3. Test Conditions. In addition to loading conditions, it is necessary to give cognizance to the condition of the material at time of test and to the ambient conditions if they affect the test results.

Depending upon the temperature at which the tests are conducted, three general classes of tests are recognized. In the first class, comprising the majority of tests, are those tests carried out at normal atmospheric or room temperatures. In the second class are tests made to determine properties of materials, such as brittleness of steel, at very low temperatures. In the third class are experiments and tests carried out at elevated temperatures, as in the development of rockets, jet engines, gas turbines, etc., to evaluate the strength, ductility, and creep of materials under such conditions.

The mechanical properties of some materials are affected by moisture conditions. For example, the strength of materials such as concrete, brick, stone, and wood is markedly influenced by the moisture in the material. The standard tests on concrete are made upon the material while it is in a saturated condition, whereas those on brick are made upon oven-dried specimens. Tests of wood may be made on specimens either in a green or in an air-dry condition, but the moisture content at time of test is always determined. Long-time tests of these materials may require the use of controlled humidity conditions. These arbitrary moisture conditions are required for a standard test so that the test results obtained by different operators will be comparable.

For special purposes, tests may be conducted which involve the use of corrosive atmospheres, salt sprays, or baths containing substances designed to ensure neutral or corrosive reactions.

In planning or specifying the details of a test, the make-up of the specimen in relation to the physical condition or nature of the material requires consideration. Some of the factors involved in the selection and preparation of specimens are discussed in Arts. 2.16 and 2.17. In the conduct of a particular test, the manner of holding, gripping, supporting, or bedding the specimen should receive attention, as well as the stability of the specimen or parts thereof.

In research investigations, the procedure in finding the effect of given variables is to maintain all conditions constant except those under investigation. In designing, conducting, or reporting tests, significant test conditions must be specified, controlled, or known.

2.4. Stress and Strain. In testing materials *loads* are applied and measured by means of testing machines such as those described in Chap. 4. Loads are usually expressed in force units, such as pounds, although for certain tests such as torsion tests, the "load" may be expressed in terms of moments, e.g., inch-pounds. Loads should be measured with an accuracy of at least 1 percent.

Stress is herein defined as the intensity of the internal distributed forces or components of forces that resist a change in the form of a body. Stress is measured in terms of force per unit area. In the United States, the units commonly used for structural materials are pounds per square inch (psi) or kips per square inch (ksi). In the metric system, kilograms per square centimeter are the units commonly used. There are three basic kinds of stress: tension, compression, and shear. It is customary to compute stresses on the basis of the dimensions of the cross section of a piece before loading, usually called the original dimensions. In simple tension and compression tests, where the specimen is subjected to uniformly distributed stress, the stress is computed by dividing the (known) load by the minimum original cross-sectional area; if the dimensions vary slightly, the area may be based upon the average of critical dimensions.* In cases where the stress distribution is not uniform, the stress at specified locations may be determined by indirect methods. In flexure and torsion

* In routine testing practice, and unless otherwise indicated, stresses and strains are calculated on the basis of the dimensions of the specimen before load is applied. These are sometimes called the *nominal* stresses and the *nominal* strains, or *conventional* stresses and strains, to distinguish them, when necessary, from the so-called "true stresses" and "natural strains," which are calculated on the basis of the instantaneous dimensions under given loads. See Art. 2.12 for further discussion of true stresses and natural strains.

tests, within the limits of elastic action, stresses may be computed by means of theoretical relations. Within the elastic range, stresses may be evaluated from measured strains, through the use of the modulus of elasticity.

The term *deformation* is used herein as a general term to indicate the change in form of a body; it may be due to stress, to thermal change, to change in moisture, or to other causes. In conjunction with direct stress, deformation is usually taken to be a linear change and is measured in length units. In torsion tests, it is customary to measure the deformation as an angle of twist (sometimes called a *detrusion*) between two specified sections; from a consideration of the dimensions of the piece, the angle of twist in a cylindrical piece may be translated into terms of shearing strain. In flexure tests, the deformation may be expressed in terms of deflection of some specified point of a beam from its original position.

Strain is defined as the change per unit of length in a linear dimension of a body, which change accompanies a change in stress. It is a unit deformation due to stress. It is a ratio, or dimensionless number, and is therefore the same whether measured in inches per inch of length, or centimeters per centimeter, etc. A convenient way of expressing (unit) strain is in terms of *millionths*; thus a strain of 0.000001 = 1 millionth, and 0.0001 = 100 millionths. Under tensile or compressive stress, unless otherwise specified, strain is measured parallel with the direction of the force and parallel with the dimension to which it is referred. Shearing deformation is measured parallel to the direction of the shearing force, but shearing strain is computed with respect to the dimension perpendicular to the direction of the force; shearing strain is therefore an angle expressed in radians. Figure 2.1 illustrates these definitions of strain. See Fig. 6.4 for strains developed by torsion.

Set, or *permanent set*, is the deformation or strain remaining in a previously stressed body after release of load.

Deformations are measured by means of a *strainometer*, a term used to denote any deformation-measuring instrument, such as an extensometer, a compressometer, a deflectometer, or a detrusion indicator. A measured deformation is reduced to strain through consideration of the *gage length*, i.e., the length over which the strainometer measures the deformation. For ordinary commercial testing, a strainometer capable of measuring to 0.0001 in. per inch of gage length is sufficiently accurate. For research on mechanical properties, however, a least reading of 0.00001 in. per inch of gage length, or smaller, may be desirable.

If a body is subjected to tensile or compressive stress in a given direction, there takes place not only strain in that direction (axial strain) but also strains in directions perpendicular thereto (lateral strain). Within the range of elastic action the ratio of lateral to axial strain under conditions of uniaxial loading is called *Poisson's ratio*. Axial extension causes

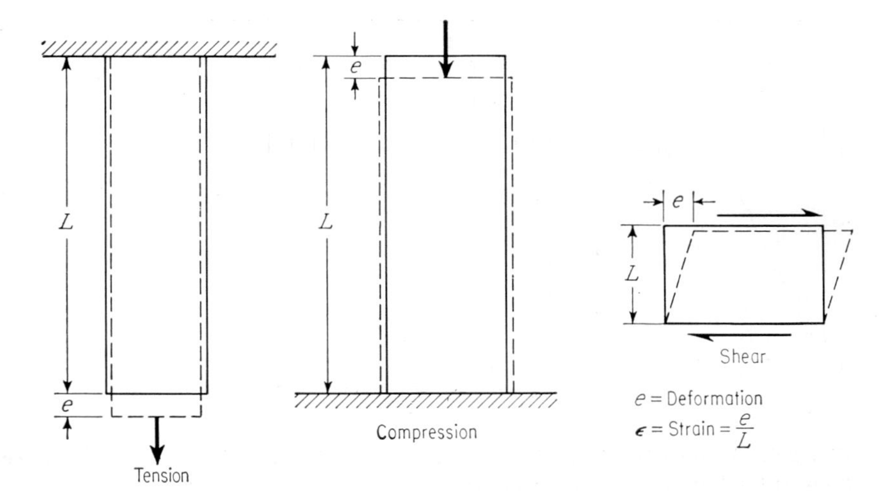

Fig. 2.1. **Relation between strain and deformation.**

lateral contraction, and vice versa. For most structural materials, Poisson's ratio has values that lie between one-third and one-sixth; hence, with ordinary measuring devices, the precision of lateral-strain measurements is not so high as that of corresponding axial-strain measurements. Poisson's ratio is rarely determined in routine commercial testing.

Occasionally volumetric deformations are determined. For solid bodies, the volumetric strain (sometimes called the cubical dilation, or the dilatation) is usually computed from measurements of linear strains. In the case of porous and highly deformable bodies, e.g., soils, the volume change is sometimes determined by the displacement of a fluid.

2.5. Stress-Strain Relations. The relation between stress and strain is commonly shown by means of a *stress-strain diagram*, which is a diagram plotted with values of stress as ordinates and values of strain as abscissas. However, the use of the term stress-strain diagram is often extended to cover diagrams in which the ordinates are values of applied load or applied moment and the abscissas are values of extension, compression, deflection, or twist.

The usual procedure in obtaining a stress-strain diagram is to plot data from a series of load readings against corresponding data from the readings of a strainometer. In some cases, stress-strain diagrams are obtained directly by an autographic attachment to the testing machine.

In planning a test requiring stress-strain data, it is necessary to select the increment of load or the increment of strainometer reading to be used between successive readings. In Fig. 2.2, which shows an idealized stress-

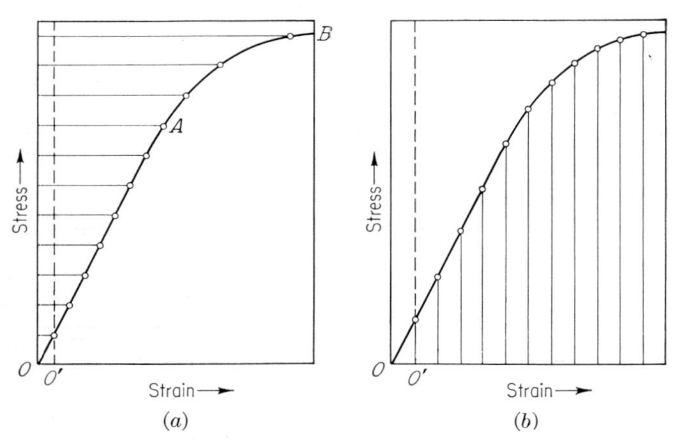

Fig. 2.2.* Determination of stress-strain relations.** (*a*) ***Equal stress increments; (*b*) ***Equal strain increments.***

strain diagram, are illustrated two methods of scheduling stress-strain readings. Figure 2.2*a* illustrates the method employing equal increments of load. It can be seen from the figure that by this method sufficient data for adequately locating the curve between points A and B may be lacking. That particular portion of the curve between A and B (the "knee" of the curve) is a part for which it is often especially desirable to locate several points on the diagram.

Sometimes smaller load increments are used in the stage of loading which corresponds to the knee of the stress-strain diagram than in the early stage of load application. This procedure is not always satisfactory, however, since it may involve a considerably greater number of readings than is really necessary, and in a test of an unknown material there is the possibility that the small-increment loading will not have been started soon enough.

Figure 2.2*b* illustrates the method employing equal increments of strain. By this method, there are located several points near the knee of the curve, and thus the shape of the diagram in this region is more definitely determined than by the method of equal load increments.

The use of increments of load rather than increments of strain has been common practice because, in general, it is simpler to schedule load increments than increments of strainometer reading. The load necessary to stress the test specimen to a value corresponding to the knee of the curve is estimated, and a fraction (frequently one-tenth) of this value is taken as the increment to be employed. The determination of a suitable increment of strainometer reading is somewhat more complicated but not difficult. The following procedure is suggested. The load corresponding to the knee of the stress-strain diagram is estimated, and one-tenth of this load is applied to the test specimen. The change in extensometer reading is then noted, and there is then chosen, for the remainder of the test, an increment of strainometer reading that is equal to some convenient interval on the scale of the strainometer and is approximately equal to this initial strain increment. In routine tests of a given material, the appropriate strain increment is known from experience. The use of equal increments of strain is considered preferable.

In connection with the practical determination of stress-strain relations it should be noted that the initial or datum strainometer reading is often taken after some small initial load has been applied. Such a procedure is desirable in order that firm gripping of or bearing on the test specimen can be attained and so that firm seating of the strainometer can take place. Data taken under this procedure, when plotted, give a stress-strain diagram that shows a finite value of stress at zero strain; this condition is shown in Fig. 2.2 provided the stress axis has its origin at O'. The effective origin of the stress-strain diagram may be obtained by shifting the stress axis to the left so that it passes through O, the intersection of the projection of the stress-strain diagram with the strain axis. For convenience, in further discussion, stress-strain diagrams will be drawn from the effective origin O.

2.6. Elasticity. *Elasticity* is that property of a material by virtue of which deformations caused by stress disappear upon the removal of stress. Some substances, such as gases, possess elasticity of volume only, but solids may possess, in addition, elasticity of form or shape. A perfectly elastic body is conceived to be one that completely recovers its original shape and dimensions after release of stress. No materials are known which are perfectly elastic throughout the entire range of stress up to rupture, although some materials, such as steel, appear to be elastic over a considerable range of stress. Some materials such as cast iron, concrete, and certain nonferrous metals are imperfectly elastic even at relatively low stresses, but the magnitude of permanent set under short-time loading is small, so that for practical purposes the material can be considered elastic up to reasonable magnitudes of stress.

If a tensile load within the elastic range is applied, the elastic axial strains result from a separation of the atoms or molecules in the direction of the loading. At the same time they move closer together in the transverse direction. For a relatively isotropic material, such as steel, the stress-strain characteristics are closely similar irrespective of the direction of loading (due to the random arrangement of the many crystals of which the material is composed), but for anisotropic materials, such as wood, these properties vary with the direction of loading.

The degree to which elastic action is exhibited is often a function of the test conditions. Some materials, which are imperfectly elastic under virgin loading, appear to become elastic after having been prestressed, and overstressing in some metals appears to raise the limit of elastic action; thus previous strain history has something to do with defining the limits of elastic action. The range of elastic action, which may be relatively great for some materials at normal temperatures, is usually reduced with increasing temperature. Also, the rapidity of loading affects the apparent elasticity of some materials; for example, with wood and concrete, a load from which practically perfect recovery takes place in a short-time test may produce considerable permanent set if long sustained.

In accordance with the concept of elastic behavior as defined above, a quantitative measure of the elasticity of a material might logically be expressed as the extent to which the material can be deformed within the limit of elastic action. However, since engineers usually think in terms of stress rather than of strain, a practical index of elasticity is the stress that marks the (effective) limit of elastic behavior.

Elastic behavior is not uncommonly associated with two other phenomena: linear proportionality of stress and strain and the nonabsorption of energy during cyclic variation in stress.* These two phenomena are not necessarily criteria of the property of elasticity and are actually independent thereof. Soft vulcanized rubber, for example, does not exhibit a straight-line relation between stress and strain, even well within limits of elastic action, and exhibits hysteresis during a cycle of loading and unloading. It is a fact, however, that the stresses at which nonlinear stress-strain behavior or actual yielding begins are useful in designating practical limits of elastic action of the common constructional materials.

2.7. Measures of Elastic Strength. In tests of materials under uniaxial loading, various criteria of elastic strength or elastic failure have been used: these are the *elastic limit*, the *proportional limit*, and the *yield strength*.

The *elastic limit* is defined as the greatest stress that a material is capable of developing without a permanent set remaining upon complete release of stress. To determine the elastic limit within the strict interpretation of this concept would require successive application and release of greater and greater loads until a load is found at which permanent set is produced. Not only is this an arduous procedure, but also the observed load at which set first develops obviously will depend upon the sensitivity and upon the nicety of operation of the strainometer used for the test. The determination of the elastic limit in this sense is not practical and therefore rarely made.

The *proportional limit* is defined as the greatest stress that a material is capable of developing without deviation from straight-line proportionality between stress and strain. It has been observed that most mate-

* The effect of permanent energy absorption under cyclic stress within the elastic range, called elastic hysteresis or frictional damping, is illustrated by the decay in amplitude of free vibrations of an elastic spring.

rials exhibit this linear relationship between stress and strain within the elastic range, and it is a matter of experience that, by ordinary methods of testing, the values of elastic limit for metals found by means of observations of permanent set do not differ greatly from values of proportional limit. The operations involved in determining the proportional limit are relatively simple; hence the proportional limit is often used as a measure of the elastic limit, and the terms have often been misused, one for the other. The proportional limit is sometimes called the "proportional elastic limit."

It is of interest to note in this connection that the concept of proportionality between stress and strain is known as *Hooke's law* because of Robert Hooke's historic generalization of the results of his observations on the behavior of springs (1678).* Although Hooke's law is a description of the action of materials over only a limited range of stress, and although in many instances it is only an approximation of actual behavior even at low stresses, from the larger view it has served as a fair and practical generalization in most cases, and it is one on which our methods of stress analysis rest.

The proportional limit is determined by use of a stress-strain diagram or modification thereof. The values of proportional limit so obtained are subject to some variation, being dependent in part upon the sensitivity of the strainometer and in part on the method of plotting; they may also be subject to considerable variation in interpretation. Up to the proportional limit the strains are usually comparatively small and require instruments of precision for their measurement. For a method of computation in which variations due to method of plotting and interpretation are reduced to a minimum, see Refs. 211 and 544.

Because of the time and care necessary to obtain significantly accurate measures of the proportional limit, its determination is of value principally in research.

It should be clear from the above discussion that it is difficult, if not impossible, to obtain the true value of stress at which inelastic action *begins*, and for commercial and routine purposes it is certainly not practical. Plastic action ("yielding") in nearly all members, even in specimens in carefully controlled laboratory tests, begins as localized actions and becomes measurable only after many local internal adjustments have occurred and after a considerable portion of the piece is affected by yielding. The limit of usefulness of many materials, especially metals, in members subjected to approximately static loading at ordinary temperatures, is therefore defined by some specified degree of plastic yielding of the material above which value the material may be considered to be damaged and below which the damaging effects may be considered negligi-

* For an interesting discussion of Hooke and the law that bears his name, see Ref. 201.

ble. This limit may in general be termed the *yield strength,* which is the practical and most commonly used measure of elastic strength.

For many materials, the stress at which a significant yielding may be said to have occurred is not readily apparent from the stress-strain relations; hence several more or less arbitrary criteria or indicators of the yield strength are in use. Although they all aim to give a relatively simple method of determination, each is conditioned by the nature of the material in connection with which it was developed, by the service for which the material is intended, by the kind of apparatus available, and by the economics of the testing problem. The various criteria for determining yield strength may be classified as follows: (1) method involving a measure of permanent set, the "offset method"; (2) methods involving the total deformation or measure thereof, up to the yield strength; (3) methods involving noticeable increase or acceleration in rate of strain.

A measure of elastic strength, no longer used to any great extent, involving the idea of increase in rate of deformation and requiring the construction of a stress-strain diagram, is Johnson's "apparent elastic limit" [141]. This is taken as the stress at which the *rate* of deformation is 50 percent greater than is the rate of deformation at the beginning of loading. This concept is also the basis of the so-called "useful limit point," a factor that has been used in connection with tests of structural members [511]. The "apparent elastic limit" is somewhat greater than the proportional limit but may be less than the yield strength.

2.8. Yield Strength by the Offset Method. Figure 2.3*a* shows a hypothetical stress-strain diagram for a material loaded to a stress (*YS*) somewhat above the proportional limit (*PL*) and then unloaded. The distance *CA* represents a deviation or offset from Hooke's law at the stress (*YS*). The set after release of load is indicated as the strain *a* on the diagram. It is an observed fact that with a number of materials the ratio of stress to strain during release of load, from a stress somewhat above the proportional limit, is constant and closely approximates the ratio of stress to strain within the elastic range. That is, the line *AB*, Fig. 2.3*a*, practically parallels the line *OC*, and the offset *CA* approximates the permanent set *a*. The offset thus approximates the inelastic deformation at a given stress. This concept is the basis for the determination of yield strength by the offset method, in accordance with the definition of yield strength given by the ASTM, namely, the stress at which a material exhibits a specified limiting permanent set (ASTM E 6).

In Fig. 2.3*b*, *OX* represents a portion of the stress-strain diagram for a material that does not exhibit a marked yield at any particular stress but yields gradually after the proportional limit is exceeded. This may be considered to be the general case. Point *B* is marked on the strain axis, at a distance *a*, the specified offset, from the intersection *O* of the stress-

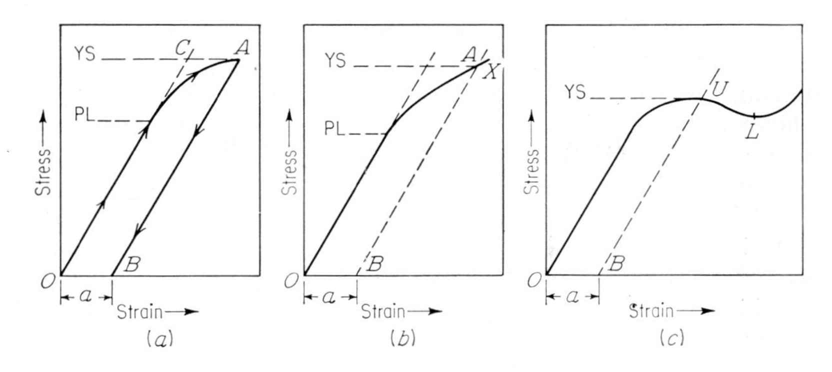

Fig. 2.3. **Determination of yield strength (diagrammatic).**

strain curve with the strain axis. The line BA is drawn parallel to the initial straight line portion of the stress-strain diagram to intersect the curve at A, thus determining the yield strength (YS) as defined by the offset method.

The precision of the determination of the yield strength becomes less exact as the magnitude of the offset decreases. Therefore too small a value of the offset should not be specified.

The magnitude of the offset a is chosen as some measurable value that is considered from experience to be of practical significance in defining a limit to the elastic range. Values of a in common use for metal are 0.001 and 0.002 (0.1 and 0.2 percent strain per inch of gage length). In reporting values of yield strength obtained by the offset method, the specified value of set used should be stated. For example, yield strength (offset = 0.1 percent) = 52,000 psi indicates that at a stress of 52,000 psi the (approximate) permanent set of the material tested was 0.1 percent per inch of the gage length.

Values of the offset that have been suggested for use with cast iron are 0.0002 to 0.0005; for use with wood parallel to the grain, 0.0005; and for use with concrete, 0.0001 or 0.0002.

The term *proof stress* found in British specifications is very similar to the term *yield strength* as determined by the offset method.

The elastic limit and proportional limit may be considered as special values of the yield strength, for which the limiting set is zero.

2.9. Yield Point of Ductile Metals. Ductile materials such as mild steel exhibit a definite *yield point*, which is defined as the stress at which there occurs a marked increase in strain without increase in stress. Only materials that exhibit this phenomenon have a yield point within this sense of the term, and the term *yield point* should not be used in connection with a material whose stress-strain diagram above the proportional limit is a line of gradual curvature with continually increasing stress. The offset method can be used to determine the yield point for materials having "sharp-kneed" stress-strain diagrams, as shown in Fig. 2.3*c*, by choice of an appropriate value of specified set, but the characteristic marked yielding in this critical range of stress permits the use of simpler methods that do not involve the taking of strain measurements.

Three methods that depend upon *total strain* or deformation as a criterion of yield are in use: (1) one in which a strainometer or even a simple multiplying lever device is employed, (2) one sometimes called the "dividers method," and (3) the so-called scaling method." These methods have the advantage that they do not require the construction of a stress-strain diagram, but they are to be considered approximate. They are used only for commercial tests of materials whose stress-strain characteristics are well-known from previous tests.

When the strainometer is to be used, the total deformation corresponding to an accepted value of set is determined and specified; the yield point is reached when the deformation as indicated by the strainometer reaches this value. This method has been applied to materials that do not have a definite yield point as well as to those which do have a yield point. The ASTM Specifications for a number of metals (e.g., ASTM A 47, Malleable Iron Castings) define yield point on the basis of 0.5 percent total elongation.

In the dividers method, the observer watches for visible elongation between two marks spaced some distance apart along the test piece. One procedure is to place one point of the dividers in a punch mark while the other point is used to scribe a very fine line on a chalked or otherwise prepared surface. In the testing of ordinary steel, to which this method is principally applicable, the yield strength is reached when the line being scribed appears to widen. Noticeable widening occurs when an elongation of probably about 0.01 in. has taken place. In a gage length of 2 in. this corresponds to a total strain of about 0.5 percent.

When marked yielding occurs in a ductile piece of metal covered with mill scale, the mill scale, being brittle and much less extensible than the metal beneath, flakes off. When there is no mill scale on the surface of the metal, a coating of brittle paint or lacquer is sometimes used. This "scaling method" or "brittle-coating method" is sometimes used as a rough indicator of yield strength. The method, though crude, is sometimes of value in tests of complicated built-up shapes or castings to indicate the regions of first occurrence of and the progress of yielding. Obviously, such a method is confined to indications of stress at the surface of the piece.

Methods that depend upon the acceleration of the rate of deformation are the strainometer method and the "drop-of-beam" method. With the strainometer attached to the specimen, the observer simply watches the strainometer until a noticeable increase in rate of deformation occurs and records the corresponding load. A strainometer reading to 0.0001 in. per inch of gage length is commonly specified. This method gives a result that is probably closer to the proportional limit than to the yield strength as previously defined.

The drop-of-beam method is applicable only to materials with a definite yield point. In this method, the load is applied to the specimen with the testing machine set to run at a steady rate. When the yield point is reached, the rate of deformation increases suddenly so that the rate of load application decreases rapidly. With a machine employing a balance arm and poise to indicate load, the operator tends to run the poise a trifle beyond the balance position as the yield is reached, and the balance beam drops for a brief but appreciable interval of time. With the machine fitted with a self-indicating load-measuring device, there is a sudden halt in the motion of the load-indicating pointer, corresponding to the drop of beam. The load at the "drop" or "halt" is recorded, and the corresponding stress is the yield point, provided that slippage in the grips has not caused a false drop or halt. This method requires only one man to conduct a test and is by far the most common method employed in the commercial testing of ductile steel.

With a ductile material it is possible, if the test is conducted in an appropriate manner, to distinguish between two critical points in the yield range, the "upper yield point" (U) and the "lower yield point" (L), as shown in Fig. 2.3*c*. The upper yield point is the one usually reported and is the one indicated by the drop-of-beam method. However, it appears that the lower yield point may possibly be of more real significance, as far as fundamental properties of the material are concerned [215].

2.10. Measures of Ultimate Strength. The term *ultimate strength* has to do with the maximum stress a material can develop. Ultimate strengths are computed on the basis of maximum load carried by the test piece and the *original* cross-sectional dimensions; these may be referred to as the nominal strengths. Ultimate strengths are usually stated in terms of the kind of stress producing the failure.

The *tensile strength* is the maximum tensile stress that a material is capable of developing and in practice is the maximum stress developed by a specimen of the material during the course of loading to rupture. Figure 2.4*a* shows, diagrammatically, stress-strain relations for a ductile metal loaded to failure in tension. The heavy line, the diagram as customarily drawn, is based on the original cross-sectional area. The ultimate strength (US) is the stress at the highest point (A) on this diagram. Beyond this point, as the specimen contracts markedly or "necks down" to final rupture, the load decreases as a result of the decreasing resisting

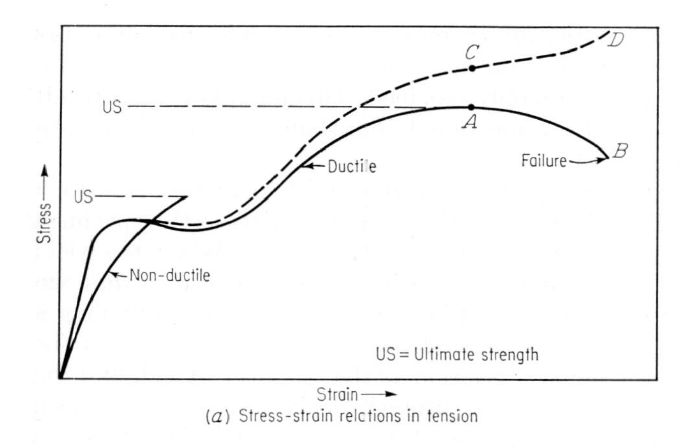

(*a*) Stress-strain relations in tension

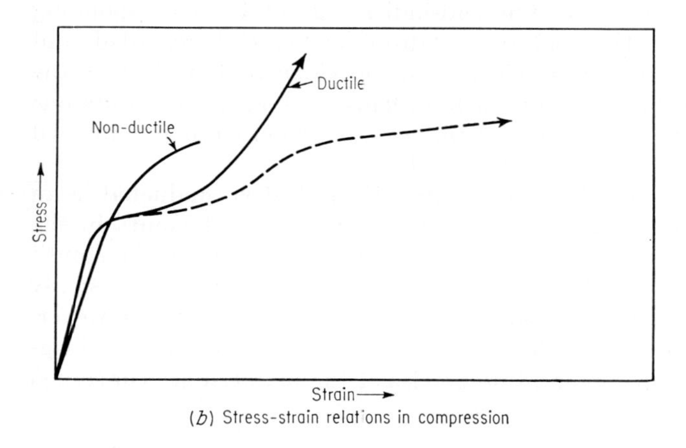

(*b*) Stress-strain relations in compression

***Fig. 2.4.* Diagrammatic stress-strain diagrams for ductile and nonductile materials tested to failure.**

area, and, if the test is conducted carefully, the nominal stress at failure (*B*) can be obtained. The stress at failure is sometimes called the "breaking stress" or "rupture stress."

The dotted line in Fig. 2.4*a* represents the true stress–conventional strain relation such as might be obtained if the load at any stage of loading were divided by the actual cross-sectional area, which decreases under tensile loading. The stress so obtained is sometimes called the "true" stress, although it is unlikely that it is the actual stress on the critical section in the range *C-D* on the diagram because the drawing of metal undoubtedly causes a complex stress distribution to be developed.* The

* See Art. 2.12 for further comment on true stress.

so-called "true stress" is not determined in routine testing. The characteristic form of the stress-strain diagram for a nonductile metal tested in tension is also shown in Fig. 2.4*a*. For a material of this type the breaking strength coincides with the ultimate strength (*US*).

The *compressive strength* is the maximum compressive stress that a material is capable of developing. With a (brittle) material that fails in compression by rupturing, the compressive strength has a definite value. In the case of materials that do not fail in compression by a shattering fracture (ductile, malleable, or semiviscous materials), the value obtained for compressive strength is an arbitrary value depending upon the degree of distortion that is regarded as effective failure of the material. Figure 2.4*b* shows characteristic stress-strain diagrams for ductile and nonductile materials in compression, the dotted line again showing the "true" stress–conventional strain relationship; in compression it is lower than the conventional stress-strain diagram owing to the increase in cross section of the specimen while under compressive loading.

The measures of ultimate strength in flexure and torsion, the *moduli of rupture* in flexure and torsion, are computed on the basis of certain assumptions that are discussed in Chap. 6.

Hardness, which is a measure of the resistance to surface indentation or abrasion, may in general be thought of as a function of stress required to produce some specified type of surface "failure." In one type of test (the Brinell) there is computed a value of stress per unit area of indentation, when the ball indenter is pressed into the material under a given load. For most hardness tests, however, inasmuch as the stress conditions are complicated and cannot be evaluated, the hardness is simply expressed in terms of some arbitrary value, such as the scale reading of the particular instrument used. Details are discussed in Chap. 7.

Under repeated loads, failure may occur because of fatigue. The capacity of a material to withstand repeated application of stress is termed its *endurance*. The *endurance limit* or *fatigue strength* is the maximum stress that can be applied an indefinitely large number of times without causing failure. For ordinary steels the endurance limit under reversed flexure is roughly half the tensile strength. The determination of the endurance limit requires the use of special apparatus and procedures, which are discussed in Chap. 9.

2.11. Plasticity. *Plasticity* is that property which enables material to undergo permanent deformation without rupture. A general expression of plastic action would involve the time rate of strain, since in the plastic state materials can deform under constant sustained stress; it would also involve the concept of limit of deformation before rupture. Evidences of plastic action in structural materials are called *yield*, *plastic flow*, and *creep*.

Plastic strains are caused by slips induced by shear stresses as shown in Fig. 2.5. Such strains can occur in all materials at high stresses, even at normal temperatures, but in Chap. 9 it is shown that plastic strains can occur in materials at relatively low stresses provided ample time is allowed and favorable high temperatures are provided. Many metals show a *strain-hardening* effect when undergoing plastic deformations, as after minor shear slips have occurred they show no further plastic strains until higher stresses are applied. There appears to be no appreciable change in volume as a result of plastic strains.

Although the maximum shear stress for tensile loading occurs on a 45° plane, the slip in a particular metallic crystal does not necessarily occur along that plane, as to do so would require a major rearrangement of the atoms. The overall plastic strain in a material depends upon (1) the number of slip planes involved, which in turn depends on the atomic arrangement, (2) the overall effects of crystal orientation, and (3) the intensity of the shear stress.

As slip does not involve any appreciable change in the spacing of the atoms, there is no tendency for them to return to their original position after the shear stress is removed. This indicates that plastic strains are irreversible and explains why they are permanent. In Fig. 2.3*a* it was shown that the stress-strain diagram during unloading is commonly a straight line. This is also explained by the irreversible character of plastic strains.

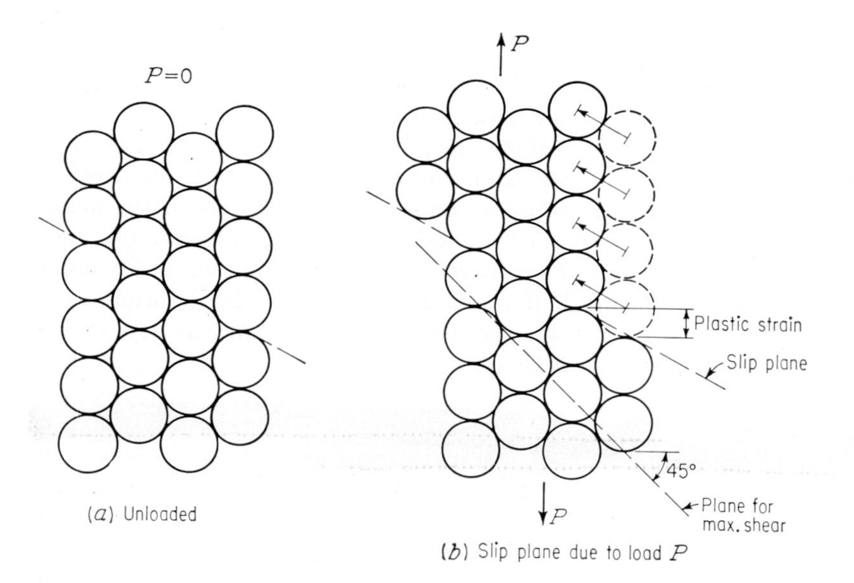

Fig. 2.5. **Plastic strain and slip plane.**

Tests show that the yield stress is increased by a prior loading into the plastic range and is an indication of the strain-hardening effect mentioned previously. This results from the fact that any plastic strains produced by a prior tensile (or compressive) loading remain unchanged until a greater load of similar type is applied.

When a material is stressed into the plastic range, some individual crystals will undergo a permanent set while adjacent crystals, more favorably oriented, may only deform elastically. This will result in some *residual stresses* in the individual crystals of the highly stressed material.

Plasticity is of importance in forming and shaping and extruding operations. Some metals are shaped cold, e.g., the deep drawing of thin sheets. Many metals are shaped hot, e.g., the rolling of structural steel shapes and the forging of certain machine parts. Metals such as cast iron are molded in the molten state. Wood is best bent while wet and hot. Malleable materials are those which can be hammered into thin sheets without fracture; malleability depends on the softness as well as the plasticity of the metal.

Of particular importance in connection with mechanical testing is one manifestation of plasticity: ductility. Ductility is that property of a material which enables it to be drawn out to a considerable extent before rupture and at the same time to sustain appreciable load. Mild steel is a ductile material. A nonductile material is said to be brittle, i.e., it fractures with relatively little or no elongation. Cast iron and concrete are brittle materials. Usually the tensile strength of brittle materials is only a fraction of their compressive strength. The usual measures of ductility are the percentage elongation and the reduction of area in the tension test; the determination of these factors is discussed in Chap. 5. Ductility is also sometimes determined by a cold-bend test, which is discussed in Chap. 6.

2.12. True Stress–Natural Strain Relations. When a ductile material is loaded beyond the yield strength, through the plastic range, the dimensions change appreciably; as the fracture load is approached, particularly after necking starts in the tensile test, the stress at the critical cross section departs more and more from the nominal stress calculated on the basis of the original cross section. Thus in studies of the behavior of materials subject to large deformations, such as metals in the plastic range, it has been found desirable to calculate stress under a given load on the basis of the instantaneous dimensions [221, 224, 502].

"True stress" is obtained by dividing the axial load P by the actual instantaneous cross-sectional area A. After necking starts in a tension test, the area is measured at the minimum section of the neck; in this range of deformation, however, the true stress is only the average stress over the area, because the actual stresses vary across the section [505].

"Natural strain," also called "true strain," is the change in gage length with respect to the instantaneous gage length over which the change occurs:

$$\epsilon_{nat} = \int_{L_o}^{L} \frac{dL}{L} = \log_e \frac{L}{L_o} = \log_e (1 + \epsilon_o)$$

where L = length of a small element under a given load
L_o = original length of element before any load is applied
ϵ_o = conventional strain

From the hypothesis that during plastic deformation the volume of material remains constant, it may be shown that

$$\epsilon_{nat} = \log_e \frac{L}{L_o} = \log_e \frac{A_o}{A}$$

where A_o = original cross-sectional area before any load is applied
A = instantaneous cross-sectional area under a given load

This relationship is particularly useful in obtaining the natural strain at the critical area of the necked section.

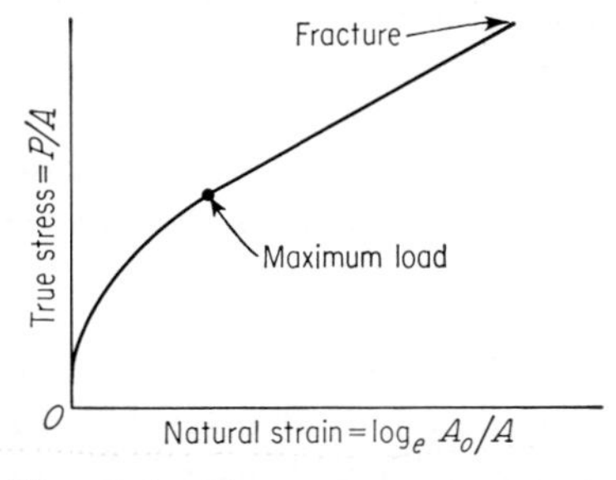

Fig. 2.6. **True stress–natural strain diagram for an axial-tension test.**

The general form of the true stress–natural strain diagram for a tension test is shown diagrammatically in Fig. 2.6. Observations indicate that the true stress is practically linearly related to the natural strain from the maximum load point to the fracture load point. This type of diagram is much more significant than the conventional stress-strain diagram in studies concerned with metal-forming operations.

2.13. Ramberg-Osgood Stress-Strain Diagram. In the ordinary stress-strain diagram the strain is dimensionless but the stress is commonly in pound and inch units. However, it is possible, and at times desirable, to make both stress and strain dimensionless as developed by Ramberg and Osgood [205].

Figure 2.7*a* shows an ordinary stress-strain diagram. If all stress values are divided by a selected *base stress* σ_0 and all strain values by a selected *base strain* ϵ_0, the resulting diagram is shown in Fig. 2.7*b*. The point representing the two base values on the original stress-strain diagram will be located at the base point (1,1) on the new Ramberg-Osgood or *one-one diagram.*

To determine the base value σ_0, draw on Fig. 2.7*a* a straight line corresponding to the secant modulus E_s equal to some selected value such as 0.7 E. The *secant yield stress* is at the intersection of the secant line and

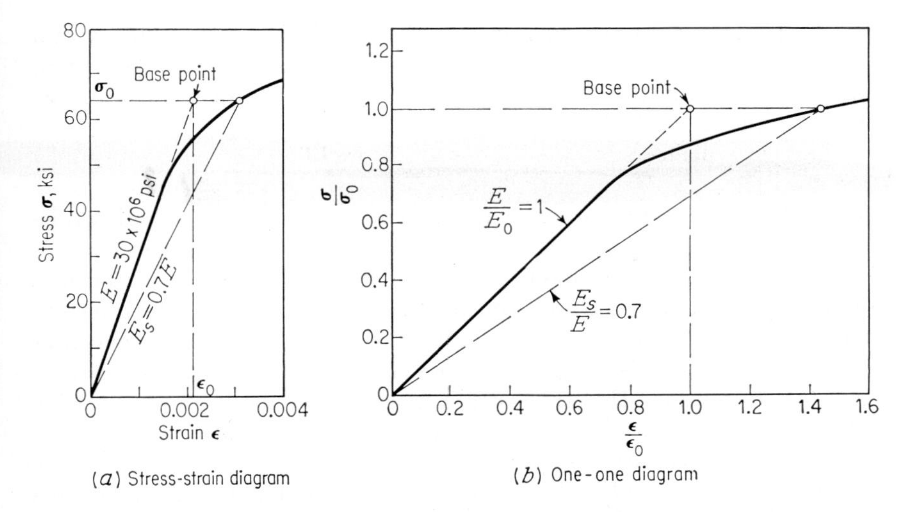

Fig. 2.7. **Development of a Ramberg-Osgood or one-one diagram.**

the stress-strain curve, and it determines σ_0. The base strain is the elastic strain corresponding to the base stress σ_0, or $\epsilon_0 = \sigma_0/E$. It can be determined as shown in Fig. 2.7*a*.

The base value of the elastic modulus (E_0) is E. Therefore E/E_0 will always be 1.0 for a one-one diagram. It is also evident that the secant yield stress corresponds to a value of $\epsilon/\epsilon_0 = 1/0.7 = 1.43$. Having determined the location of the two common points (1,1) and (1,1.43) all one-one diagrams are alike except for their shape.

It has been shown [205] that

$$\frac{\epsilon}{\epsilon_0} = \frac{\sigma}{\sigma_0} + \frac{3}{7}\left(\frac{\sigma}{\sigma_0}\right)^n$$

This is the Ramberg-Osgood equation which simplifies the stress-strain diagrams so that the difference between a variety of materials is expressed solely by the exponent n, which is a function of the shape of the diagram. The equation is valuable in comparative studies of materials as it permits the studies to be conducted on a more generalized basis.

The tangent-modulus ratio in dimensionless form is

$$\frac{E_t}{E} = \frac{1}{1 + \frac{3}{7}n(\sigma/\sigma_0)^{n-1}}$$

This ratio is useful in problems involving inelastic buckling. A method for determining the exponent n in the above equations from any stress-strain diagram has been developed [205]. Values of n for a few metals are shown in Ref. 314, p. 594.

2.14. Stiffness. Stiffness has to do with the relative deformability of a material under load. It is measured by the rate of stress with respect to strain. The greater the stress required to produce a given strain, the stiffer the material is said to be.

Under simple stress within the proportional limit, the ratio of stress to corresponding strain is called the *modulus of elasticity* (E). This term is somewhat of a misnomer, since it refers to stiffness in the elastic range rather than to elasticity. Corresponding to the three fundamental types of stress, there are three moduli of elasticity, the modulus in tension, compression, and shear. Under tensile stress, this measure of stiffness is sometimes called *Young's modulus*, after the English physicist who first defined it. Under simple shear the stiffness is sometimes called the *modulus of rigidity*. In terms of the stress-strain diagram, the modulus of elasticity is the slope of the stress-strain diagram in the range of linear proportionality of stress to strain.

Because many materials are imperfectly elastic, special definitions of the modulus of elasticity may be necessary. Figure 2.8 shows a stress-strain diagram that is continuously curved, a type of diagram such as is obtained for concrete or cast iron. The slope of a line (OA) drawn tangent to the curve at the origin is the *initial tangent modulus*. The slope of the curve at, say point B, is the *tangent modulus* at a stress of b psi. The ratio of any given stress to the corresponding strain, which is equivalent to the slope of the line OC, is the *secant modulus* of elasticity at stress c psi. Some column formulas, especially those for long columns which fail by bending, involve the modulus of elasticity. In such cases the tangent modulus at the stress causing yielding of the material is the critical value, as shown in Art. 3.9.

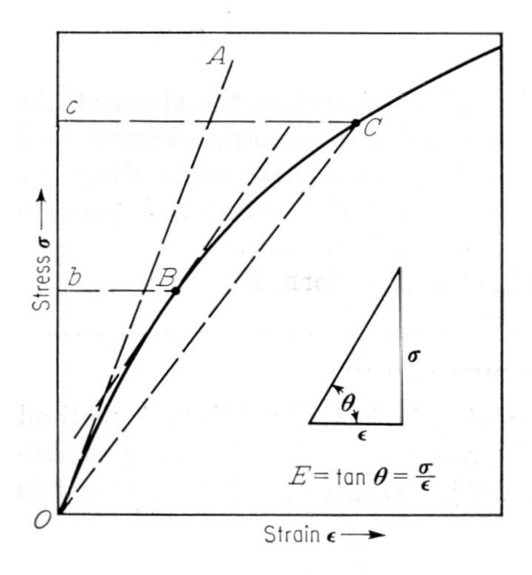

Fig. 2.8. **Moduli of elasticity.**

The modulus of elasticity of materials is ordinarily determined directly by tests involving stress-strain measurement of specimens subjected to simple stresses. However, modulus determinations are sometimes made through the use of special properties and theoretical relationships. For example, the modulus is computed from observations of the deflection of beams, the period of vibration of wires and rods, and the sound emitted by vibrating bars. The modulus of elasticity is expressed in units of force per unit area, as pounds per square inch, kips per square inch, or kilograms per square centimeter. The modulus under axial stress is of the order of several million pounds per square inch for most metals. It becomes less at higher temperatures; for low-alloy steel it reduces from 30,000,000 psi at normal temperatures to about 12,000,000 psi at 1200°F. The modulus in shear for metals is about two-fifths of the modulus under axial stress.

There is no established measure of stiffness in the plastic range.

The term *flexibility* is sometimes used as the opposite of stiffness. However, flexibility usually has to do with flexure or bending; also, it may connote ease of bending in the plastic range. The effective or overall stiffness or rigidity or flexibility of a body or structural member is obviously a function of the dimensions and shape of the body as well as of the characteristics of the material.

The measure of resistance to change in volume is called the *coefficient of compressibility* or the *bulk modulus* and is taken as the ratio of (hydrostatic) stress to corresponding unit change in volume.

2.15. Energy Capacity. The capacity of a material to absorb or store energy is of importance in connection with problems of shock resistance, impact loading, etc. The basic principle involved is that work or energy is equal to force times distance.

The amount of energy absorbed in stressing a material up to the elastic limit, or the amount of energy that can be recovered when stress is released from the elastic limit, is called the *elastic resilience*. The energy stored per unit of volume at the elastic limit is the *modulus of resilience*. For a unit volume, say a 1-in. cube, the resilience is the product of average stress times strain: $\frac{\sigma}{2}\epsilon = \frac{\sigma}{2}\frac{\sigma}{E} = \frac{\sigma^2}{2E}$. In terms of the stress-strain diagram, energy absorption is represented by the area under the diagram. In Fig. 2.9 which shows a typical diagram for mild steel, the elastic resilience is represented by area I. If the load is released from some point A in the plastic range, the recovery diagram is approximately a straight line (AB) and the energy released is represented by area II; this has been called the *hyperelastic* resilience.

The modulus of resilience is a measure of what may be called the "elastic energy strength" of the material and is of importance in the

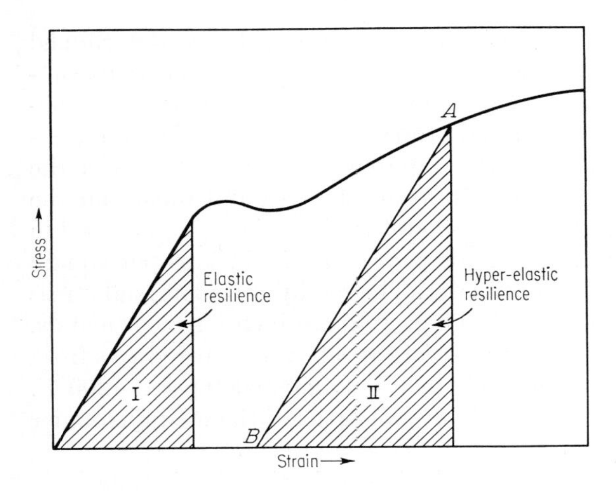

Fig. 2.9. **Resilience in terms of stress-strain diagram.**

selection of materials for service where parts are subjected to energy loads but where the stresses must be kept within the elastic limit [135]. It should be noted that for a high modulus of resilience a material should have a relatively high elastic strength, low modulus of elasticity, or both. It is commonly expressed in units of inch-pounds per cubic inch. For various grades of steel it has values ranging from about 15 to 650 in.-lb per cu in. The higher the carbon or alloy content, the higher the modulus of resilience. For steel the modulus of resilience for shear is approximately equal to that for axial stress.

When a material is subject to repeated loading, during any cycle of loading and unloading, or vice versa, some energy is permanently absorbed or lost. This is true even in the elastic range as is evidenced by the decay in the free vibrations of rods and springs. For metals such as steel it is true that the energy lost per cycle is small, but to this extent the idea that the modulus of elastic resilience represents recoverable energy is approximate. This phenomenon of lost energy is called, in general, *hysteresis*, and within the elastic range, *elastic hysteresis*. In terms of the stress-strain diagram, the hysteresis loss is represented by the area enclosed by the loop formed by consecutive segments of the diagram. In Fig. 2.10*a* is shown a hypothetical diagram for a material such as soft rubber, which is elastic but does not follow Hooke's law, nor release all the energy expended during loading. The shaded area represents the energy lost or the hysteresis loss. Figure 2.10*b* shows typical hysteresis loops for a material like steel when unloaded and reloaded to stresses above the elastic limit.

In general, the determination of hysteresis loss, at least within the elastic range, is not practicable by means of a direct static loading test using a strainometer. Strainometers of appropriate precision are too

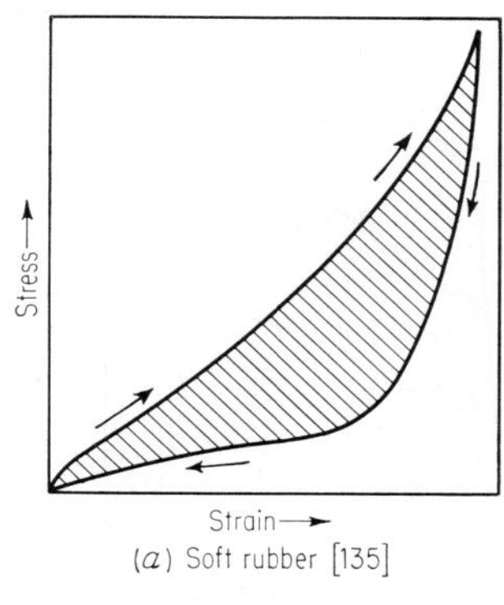

(*a*) Soft rubber [135]

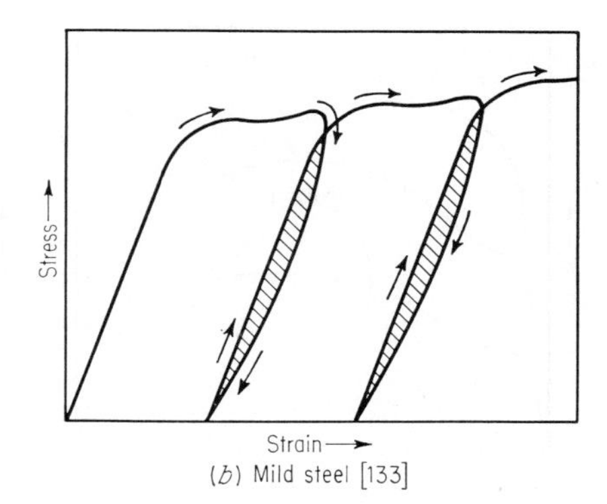

(*b*) Mild steel [133]

Fig. 2.10. **Hysteresis loops.**

cumbersome to operate, and, in practically all strainometers, errors due to backlash, when the direction of loading is reversed, are of serious magnitude. The most feasible means for determining the hysteresis loss in the elastic range is observation of the damping of vibrations.

Toughness involves the idea of energy required to rupture a material. It may be measured by the amount of work per unit volume of a material required to carry the material to failure under static loading, called the *modulus of toughness*. Under this criterion it may be represented by the area under the complete stress-strain diagram, as illustrated in Fig. 2.11.

Toughness is a measure of what may be called the "ultimate energy strength" of a material and is of importance in the selection of a material for types of service where impact loads are applied which from time to time may cause stresses above the yield point [135]. Toughness may be expressed in units of inch-pounds per cubic inch. For carbon steels, values range from about 5000 to 17,000 in.-lb per cu in.; medium carbon steels are the toughest, as may be noted from Fig. 2.11.

For ductile materials, an approximate measure of toughness, taken as the product of the ultimate strength and the elongation, is used for purposes of comparison and is called the *merit number* [135]. In reporting or interpreting toughness data for ductile materials in tension, the gage length over which the deformation measurements are made should be known, and the L/d ratio should be reported, because, owing to localized drawing or necking of the specimen near the point of failure, the percentage elongation is greater when measured over shorter lengths.

Direct impact tests are often made in which the work required to rupture a specimen is determined from the energy of a falling weight or pendulum. Such tests are of value principally to indicate the effects of variations in a given material, or the sensitivity of the material to notching. They do not appear to give direct measures of toughness in the sense previously discussed. Impact tests are discussed in Chap. 8.

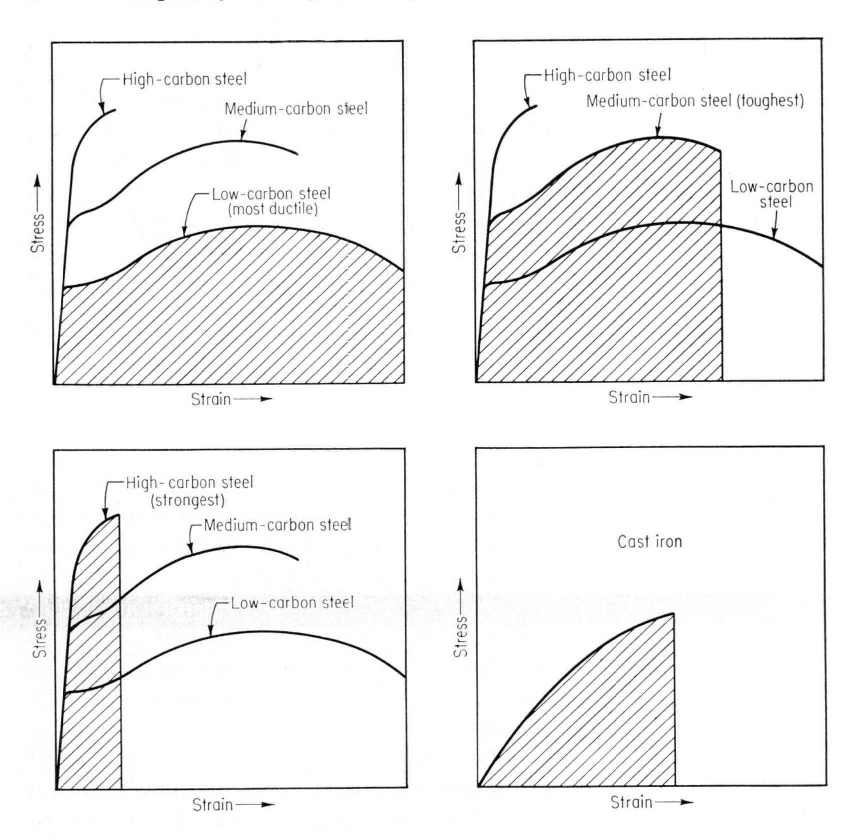

Fig. 2.11. **Toughness in terms of area under stress-strain diagram.**

2.16. Failure. In connection with tests to failure of materials and of structural parts or members, it is important to observe and to record the type of failure and the characteristics of the fracture. Such observation should include not only the phenomena associated with final rupture but also all evidences of change of condition such as yield, slip, scaling, necking down, localized crack development, etc. Although observations of failure are necessarily qualitative, much can be learned from a study of failures, and with experience it is possible to recognize from a break the kind of stress that caused failure and something about the kind and condition of the material. In this connection, it is important to be alert to discover the presence of flaws and defects, for premature failure is frequently caused by defects. Defects may be cause for invalidating the test; also, with any material it is desirable to form an estimate of the frequency of occurrence of defects. Further, distinction should be made between failure of the *material* and failure of the test piece owing to conditions inherent in the setup, as by buckling or elastic instability.

The characteristic fractures under the various kinds of loading are discussed in detail in the chapters on particular types of tests.

For the testing engineer engaged in research work, the mechanism of failure is of interest. A fairly adequate picture of the nature of both elastic breakdown in metals and rupture in brittle materials has been developed and is discussed in Chap. 3. For discussions bearing on some of the broader aspects of the strength of materials and the significance thereof, see Refs. 163, 202, 231, 240 to 245, 250, 260.

2.17. Auxiliary Measurements and Tests. In conjunction with mechanical tests a number of auxiliary measurements and tests may be necessary, usually for furnishing data on dimensions, area, volume, weight, density, etc. These are mentioned here for the purpose of directing attention to the desirability of making such measurements with a precision consistent with that of load and strain measurements.

2.18. Selection of Test Specimens. Specimens must be selected and prepared so as to give a reliable indication of the properties of the materials or parts they represent. There are two problems involved in the selection of specimens for testing: (1) the setting up of physical procedures for securing samples and (2) the determination of the number of specimens or frequency of tests necessary. As regards the first problem, standard specifications for sampling have been especially prepared by the ASTM for a number of engineering materials, e.g., cement, C 183; lime, C 50; brick, C 67; tile, C 112; concrete masonry units, C 140; concrete aggregates, D 75; wood, D 143; and rubber, D 15. In addition, many specifications for particular materials contain sampling requirements. Some of the considerations involved in the selection of specimens are outlined below. As regards the second problem, a guide is furnished by sampling theory, which is outlined in Refs. 1105 and 1130 to 1132, although in many instances the number of specimens or tests to be used is based on custom or experience. Although for routine testing and inspection the sampling procedure is usually definitely prescribed, nevertheless it is desirable that the inspector have sufficient background knowledge of the production of a particular material to know what constitutes a really representative sample. Above all, common sense should be used in both selection and preparation of specimens.

In testing material from metal plate, due regard should be given to the direction of rolling; sometimes tests are made on specimens cut both parallel with the direction of rolling and at right angles to the direction of rolling. Directional effects are particularly important in wrought iron, but not so important in steel; there is apparently little directional effect in rolled brass and copper. The strength and ductility of metal cut from rolled structural shapes appear to be influenced to some extent by the working under the rolls; the thinner parts tend to be slightly stronger and less ductile. The properties of metal cut from castings are influenced by the rate of cooling and by shrinkage stresses at changes in section; generally, specimens near the surface of iron castings are stronger. If specimens are cast separately, the size of casting requires consideration. With some materials (e.g., bronze) the composition of the molten mixture may change during the time required to cast the metal, so that a specimen taken at the beginning may not correspond to the average composition of the final product. To obtain representative specimens from forgings or heat-treated parts may offer real difficulties; each case must be treated on its own merits.

With specimens that are molded (mortar, concrete, rubber, plastics, etc.), care must be taken that molding conditions do not cause defects in the specimen; e.g., restraint to shrinkage in the molds may induce cracks in brittle materials. Attention should be directed to the maintenance of known or standard "curing" conditions (moisture and/or temperature) if they tend to influence the results of the tests.

In the selection of specimens of brick, tile, and other ceramic products, consideration must be given to variation in degree of burning, defects, and the like. Specimens of stone should be selected with regard to homogeneity of the deposit and the piece, as well as with regard to the direction of bedding planes.

Specimens of wood should be selected with due regard for the direction of the grain, the density of the wood as influenced by rate of growth, the proportion of springwood to summerwood, the degree of seasoning, and the presence of defects.

2.19. Preparation of Test Specimens. In preparing specimens of metal, if a rough "blank" is taken by shearing, punching, or flame-cutting, the finished specimen should not contain any of the damaged metal. The finished surface of specimens from sheared blanks should be at least $\frac{1}{8}$ in. from sheared faces and at least $\frac{1}{4}$ in. distant from flame-cut faces. Care should be taken not to bend the piece, because the working of the metal tends to change its properties; in tests of specimens cut from tubing, flattening of specimens is sometimes prohibited for this reason.

The finish cut on machined metal specimens should be made by turning, planing, or milling and should give a surface fine enough not to influence the break. If the ends of a specimen are to be threaded, U.S. Standard, or slightly rounded threads, rather than sharp V threads should be cut, particularly if the specimen requires heat-treatment.

Specimens of concrete, mortar, and several other types of materials have to be molded while in the plastic state. Special consideration must be given to each type of material. For specifications for molding concrete, see ASTM C 31.

The sawing, coring, or grinding of stone, concrete, and ceramic materials should receive attention with respect to flatness and squareness of bearing surfaces and adequate smoothness of lateral surfaces. The axes of stone specimens should be so oriented that bedding planes do not form planes of weakness for the particular test specimen.

Wood specimens should be sawed or planed or turned so as to avoid sharp saw cuts or nicks or split fibers on critical surfaces.

The size of the finished specimen is in general governed by the size of the piece or product from which it is taken and by the capacity of the testing machine available to test it. In many materials, the degree of homogeneity or uniformity of structure of the material may dictate the size of specimen that can be used. For example, the diameter of concrete specimens should be three or four times the diameter of the largest aggregate particles.

Dimensions and tolerances for standardized specimens should be noted and adhered to; these are discussed in connection with particular types of tests.

Finally, attention should be given to the marking and identification of test specimens and the method of relating test samples to the lot or lots of material they represent.

2.20. Selection of Testing Apparatus. The selection of apparatus for a particular test involves considerations of (1) purpose of the test, (2) accuracy required, (3) convenience or availability, and (4) economy. In a number of instances, the final choice represents a compromise between the last three.

For commercial testing, the accuracy required should be readily attainable with existing equipment, but this accuracy should be known and should be maintained. The usual requirement for ordinary testing machines is that they be accurate to within 1 percent in the loading range. If the final result is to have the same order of accuracy, then measurements of dimensions and of strains must also be accurate to at least 1 percent and preferably less than 1 percent. For work of higher precision, or of any precision, it should be emphasized that consistent accuracy in all measurements is to be desired. It may require some study of the performance of various pieces of apparatus in order to select the right equipment for a particular test.

Considerations of convenience and economy are generally dictated by the equipment available in a particular laboratory. Certainly, however, within the limits of required accuracy the procedure that is simplest and least time consuming should be selected.

Chapter 3.

THE PROBLEM OF FAILURE

3.1. Aspects of Failure. In the experimental study of materials, the attainment of failure may sometimes be counted as success; a full understanding of why and how it occurred can be a triumph. Confident utilization of the potential strength of materials and structures is aided by insight into the processes of failure.

While complete rupture or disintegration may provide an obvious demonstration that some end point has been reached in the course of subjecting a material or a structure to some treatment, it is only one aspect of the question, "What constitutes failure?" In a sense, it is only a superficial aspect.

The manifestations that failure has occurred sometimes appear after basic processes by which failure was initiated have run their course, so that the real mechanism of failure may be masked or pass undetected. The initial conditions in the microstructure of the material, particular conditions of loading, macro- and microdistortions occurring during the later stages of loading, or shape or restraint of a test piece or structure may critically influence the mode of failure. Astute observation of ambient conditions, as well as a comprehension of the fundamental behavior of materials, is often important in order to avoid erroneous conclusions about the "causes of failure."

With respect to the performance of a material per se, an alteration in characteristic behavior as governed by some basic physical property may be said to constitute failure. For example, if a material is stressed (or strained) beyond the elastic limit (i.e., it does not recover its original shape or length on release of load), it may be said to have reached elastic failure; this does not mean that the ability of the material to exhibit some elastic recovery has been obliterated—rather it means that relative to its initial condition a nonelastic or permanent strain has been induced. In a sense, then, the *concept* of failure is elusive.

On a macro scale, a degree of deformation (elastic or plastic) that is excessive in relation to the acceptable performance of a member of some structure or machine may also define *failure*, even though a considerable reserve of load resistance remains. The choice of what constitutes tolerable deformation may involve subjective judgment, and in such a situation the limit of deformation beyond which failure is judged to have taken place is arbitrary. Further, since the same material may be employed in structures or machines whose tolerable deformations differ, the criterion of failure here is relative to the *use* of the material.

A question that has often been posed is, "Is it stress or strain that causes failure?" A simple, direct answer does not appear to be an answer at all—in part, because the criteria of failure may differ, but more importantly, because there are several possible modes of failure.

Our interest in this chapter is in laying a basis for understanding some of the processes that lead to failure, as well as the nature of failure as a phenomenon of materials under load. We are, then, as much concerned with the nature of *resistance* to deformation as with the conditions under which an end point or instability is reached.

During the past three decades, important advances in, and consolidation of, knowledge concerning the composition and microstructure of crystalline materials have provided greatly improved insight into the resistance of all materials and the mechanisms of failure. The serious student of materials should have, as a background, an appreciation of the fundamental nature of materials such as that provided in Refs. 301 and 302.

3.2. Modes of Failure. It is conceivable (and it may be observed) that failure can occur in three fundamental ways: by slip or flow,* by pulling apart or separation,* or by buckling, as does a column. A combination or

* See Ref. 303 for one of the early statements concerning the distinction between slip and separation as fracture modes.

succession of these actions may take place during the course of loading some test pieces or structures to final rupture.

Slip or flow occurs under the action of shearing stresses. Essentially, parallel planes within an element of a material move (slip or slide) in parallel directions. Continuous action in this manner, at constant volume and without disintegration of the material, is termed *creep*, or *plastic flow*. Slip may be terminated by rupture when the molecular forces (or forces of similar scale) are overcome. An ideal plastic or a fluid could flow continuously and maintain its integrity. Most constructional materials, however, exhibit a relatively *limited range* of plastic action before physical rupture occurs. Tensile or compressive loadings, as well as direct shear, torsional, or flexural loadings, may induce states of stress in which the shearing stresses cause slip. Thus, in a simple tension test, failure may occur due to slip, which is a shear-induced action.

Slip and its end point, shear fracture, may take place through intercrystalline material, or within the crystal lattice along certain critical planes. The planes susceptible to slip depend on the arrangement of the atoms in the lattice.

Separation is an action induced by tensile stresses. It takes place when the stress *normal* to a plane exceeds the internal forces that bind the material together. Failure by separation is often called *cleavage fracture* and is exemplified by the pulling apart of a crystal of mica by forces acting perpendicular to the prominent cleavage planes of such a mineral. States of stress which involve tensile stresses sufficiently critical to cause cleavage fracture may be induced by loadings other than primary tensile loadings; for example, in torsional loading of a material such as cast iron, the induced tensile stresses are responsible for the type of helicoidal fracture surface that normally is observed (see Fig. 6.9*b*). This can be shown by twisting a piece of chalk until it ruptures. Cleavage may occur at grain boundaries, through intercrystalline material, or across susceptible planes within the crystal lattice. The planes susceptible to cleavage depend upon the arrangement of the atoms in the lattice.

Buckling is a compression phenomenon. A complex system of stresses, exemplified by those in an end-loaded slender column or strut after it begins to deflect laterally, is involved; after some critical level of load the resistance of the composite mass drops and the system of forces is unstable. A buckling failure may be induced by loadings other than primary compressive loading; for example, the torsional loading of a thin-walled tube may result in buckling caused by the induced compressive stresses. Or, in a wood beam, under flexural loading, failure may be initiated by the localized buckling of the wood fibers at the compression surface of the beam, as well as by lateral buckling of the compression flange as a whole. The buckling phenomenon is not confined to, or limited by, the range of elastic action.

Several terms have been used to describe observable characteristics associated with failure and fracture. To avoid confusion, distinction should be made among them. The terms *shear* and *cleavage* refer to the crystallographic mode of fracture. A shear fracture results from excessive slip or flow. A cleavage fracture results from separation across a cleavage plane under tensile stress. The terms *ductile* and *brittle* refer to the relative strain that occurs before fracture occurs. A ductile material is one which can undergo appreciable slip or plastic flow before fracture occurs. In particular instances, ductile behavior might be followed by fracture in the cleavage mode. A brittle material is one which undergoes very little or no slip before fracture occurs. A brittle material usually fractures by the cleavage mode, but some slip or flow will usually also have taken place.

For metals, the terms *fibrous* (or sometimes, *silky*) and *granular* refer to the appearance of a fractured surface. Generally, the fracture surface appears fibrous or silky when the fracture is predominantly in the shear mode, and the fracture surface appears granular when the fracture is predominantly in the cleavage mode. However, as pointed out by Gensamer [304], a granular-appearing fracture surface is seldom a fracture that is purely in the cleavage mode, and fibrous-appearing fracture surfaces may contain some grains that have not fractured by shear.

3.3. Mechanism of Slip. Permanent or plastic deformation in most metals* is attributed primarily to the slip of an array of atoms in an intracrystalline plane over another such array in an adjacent plane. Such a slip occurs most easily along planes containing the greatest number of atoms per unit area. The stresses required to cause slip are observed to be much less than those calculated from what is known about interatomic forces. This is attributed to the presence of imperfections in the crystal lattice, called *dislocations*. The presence of dislocations permits the slip of groups of only a few atoms at a time; hence, some movement can take place at a lower stress than would have been required to cause all the atoms in all planes in the lattice to slip at the same time.

As localized slips take place, adjacent material in some metals can now offer greater resistance, so that additional slip can occur less readily. This action is probably accentuated in polycrystalline materials, in which the preferred slip planes are randomly oriented. The increase in strength and hardness observed in some metals, called *strain hardening*, is explained by this mechanism (see Ref. 301, Chap. 4).

In a viscous liquid, flow is also a shear phenomenon, but without the presence of an ordered atomic arrangement and with the condition of relatively weak interatomic forces, any small shear force causes flow.

* Studies of the structure and behavior of ceramic materials, especially in connection with efforts to develop ductile ceramics, are providing additional important confirmation of the concepts outlined here (see Ref. 305).

Viscosity of a liquid is the resistance to continuous flow. The deformation of materials described as plastics, which may have amorphous or microcrystalline structures, can be visualized as a characteristic intermediate between that for a crystalline metal and a viscous liquid (see, for example, Ref. 301, Arts. 5.11 and 11.5).

The *rate* at which slip (plastic deformation) can occur is affected by temperature (higher temperature, greater strain rate with time). For materials such as metals, in which slip of sufficient magnitude results in rupture of the interatomic bonds, the force (or slip) required to cause such shear failure increases, sometimes markedly, at the lower temperatures. High rates of load application also appear to raise the shear resistance governed by slip.

3.4. Mechanism of Cleavage. The concept of cleavage is fairly straightforward. But complete understanding and agreement on how it is initiated, influenced, and propagated in real materials have yet to be achieved (see, for example, Orowan, "Fundamentals of Brittle Behavior in Metals," Ref. 304, p. 139; also Ref. 305).

The critical normal stress which would cause fracture by cleavage would appear to be somewhat increased by decreasing temperature, but it is surmised that the increase in cleavage resistance as temperature is lowered is much less than the increase in resistance to slip. It is also thought that the cleavage resistance is not greatly affected by rate of load application or rate of straining. It is believed that prior plastic deformation may affect the resistance to cleavage [245].

3.5. Shear versus Cleavage Fracture during the Course of Loading. It is conceivable that, for some material at some temperature, a state of stress can be induced such that the maximum shear stress reaches the level of slip resistance before the major principal tensile stress reaches the level of cleavage resistance, or vice versa. Thus, the exhibition of ductile or brittle behavior depends, at least in part, on the ratio of critical shear to tensile stresses. For example, if a state of equal triaxial tension could be induced (no shear stress on any plane), we would expect no slip (brittle behavior) and a cleavage fracture. In the simple tension test of a mild steel at room temperature, we observe a considerable degree of ductility, meaning that the critical shear stress for plastic deformation is attained before the tensile stress reaches the cleavage resistance, and also, judging by the fracture surface, that the shear stress reaches the level of shear fracture resistance before the tensile stress reaches the potential cleavage resistance. In contrast, if this material is subject to simple tension at, say, the temperature of liquid air, it has been observed that the fracture takes place by the cleavage mode [306, 505], although some very small plastic deformation still occurs at this low temperature. This indicates that the tensile stress reached the level of the cleavage resistance at that temperature before the shear stress reached the level of resistance to fracture by the shear mode, but that the shear stresses attained a value sufficient to cause some slip before the test was terminated by the cleavage fracture.

From evidence of this kind, Gensamer [245] postulated a model of the variation of shear and cleavage resistance with temperature, as shown in Fig. 3.1. The model is qualitative and idealized, but helpful in indicating the kind of behavior that may be expected from different loadings at various relative temperatures.

In Fig. 3.1, Curve A represents the possible variation of cleavage resistance with temperature. Curve B represents the variation of slip resistance (in this case, the *onset* of plastic deformation, not shear-fracture resistance) with temperature. For a bar of material subjected to simple tensile loading ($\tau_{max} = \frac{1}{2}\sigma_1$) below some temperature T_2, we should expect completely brittle behavior (bar fails by cleavage, before any plastic deformation can take place); above this temperature, some plastic deformation would take place regardless of the mode of fracture.

If the bar were subjected to torsion ($\tau_{max} = \sigma_1$), the temperature T_1 below which the material would be completely brittle is much lower. On the other hand, if a state of stress were induced such that τ_{max} were *less* than $\frac{1}{2}\sigma_1$, such as might be developed in the plane of the notch of a notched bar in a bend test, a higher so-called "transition temperature" T_3 is noted. It has been observed that some steels which are fairly ductile in the tension test become fairly brittle in a notched bend test at temperatures in the lower atmospheric range.

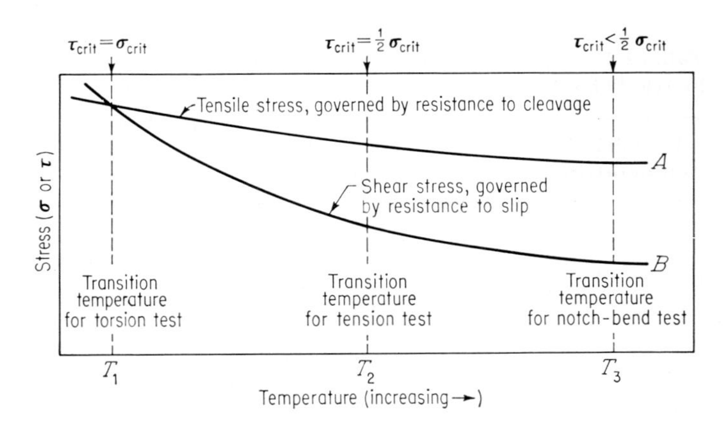

Fig. 3.1. Schematic representation of variation of resistance to cleavage and slip with temperature, and the influence of state of stress on nature of failure. (*After Gensamer, Refs.* 245, 304.)

If curve *B* in Fig. 3.1 were to represent the variation in resistance to *fracture* as governed by shear, then it would indicate similarly the temperatures for transition from shear *fracture* to cleavage fracture. We should expect the transition temperature based on the criterion of the mode of *fracture* to be higher than the temperature for transition from complete brittleness to the presence of some degree of ductility, since the stress required to cause a shear *fracture* is obviously higher than the stress required to initiate slip.

A notch induces a complex state of stress (variable across the plane through the base of the notch) because of the variation in strain in the vicinity of the notch. An increase in both the sharpness of the notch and the thickness of the material parallel to the base of the notch increases the ratio of the critical maximum tensile stress to maximum shear stress. Thus the more "severe" the notch geometry, the higher the transition temperature to be expected.

Although less is known of the variation of slip and cleavage resistances with rate of strain, there seems to be a general agreement among investigators of the subject that increasing the rate of strain has an effect similar to that of lowering the temperature [308, 309, 310]. Thus, if notched bars are subjected to impact tests, we should expect higher transition temperatures than for a bend test in which the load is applied slowly. In the notched-bar impact test, the transition from ductile to brittle behavior is usually brought into the atmospheric temperature range because of the effect of the notch, combined with the rapid strain rate, in reducing plastic deformation. As the condition of the material at the base of the notch becomes a critical factor in determining the type of failure, the transition temperature, as well as the amount of energy

Table* 3.1. Factors influencing temperature of transition from ductile to brittle behavior

Transition temperature	Mechanical factors	Metallurgical factors
Increased by:	Increasing triaxiality of stress Increasing strain rate Increase in size of specimen Increasing prestrain Prior cycle of fatigue	Increase in strain aging Increase in final grain size Increase in carbon content Quenching and tempering
Decreased by:	Shot peening of notched specimens	Normalizing Certain alloying additions such as nickel and copper

* C. W. MacGregor, "Significance of Transition Temperature in Fatigue," Murray, W. M. (ed.): *Fatigue and Fracture of Metals,* Wiley, New York, 1952.

absorbed in causing the bar to fracture, is an index of the effect of composition and structure on toughness and brittleness.

In summary, then, factors influencing the temperature at which transition from ductile to brittle behavior occurs are outlined in Table 3.1.

Or, stated another way, factors which inhibit plastic slip and tend to cause brittle behavior, with a possibility of shifts from shear to cleavage fracture, in otherwise ductile metals are: (1) a state of stress which holds shearing stresses to a small magnitude relative to tensile stresses, (2) localization of the deformation by the presence of discontinuities or notches, (3) very rapid application of load, (4) lowered temperatures, and (5) certain types of structure or composition.

3.6. Tensile Instability in the Plastic Range. In Art. 3.3, reference was made to the phenomenon of *strain hardening*. Because of this, in a tension test after yielding occurs, the stress must be increased to produce increased strain. At the same time, the bar under load decreases its cross-section as it stretches, so that to maintain a given stress level, the load should be decreased. During the earlier stages of the loading in the plastic range, the strain-hardening effect predominates, such that a given load corresponds to a given degree of strain; to increase the strain, the load must be increased. Eventually, however, the tendency to decrease in cross section offsets the strain-hardening effect, so that the load passes through a maximum and must decrease to maintain equilibrium. If it is not reduced, a condition of instability exists and the bar stretches to failure.

At the maximum load point, some segment that is slightly weaker than all other segments along the length of the bar stretches more and decreases in cross section more, because the other segments would require an increase in load to cause additional strain. The condition of instability is thus concentrated in a localized segment of length which forms a "neck."

3.7. Action in Compression. Unless instability develops in the form of buckling, the basic mode of failure of continuous solids under compression loading is by slip. For materials which can undergo appreciable plastic flow, large distortion of cross section can occur without fracture. Those materials which undergo slip will fracture along a plane or a surface on which the shear stress becomes equal to the shear resistance. Some materials, such as concrete, mobilize frictional resistance as well as sliding cohesion on the potential failure plane.

3.8. Elastic Buckling of Columns. Several excellent texts [e.g., 313, 314, 315] cover the general concepts and theory of instability of columns and their tendency to buckle, so only a brief résumé of this subject will be presented here.

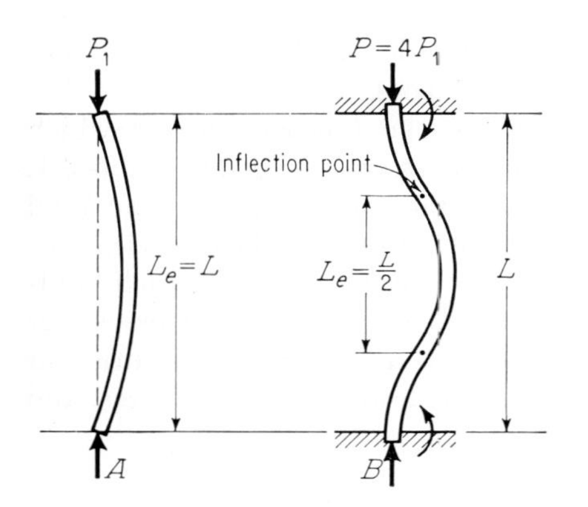

Fig. 3.2. **Effective length of Euler columns.**

The Euler equation for long pin-ended columns shows that the load capacity P_E at buckling is a function of the lateral stiffness of the column designated by EI, rather than of the critical compressive stresses in the material. The Euler equation is

$$P_E = \frac{\pi^2 EI}{L_e{}^2}$$

where L_e = effective length of column

This equation is not applicable to short or intermediate-length columns, since their load capacity depends more on the strength of the material than on the stiffness of the column.

If the ends of a Euler column are fixed against rotation, the effective length L_e is one-half the total length, and the load capacity is four times as great as for a pin-ended column, as shown in Fig. 3.2.

A modified form of the Euler equation is obtained by dividing the critical load by the cross-sectional area of the column to obtain the aver-

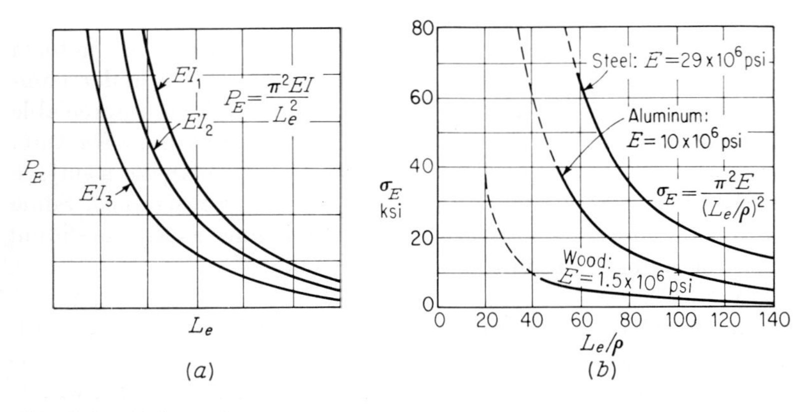

Fig. 3.3. **Euler column curves.**

age Euler stress

$$\sigma_E = \frac{P_E}{A} = \frac{\pi^2 EI}{L_e{}^2 A} \qquad \text{or} \qquad \sigma_E = \frac{\pi^2 E}{(L_e/\rho)^2}$$

where ρ = radius of gyration = $\sqrt{I/A}$
The two forms of the Euler column curves are shown in Fig. 3.3.

3.9. Inelastic Column Action. The Euler equations apply only if buckling failure occurs before the average stress P/A exceeds the proportional limit; that is, they are restricted to long columns. However, if E be replaced by E_t, the tangent modulus, as suggested by Engesser, the critical stress σ_{cr} causing failure is

$$\sigma_{cr} = \frac{\pi^2 E_t}{(L_e/\rho)^2}$$

This tangent-modulus equation involves the effects of different materials, inelastic effects, column cross-section, column length, and end conditions.

By modifying the above equation to

$$\left(\frac{L_e}{\rho}\right)_{cr} = \pi \sqrt{E_t/\sigma}$$

it is possible to obtain the critical slenderness ratio L_e/ρ for a given stress and corresponding tangent modulus as determined from a stress-strain diagram, even though the stress is in the inelastic range. A typical column

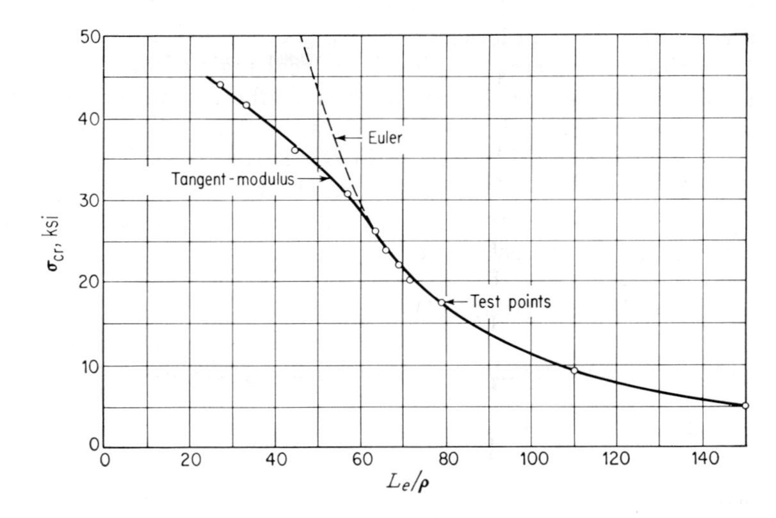

Fig. 3.4. **Tangent-modulus curve determined from stress-strain diagram.** (*From Shanley, Ref.* 314.)

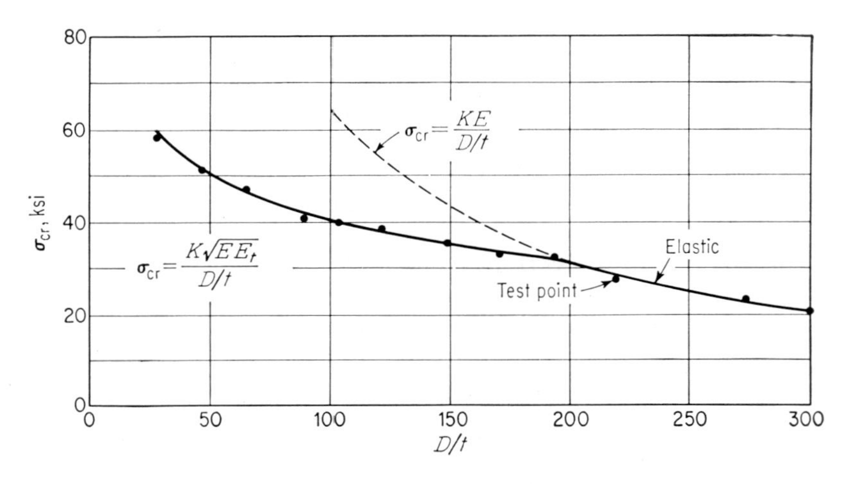

***Fig. 3.5.* Typical test results for local buckling of aluminum alloy tubes in compression.** (*From Shanley, Ref.* 314.)

curve is shown in Fig. 3.4. Tests of actual columns confirm the validity of this procedure, as shown by the test points. When L_e/ρ is less than about 12 the column is no longer affected by the slenderness ratio but instead fails by direct compression as for a short block.

3.10. Local Buckling of Round Tubes in Compression. The *primary* buckling of columns considered in previous sections resulted from excessive bending of the column as a whole while the shape of the cross section remained practically unchanged. *Local* buckling alters the configuration of the cross section in a local region but does not involve a change of the general shape of the member. It usually occurs in thin-walled members, such as tubes, I beams, etc.

The diameter-thickness ratio D/t is the shape factor affecting local buckling of short cylindrical tubes. The result of actual tests plotted in Fig. 3.5 shows that for large D/t values elastic buckling is reasonably represented by the equation

$$\sigma_{cr} = \frac{KE}{D/t} \qquad \text{(Ref. 314)}$$

but for smaller D/t values inelastic buckling occurs at stresses indicated by

$$\sigma_{cr} = \frac{K\sqrt{EE_t}}{D/t}$$

where σ_{cr} = critical local-buckling (crushing) stress
E_t = tangent modulus at σ_{cr}
K = constant, ranging from 0.4 to 0.65 for various conditions

3.11. Lateral Buckling of Beams in Bending. The compression flanges of long beams having no lateral support tend to buckle laterally, in much the same manner as columns. This causes the interior sections to rotate with respect to the end sections, so that the lateral bending stiffness of the compression flange will be amplified by some part of the torsional stiffness of the entire beam. The end sections are usually held in a vertical position, but many loading conditions and many types of beam sections complicate the analysis.

A common semiempirical formula for lateral buckling of I beams and wide-flange beams is

$$\sigma_{cr} = \frac{KE}{Ld/bt} \qquad \text{(Ref. 314)}$$

where d = beam depth
b = flange width
t = flange thickness
K = 0.7 (based on average test values)

The AISC code for steel structures considers KE = 12,000,000 psi after reduction by a factor of safety. Furthermore, it is specified that the safe stress for design shall never exceed 0.6 of the yield strength.

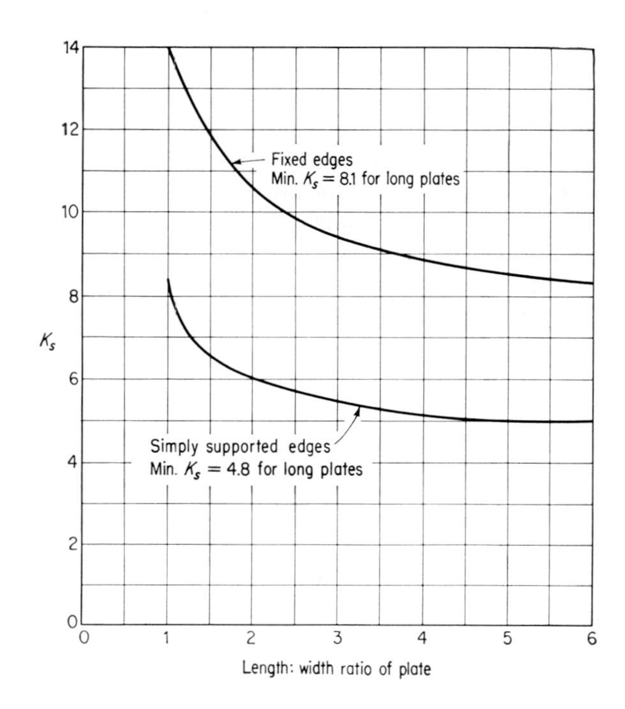

Fig. 3.6. **Buckling coefficient for flat plates in shear.** (*From Shanley, Ref.* 314.)

3.12. Shear Buckling of Flat Plates. Pure shearing stresses cause diagonal compressive and tensile stresses, as shown in Fig. 6.2*a*, so that thin plates subject to pure shear will buckle in compression when the critical shear stress τ_{cr} is reached. The equation which covers both elastic and inelastic buckling is

$$\tau_{cr} = K_s n E (t/b)^2$$

where K_s = shear-buckling coefficient (see Fig. 3.6)
n = inelastic factor = $\sqrt{E_t/E}$ from the compressive stress-strain diagram at values of $\sigma = 2\tau$
E = elastic modulus
E_t = tangent modulus
t = thickness
b = shorter dimension of rectangular plate

3.13. Theories of Failure. A number of theories or mathematical models have been devised which purport to equate the conditions of failure under a complex state of stress to a function or property measured under a simple state of stress, such as that produced by uniaxial tension. For example, according to this theory failure should occur when the loading induces a value of principal stress at failure in simple tension (or compression). Other theories have utilized maximum shear stress, maximum strain, maximum strain energy, etc. Generally, theories of this type have failed to correlate with results observed over the full gamut of possible states of stress, although one theory may "fit" for some materials over a limited range of stress combinations. These theories do not take into account the fact that, depending on ambient conditions, or for some states of stress, the mode of failure may shift. While they may have some circumscribed usefulness for design, they do not appear to contribute insight into physical behavior.

Chapter 4.

MEASUREMENT OF LOAD, LENGTH, AND DEFORMATION—COMMON TESTING APPARATUS

4.1. Measurements in Materials Testing. Although the determination of quantitative measures of the many properties of materials calls for a wide diversity of observations, the basic quantities actually to be measured are relatively few; some of the more important are lengths (including changes in length), angles, volumes, forces (including weights and pressures), time intervals, temperatures, and electrical currents, voltages, and resistances.

In mechanical testing most measurements ultimately have to do with the determination of stress and strain. Although direct comparison with known weights and distances often is used as the means for determining force and length, in general, a variety of physical principles and phenomena is employed in the numerous types of apparatus used to determine load and deformation. In addition to mechanical devices that multiply or magnify load and length changes, there are instruments that take advantage of phenomena such as elasticity, reflection of light, interference of light waves, electrical resistance, magnetism, inductance, and sonic vibrations.

In order to control the accuracy of numerical data, it is necessary to know the error, or the limit of error, of the contributing measurements. The error (i.e., the difference between an observed value and what is believed to be the true value) in the indicated readings of a measuring instrument is normally determined by a process of calibration. The true or correct value is obtained through some more fundamental method of measurement or by comparison with observations on apparatus of known accuracy.

Intimately related to the *accuracy* of an instrument are the *sensitivity* and the *least reading* of the instrument. The sensitivity is expressed in terms of the smallest value of the quantity to be measured corresponding to which there is a response on the indicating device of the measuring instrument; an instrument that requires a relatively large change in magnitude of the thing being measured in order to actuate the instrument is said to lack sensitivity. The least reading is the smallest value that can be read from an instrument having a graduated scale. Except on instruments provided with a vernier, the least reading is that fraction of the smallest division which can be conveniently and reliably estimated; this fraction is ordinarily one-fifth or one-tenth, except where the graduations are very closely spaced. On vernier instruments the least reading is the least count of the vernier. Obviously, the accuracy of a measuring device is limited by its sensitivity and least reading, although a sensitive instrument is not necessarily accurate.

All measurements, except the counting of individual objects, are subject to accidental variation, which must be controlled or known if the final results of a test are to be of known *precision*. The accuracy and precision of measurements are discussed in Chap. 11.

In this chapter there are discussed some of the more common instruments for determining load, length, and deformation and the principles on which they operate. The discussion is confined primarily to the subject of measurement of mechanical properties, since most of the effort in the ordinary materials-testing laboratory is concerned with these. However, the principles developed here apply directly to measurements of many other physical properties, e.g., the determination of thermal expansion or of contraction due to drying involves principally length-change measurements.

4.2. Fulcrums. In any device employing levers, whether it is a testing machine, a balance, or a strainometer, fulcrums or pivots are important details. It is necessary that they operate with a minimum of friction and

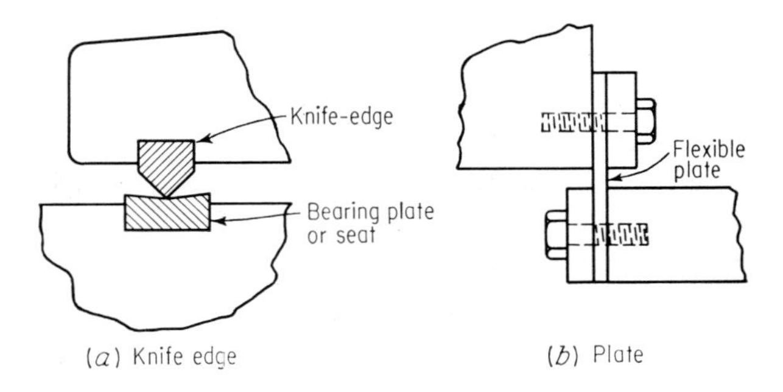

Fig. 4.1. **Fulcrums.**

without lost motion and that they maintain a constant position (lever arm). Further, they should be designed so as to be stable and remain in alignment under load.

In testing machines using a lever weighing system, the fulcrums are usually hardened-steel "knife-edges" in which two ground surfaces meet at a 90° angle to produce a straight line, which is the bearing edge. In small instruments, the angle between the surfaces meeting at the bearing edge is often much less than 90°. The bearing plate or knife-edge seat as shown in Fig. 4.1*a*, which is usually made of hardened steel, also has ground surfaces commonly meeting at a reentrant angle somewhat less than 180°.

In testing machines the allowable compressive load per linear inch on knife-edges is about 7000 lb, although bearing values of 11,000 lb per in. have been used.

An alternate design shown in Fig. 4.1*b* places the fulcrum in tension by using thin sheets of steel with a short unsupported length. Allowable tensile loads of 50,000 lb per lin in. have been used in plate fulcrums. Translatory motion of the joint can be prevented by using two adjacent fulcrum plates at right angles to each other. Crossed fulcrum plates are frequently used in instrument design, as shown in Fig. 4.26*b* [496, 497; see also 101, p. 64].

LOAD MEASUREMENTS

4.3. Determination of Load. In the following paragraphs are described some of the methods of measuring load in materials-testing practice. Certain of these methods may be used alone or in combination with other methods.

Weights. When weights of known magnitude are used directly as the means of applying load, they also serve to measure the load. The procedure is limited in application.

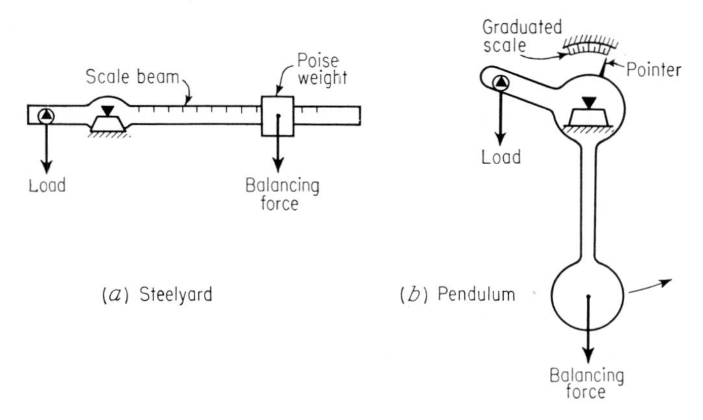

Fig. 4.2. **Weighing by variable lever.**

Weights with Fixed Lever. By means of a horizontal lever, the arms of which are of fixed but not necessarily equal length, a given load in one arm may be balanced by some combination of weights on the other arm. This principle is sometimes used to bring a lever weighing system within a desired *range* of load, but because the process of balancing by continually adding separate weights is a slow one, it is rarely used alone in testing machines. It is, of course, commonly employed in weighing scales of the "balance" type.

Weight with Variable Lever. For testing purposes one of the useful principles of weighing is that of the steelyard, whereby the load applied to the short arm is balanced by a weight of constant magnitude placed at the appropriate point on the long arm. The long arm, or scale beam, is graduated to indicate the load corresponding to the position of the movable weight (sometimes called a rider, poise, or jockey weight) (see Fig. 4.2*a*). Another form of the variable-lever principle is the pendulum, illustrated in Fig. 4.2*b*. The steelyard method of weighing load requires manual operation for balancing; on the other hand, the pendulum method together with the use of a suitable scale is self-indicating.

The actual load to be balanced by the elementary weighing device is often reduced or stepped down from a given full load by an intermediate compound- or multiple-lever system. This is necessary, when large loads are to be measured, in order to keep the weighing device within convenient usable proportions.

Hydraulic Devices. Liquid pressures are commonly measured by means of manometers or bourdon tubes. A manometer is simply a glass tube, usually placed vertically, in which a liquid (say mercury) can rise to such a level that it balances the applied pressure; the level of the liquid is read from a graduated scale. Obviously the manometer is limited to the measurement of relatively low pressures, so that its use for large loads would require an intermediate transmission device to step down the load.

The bourdon tube is essentially a closed-end curved metal tube which tends to straighten out as the pressure is increased in the liquid in the tube. In the usual bourdon gage the motion of the end of the tube is mechanically magnified to rotate a pointer over a scale, as indicated schematically in Fig. 4.3. The accuracy of the ordinary bourdon gage may be considerably affected by temperature changes, hysteresis, and the friction of its moving parts.

The load to be weighed may be transmitted hydraulically through the use of either a hydraulic cylinder and piston or a closed flexible capsule, both of which are shown schematically in Fig. 4.3.

Interconnected hydraulic devices of different piston areas may be used in place of an intermediate lever system to step down the load, and the small piston can be made to actuate a steelyard or pendulum weighing device; this is simply the reverse of the usual hydraulic-jack principle.

The hydraulic cylinder has two marked disadvantages when used in load-weighing systems: the leakage of the liquid in loosely fitted pistons and the variable friction on the piston when packing is used. The friction can be reduced by the use of cylinders fitted with carefully ground and lapped pistons, and it can be still further reduced by rotating the piston during operation of the unit; these devices do not, however, fully eliminate the difficulties, and they complicate the manufacture of the apparatus.

The hydraulic capsule, which operates without appreciable friction, has proved to be a very satisfactory means of transmitting load and has come into wide use. Figure 4.4 shows a schematic cross-sectional view of the Emery capsule, which consists essentially of a thin, flexible metal diaphragm and a heavy recessed block, tightly clamped together to seal in the liquid, usually oil, to form a closed hydraulic system. Because the

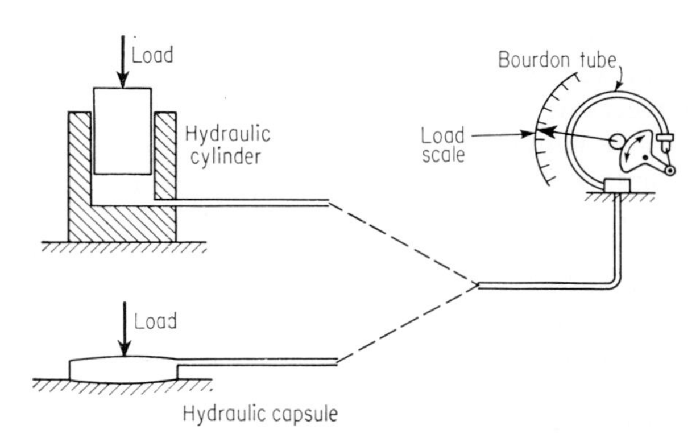

Fig. 4.3. **Weighing by hydraulic pressure.** (*From Gibbons, Ref.* 411.)

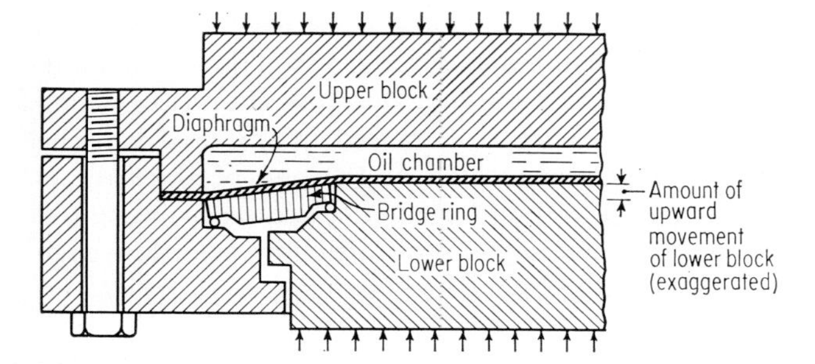

Fig. 4.4. **Partial section of Emery capsule.**

liquid is practically incompressible, the load on the block is transmitted to the oil in the shallow chamber with but little movement of the diaphragm. The oil shown is only in the load-measuring system. Another, separate oil system is used in testing machines which apply the load hydraulically (see Art. 4.6).

Dynamometers. In general, dynamometers are a class of devices by means of which power output or power transmission can be measured. Because the mechanical measurement of power usually resolves itself into the determination of a force (along with other quantities), the term *dynamometer* is often applied to self-contained load-measuring instruments (usually portable).

Many dynamometers (in the restricted sense of a load-measuring instrument) utilize the deformation or deflection of an elastic member as the basis for determining the force applied to the device, although pressure developed in a hydraulic capsule has also been used as the basis for indicating applied force. In use, a dynamometer is inserted in the force circuit and the force to be measured (or a known fraction of the force to be measured) is transmitted through the dynamometer. By calibration under known forces the deflection of the elastic element can be translated directly into terms of force transmitted, using a suitably graduated scale or by applying a calibration factor to the indicated deflections.

In materials testing, two types of dynamometer are commonly used. One type is the spring balance made with a closely wound helical spring which can be used directly to measure the loads on a small specimen or be used in conjunction with a multiple-lever or hydraulic transmission system. In the other type, instead of a helical spring, the elastic deflection of a beam or frame or ring may be used to measure load. The "calibration ring" is such a device; it is described in Art. 4.8 in connection with the calibration of testing machines. Some elastic devices have electric resistance-wire gages (as described in Art. 4.15) mounted permanently on them to measure strains so they may serve as dynamometers.

4.4. Testing Machines. Two essential parts of a testing machine are (1) a means for applying load to a specimen and (2) a means for balancing and measuring the applied load. Depending upon the design of the machine, these two parts may be entirely separate or they may be superimposed one on the other. In addition to these basic features, there are a variety of accessory parts or mechanisms, such as devices for gripping or supporting the test piece, the power unit, controllers, recorders, speed indicators, and recoil or shock absorbers.

The load may be applied by mechanical means, through the use of screw-gear mechanisms, in which case the machines are referred to as "screw-gear" or "mechanical" machines. When the load is applied by means of a hydraulic jack or press, the device is called a *hydraulic* machine. The power may be supplied by hand or by some prime mover (usually an electric motor) to a pump or gear train, depending upon the design of the machine and its capacity.

Some machines are designed for one kind of test only, such as a tension machine made for testing chain and wire; and others are made for compressive tests only. However, if a machine is designed to test specimens in tension, compression, and flexure, it is called a *universal testing machine*. There are also special machines for torsion, hardness, impact, endurance, cold-bending and other tests. In some of these special machines, the load is not measured.

Sometimes it is advantageous to have the specimen horizontal, as when testing chain or long specimens of wire rope. On the other hand, vertical machines are preferable when testing columns so as to prevent the bending of the column due to its own weight, as would occur if it was in a horizontal position.

The various features mentioned above, together with the type of load-indicating mechanism and the size or capacity, serve to classify a testing machine.

Except for the direct use of weights, in its elementary (and earliest) form the testing machine consisted of a single lever, which was used both to apply and balance the load (see Fig. 4.5*a*). In such a machine, the lack of means to compensate for the deformation of the specimen and movement of the machine parts was a serious disadvantage, so that the next step in the development of the testing machine was to provide a means of loading independently of the means of weighing (see Figs. 4.5*b* and 4.5*c*) [411]. In Fig. 4.5*b* the load is shown to be applied hydraulically, whereas in Fig. 4.5*c* a mechanical means is employed. Either system could have been used in both machines. Such machines are of interest in demonstrating fundamental action, although single-lever machines are now rarely used in this country.

Two principal types of universal power-driven machines are now in common use in this country: (1) screw-gear machines with multiple-lever

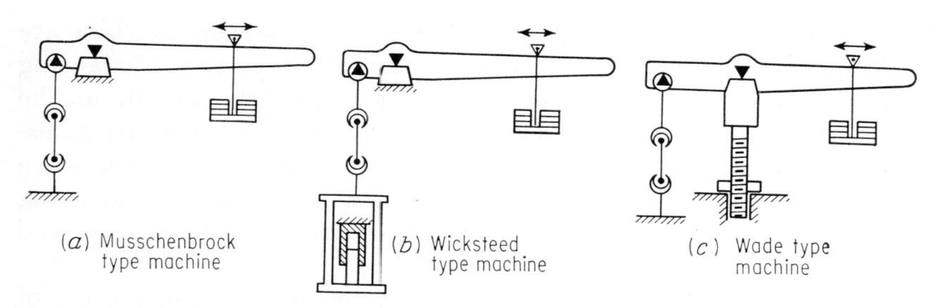

Fig. 4.5. **Early types of testing machines.** (*From Gibbons, Ref.* 411.)

and movable-poise or pendulum weighing devices or with electronic load-measuring devices, and (2) hydraulic machines, which in the more precise types use the Emery capsule and a modified bourdon tube, or a bourdon tube in combination with an isoelastic spring or an electronic device, to weigh and indicate the load.

In a mechanical machine the load ordinarily is applied to a specimen through the "movable crosshead" (see Fig. 4.6). In the case of a specimen in tension, the load is resisted by a "fixed crosshead," which may, however, be placed in any one of several positions. In a compression or cross-bending test the load is resisted by the "bed" or "platen" of the machine. In screw-gear lever-type machines, the fixed head or the platen then transmits the load to the compound-lever weighing system. In a hydraulic machine the load ordinarily is applied by the movement of the piston of the hydraulic system, which is connected either to the platen of the machine or to a movable crosshead. The load-weighing mechanism may originate in either the fixed or movable portions of such machines.

The capacities of screw-gear machines are generally less than 300,000 or 400,000 lb, although one with a capacity of 3,000,000 lb has been built.

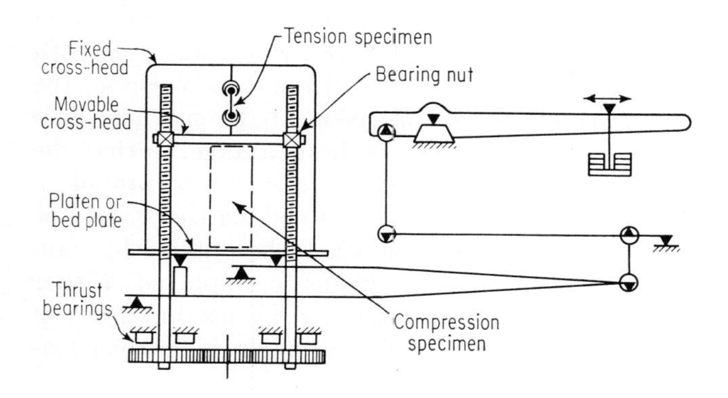

Fig. 4.6. **Schematic diagram of screw-gear testing machine. The movable crosshead always moves downward during any test.** (*From Gibbons, Ref.* 411.)

The difficulty of providing adequate knife-edge fulcrums that must carry the full load developed by the machine places a practical upper limit on the size of machines using levers to measure the load. The American screw-gear lever-type machine of ordinary capacity (say 30,000 to 300,000 lb) can be made to be accurate and sensitive. It has proved to be very serviceable and, until recently, has found wide acceptance, although it has the disadvantage of requiring a large amount of floor space and may produce some noise and vibration.

The hydraulic machine offers a means of attaining very large capacities. The largest machine, used for compression only, is the 10,000,000-lb machine at the National Bureau of Standards [410]. A number of universal machines in the range from 1,000,000 to 5,000,000 lb are in use. In recent years, hydraulic machines of ordinary capacity have come into wide favor. In the modern hydraulic machine the load can be applied quickly, easily, with little noise or vibration, and with good control of the loading rate. The cheaper hydraulic machines that use the pressure in the loading cylinder as an indication of the load on the specimen may be susceptible to larger inaccuracies, but the better machines, especially those which incorporate the Emery capsule, can be very accurate. The cost of a good hydraulic machine of ordinary capacity is not greatly different from that of a good screw-gear machine.

Some of the general requirements for testing machines are as follows:

1. Required accuracy must be obtained throughout the loading range; errors are ordinarily required to be less than 1 percent, but 0.5 percent or less is desirable.
2. Should be sensitive to small load changes.
3. Jaws in the crossheads should be in alignment.
4. Movable heads should not rock, twist, or shift laterally.
5. Load application should be uniform, controllable, and capable of considerable range in speed.
6. Should be free from excessive vibration.
7. Recoil mechanism should be adequate to absorb the energy of rupture of test pieces breaking suddenly so as to avoid injury to the machine when loaded to capacity.
8. Should be capable of easy and rapid manipulation and adjustment and should permit easy access to specimens and strainometers.

Sometimes autographic or semiautographic stress-strain recorders are used. Gripping devices, bearing blocks, and specimen supports are described in connection with the tests to which they apply.

4.5. Screw-gear Machines. In some universal machines, a motor-driven screw-gear mechanism actuates the movable head, which transmits the load through the test piece directly to the platen or to the fixed

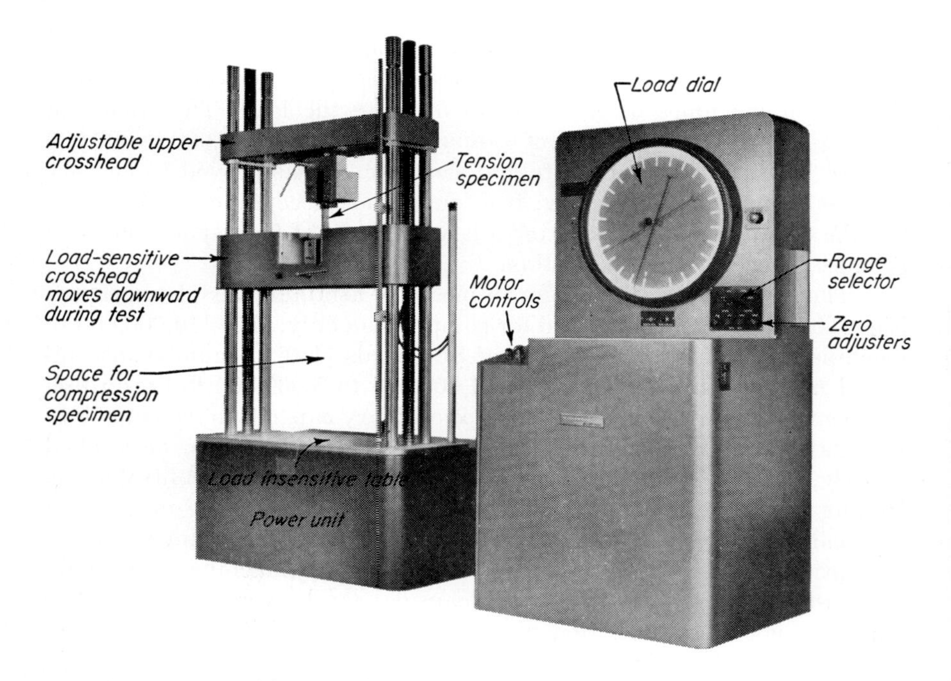

Fig. 4.7. **Mechanical universal machine having a universal electronic load cell in movable crosshead.** (*Courtesy of Wiedemann Machine Co.*)

head and then indirectly to the platen. The load on the platen may in turn be balanced by a multiple-lever system that terminates with the scale beam and poise weight, as shown schematically in Fig. 4.6; however, some screw-gear machines built in recent years weigh the load by a direct-reading pendulum weighing system, the principle of which is illustrated in Fig. 4.2*b*. One of the newer types of mechanical testing machines (Wiedemann-Baldwin, shown in Fig. 4.7) differs from all others in that the load is measured by means of an SR-4 universal load cell (see Art. 4.15) which electronically operates the load indicator [403 *n*].

In some testing machines the screws themselves rotate within bearing nuts mounted in the movable head as shown in Fig. 4.6; in other machines the screws are fixed to the movable head, and the bearing nuts are in the gears below the bedplate. Either system serves satisfactorily to move the head.

Machines having two, three, or four screws are used. The two-screw machines are well adapted to tension and transverse tests, but when used for compression tests, to avoid bending the screws care should be exercised to place the specimen in the plane of the screws and midway between them. The specimen is not so accessible in the three- and four-screw machines as it is in those having only two screws, but the former are not so readily damaged by accidental eccentricity or off-center loads.

Single-screw machines are sometimes used for tension tests of wire, rubber, fabric, or cement briquets.

The knife-edges of a machine should be checked from time to time, because if they become dulled, chipped, clogged with debris, or displaced on the seat, the accuracy and sensitivity of the machine may be appreciably reduced.

To prevent the platen from jumping off the knife-edges when a specimen ruptures suddenly, hold-down bolts with rubber-insulated recoil nuts are used. These nuts should *not* bear on the platen and should always be checked at the beginning of a test to see that they are just barely loose.

At the beginning of any new test, or series of similar tests, the scale beam or indicating mechanism should be balanced at zero load, with all the equipment on the platen that is to be used in the test but that will not stress the specimen. The scale beam or indicating mechanism is usually balanced by adjusting small counterweights.

Some screw-gear machines are designed so that different poise weights or pendulum weights may be used. By using a weight that is smaller than that required for full-capacity loads, the machine can be used for small loads with a precision somewhat comparable to that for large loads, provided adequate sensitivity can be maintained.

4.6. Hydraulic Machines. The main features of two types of hydraulic machine are shown diagrammatically in Fig. 4.8. In type A, load is applied by a hydraulic press and is measured by the pressure developed within the hydraulic cylinder. The main piston is usually carefully fitted and lapped; to reduce the friction of the small piston used in the weighing system, the latter piston is rotated during operation of the machine. In

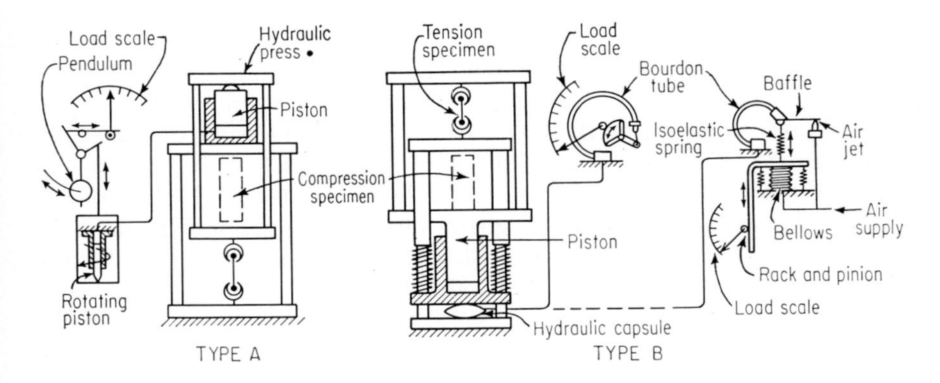

Fig. 4.8. **Essential features of two types of hydraulic testing machines.** (*From Gibbons, Ref.* 411.)

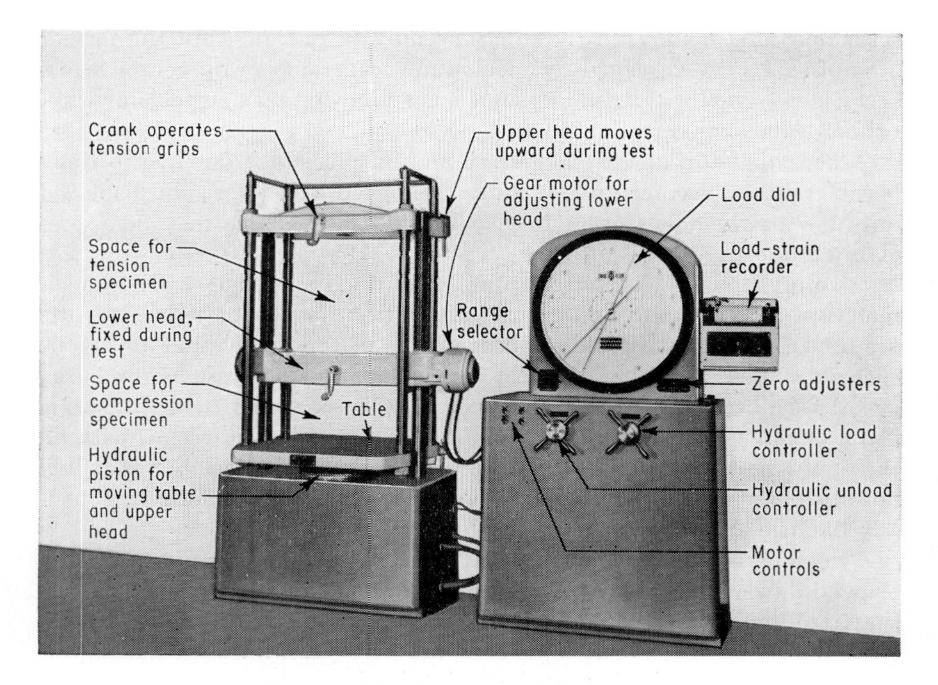

Fig. 4.9. **Hydraulic universal machine having electronic load indicator connected to tip of bourdon tube.** *(Courtesy of Tinius Olsen Testing Machine Co.)*

the diagram shown, the load is finally weighed by a pendulum device, although a bourdon tube is used in some machines.

In machines of type *B*, the load is applied by a hydraulic press independently of the weighing system, which is actuated by a hydraulic capsule. In some machines, such as that shown in Fig. 4.9, the very slight motion of the tip of the bourdon tube operates electronic units which in turn operate the load pointer (see Art. 4.25). In other machines the direct use of the bourdon tube has been replaced by a mechanism, operating on the "null" method, indicated at the right of Fig. 4.8. In this method, a small motion of the tip of the bourdon tube moves the baffle on an air jet and allows the air pressure in the jet and hence in the bellows to decrease. The springs to the right and left of the bellows collapse the latter and extend the isoelastic spring (a spring of constant modulus) which is attached to the bourdon-tube tip. This movement restores the baffle on the air jet to its original position and actuates the pointer on the load scale. This method overcomes the well-known disadvantage of the ordinary bourdon tube, namely, that it does not give a straight-line relation between pressure and motion of the tip. In some large machines the hydraulic capsule is located in the movable crosshead.

In large-size, well-designed machines, such as the one illustrated in

Fig. 4.10, it is possible readily to change the location of the movable crosshead. This is accomplished by rotating the two screws that turn in bearing nuts in the crosshead. These screws do not rotate during application of the load, however. The movable crosshead can be used to place the fixed crosshead in one of several positions on the side columns or main framework, at which points the fixed head can be fastened by keys or shear pins. These adjustments of the crossheads are necessary to accommodate various lengths of test specimens.

In machines of small to moderate capacity, the recoil energy may be absorbed by rubber pads, but on large machines hydraulic recoil cylinders are used.

In a closed hydraulic load-indicating system, the presence of air in the system causes erratic and inaccurate measurements. Whenever repairs

Fig. 4.10. **A 5,000,000-lb Baldwin universal hydraulic testing machine in the Naval Air Material Center at Philadelphia. A compression panel is ready for test, and a tension bar is placed in the upper crosshead for a subsequent test.** (*Official photograph, U.S. Navy.*)

have been made, or if there is any possibility of leaks, the reliability of the load-indicating system should be checked.

Most hydraulic machines are equipped with two or more load-indicating dials to serve for different ranges of load, or they have one dial with a mask that can be rotated to expose different groups of figures and thus permit the one dial to serve for various ranges of load. Suitable load-measuring mechanisms are used for each load range so that small loads can be observed with a precision comparable to that for large loads.

4.7. Speed Adjustment. The driving mechanisms for screw-gear testing machines are usually made to operate the head at four or more speeds. The several speeds may be obtained by the selective use of different gear ratios, by the use of various fixed motor speeds, or by electronically controlled drives which permit the use of any desired speed of testing [420, 421]. In most modern hydraulic machines, any desired rate of load application can be obtained by the use of an appropriate pump speed or valve setting controlling the flow of oil from pump to loading cylinder. In such machines, the load rate is often controlled through the use of an auxiliary pacing arm or disk on the load-indicating dial; to apply load at a desired rate, the operator sets the pacer to operate at the given rate and then adjusts the motor or pump controls to make the load-indicating pointer follow the pacer.

4.8. Calibration of Testing Machines. Three methods commonly used to calibrate testing machines are (1) use of weights alone, (2) use of levers and weights, and (3) use of elastic calibration devices (ASTM E 4).

When they can be used, standardized weights afford a simple means of calibration. The usual weight is a 50-lb unit. Such weights are often made accurate to the nearest 0.01 lb or less by comparison with a known standard. Weights alone are suitable only for use with vertical-type testing machines that operate the weighing mechanism by a downward pressure on the platen. The use of weights is limited by the available space on the platen of the machine and by the number of weights available; often twenty 50-lb weights, a total of 1000 lb, are used. In special cases, standardized weights weighing 10,000 lb have been used [414].

The range over which calibrated weights can be used may be increased by the use of a pair of levers, which are usually made with a 10:1 lever ratio, so that 20 weights give an effective load of 10,000 lb. A common arrangement of these so-called "proving levers" is shown in Fig. 4.11. For calibrating horizontal testing machines, "bell-crank" levers (i.e., levers the two arms of which are at right angles) are sometimes used. The lever ratio of any lever system should be determined by a loading test rather than by a direct measurement of the lever arms.

General limitations to the use of weights, or levers with weights, are

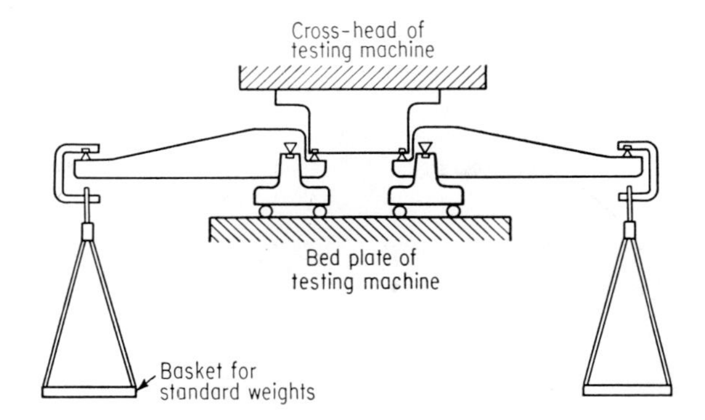

Fig. 4.11. **Proving levers (ASTM E 4).**

that they are inconvenient to transport and they can be used only for a relatively small range of load—usually less than the usable loading range of intermediate- and large-capacity testing machines.

Probably the simplest and most common method of calibrating the largest capacity machines is by use of an elastic calibration device, which consists of an elastic metal member or members, combined with a mechanism for indicating the magnitude of deformation under load. Two forms of this device are (1) a steel bar together with an attached strainometer and (2) a "calibration ring," which is a steel ring or loop combined with some type of deflection indicator. The steel bar is suitable principally for use in tension, although some bars are used in compression. The ring or loop devices are made for use either in tension or in compression. A calibration ring for use in compression is illustrated in Fig. 4.12. A compressive load shortens the vertical diameter, and this change is measured by the micrometer. From this change and the calibration data for the ring, the applied load can be determined. Calibration rings of this sort are available in capacities up to 300,000 lb, but compression bars having capacities up to 3,000,000 lb, which are equipped with electronic strain gages, are available at the National Bureau of Standards. Also, for calibrating very large machines in compression, several calibration rings or bars can be used in parallel.

Following are three important requirements of an elastic calibration device (ASTM E 74):

1. It should be so constructed that its accuracy is not impaired by handling and shipping and that parts subject to damage or removal can be replaced without impairing the accuracy of the device.

2. It should be provided with shackles or bearing blocks so constructed that the accuracy of the device in use is not impaired by imperfections in the shackles or blocks.

3. It should be calibrated in conjunction with the strainometer that is to be used with it, and the strainometer should be used in the same range as that covered by the calibration.

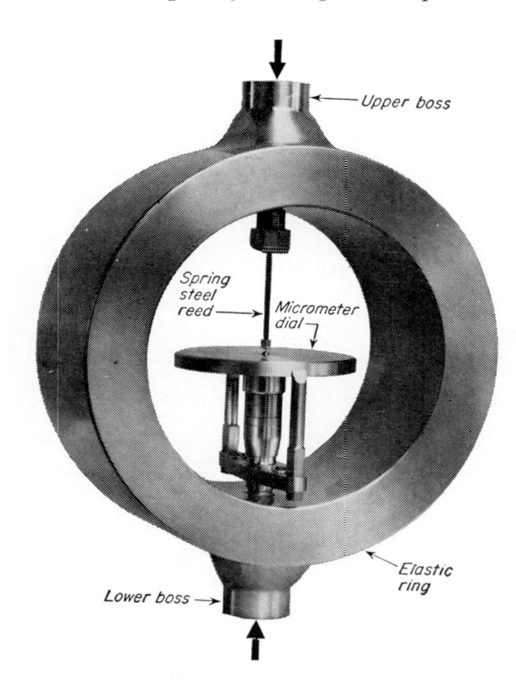

***Fig. 4.12.* Calibration ring for use in compression.** (*Courtesy of Morehouse Machine Co.*)

Care must be taken to minimize any temperature changes during the use of an elastic calibration device. Furthermore, the actual temperature at time of use and at time of its own calibration must be known, since the elastic properties of the device change with temperature. In general, the reading of a ring-type device changes by about 0.015 percent for each degree Fahrenheit change in temperature from the standard.

Devices of this sort may be calibrated directly by standard weights. A testing machine having a capacity of 111,000 lb, produced by the use of standard weights, is available at the National Bureau of Standards [412]. For loads in excess of this, calibration is performed by the use of combinations of previously calibrated devices. The ASTM requires that the calibrating loads should be accurate to 0.02 percent for loads up to 100,000 lb and to 0.1 percent for loads from 100,000 to 300,000 lb, but allows somewhat larger errors for greater loads (ASTM E 74). Before calibrating an elastic device it is desirable to subject it to a series of cyclic loadings over a range about 20 percent greater than that of its proposed use. The usable loading range for ordinary use is defined as that range in indicated loads for which the percentage error does not exceed specified tolerances (ASTM E 74).

In all ordinary calibration work, the calibrating loads should be applied so that the resultant load acts as nearly as possible along the axis of the weighing head. In special instances, calibrations may be made with the load applied at known eccentricities.

Distinction should be made between the *calibration* of testing machines, or the procedure of determining the magnitude of error in the indicated loads, and what the ASTM (ASTM E 4) calls the *verification* of testing machines. Verification has to do with ascertaining whether or not the errors are within a stated permissible range, and it implies certification that a machine meets stated accuracy requirements. The "permissible variation," or maximum allowable error in indicated load of a testing machine, is 1 percent. The "loading range" is the range of indicated loads for which the machine gives results within the specified permissible variation. The allowable loading range should be stated in any verification certificate. It is recommended that no correction be used with machines tested and found to be "accurate" within the limits prescribed (ASTM E 4).

It is specified that calibration corrections shall *not* be applied to indicated loads to obtain values within the required range of accuracy. Obviously this implies that the machine must be adjusted or modified until the calibration shows it to be within specified limits. Subsequent calibrations that establish the fact that errors are within prescribed limits are called *verifications.*

The adjustment of machines having a poise weight or a pendulum is readily accomplished by changing the weight of these elements. For hydraulic machines having an isoelastic spring, adjustment is accomplished by changing its effective length; for those using an electrical gage at the tip of the bourdon tube, adjustment is accomplished in the tube connection; for those having a simple bourdon-tube load gage, adjustment must be made in the linkage of the gage.

Temperature changes do not affect the accuracy of a mechanical machine but do have a slight effect on all hydraulic machines using a bourdon tube. However, for normal temperature changes, the errors so introduced are usually less than about 0.1 percent.

Directions for the calibration of a testing machine by use of weights alone and by use of an elastic calibration device are given in Part 2.

STRAIN MEASUREMENTS

4.9. Measurement of Length. With a few notable exceptions, the operation of making quantitative linear measurements ultimately reduces to the making of readings on a graduated scale, and the latter essentially consists in estimating the position of some mark (index line, pointer, or the like) along the scale. To obtain an accurate estimate, it is necessary to eliminate parallax, which is usually done in one of two ways, by "edge coincidence" or by use of the "mirror-scale" principle. In the former method, the mark is made to lie in the plane of the scale graduations. One form of a mirror-scale device is shown in Fig. 4.13. When the cross

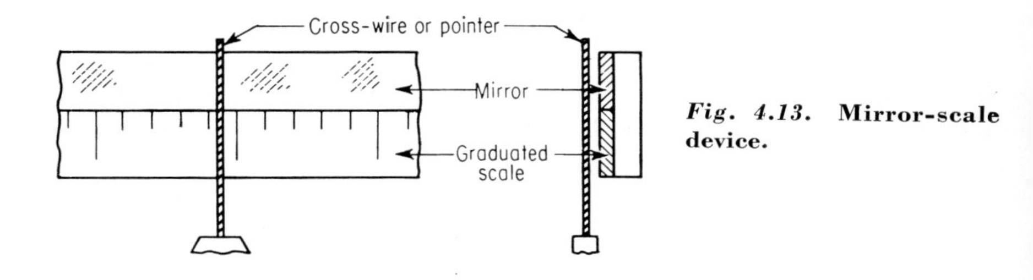

Fig. 4.13. **Mirror-scale device.**

wire or pointer appears to coincide with its image, the line of sight through the cross wire is perpendicular to the scale and mirror.

In the simplest case, the position of a mark along a scale is obtained by estimating its distance from an adjacent graduation. The least reading of a scale depends upon the spacing of the graduation marks, and wherever possible it is desirable to estimate tenths of divisions. For greater refinement in reading fractions of a division, a vernier may be used.

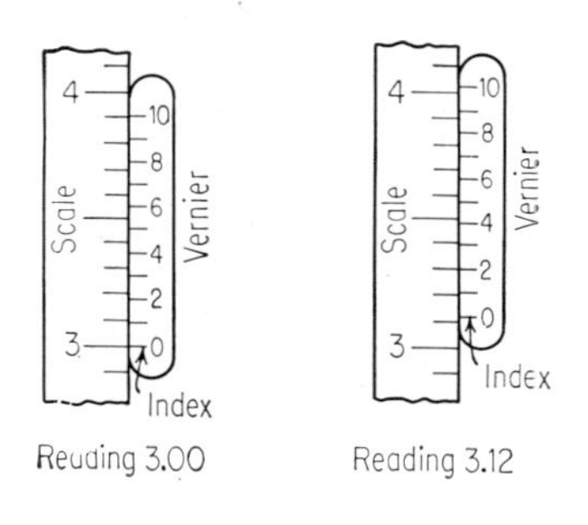

Fig. 4.14. **Direct vernier.**

A direct-reading single vernier, perhaps the most common type, is shown in Fig. 4.14. A distance equal to nine divisions of the scale is divided into ten equal divisions on the vernier. Then each division on the vernier equals nine-tenths of a division on the scale. Hence, if the first mark of the vernier beyond the zero or "index" mark matches up with any mark on the scale, the index is one-tenth of a division beyond the preceding scale mark; if the second mark on the vernier matches up, the index is two-tenths of a division beyond the preceding scale mark; etc.

Measurement of the distance between two points can be made directly by comparison with a graduated steel scale or tape. The distance between opposite surfaces of a solid object is commonly determined by the use of a caliper, the separation of the points of which can be measured directly with a scale. For small distances, the direct use of the graduated scale yields results of limited accuracy, because the practical least reading of a scale with the unaided eye is about 0.01 in. Resort then is made to a micrometer (i.e., a small-distance measurer) for making finer measurements.

4.10. Micrometers. In principle, a micrometer is simply an instrument for giving a magnified indication of small distances. In many micrometers, the distance is, in effect, traversed by some moving part, and the resulting movement is magnified and measured. The determination of distances

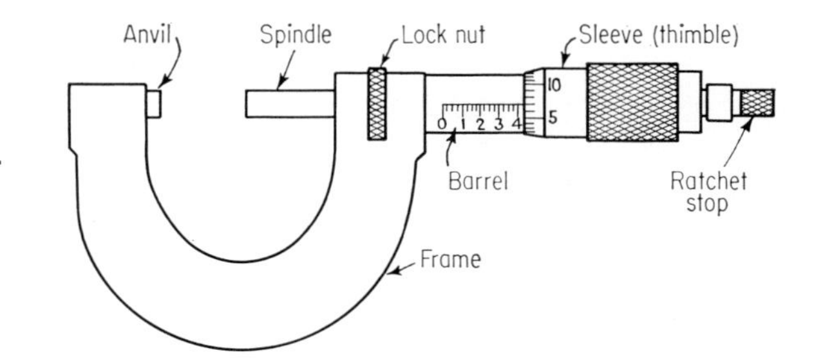

Fig. 4.15. **Screw micrometer caliper.**

greater than the range of motion of the micrometer device requires that measurements be made with respect to some fixed point whose position is accurately known.

Screw Micrometer. Perhaps the simplest form of micrometer is the screw micrometer. A common illustration of the screw micrometer is found on the ordinary micrometer caliper, shown in Fig. 4.15. Here the screw generally has 40 threads per inch (thread pitch = 0.025 in.), and the barrel has 25 divisions, so that a $\frac{1}{25}$ turn gives a motion of the spindle (and a corresponding reading) of 0.001 in. More precise screw micrometers are made which are graduated to 0.0001 in. to give a practical least reading of 0.00001 in. The range of travel of the spindle in micrometer calipers is usually not more than 1 in., but such devices are available for measuring lengths of 2 ft or more.

In many uses of the screw micrometer, the end of the spindle or screw must make contact with the piece with reference to which measurements are being made. Some method of controlling the contact pressure is necessary if consistent results are to be obtained. The machinists' micrometer caliper (see Fig. 4.15) is often equipped with a spring ratchet that releases at a definite contact pressure. In some early uses of the screw micrometer on strain-measuring instruments, contact of the spindle closed an electric circuit and caused a bell to ring or a light to flash. In some recent uses of the screw micrometer (e.g., to measure deflection in a proving ring), the spindle is positioned accurately by the use of a vibrating reed arranged as indicated in Fig. 4.16. The reed is set in motion, and the screw is advanced until it comes in contact with the reed and alters its tone.

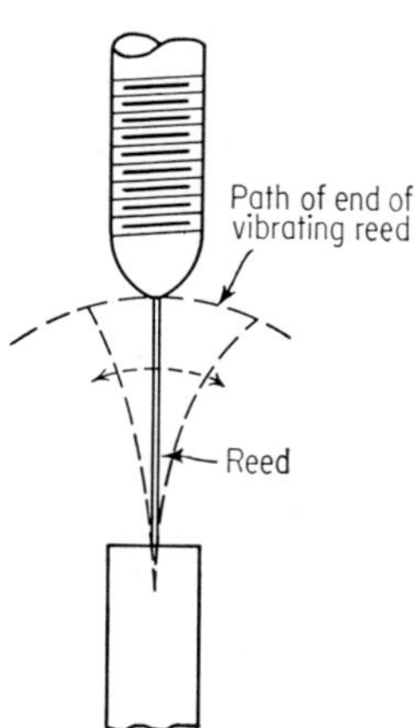

Fig. 4.16. **Vibrating-reed principle of setting screw micrometers.**

Micrometer Microscope. Another device is the micrometer microscope. In one form, the magnified image of the points or marks (with

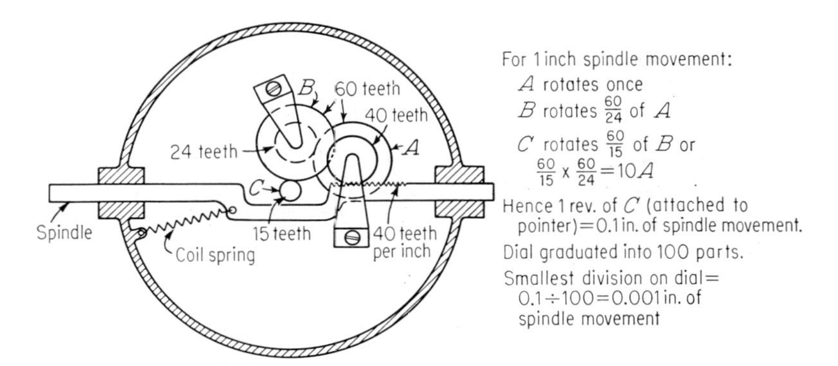
60 teeth
40 teeth
24 teeth
B
A
C
Spindle
15 teeth
40 teeth per inch
Coil spring
For 1 inch spindle movement:
A rotates once
B rotates 60/24 of A
C rotates 60/15 of B or
60/15 x 60/24 = 10A
Hence 1 rev. of C (attached to pointer) = 0.1 in. of spindle movement.
Dial graduated into 100 parts.
Smallest division on dial = 0.1 ÷ 100 = 0.001 in. of spindle movement

Fig. 4.17. **Typical dial-indicator mechanism (Ames, Federal, Starrett).**

respect to which the measurement is to be made) is superimposed on a scale within the instrument. In another form, the magnified distance to be measured is traversed by a movable cross hair the motion of which is indicated by a screw micrometer. The practical least reading of an ordinary micrometer microscope is about 0.00005 in. with a range of about 0.1 to 0.2 in.

Dial Micrometer. A type of micrometer now in wide use is the dial micrometer or "dial indicator." In these instruments, the motion of the spindle actuates a lever or gear train, which in turn operates a pointer on a graduated dial. The dial indicator has the great advantage of being self-indicating.

The internal mechanism of one form of dial indicator (Ames, Federal, Starrett), which uses trains of gears, is shown in Fig. 4.17. It should be noted that in this device the rack drives a pinion, which in turn drives a gear. This is the reverse of normal gear-train operation and makes bearing friction important, so that, on the better grades of dial indicators, jewel bearings are used. In the ordinary dial indicator, the smallest division on the dial corresponds to a spindle movement of 0.001 in., giving a least reading by estimation of 0.0001 in. Indicators are available, however, which are graduated to 0.0001 in. For an indicator fixed in position, movement of the spindle is used to measure a strain or other value such as thickness, height, etc. These indicators are constructed for various ranges of spindle motion, a common range being 0.2 in.; however, ranges of ½ or even 1 in. are available. Over any considerable range, most indicators of this type are reliable to one or two divisions on the dial. Over a restricted range, however, or by calibration, they can be made to give measurements accurate to a value corresponding to perhaps one-fifth of a division.

Two indicators using lever magnification are shown in Fig. 4.18. In one of these (the Starrett Last-Word indicator), the motion of the lever is

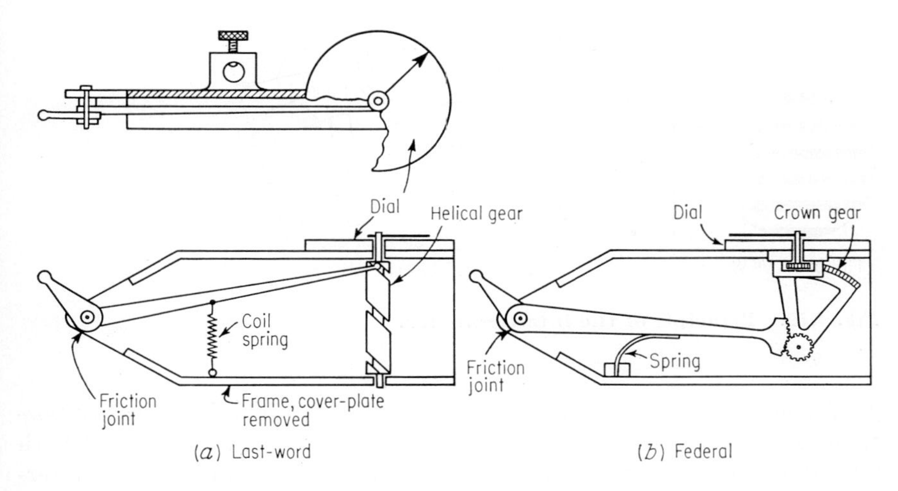

Fig. 4.18. **Lever dial indicators.**

transformed into rotary motion through the use of a shaft cut with a helical groove. In the other (the Federal indicator), an arrangement of pinions and crown gears is used. One main division on these indicators corresponds to 0.001 in., giving a least reading by estimation of 0.0001 in. They are of limited range, usually about 0.03 in. More sensitive models of these indicators are available which are graduated directly to 0.0001 in., but their range is only 0.016 and 0.008 in., respectively, for the two types described.

Light Lever. A means of magnifying angular motion with high precision, which is sometimes adapted to strain measurements, is the "light-lever" method. A beam of light is reflected from a mirrored surface on the movable element, and the movement of the beam (a long lever) is observed on a scale at some desired distance from the mirror (see Art. 4.11).

Interferometer. The most accurate known means of measuring small movements is an adaptation of the phenomenon of interference of light waves. An interferometer is so constructed that, when monochromatic light is reflected from two nearly parallel plane surfaces, reflected rays that are out of phase by one-half wavelength of the light interfere, causing a series of light and dark bands to be seen (see Fig. 4.19). As the distance between the reflecting surfaces changes, the bands appear to move across the field of view. The shift of one band (across a reference mark) corresponds to a movement of the plates of one-half wavelength of the light used.

The interferometer carries its own calibration, as the wavelengths of light are known to a high degree of accuracy. Such an instrument is extremely sensitive and is suitable for very fine measurements. Under

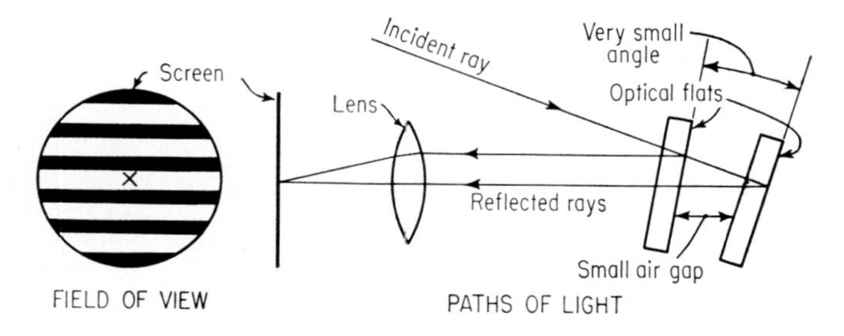

Fig. 4.19. **Principle of the interferometer.**

good conditions, measurements to one-tenth wavelength are easily possible, which for green light from a mercury-vapor lamp, of wavelength 0.0000215 in., corresponds to 0.000002 in. The total range of an interferometer device is small, being about 1000 bands (often less), or 0.01 in.

The interferometer principle finds important use in the calibration and checking of gage blocks where extreme accuracy is required. In this application the bands do not move, and the data are quickly and easily obtained. However, when the interferometer principle is used as the measuring device in a strainometer [456], it makes a delicate instrument that requires skillful handling to secure reliable data, because the relative positions of the separate parts, namely, the light source, reflecting mirror, viewing telescope, and strainometer, must be maintained in good adjustment. Furthermore, the counting of the bands as they pass the reference mark requires constant attention and, with large numbers, is a tedious operation even for the trained observer. It is not recommended for ordinary use.

4.11. Measurement of Change in Length: Mechanical Strainometers. A strainometer has been defined as any deformation-measuring instrument (Art. 2.4). The deformation may be a change in length resulting from linear strains, it may be the deflection of a beam, or it may be an angular twist as in a shaft. The instruments described in this article apply principally to the measurement of linear strains; apparatus for measuring deflection and angle of twist are described in connection with the types of test to which they apply. The length-measuring devices described in the previous articles are the basis on which many of the commonly used strainometers operate.

The majority of the instruments for measuring linear strains are applied to the surface of the test piece—surface strainometers. For measuring internal strains, a few remote-reading instruments have been devised—sometimes called "telemeters" or "strainmeters." Most strainometers remain attached to the test piece during the course of a test, but certain portable instruments, sometimes called "strain gages," can be removed from the test specimen and applied again only when a strain observation is to be made. Depending upon whether it is to measure tensile or compressive strains, an instrument may be called an extensometer or compressometer.

The points between which the deformations are measured are called the *gage points,* and the initial or nominal distance between gage points is called the *gage distance* or *gage length.* The measurement of the overall deformation does not depend upon the gage length if a micrometer type of device is used.

Averaging-type Collar Extensometer. For ordinary strain measurements of axially loaded test specimens, perhaps the most generally useful type of strainometer is the "averaging" type of extensometer or compressometer, shown schematically in Fig. 4.20. This type of instrument gives very closely the average strain in the bar, even though slight bending may occur. The deformation is first magnified (usually about twice) by the lever action of the collars and then measured by some type of micrometer. In early forms, a screw micrometer was used. The "Ewing" instrument used a micrometer microscope. In many instruments in present-day use a dial indicator is used. With an 8-in. gage and the use of an ordinary dial indicator reading to 0.0001 in. by estimation, this type of instrument is capable of measuring strains to about 0.000006 in. per in.

To obtain the average strain in a prismatic bar, strains must be measured on at least two gage lines, which must be diametrically opposite. To determine the complete strain condition in a cylindrical bar, subject

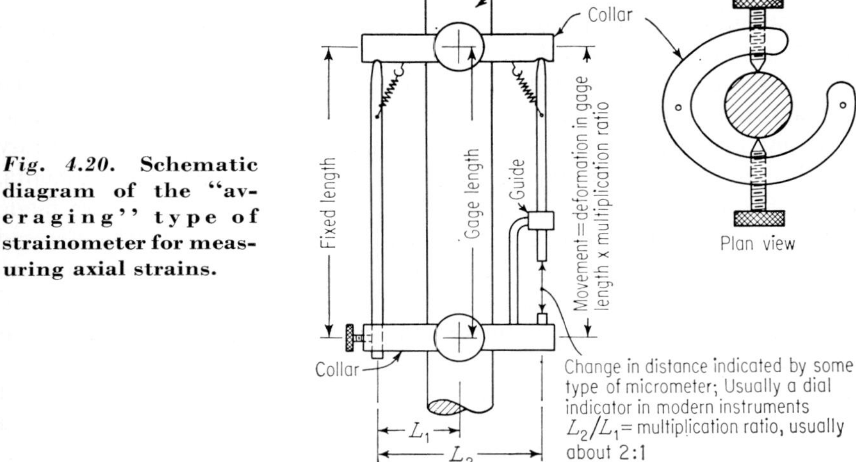

Fig. 4.20. **Schematic diagram of the "averaging" type of strainometer for measuring axial strains.**

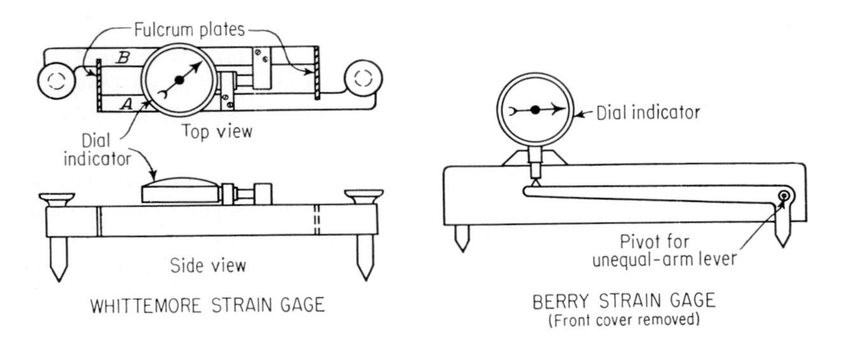

Fig. 4.21. **Strain gages (diagrammatic).**

to bending as well as direct stress, strains must be measured on at least three gage lines [540].

Strain Gages. For measurements of strain along a single gage line, a number of methods are used, the principles of some of which are explained in the following paragraphs. Obviously, a variety of particular instruments may be and have been designed on these principles.

The principal features of the Whittemore and Berry strain gages are indicated in Fig. 4.21. These devices have two pointed legs that are seated into small holes drilled into the object tested. In the Whittemore gage, the side bars are connected by flexible fulcrum plates. The dial indicator is connected to side bar A, and the dial spindle bears against a lug attached to side bar B so that any change in the gage length causes a change in the relative position of the side bars which is indicated by the dial indicator. The Berry gage actuates a dial indicator through the use of a bell-crank lever, one arm of which constitutes one leg of the gage. The lever ratio is usually 1:5.

These gages are called *portable* instruments because they may be applied to a gage line only at the time an observation is to be made, and thus one instrument may be used for measuring deformations along any number of gage lines. When they are used as portable instruments, readings on an auxiliary unstressed bar, called a *standard bar*, are made at intervals to serve as a base to which the observations on the stressed specimen are compared. Sometimes they are clamped to the test piece to give a continuous indication of change in deformation. The precision of the Whittemore gage is a function of the accuracy of the "ten-thousandth" dial indicator used. Over a 10-in. gage length, strain measurements are probably reliable to about 0.000005 in. per in. The precision of a 10-in. Berry gage is probably close to this figure when the movable leg is close to being perpendicular to the surface of the test piece; however,

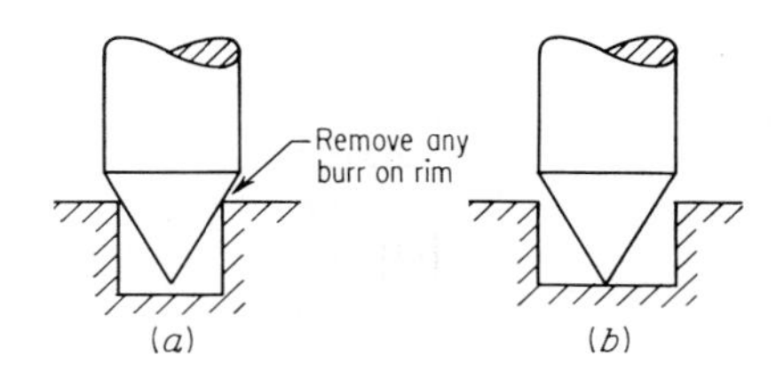

***Fig. 4.22.* Strain-gage holes. Hole a is satisfactory. Hole b is too shallow.**

when the leg is inclined to any extent, it does not fit the gage hole properly, and relatively large errors may be introduced.

For accurate readings, the gage holes must be made with care. As shown in Fig. 4.22 the shallow hole does not accurately position the pointed leg of the strain gage; so large errors may occur. In preparing gage holes, care must also be taken to remove any burr, produced in drilling or punching the hole.

Possible errors in strain-gage determinations owing to temperature changes occurring during the progress of the work are eliminated by using a standard bar of the same material and coefficient of thermal expansion as the object tested, or by observations of temperature of the standard bar and the object tested, if both have known expansion coefficients. In the latter case, all readings are corrected to what they would have been at some constant temperature, say 70°F; the corrected readings on the standard bar and the object are then compared in determining strains.

Instructions for the determination of stress by use of a strain gage are given in Part 2.

Huggenberger Tensometer. The principal features of the Huggenberger "tensometer," which operates on the multiplying-lever principle, are shown in Fig. 4.23. Each scale division on this instrument represents 0.0001 in. The range of the instrument is about 0.008 in. [440, 441].

Marten Mirror Strainometer. Mechanical-lever systems are limited by friction in the joints and by the weight of the levers. By using jeweled bearings, plate fulcrums, or knife-edges, the bearing friction can be greatly reduced, but the levers still have weight; the latter objection can be removed by the use of beams of light for magnification [447, 450–452]. One such unit as used in Marten's extensometer is illustrated in Fig. 4.24. A plane mirror is attached to a double knife-edge or lozenge so that the mirror is rotated as the specimen changes in length. The degree of magnification of this movement depends upon the length of the long lever, which is the distance between mirror and scale, and the length of the short lever, which is the distance between opposite edges of the lozenge to which the mirror is attached. Sometimes a roller is used instead of knife-edges. The complete strainometer consists of two such units attached to opposite sides of the specimen.

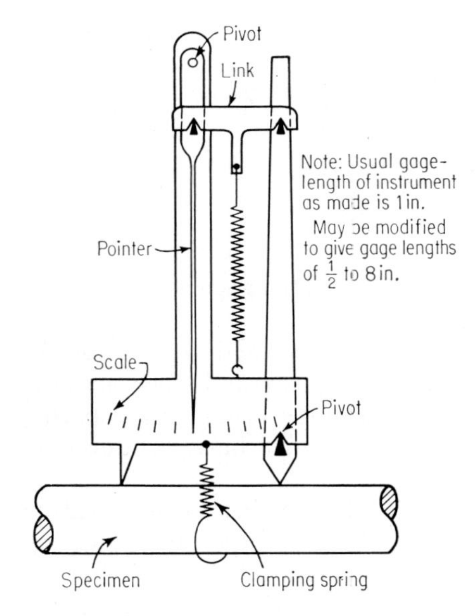

Fig. 4.23. **Schematic diagram of Huggenberger tensometer.**

Tuckerman Optical Strainometer. One of the most precise strainometers is Tuckerman's optical strain gage, which is also illustrated in Fig. 4.24. One surface of the lozenge, which has been polished to optical flatness, is used as a rotating or rocking mirror. Its rotation with respect to another mirror fixed in position with respect to the frame is measured with an optical device called an *autocollimator* (see Fig. 4.24). The standard gage length is 2 in., but it can be connected to one as short as $\frac{1}{4}$ in. With the 2-in. instrument, the smallest measurable strain is 0.000002, and the range is 0.0025 in. per in. [451, 452].

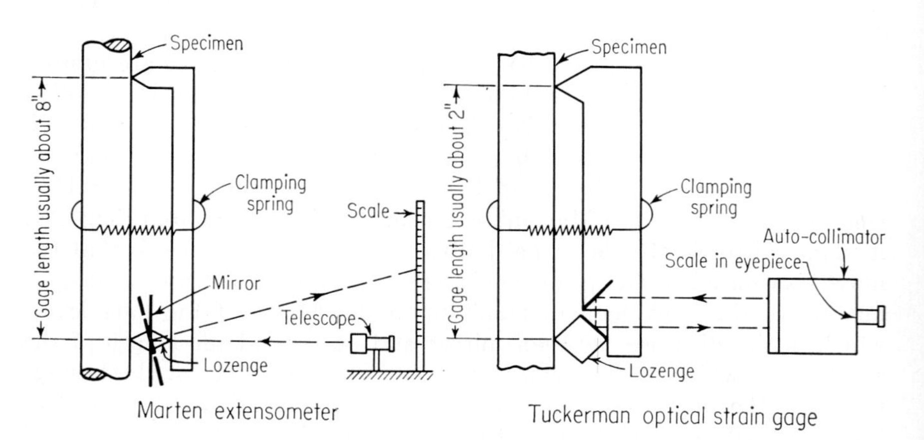

Fig. 4.24. **Mirror or "light-lever" extensometers.**

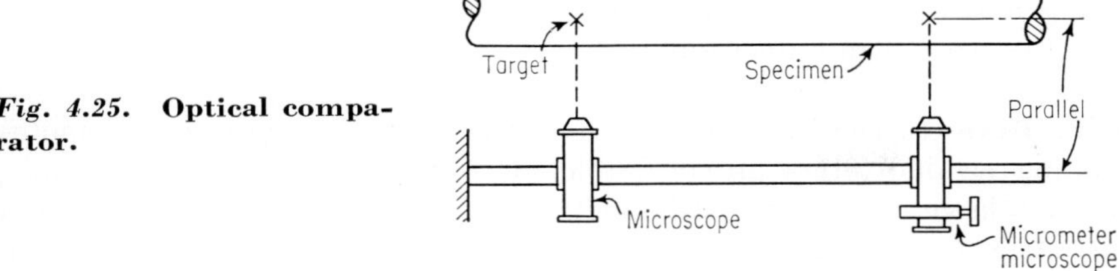

Fig. 4.25. **Optical comparator.**

Some Other Types. For some purposes, a "comparator" type of arrangement is convenient. The principal features of this device are shown schematically in Fig. 4.25. Two microscopes are mounted on a fixed base or bar. One or both of the microscopes are fitted with micrometer eyepieces so as to measure the movement of the targets on the specimen. Except for the fact that the microscopes are replaced by telescopes, this general type of device is often used for observing the deformations in specimens tested at elevated temperatures.

In connection with Poisson's ratio determinations, lateral strains are determined by means of "lateral strainometers." Many lateral strainometers are constructed along the lines indicated in Fig. 4.26*a*. Dial indicators are often used as a means of measurement. However, if precision is to be obtained, extremely fine measurements are necessary. Vose's adaptation of the interferometer principle to measuring lateral strains is illustrated in Fig. 4.26*b* [457]. The mirror principle has also been used.

4.12. Stress-Strain Recorders. Some testing machines are equipped with autographic stress-strain recorders which automatically draw a stress-strain diagram. In one type, the electric strainometer clamped to the specimen has a strain-actuated lever which moves a core inside an electrical coil of a miniature transformer. The movement of the core is transmitted electronically to a similar transformer which actuates a servomotor that rotates the drum of the recorder. The load-actuated

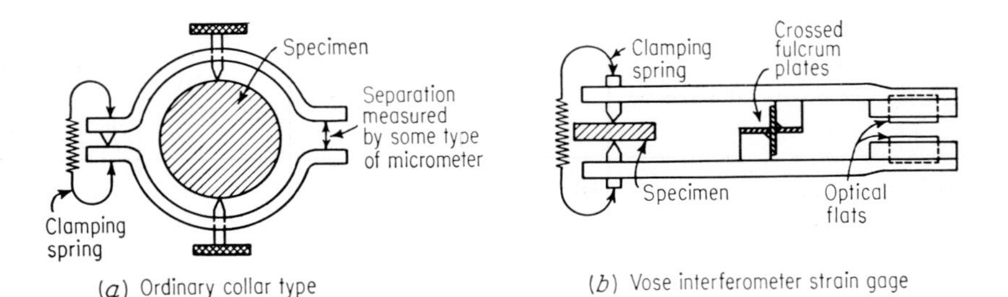

Fig. 4.26. **Lateral strainometers.**

stylus and the strain-actuated drum control the two motions needed to plot the stress-strain diagram.

4.13. Remote-reading Strainometers. For many purposes, remote-reading strainometers are desirable, or even necessary. For example, where a large number of strainometers are located in different parts of a test structure, it is convenient to make readings at some central instrument. Where strains are to be observed up to the point of rupture, under high loads, it is prudent for the observers to remain at a safe distance. Where strains are to be determined at some inaccessible position, as in the interior of a large concrete dam or the interior of a pressure vessel, a remote-reading instrument is necessary. Where strains due to rapidly fluctuating stresses are to be measured, a remote-reading strainometer connected to some recording device may be used. Many of the commonly used remote-reading strainometers are of the electric-resistance type [462–471].

4.14. The Strain-Resistance Change Relationship. Lord Kelvin discovered that the electric resistance of a given wire is a function of the strain to which it is subjected, tensile strains usually increasing the resistance and compressive strains decreasing it. For strain-gage work it is common to express the change in strain in terms of change in resistance, giving a ratio called the strain sensitivity or gage factor K. This gage factor $K = \Delta R/R \div \Delta L/L$, where ΔR represents resistance change in the total gage resistance R, and ΔL is the corresponding change in length in the total length L of the conductor. The strain ϵ in microinches per inch $= \Delta R/RK$.

The strain sensitivity is markedly influenced by the type of resistance wire, as shown in Table 4.1.

The various points to be considered in the selection of a resistance wire, in the order of their importance, are (1) gage factor, the higher the better; (2) resistance, the higher the better; (3) temperature coefficient

Table 4.1. **Characteristics of resistance wires**

Trade name of wire	Composition	Strain sensitivity	Temp. coef. of resistance
Nichrome	80% Ni; 20% Cr	2.0	High
Manganin	4% Ni; 12% Mn; 84% Cu	0.47	Very low
Advance, Copel, or Constantan	45%; Ni 55% Cu	2.0	Negligible
Isoelastic	36% Ni; 8% Cr; 0.5% Mo	3.5	High
Nickel	Ni	−12.1	Unstable

of resistance, the lower the better so that the gage will not be too sensitive to temperature; (4) coefficient of linear expansion, the lower the better; (5) thermoelectric behavior, or the tendency to generate a thermal emf at the connections, the lower the better; (6) physical properties, soft wire easily formed and soldered is preferable to hard, springy wire; and (7) hysteresis behavior, undesirable.

Of the kinds of wire shown in Table 4.1, wire of the Advance type is preferred for most gages as it has a good enough gage factor, a negligible temperature coefficient of resistance, and it works easily. For dynamic measurements, where a high temperature coefficient of resistance is not of much importance, it is desirable to use isoelastic wire because of its high strain sensitivity. This provides a higher output, requires less amplification for the measuring instruments, and results in a lower cost of instrumentation. Some gages are made by etching thin sheets of metal foil to produce the equivalent configuration of the wire gages.

4.15. Bonded Resistance-wire Gages. The usual commercial resistance-wire gage used in the United States is made by the Baldwin-Lima-Hamilton Corp. and is known as the SR-4 gage. A few of the common types are shown in Fig. 4.27. In many cases the sensitive elements are made up from a continuous length of wire looped back and forth so that all loops are in the same plane. This wire is then cemented to a paper or other type of carrier material. In other cases the sensitive elements are made of a continuous length of wire wrapped in a helical pattern around a thin flat paper core. This sensitive element is then sandwiched between two cover papers for protection. The effective gage lengths range from $\frac{1}{8}$ to 8 in. or more, but this range is not available in all types of gages.

A few special types of gages are made for particular service conditions. One such gage, called a *post yield* strain gage, is used for measuring strains up to 10 percent, which is far beyond the 1 to 2 percent strain range of other SR-4 gages.

A special arrangement of ordinary SR-4 gages to permit the measurement of large elastic or inelastic strains is the clip gage shown in Fig. 4.28. It consists of two SR-4 gages cemented to a U-shaped clip of spring bronze so that, when the test piece to which it is attached is subjected to strain, one gage is elongated while the other is compressed, resulting in a large change in resistance between the two. Each clip gage must be calibrated, but this is a relatively simple procedure. One advantage of this gage is that it may be used repeatedly on many different specimens.

Some special gages are temperature-compensated so that they cancel out unwanted temperature effects, allowing measurement of only the strains due to stress. Other gage patterns have been developed to eliminate the effect of transverse stresses from the readings—these are known as

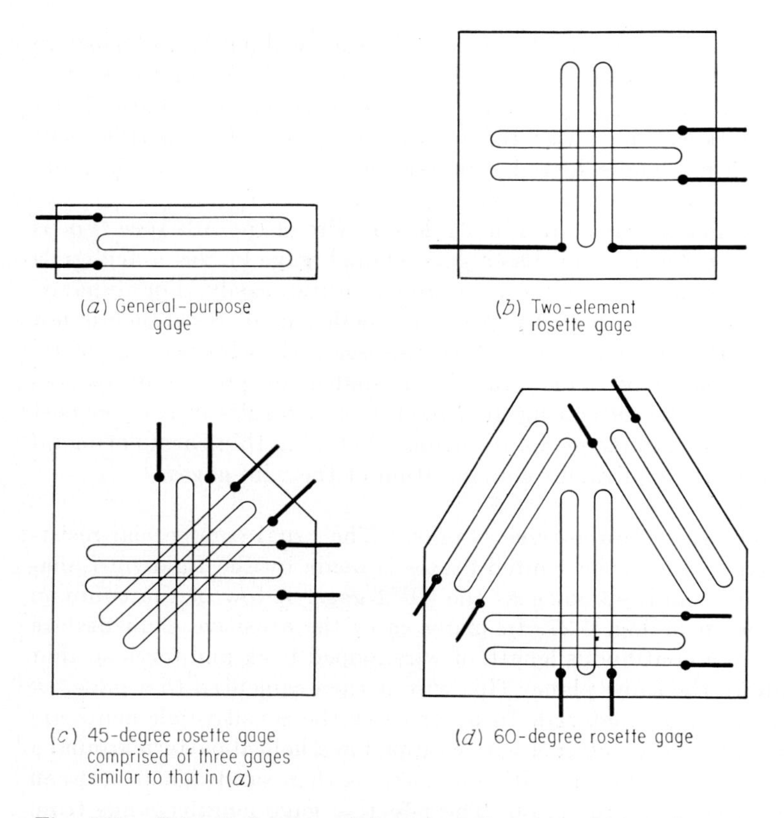

Fig. 4.27. **Principal types of electric resistance-wire strain gages.**

stress gages. A special gage made of nichrome foil can be used at temperatures up to about 1200°F if it is suitably bonded to the test piece. Another gage encased in a plastic cover is made for embedment in fresh concrete without danger of interference from the moisture in the concrete.

The wire used in SR-4 gages is only 1 mil in diameter and is carefully controlled to give dependable results. For overall control in manufacture the resistance of every gage is measured, and at frequent intervals during

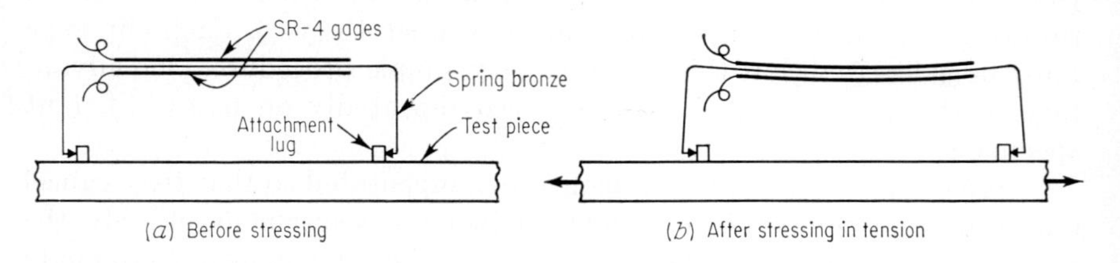

Fig. 4.28. **Clip gage.**

production some gages are cemented to steel specimens and tested under actual service conditions to determine the exact gage factor and to check performance. If the gages are used on materials other than steel, correction factors must be applied to the gage readings because of the different value of Poisson's ratio.

The outstanding advantages of bonded electric resistance-wire gages, compared to mechanical gages, are

1. Ease of installation
2. Relatively high accuracy
3. Adjustable sensitivity (by changing the gain of the amplifier used)
4. Remote indication (making possible the observation of strains at distant inaccessible points)
5. Very short gage lengths
6. Measurement of strain at the surface of the test member
7. Response to dynamic strain

4.16. Attachment of Gages. The surface to which a gage is to be attached must be thoroughly clean and even, but preferably not too smooth or highly polished. Cleaning is done satisfactorily with acetone or carbon tetrachloride. Very smooth surfaces should be roughened with a medium-grit emery paper. A liberal amount of cement should be used in bonding the gage to avoid the formation of air pockets in the cement. The use of moderate weights, or spring clamping pressure applied through a rubber pad, will squeeze out excess cement. After air-drying for an hour, infrared heat lamps judiciously used will hasten the hardening of nitrocellulose and epoxy cements. The customary drying period of 24 hr without heat may be reduced to perhaps 8 hr with heat. In any case, thorough drying is essential to avoid yielding of the bonding material.

For bonding Bakelite gages, it is necessary to use a Bakelite cement. A neoprene or felt pad is placed over the gage, a metal plate is positioned, and spring pressure is then applied. Baking at a temperature of 300 to 400°F is essential to proper bonding of these gages. Bakelite cements may be used for high-temperature applications. In common with epoxy cements, they undergo less creep than nitrocellulose cements when the test specimen is subjected to long-time loading.

4.17. Moistureproofing. A high degree of stability of gage resistance is essential since the measurement of strain involves the determination of very small changes of resistance. Errors due to instability or "drift" should not exceed a small fraction of the resistance change due to the strain being measured.

Gage instability is caused primarily by moisture absorbed by the gage. Moisture produces (1) changes in conductivity and (2) dimensional changes of the bonding cement which result in strains in the gage wires and thus in turn cause resistance changes. Also, long-time aging of the bonding cement may cause some instability. The latter is largely corrected by thorough drying or baking of the gage, and the effect of moisture is minimized by thorough moistureproofing. However, only a few types of service require this protection as in most applications the effect of moisture changes is not significant during the course of a test. In general, the Bakelite and epoxy bonded gages are less sensitive to aging and to moisture than the nitrocellulose cemented gages and are therefore advantageous, when properly waterproofed, where high stability is required.

Before moistureproofing any gage it is necessary that the cement be thoroughly dried or cured and that the gage be free of all absorbed or surface moisture. The gage may then be waterproofed by coating with (1) grease, (2) melted sealing compounds, (3) neoprene, or (4) self-vulcanizing synthetic-rubber compounds.

Petroleum jelly is an effective coating but is suitable only for temporary installations. Melted sealers such as Cerese wax, petrolastic asphalt No. 155, Petrocene wax, or Ozite A have been commonly used for various degrees of protection. The neoprene coating requires several applications which must be baked at 190°F for several hours. It is a laborious method but will give protection for underwater service. The self-vulcanizing rubbers compounded by the Minnesota Mining & Mfg. Co. require a precoat such as Petrocene wax over the gage and a primer over metal surfaces, but the entire operation can be completed more quickly than with neoprene.

In all cases the moistureproofing material must be carefully applied, covering the gage and cement area, extending well beyond the cement on to clean metal. It must extend well up the lead wires, and these wires should be arranged so that they will not be flexed in such a way as to break the seal. The coating must be thick enough over the gage (at least 1/16 in. and sometimes much more) to block completely all moisture paths leading to the gage.

For the moistureproofing of gages attached to metal surfaces under water, a successful method involves the use of hemispherical plastic cups placed over the gage and bonded to the metal surface. The lead wires are sealed where they emerge from a hole in the plastic cup [468].

4.18. Wheatstone-bridge Hookup. Static strains may be determined by placing a wire gage *A* in a four-arm Wheatstone-bridge d-c circuit, as shown in Fig. 4.29. When the gage resistance is changed by deformation of the gage, the bridge circuit is unbalanced. To compensate for strains caused by temperature and humidity variations, a so-called "dummy gage" *D* is connected into the Wheatstone-bridge circuit. The dummy gage is a duplicate of the active or strain-measuring gage A. It is attached to a piece of unstressed material of the same kind and subject to the same temperature and humidity conditions as the member to which A is attached. The active gage measures strain due to stress, plus deformation due to temperature and humidity effects, while the dummy gage measures deformation due to temperature and humidity only. The circuit is arranged so that the resultant bridge unbalance is a function of strain due to stress, free from deformation due to temperature or humidity.

The Wheatstone bridge consists of resistances A plus D and R plus R' in parallel, with a means of determining points of zero potential by moving the lower lead of the high-resistance exploring galvanometer G along the resistances R and R' until it reads zero, thus using R and R' as a slide-wire resistance. In this way the resistances are divided between the four arms so that $A/D = R/R'$. The resistances R and R' are commonly made variable to avoid the need for moving the galvanometer lead. From the above relationship the resistance A can be measured at any stage, and thus the change is A and the corresponding strain can be determined (as shown in Art. 4.14).

Although the simple hookup shown in Fig. 4.29 can be used satisfactorily, in practice it is much more convenient to use a specially designed SR-4 strain indicator described in Art. 4.19, which is calibrated to give strains directly instead of resistances.

To obtain direct or axial strain averaged from two sides of a tension or compression member and at the same time to cancel out unwanted bending strains, two active gages A_1 and A_2 should be mounted back to back on opposite sides of the specimen and connected with two dummy gages, as shown in Fig. 4.30. In this case the strain in one gage, say A_1, will be greater than if carrying axial load alone and the strain in the other will be less by exactly the same amount. Thus the circuit

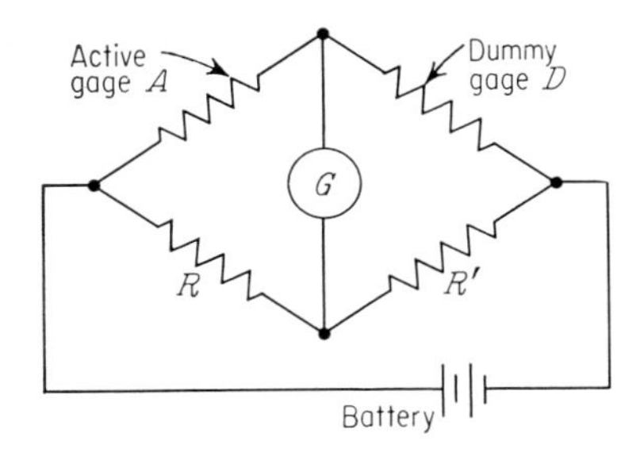

Fig. 4.29. **Wheatstone-bridge hookup.**

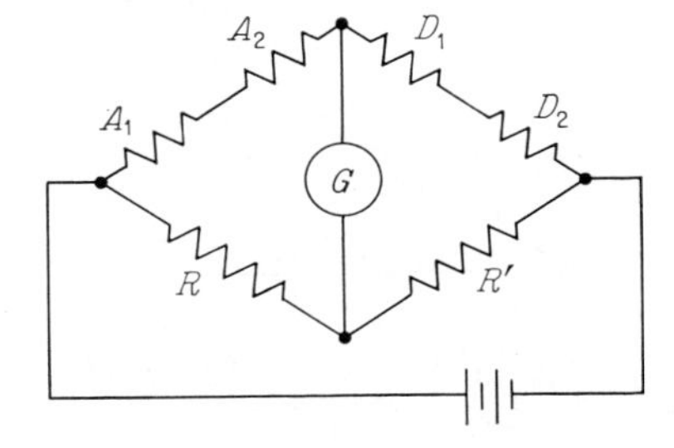

Fig. 4.30. **Bridge arrangement for measuring axial strain and eliminating bending strain.**

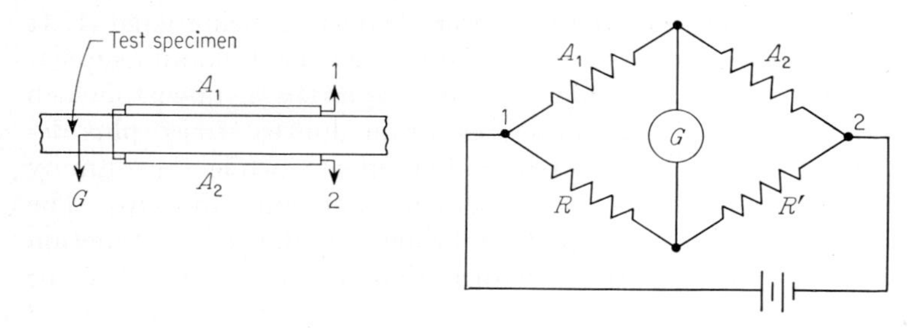

***Fig. 4.31.* Bridge arrangement for measuring bending strain and eliminating axial strain.**

adds their results and divides by two, which gives the true axial strain irrespective of any bending.

To measure bending only and to cancel out axial strains, gages A_1 and A_2 should be mounted on opposite sides of the specimen as before, but connected in adjacent arms of the bridge, eliminating the dummy gages as shown in Fig. 4.31. This circuit not only compensates for temperature but doubles the sensitivity as well because with A_1 in tension and A_2 in compression, located in adjacent arms, the resistance changes of opposite signs in effect add together. Whenever possible, in measuring bending strains, gages should be used in this manner, as this hookup provides the most efficient circuit with the least number of gages.

4.19. SR-4 Strain Indicator. The instrument most commonly used in conjunction with SR-4 gages is a strain indicator, a simplified form of which is shown schematically in Fig. 4.32. It is a battery-powered d-c unit which gives optimum accuracy and sensitivity. In this instrument the active gage A and the dummy gage D are external to the indicator. G represents the galvanometer; the decade resistance is provided for a coarse balance of the bridge. The variable resistances R and R' are gages mounted back to back on a slender cantilever beam; a micrometer screw

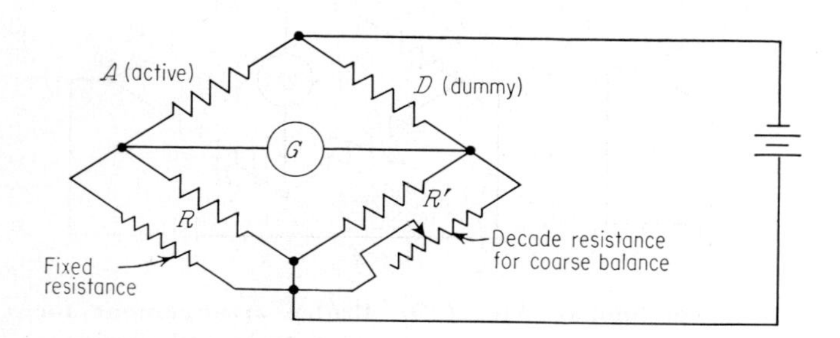

***Fig. 4.32.* SR-4 strain indicator without amplification.**

is provided to deflect this beam and change the resistances R and R' to accomplish the fine balance. The micrometer also serves to measure the unbalance in the circuit when the bridge is used as a "null" indicator.

By the null method the galvanometer is balanced back to zero by turning the micrometer screw which is graduated so that changes in reading give strains directly in microinches per inch. The null method is slow, but it is the more accurate way of measuring the strain. A quicker method is to note the amount of deflection of the galvanometer from the zero position when provision is made to impress a definite voltage on the indicator, because the deflection is proportional to voltage. The null method is independent of the voltage used.

The indicating device used with a d-c circuit is almost always a galvanometer because it is one of the few instruments which will operate over a significant range without amplification, although some oscillographs with supersensitive elements require no amplification.

The strain-gage signal is actually very low as a stress change of 90,000 psi in steel will change the resistance of a 120-ohm gage by only 0.7 ohm, and the bridge output for one active A-1 gage is only 10 mv for a strain

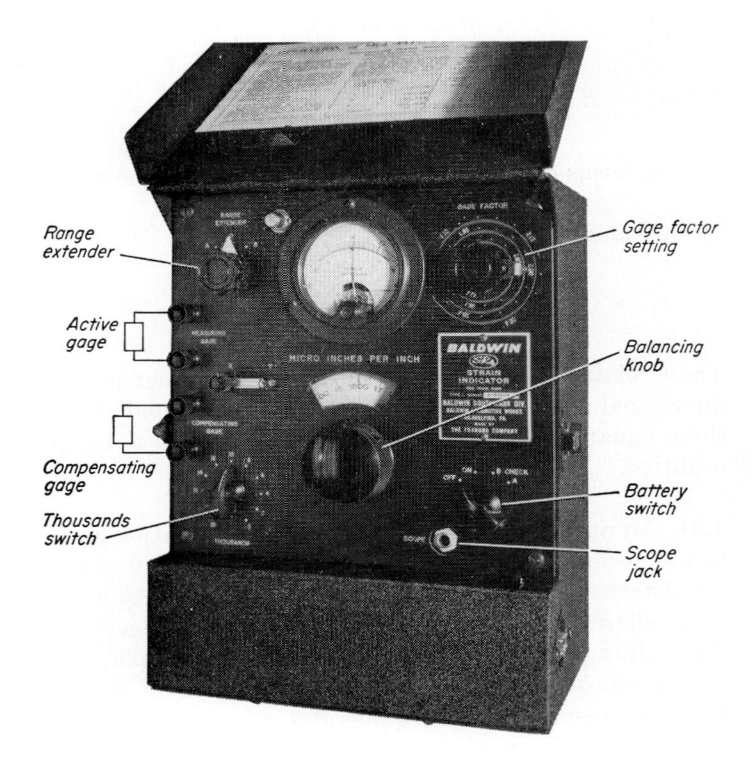

Fig. 4.33. **SR-4 strain indicator with amplification.**

of 0.001 in. per in., even though a mechanical advantage of 2 is provided by the gage factor. Therefore, the actual SR-4 strain indicator includes electronic amplification to obtain a stronger signal. Figure 4.33 shows one of these instruments.

4.20. Principal Stresses Using Strain Rosette. When using a 45° rosette (Fig. 4.27*c*), if strains ϵ_1, ϵ_2, and ϵ_3 are observed at 0°, 45°, and 90° angles, respectively, with the X axis, then

$$\epsilon'_X = 0.5[(\epsilon_1 + \epsilon_3) + 1.41 \sqrt{(\epsilon_1 - \epsilon_2)^2 + (\epsilon_2 - \epsilon_3)^2}]$$

where ϵ'_X = principal strain most nearly in the direction of the X axis. The other principal strain occurring at right angles to ϵ'_X is

$$\epsilon'_Y = 0.5[(\epsilon_1 + \epsilon_3) - 1.41 \sqrt{(\epsilon_1 - \epsilon_2)^2 + (\epsilon_2 - \epsilon_3)^2}]$$

Converting strains to stresses, based on the relationship $\sigma = E\epsilon$, and correcting for the transverse strain with Poisson's ratio μ, the principal stresses are

$$\sigma'_X = E\left[\frac{0.5(\epsilon_1 + \epsilon_3)}{1 - \mu} + \frac{0.71 \sqrt{(\epsilon_1 - \epsilon_2)^2 + (\epsilon_2 - \epsilon_3)^2}}{1 + \mu}\right]$$

$$\sigma'_Y = E\left[\frac{0.5(\epsilon_1 + \epsilon_3)}{1 - \mu} - \frac{0.71 \sqrt{(\epsilon_1 - \epsilon_2)^2 + (\epsilon_2 - \epsilon_3)^2}}{1 + \mu}\right]$$

In terms of the principal strains ϵ'_X and ϵ'_Y, the principal stresses are

$$\sigma'_X = (\epsilon'_X + \mu\epsilon'_Y)\frac{E}{1 - \mu^2}$$

$$\sigma'_Y = (\epsilon'_Y + \mu\epsilon'_X)\frac{E}{1 - \mu^2}$$

These relationships hold only for an isotropic material (E the same in all directions) and for stresses below the proportional elastic limit. In using these equations a tensile strain is positive and a compressive strain is negative.

4.21. Dynamic Testing. For rapidly fluctuating strains it is essential to use pen-and-ink recorders, oscillographs, or cathode-ray oscilloscopes for determining the strains produced, the selection of the instrument best adapted to a given problem being contingent upon the rate of fluctuation of the strains. It is necessary to amplify strongly the weak but precise signal produced by the unbalance of the strain-gage bridge to develop enough power to operate any of these instruments. For these applications a-c amplifiers are essential, but when using alternating current it is necessary to balance the capacitance and the inductance as well

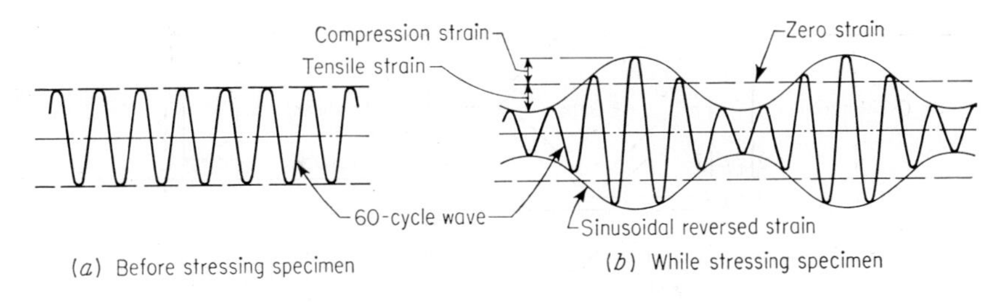

Fig. 4.34. **Record of dynamic test.**

as the resistance of the circuit. Alternating-current circuits permit amplification very simply and eliminate the contact potentials or thermal emfs in the gage because of the constantly changing polarity.

The type of record produced on an oscillograph film or on a Brush pen-and-ink recorder chart by the amplified output of a 60-cycle a-c bridge is shown in Fig. 4.34. The strain variations picked up by the SR-4 gage appear as a modulation of the 60-cycle output of the bridge produced when the gage is under no strain. The result is an envelope or unrectified record, and the pattern above the center line duplicates that below. Tension increases the gage resistance and correspondingly reduces the galvanometer swing; conversely, compression decreases the resistance and increases the swing. This serves to distinguish between tension and compression on the record. If the output is rectified by a copper-oxide rectifier before recording, a single line trace or the envelope of Fig. 4.34 is obtained as in Fig. 4.35. This curve is easier to interpret.

A 60-cycle a-c source is satisfactory to power the bridge so long as the frequency of fluctuation of the strain picked up by the gage does not exceed about 6 cycles per sec or one-tenth of the frequency of the impressed wave, commonly called the carrier wave. As the strain frequency approaches the carrier-wave frequency, the strain-wave pattern becomes less well defined. For high-frequency dynamic strain recordings it is desirable to have a carrier frequency at least ten times the frequency of the gage pickup. This can be done by electronic oscillators, which are vacuum-tube generators of high-frequency currents. A typical hookup for high-frequency dynamic testing is shown in Fig. 4.36.

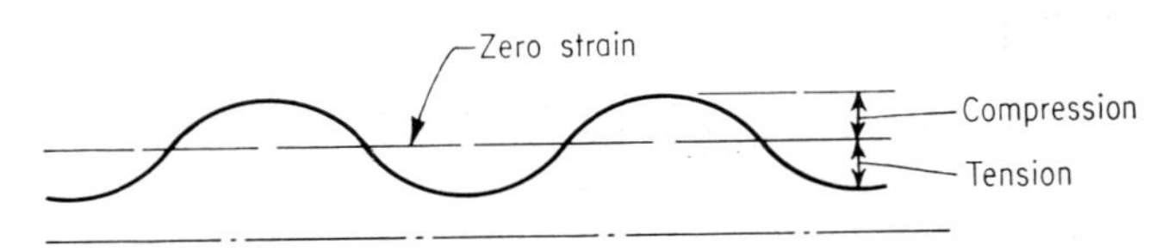

Fig. 4.35. **Rectified trace of dynamic-test record.**

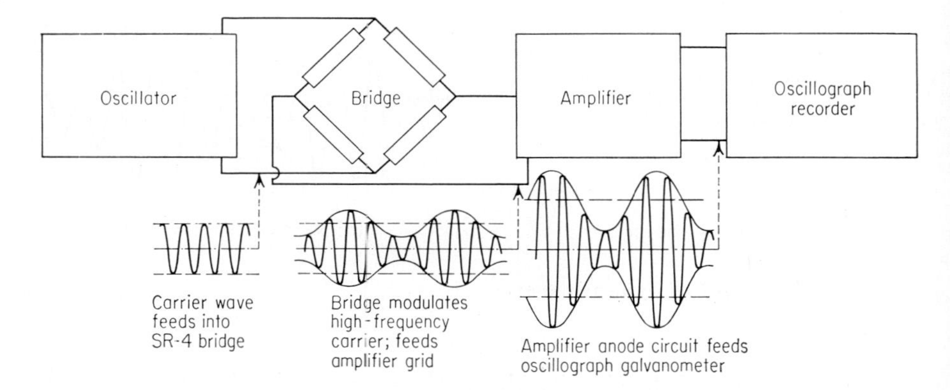

Fig. 4.36. **Hookup for high-frequency dynamic testing showing waves transmitted between units.**

4.22. Carlson Strain Meter and Stress Meter. An entirely different type of resistance-wire strain gage is the Carlson electric strain meter, illustrated in Fig. 4.37 [470]. Steel wire, about 0.004 in. in diameter, is wound with an initial tension of about 100,000 psi over porcelain spools, so arranged that elongation of the instrument increases the tension in coil A and reduces the tension in coil B. The ratio of the resistances of the two coils is measured by a portable testing set, which is essentially a Wheatstone-bridge circuit. With a 10-in. gage length, it is possible with this instrument to obtain strain readings to about 0.000004 in. per in. The range of the instrument is about 0.02 in. By connecting the two coils in series and measuring their over-all resistance, the strain meter becomes an electric resistance thermometer for determining temperatures. Many of these instruments have been placed in concrete dams during

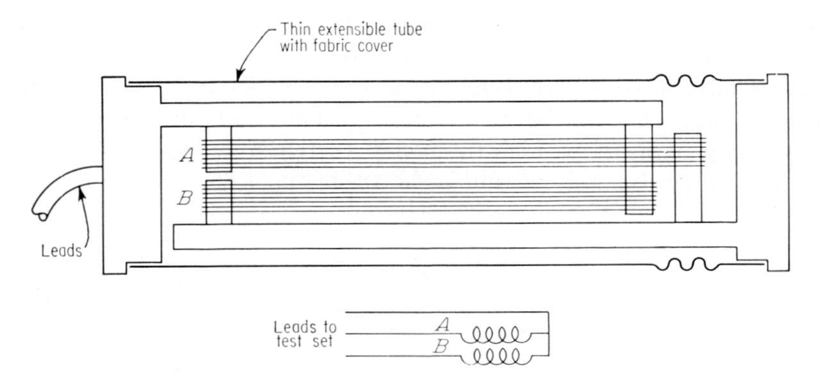

Fig. 4.37. **Schematic diagram of Carlson electric strain meter.**

the past decade to determine long-time deformations due to load and temperature changes.

The stress meter [471] is used for determining the compressive stress within a mass of material such as concrete or earth. As shown in Fig. 4.38 the stress meter consists of a disklike assembly connected to a special strain meter, basically similar to that shown in Fig. 4.37. The responsive part of the device is an internal diaphragm which detects stress applied normal to the faces of the external disk. This external disk consists of a pair of spaced plates welded around their rim and separated by a thin film of mercury. Pressure applied normal to the disk is transmitted to the mercury, which in turn defects the internal diaphragm. The small elastic deflections of this diaphragm actuate the strain-meter unit, which can be read remotely by use of a Carlson test set. By a suitable calibration, the readings of the test set can be converted to unit stresses. The unique feature of the stress meter is that its response is substantially a straight line function of the applied stress. In contrast, strains do not always vary as linear functions of stress, so that the interpretation of strain-meter observations in terms of stress is sometimes difficult, if not impossible. For example, concrete may exhibit creep under a constant sustained load. Furthermore, for a material such as concrete which develops a higher modulus of elasticity with age, a given stress applied at an early age will produce a greater strain than when applied later.

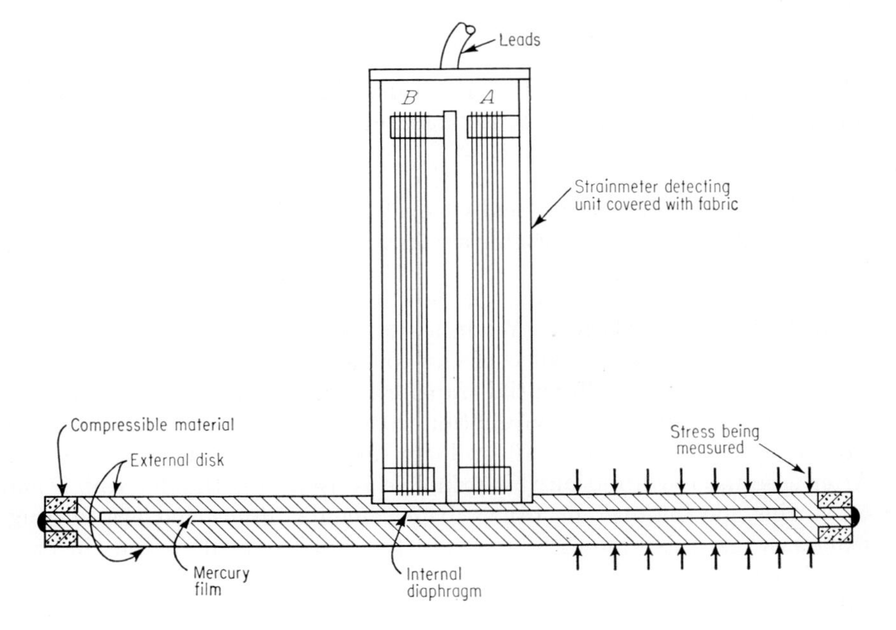

Fig. 4.38. **Schematic diagram of Carlson electric stress meter.**

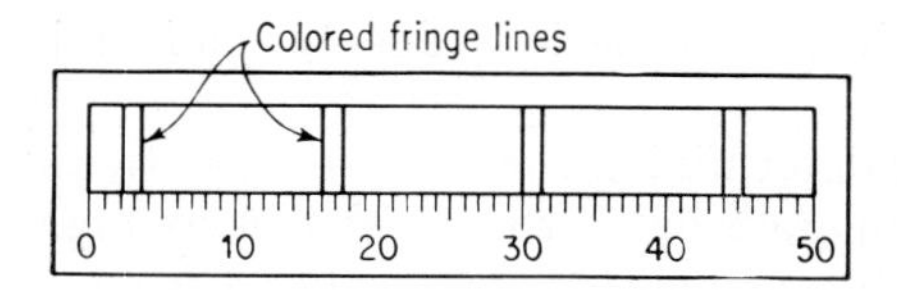

Fig. 4.39. **Strainline gage.**

4.23. Photoelastic Strainometers. The Strainline photoelastic gage permits the determination of strains along the axis of the gage using ordinary light as the only power source [469] (see Art. 10.36 for a description of the photoelastic method). It is made of a thin ¼ by 2-in. photoelastic material, schematically illustrated in Fig. 4.39, which is bonded to the specimen by means of an epoxy cement. Stress applied to the gage causes colored bands, or *fringes,* to move with respect to the gage scales and thus permits a direct reading of the strain. Tension causes the fringe lines to move toward higher numbers on the scale and compression causes a reverse motion. Then

$$\epsilon = X(SF)$$

where ϵ = unit strain along the gage
X = the movement of a given fringe in the gage, in divisions
SF = strain factor of the gage in μ in./in./division*

The fringe factor FF of the gage is the strain required to cause one fringe to be replaced in position by the adjacent fringe. For Strainline material

$$FF = 700\mu \text{ in./in./fringe}$$

The strain factor, SF, is obtained by dividing the fringe factor by the number of divisions, d, between fringes, or

$$SF = FF/d$$

These gages have practically constant strain factors between 50° and 100°F. The maximum usable strain when bonded with epoxy cement is about 7000μ in./in.

4.24. Inductance Gages. When a steel core is moved in or out of an electrical coil, the inductance of the coil will change. This principle is used in the Atcotran differential transformer and in the General Electric deflection gages, which in conjunction with a suitable electronic amplifier can be used to measure small or large displacements quite accurately. After a suitable calibration this principle is useful for the determination of strains, deflections, or other displacements. Some modern testing machines use this system to measure the displacement of the tip of a bourdon tube connected to the hydraulic load cell, and by a suitable calibration this displacement indicates the load acting on the system.

* μ means "micro."

4.25. General Use of Brittle Coatings. It has long been known that fine cracks develop in the mill scale of hot rolled steel at stresses slightly beyond the yield point. This same effect can be observed, even at strains well below the yield point, when the surface is covered with a brittle coating. One type of commercially available brittle coating, called *Stress-coat,* is basically a limed-wood rosin K and dibutyl phthalate, with carbon disulfide as a solvent. As load is applied to a coated specimen, the brittle coating will crack along lines normal to the direction of the maximum tensile stresses as soon as the tensile strains reach the level of strain sensitivity of the coating. For Stresscoat this strain is about 0.0005 to 0.0009 in. per in., corresponding to a stress in steel of 15,000 to 27,000 psi. The actual level of strain sensitivity for each test is determined by calibration strips coated at the same time as the test specimen [477–479]. By applying increments of load to the specimen, and noting the load at which cracking of the coating begins at various locations, it is possible to determine the critically strained parts of the specimen and to note how other parts reach critical strains and corresponding stresses as the load is increased.

Calibration of the coating on each specimen is conducted using the apparatus shown in Fig. 4.40. The calibration strips are of spring steel 12 by 1 by ¼ in., coated on one side, and loaded as a cantilever beam using a known deflection. This deflection is such that the calibration strip will undergo a varying strain along the strip as shown on the scale accompanying the loading rig. After the strip is deflected, a mark can be made on it at the location where cracking begins. The strip is then

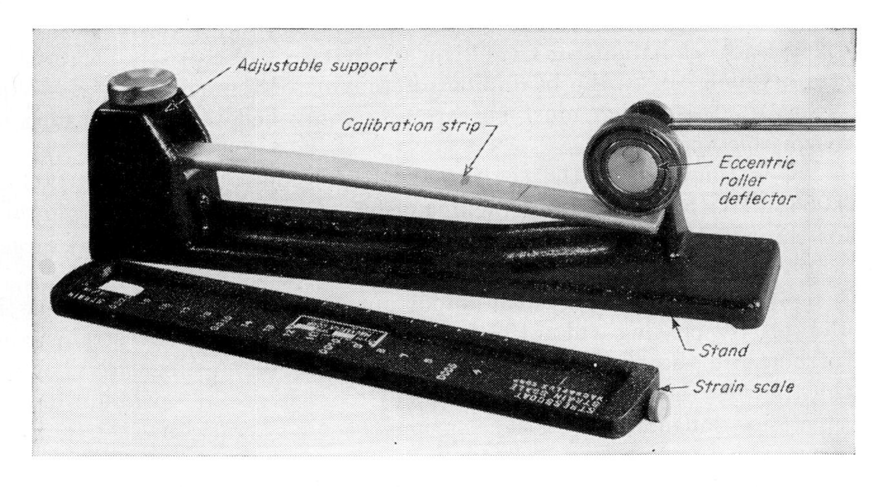

***Fig. 4.40.* Method of loading calibration strip in calibrator.** *(Courtesy of Magnaflux Corp.)*

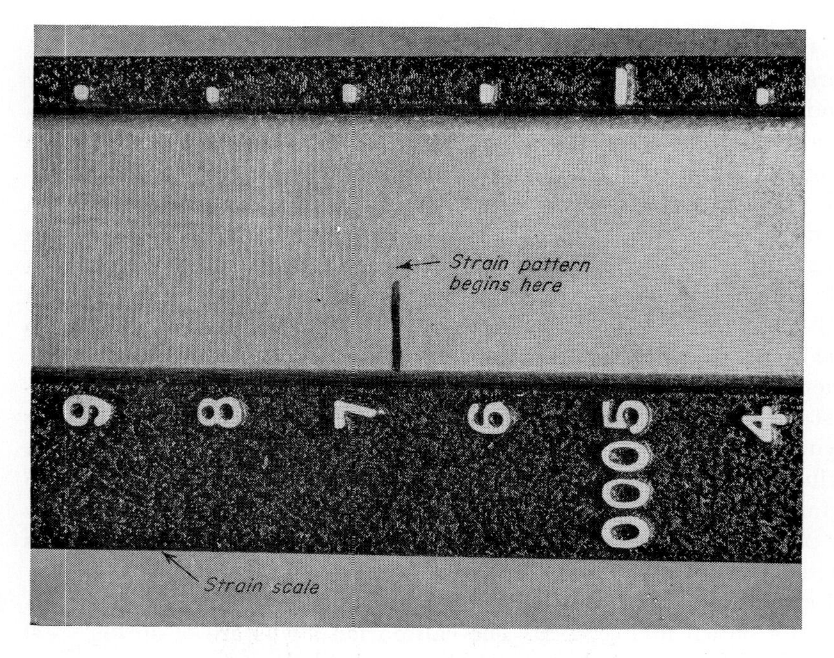

Fig. 4.41. **Typical strain pattern on calibration strip held in special scale. A strain sensitivity of 0.00067 is indicated.** (*Courtesy of Magnaflux Corp.*)

placed in the special scale provided with the apparatus, from which the amount of strain causing the cracking is noted, as shown in Fig. 4.41.

4.26. Factors Affecting Cracking of Brittle Coatings. The critical strain which causes the beginning of cracking of the coating varies with several factors which must be carefully controlled to obtain results of reasonable accuracy.

The thickness of the coating should be controlled within 0.003 to 0.008 in. Within this range, cracking occurs at a fairly uniform strain, but for thinner and thicker coatings the initial cracking will occur at larger strains.

The drying period is an important factor; if insufficient time is allowed, the coating will not be brittle enough to crack at low strains, and furthermore the strain at initial cracking will be supersensitive to the thickness of the coating. In general, 15 to 24 hr is the optimum drying time. For longer periods the coatings become erratic in their pattern formation.

Atmospheric conditions during the drying and testing period have definite effects on the strain sensitivity of the coating. Gradual changes

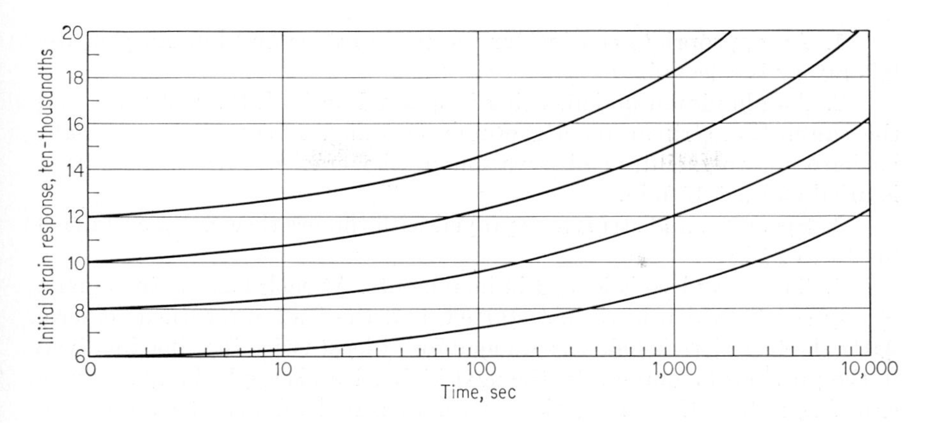

Fig. 4.42. **Variation of strain sensitivity with time of loading (constant rate of loading).** (*From de Forest and Ellis, Ref.* 477.)

of temperature during the *drying* period are not so critical, but changes during the *test* period should not exceed plus or minus 1°F and the temperature of the calibration strip should not vary more than plus or minus ½°F from that of the specimen.

Sensitivity also varies with relative humidity but ordinarily this does not change much during a given test. In order to obtain the optimum coating for given atmospheric conditions, a series of 12 coatings which vary in brittleness are available.

Although the coatings are called *brittle,* they are actually somewhat plastic. Figure 4.42 shows how the strain necessary to crack a coating varies with the time taken to cause cracking. This influence of creep may be corrected by checking a calibration strip at the beginning and then, by means of Fig. 4.42, correcting all observed strains on the specimen to the values that would have been obtained if the load rate had been the same as for the calibration strip.

If temperatures are controlled properly, the Stresscoat method is capable of producing results with an accuracy of about plus or minus 10 percent. Even if temperatures are not controlled, the method will indicate satisfactorily the location and the direction of the maximum stress developed. To avoid the difficulties involved in proper temperature control, some experimenters use Stresscoat for qualitative determinations and then in later tests measure actual strains by use of strainometers applied at the critical points, as indicated by the tests with brittle coatings.

4.27. Method of Conducting Test with Brittle Coating. The test procedure is generally as follows:

1. An appropriate coating for the temperature and humidity in the laboratory is selected.

2. An aluminum undercoating is sprayed on both the test piece and the calibration strips after a thorough cleaning. This provides a bright background to facilitate observation of the cracks. The undercoating is dried at least 15 min.

3. Specimen and strips are sprayed with the brittle coating and dried for 15 to 24 hr.

4. The test piece is loaded in increments. At each load, note is made of any cracks which have just started, and the load is returned to zero. After the test piece is held at zero load for at least twice the time involved in the previous loading cycle, the next higher load is applied and any new cracking noted. If desired, the return to zero load may be omitted if proper corrections for creep are applied.

5. The coating is calibrated by loading a calibration strip in the calibrator in the same length of time as that used in loading the specimen. The point of initial cracking is marked, and the strain in the strain scale is read.

6. After completion of the test, the crack patterns may be more clearly shown for photographing by dye-etching.

This procedure applies to tensile strains only, but brittle coatings can also be used for analyzing compressive stresses. The difference in the procedure is that the specimen is coated while under a compressive load. After drying, the load is gradually released, and the crack pattern formed when the compression load is released is analogous to that produced when a coated specimen is subjected to tension. A similar condition is obtained by coating the nonloaded specimen, slowly loading in compression for an hour without cracking the coating, holding the load for at least another hour until the coating creeps sufficiently to become stress free, and then releasing the load in increments to cause cracking of the coating.

4.28. Calibration of Length- and Strain-measuring Devices. In so far as linear measurements are concerned, the principal devices that require calibration in the ordinary materials-testing laboratory are micrometers, especially those for use on strain-measuring devices, and assembled strainometers. The calibration can be made by the use of high-precision micrometers of known accuracy, as determined, say, by measurements made at the National Bureau of Standards, by the use of gage blocks, or by the use of an interferometer whose calibration is self-determined by the wavelength of the light used.

One form of calibrator for use with dial indicators is shown in Fig. 4.43. Built on the screw-micrometer principle, an instrument of this type

Fig. 4.43. **Calibrator for dial indicator.**

with an 8-in.-diameter graduated circle can measure accurately to about 0.00001 in.

Gage blocks are now available which are accurate to two-millionths inch per inch. They are made in sizes ranging from 0.01 to 20 in., and by combining blocks thousandth-inch intervals (above 0.04 in.) can be obtained. The principal manufacturers of blocks are (1) Pratt and Whitney Co., (2) Brown and Sharpe Co., and (3) DoAll Co.

One form of calibrator for strainometers is described in a report by Stang and Sweetman [435]. The principal features of this device are shown in Fig. 4.44. The instrument to be calibrated is attached to the coaxial spindles which are moved with respect to one another by rotating a worm gear into which the movable spindle is threaded. The spindle motion is under good control since one revolution of the worm will move it 0.00014 in. The amount of motion is measured by gage blocks that are clamped in a vise at the upper end of the spindle. The pointer on the dial indicator is always brought to the same reading by operating the worm drive after each change in gage-block arrangement required to give a desired spindle movement. The range in spindle movement is 1 in., and strainometers with gage lengths up to 24 in. can be accommodated.

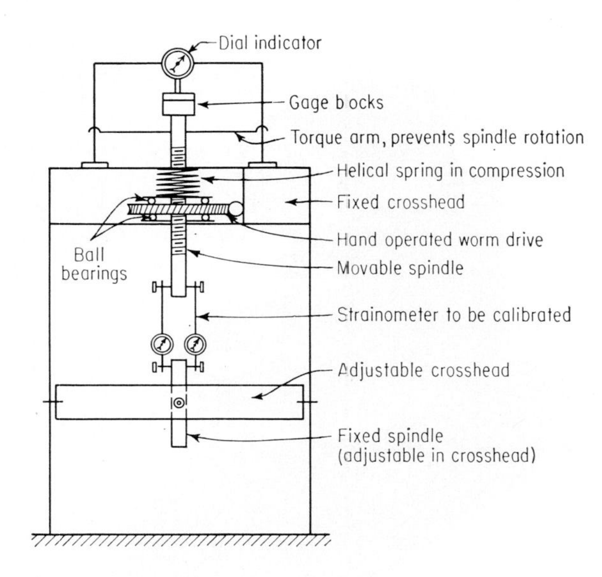

***Fig. 4.44.* National Bureau of Standards strainometer comparator.** (*From Stang and Sweetman, Ref.* 435.)

Directions for conducting the calibration of a strainometer are given in Part 2.

GAGING

4.29. Use of Gages. In connection with the inspection of materials, it is often required that not only the quality of the material but also the dimensions and tolerances be checked. In the case of manufactured machine parts, this is done by the use of "gages." Although gaging is not an essential part of materials testing, a brief outline of common gaging procedures is given in the following articles for the benefit of the materials inspector whose duties call for some knowledge of this subject. Some of the treatises on this subject should be examined for more complete details [482-485].

Gages are devices used for determining whether one or more dimensions of a single part or assembly are within permissible limits. A simple gage is an individual instrument and can be used to check only one dimension, whereas a micrometer or scale, where it can be used, will serve for any measurement within its range. The gage, however, assures a positiveness and rapidity of measurement not obtainable with a micrometer and, when proper limits are established, assures interchangeability of parts so that any of those accepted will fit into the same place without adjustment of any kind. Reference gages may be used to check service gages that are used in manufacturing operations or inspections. In many cases, manufacturing or inspection gages are calibrated directly by comparison with gage blocks of known accuracy or by the use of an interferometer instead of using check or reference gages.

4.30. Simple Gages. There are many types of gage varying from the simplest snap gage to the most complex arrangement, which corresponds to a fixture or jig in the manufacturing process, made to check simultaneously such items as parallelism of surfaces, squareness, and distances between holes.

Simple gages are made in "go" and "no go" or "not go" sizes. The "go" gage will fit over or enter into the part being measured as shown in Fig. 4.45, which illustrates two simple types of gage, the receiving and the entering or plug gage. The entering type may be straight or tapered and of any cross section, such as circular, square, rectangular, or irregular. When circular and straight, they are designated as plain cylinder plug gages and are used to check inside diameters. In using these gages the work is satisfactory if its dimensions are between the "go" and "no go" gages.

Simple plug gages may be made solid, i.e., of a single piece, or may be fitted into separate handles. The ASA has three different types for small (0.059 to 1.510 in.), intermediate (1.510 to 8.010 in.), and large (8.010 to 12.010 in.) plug gages. The separable types are designed and constructed so as to minimize the possibility for any "shake" that would interfere with the sensitive "feel" so essential in the proper manipulation and use of these gages. They require delicate handling and should never be "forced." In the large sizes especially, skill and experience are necessary to secure reliable results.

The small sizes, when not solid, are usually made with a tapered shank that is driven into the end of a hollow handle—the "go" gage in one end and the "no go" in the other. So as quickly to distinguish between them, the "no go" gage is usually made shorter than the "go" gage. Intermediate sizes have a hole in the gage into which a shouldered handle

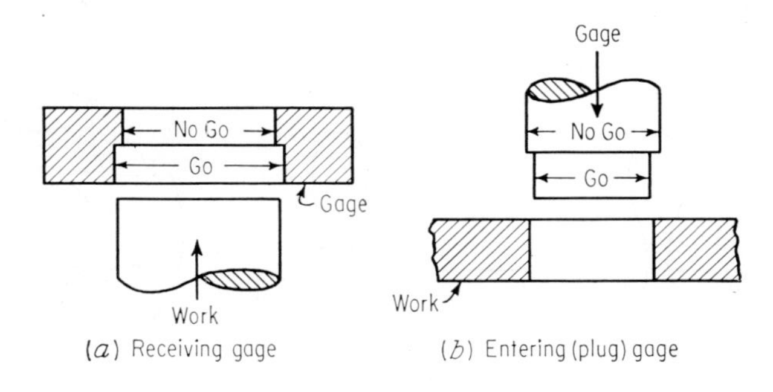

Fig. 4.45. **Simple gages.**

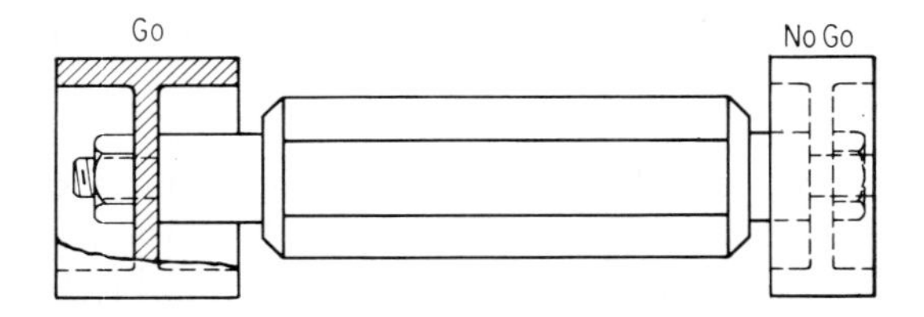

Fig. 4.46. Intermediate-size plug gage.

is fitted and held by a nut. They are usually single ended, although they could be made up as double-ended limit gages, as shown in Fig. 4.46. Large-sized plug gages have two handles, as shown in Fig. 4.47.

When the plug is threaded instead of being plain, it is frequently referred to as a screw, and the gage is called a screw-plug or a thread-plug gage.

The receiving type, when circular, is commonly called a ring gage and may be cylindrical or tapered and plain or threaded. However, when not threaded, the opening may be of any shape such as required for a spline gage. Ordinary ring gages are used to measure outside diameters. The "go" and "no go" openings may be made in the same plate as in Fig. 4.45*a*; usually, however, each ring gage is a separate unit knurled on the outside with the "no go" gage given a heavy chamfer or a deep annular groove to distinguish it from the unmarked "go" gage. "No go" ring gages can obviously detect errors only at the ends of cylindrical parts.

The usual thread ring gage is made adjustable. The screw shown in Fig. 4.48 is in two parts, one a slotted sleeve that is threaded both internally and externally so that when the end *A* is turned it opens or closes the adjusting slot, while turning the end *B* locks the screw in position.

Snap gages are made both fixed and adjustable, as shown in Fig. 4.49. The fixed or plain type is cheaper but not so serviceable as the adjustable or built-up type. Snap gages are used for checking outside diameters, thicknesses, and lengths. In general, snap gages permit more rapid inspection than do ring gages, but where tolerances are close, "go" ring gages will quickly check roundness and diameter.

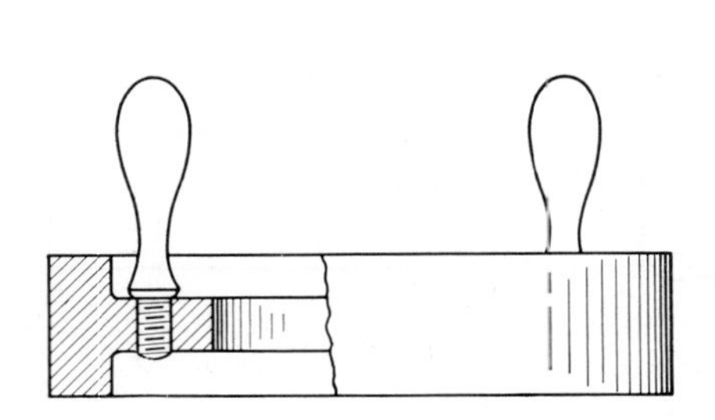

Fig. 4.47. Large-size plug gage.

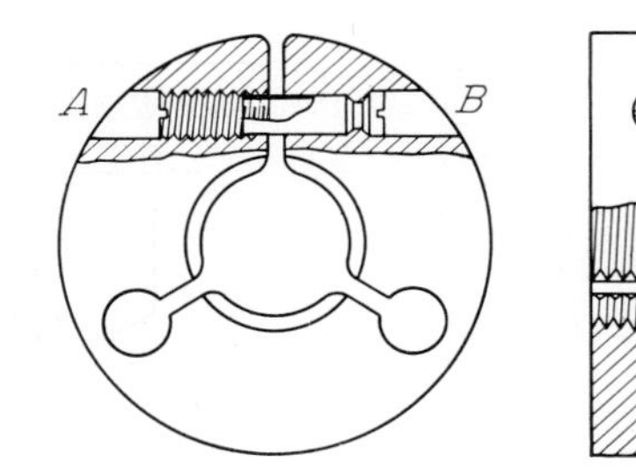

Fig. 4.48. Thread ring gage.

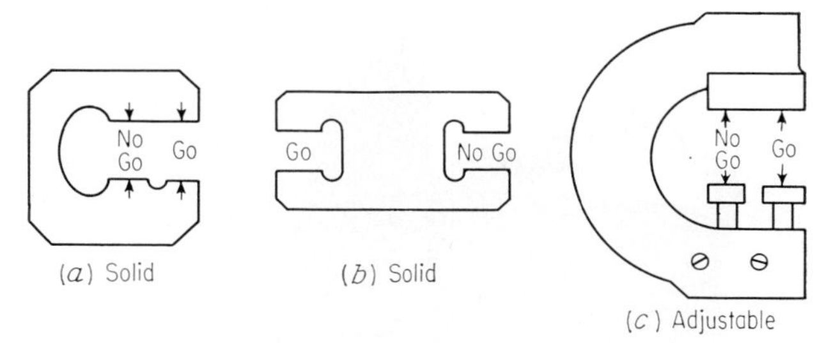

Fig. 4.49. **Typical snap gages.**

4.31. Contour Gages. Contour, profile, and template gages to determine the contour or outline of a part may be made of the "go" or "no go" type; however, instead of using such gages, an enlarged shadow of the part being examined or inspected may be projected on an enlarged or "blown-up" drawing. The latter method is used both in the manufacture and inspection of such parts as gears, threads, and cams. Even in a relatively low magnification such as 20, a difference between the projected shadow and the master drawing, which represents one-thousandth of an inch on the original part, can readily be measured. Special instruments made to test screw threads by this method make it possible both to rotate and shift the viewing screen carrying the master pattern or drawing so as accurately to measure variations in angle and pitch. These optical projectors have a wide range of magnification varying from four to several hundred times. However, for measuring the pitch diameter of a thread gage, the National Screw Thread Commission still recommends the so-called "three-wire method," which involves placing three wires in the grooves, two on one side and one on the other side. An overall diametral measurement is taken, using an outside micrometer caliper, and the pitch diameter is then calculated, using equations given in various handbooks.

4.32. Indicating Gages and Comparators. Indicating gages employ a device such as a lever and graduated scale or a dial indicator to show the variations in dimensions being checked. These are more expensive units than the ordinary "fixed" gages described previously; however, the elements of wear and feel are not involved in making the measurements. Dial indicators as described in Art. 4.10 which read to 0.001 or 0.0001 in. or other types of indicating devices may be locked to the column of the instrument (sometimes called a comparator) in any position, as shown in Fig. 4.50. The zero setting of the dial can be adjusted by a gage block or some other unit and "go" and "no go" marks made on the dial or actual plus-minus readings from the zero position observed.

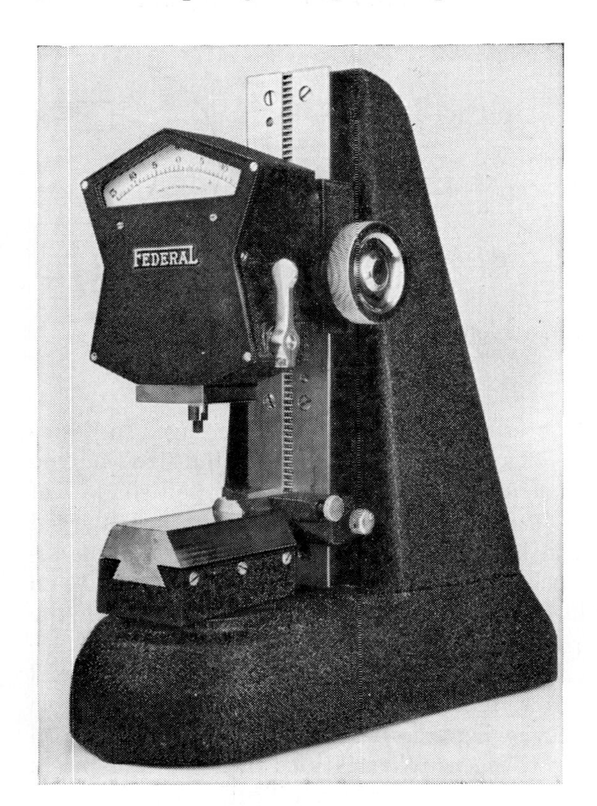

***Fig. 4.50.* Indicating comparator.**

A novel use of flexible metal strips to produce a mechanical lever effect is found in the Sheffield gage, which is shown diagrammatically in Fig. 4.51. The final step-up to yield a sensitiveness of 0.00001 in. is secured by casting the shadow of the pointer on a translucent scale. The range of this indicator is about 0.0005 in. Its use has been largely confined to shop gaging work.

Another type of comparator is the electrolimit gage. In this instrument, the spindle or plunger, which is in contact with the piece to be measured, does not move a pointer through a series of gears, as in the ordinary dial indicator, but instead actuates an armature that is balanced between two electromagnetic coils. Any movement of the armature will produce a current in the coils, which can be read on a microammeter whose scale is graduated to decimals of an inch of movement of the spindle. Usually a 0.0001-in. increment of the spindle deflects the end of the ammeter needle 1 in. Since the range of any comparator is usually decreased when the sensitivity is increased, the full range of plunger movement in the electrolimit gage is only 0.0005 in.

The Zeiss optimeter, instead of using an electrical current to amplify

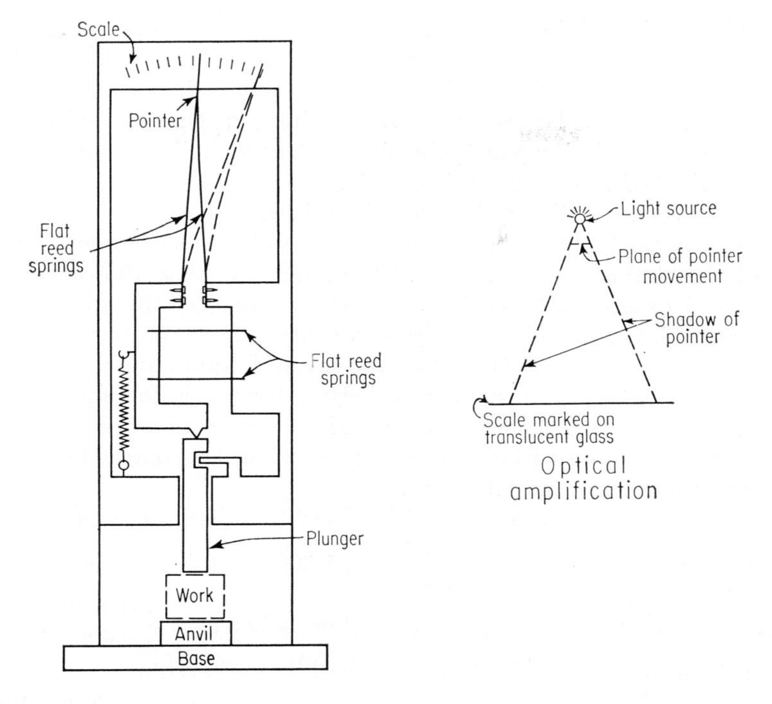

Fig. 4.51. **Sheffield visual gage.**

the motion of the plunger, employs a combined mechanical (hinged mirror) and optical amplification that gives a range of 0.0035 in. on both sides of the selected datum on a scale permitting estimated readings to 0.00001 in.

When making such precise measurements, it is well to remember that a temperature change of 1°F produces a change of length of 0.0000065 in. for each inch of length of any object of steel. Since the standard temperature when checking gages is 68°F (20°C), all important measurements should be made at this temperature.

4.33. Tolerances. It is quite generally agreed that interchangeable parts were first used about 1790, but the concept of a "go" tolerance limit was not introduced until 1840 and the "no go" limit some thirty years later.

In the case of a shaft and bearing, a set of "go" gages might bring together a shaft of maximum diameter (one that will just pass through the "go" ring gage) and a bearing having the minimum or smallest allowable diameter (one that will just fit over the "go" plug gage).

The mating of these parts will produce the minimum clearance. In order to control the maximum clearance which is encountered when the smallest allowable shaft (one that just fails to enter the "no go" ring gage) is mated with the largest allowable bearing (one that just fails to be entered by a "no go" plug gage), the size of the "no go" gages must be based on the desired fit of the mating parts.

"Clearance" may be defined as the difference in dimensions between mating parts, but when clearance is specified as a definite amount, it is designated as an "allowance." We might say that allowance is an intentional clearance. Since it is impossible intentionally to make a part to exact size, a definite leeway or variation must be permitted. This difference or range between maximum and minimum acceptable sizes is known as tolerance, or more briefly, tolerance is the allowable variation in a dimension. Tolerances for various classes of work are nonprecision, 0.01 in.; semiprecision, 0.001 in.; precision, less than 0.001 in.; and high precision, less than 0.0001 in.

Tolerances are expressed in either the bilateral or the unilateral system. For example, $1\begin{smallmatrix}+0.0006\\-0.0000\end{smallmatrix}$, i.e., 1.0006 to 1.0000, expresses a tolerance of 0.0006 in. in the unilateral system, whereas $1\begin{smallmatrix}+0.0003\\-0.0003\end{smallmatrix}$ also expresses a tolerance of 0.0006 in., but in the bilateral system.

In the unilateral system, a shaft and hole each with a thousandth of an inch tolerance could be dimensioned

$$\text{Shaft} = 0.774 \text{ in.} - 0.001 \text{ in.}$$
$$\text{Hole} = 0.775 \text{ in.} + 0.001 \text{ in.}$$

In this case, the clearance would vary from 0.001 in., giving the maximum tightness, to 0.003 in., which would produce the greatest looseness.

The American Standards Association has published a list of recommended tolerances and allowances for eight classes of fits, ranging from loose, which has a large positive allowance, to a heavy shrink fit, which has considerable negative allowance. Detailed information on these is readily available in engineering handbooks and various texts; so it will not be repeated here.

4.34. Inspection and Manufacturer's Gages. When a drawing calls for a shaft $D\begin{smallmatrix}+0.000\\-d\end{smallmatrix}$, the maximum allowable size of the shaft is D and minimum is $(D - d)$. The dimensions for the "go" part of an inspection ring gage are $(D - a)\begin{smallmatrix}+0.000\\-b\end{smallmatrix}$, where a is a wear allowance and b is the

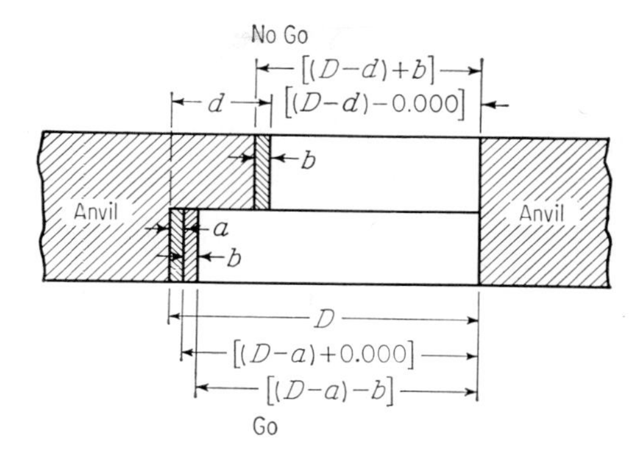

Fig. 4.52. Dimensions of inspection gages.

gage maker's tolerance. The "no go" element would be $(D - d) \begin{smallmatrix} +b \\ -0.000 \end{smallmatrix}$ (see Fig. 4.52).

When the "go" element of a solid snap gage having these dimensions enlarges by wear to a dimension D, it should be taken out of service, and if the gage is adjustable it should be reset.

The "go" dimension for a new manufacturer's gage for the small shaft would be $(D - a - b - c) \begin{smallmatrix} +0.000 \\ -e \end{smallmatrix}$, where the wear allowance is c and the gage maker's tolerance is e.

The "no go" dimension would be $(D - d + b) \begin{smallmatrix} +e \\ -0.000 \end{smallmatrix}$, and when this element has worn to $(D - a - b)$ it becomes unserviceable, because a part made to that size would not pass the inspector's gage if the latter happened to be at its allowable minimum (see Fig. 4.53).

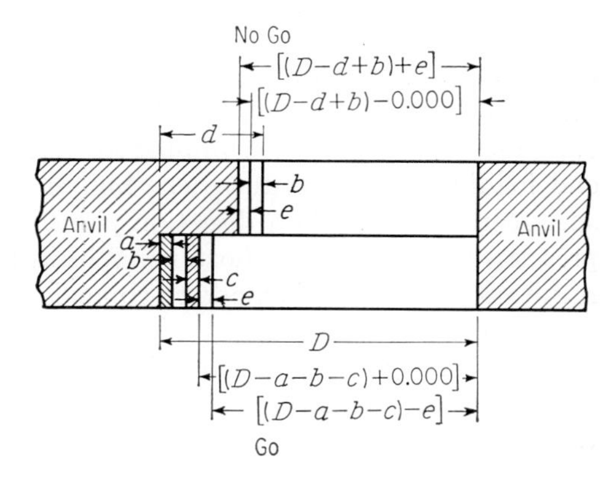

Fig. 4.53. Dimensions of manufacturer's gages.

Chapter 5.

STATIC TENSION AND COMPRESSION TESTS

5.1. Scope and Applicability. The terms *tension test* and *compression test* are usually taken to refer to tests in which a prepared specimen is subjected to a gradually increasing (i.e., "static") uniaxial load until failure occurs. In a simple tension test, the operation is accomplished by gripping opposite ends of the piece of material and pulling it apart. In a compression test, it is accomplished by subjecting a piece of material to end loading which produces crushing action. In a tension test, the test specimen elongates in a direction parallel to the applied load; in a compression test, the piece shortens. Within the limits of practicability, the resultant of the load is made to coincide with the longitudinal axis of the specimen.

Except for certain arbitrarily formed test pieces, the specimens are cylindrical or prismatic in form and of approximately constant cross section over the length within which measurements are made. Compression specimens are limited to such a length that bending due to column action is not a factor. Thus (with certain exceptions), attempt is made to obtain a uniform distribution of direct stress over critical cross sections normal to the direction of the load. The attainment of these ideal conditions is limited by the form and trueness to form of the test piece, by the effectiveness of the holding or bearing devices, and by the action of the testing machine.

The static tension and compression tests are the most commonly made and are among the simplest of all the mechanical tests. Since what may be considered the beginning of scientific testing, tension tests, at least, have occupied a large share of attention, and during the past fifty years great and probably well-merited confidence has been placed in the value and significance of both the tension and compression tests.

When properly conducted on suitable test specimens, these tests, of all tests, come closer to evaluating fundamental mechanical properties for use in design, although it should be observed that the tensile and compressive properties are not necessarily sufficient to enable the prediction of performance of materials under all loading conditions. When standard methods of test are employed, the results are acceptable criteria of quality of materials with which sufficient experience has accumulated to provide assurance that a given level of quality means satisfactory behavior in service. Such tests imply the standardization of specimens with regard to size, shape, and method of preparation and the standardization of the procedures of testing. As with any test, however, for newly developed materials, the tension and compression tests should be used with caution as quality-level indicators because the significance of such tests is limited by their correlation with performance.

Properly conducted tests on representative parts can be valuable in indicating directly the performance of such parts under loads in service. Suitable tests of specimens or fabricated parts subjected to specific treatments can be of use in evaluating quantitatively the effect of such treatments.

Although, in so far as the sense or direction of stress is concerned, compression is merely the opposite of tension, there are several factors that make the tension or the compression test the more desirable in a specific case. The most important of these factors are as follows: (1) suitability of the material to perform under a given type of loading, (2) differences in properties of a material under tensile and compressive loading, and (3) relative difficulties and complications induced by the gripping of or end bearing on the test pieces.

The use of the tension as against the compression test is probably largely determined by the type of service to which a material is to be subjected. Metals, for example, generally exhibit relatively high tenacity and are therefore better suited to and are more efficient for resisting tensile loads than materials of relatively low tensile strength. The tension test is thus very commonly employed with and is appropriate for general use with most cast, rolled, or forged ferrous and nonferrous metals and alloys.

With brittle materials, such as mortar, concrete, brick, and ceramic products, whose tensile strengths are low compared with their compressive strengths, and which are principally employed to resist compressive forces, the compression test is more significant and finds greater use.

The tensile strength of wood is relatively high, but it cannot always be effectively utilized in structural members because of low shear resistance, which causes failure at the end connections using bolts, split rings, or shear plate connectors before the full tensile resistance of a member can be developed. Thus, in so far as direct stresses are concerned, the compression test of wood is of greater practical significance than the tension test. Some materials, like cast iron, although having a lower tensile than compressive strength, are used to resist either type of stress, and both types of test are sometimes made.

The use of tension and compression tests is not confined to the determination of the properties of the material in the form of prepared (shaped) specimens. Full-size tests of manufactured materials, fabricated parts, and structural members are commonly made. The variety of full-size fabricated parts and members to which tension or compression tests may be applied is large. In many cases, tests of this sort are essentially the same as tests on prepared specimens. For example, more or less standardized apparatus and test procedures are used to determine the properties of selected lengths of wire, rod, tubing, reinforcement bars, fibers, fabrics, cordage, and wire rope in tension and brick, tile, masonry blocks, and certain types of metal castings in compression. The important feature of other full-size tests is the duplication, as nearly as possible, of the load conditions of service and the observation of the development of localized weaknesses as well as critical loads. A few of the full-size pieces on which tests are not uncommonly made may be mentioned: (in tension) eyebars, anchor chain, crane hooks, drawbars, and riveted and welded joints; (in compression) cast-iron and concrete pipe, built-up piers, pedestals, columns, and wall sections. Compression tests on columns, thin-walled fabricated parts and structures, and the like involve problems of elastic stability and usually require special procedures.

THE TENSION TEST

5.2. Requirements for Tension Specimens. Although certain fundamental requirements may be stated, and certain shapes of specimens are customarily used for particular types of tests, prepared specimens for tension tests are made in a variety of forms. The cross section of the

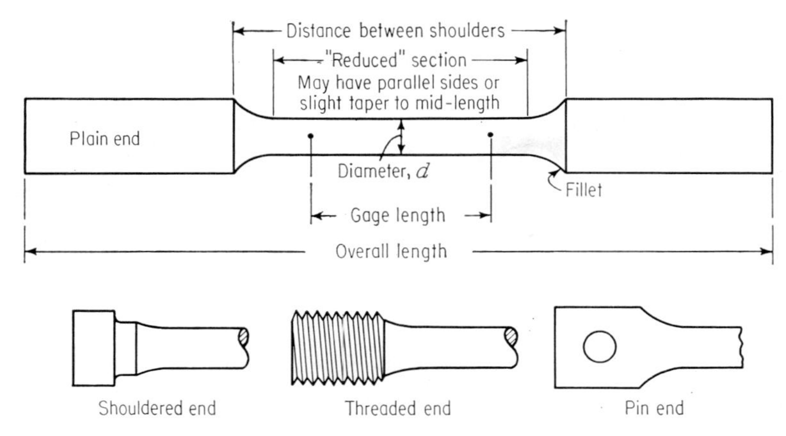

Fig. 5.1. **Typical tension specimen.**

specimen is usually round, square, or rectangular. For metals, if a piece of sufficient thickness can be obtained so that it can be easily machined, a round specimen is commonly used; for sheet and plate stock, a flat specimen is usually employed. The central portion of the length is usually (but not always) of smaller cross section than the end portions in order to cause failure to occur at a section where the stresses are not affected by the gripping device. Typical nomenclature for tension specimens is indicated in Fig. 5.1. The gage length is the marked length over which elongation or extensometer measurements are made.

The shape of the ends should be suitable to the material and such as to fit properly the gripping device to be employed. The ends of round specimens may be plain, shouldered, or threaded. Plain ends should be long enough to accommodate some type of wedge grip. Rectangular specimens are generally made with plain ends, although sometimes they may be shouldered or contain a hole for a pin bearing.

The ratio of diameter or width of end to diameter or width of reduced section is determined largely by custom, although for brittle materials it is important to have the ends sufficiently large to avoid failure due to the compounding of the axial stress and the stresses due to action of the grips. If a specimen is machined from larger stock, the reduction should be at least enough to remove all surface irregularities. The transition from end to reduced section should be made by an adequate fillet in order to reduce the stress concentration caused by the abrupt change in section; for brittle materials, this is particularly important. The effect of change of section on stress distribution is practically inappreciable at distances greater than about one or two diameters from the change.

To obtain uniform stress distribution across critical sections, the reduced portion of the piece is often made with parallel sides for its entire

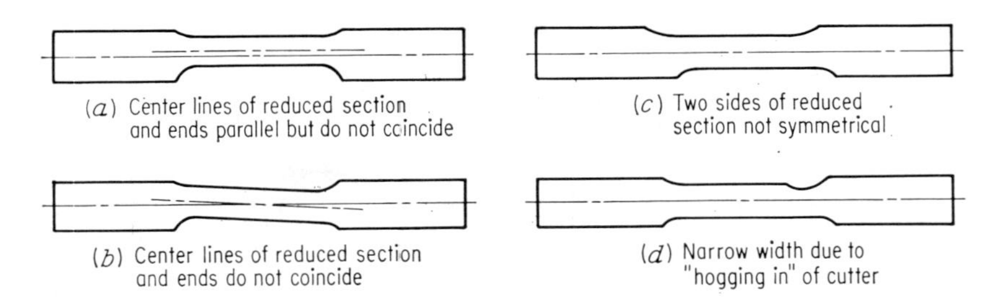
(a) Center lines of reduced section and ends parallel but do not ccincide
(c) Two sides of reduced section not symmetrical
(b) Center lines of reduced section and ends do not coincide
(d) Narrow width due to "hogging in" of cutter

Fig. 5.2. **Common faults of flat test specimens.**

length, although many types of specimens are made with a gradual taper from both ends of the reduced section to its mid-length. Specimens of some materials (e.g., mortar briquets as shown in Fig. 5.6) are curved throughout the central portion of their length in order to prevent breakage at or near the grips; in such specimens, the stress is not uniform on the critical cross section, and all dimensions of the specimen must be standardized to obtain comparable results.

A specimen should be symmetrical with respect to a longitudinal axis throughout its length in order to avoid bending during application of load. Figure 5.2 illustrates common faults in the preparation of flat specimens.

The length of the reduced section depends on the kind of material to be tested and the measurements to be made. With ductile metals, for which elongation or reduction of area is to be determined, the length must be sufficient to permit a normal break, i.e., the drawing out or necking down should not be inhibited by the mass of the ends. With brittle materials for which the elongation is very small and is not measured and for which fracture is plane, the length of the reduced section may be relatively short.

The gage length is always somewhat less than the distance between shoulders, but practice with regard to the ratio between these two lengths is not uniform. If extensometer measurements are to be made, it is considered desirable to have the gage length short of the distance between shoulders by at least twice the diameter of the test piece. The gage points should be equidistant from the center of length of the reduced section.

The *percentage* elongation of a ductile metal specimen of given diameter depends upon the gage length over which the measurements are made. It has been established by many tests that the elongation is practically constant for pieces of various sizes if the pieces are geometrically similar. For small cylindrical specimens of ductile metals, the ASTM (ASTM E 8) calls for a gage length of four times the diameter. For larger specimens of ferrous metal, various ASTM Specifications (ASTM A 7, A 15) use some specified gage length and thickness or diameter as a base, and the effect of different thickness or diameters is provided for by deductions from the permissible elongation in accordance with a stated rule.

Some general requirements with regard to the selection and preparation of specimens have been outlined in Chap. 2.

5.3. Standard Test Specimens. The dimensions of various standardized test pieces, with permissible tolerances, are given in Figs. 5.3 to 5.7.

The ASTM Standard round tension specimen for ductile metals shown in Fig. 5.3*a* is often made 0.505 in. in diameter in order to have an

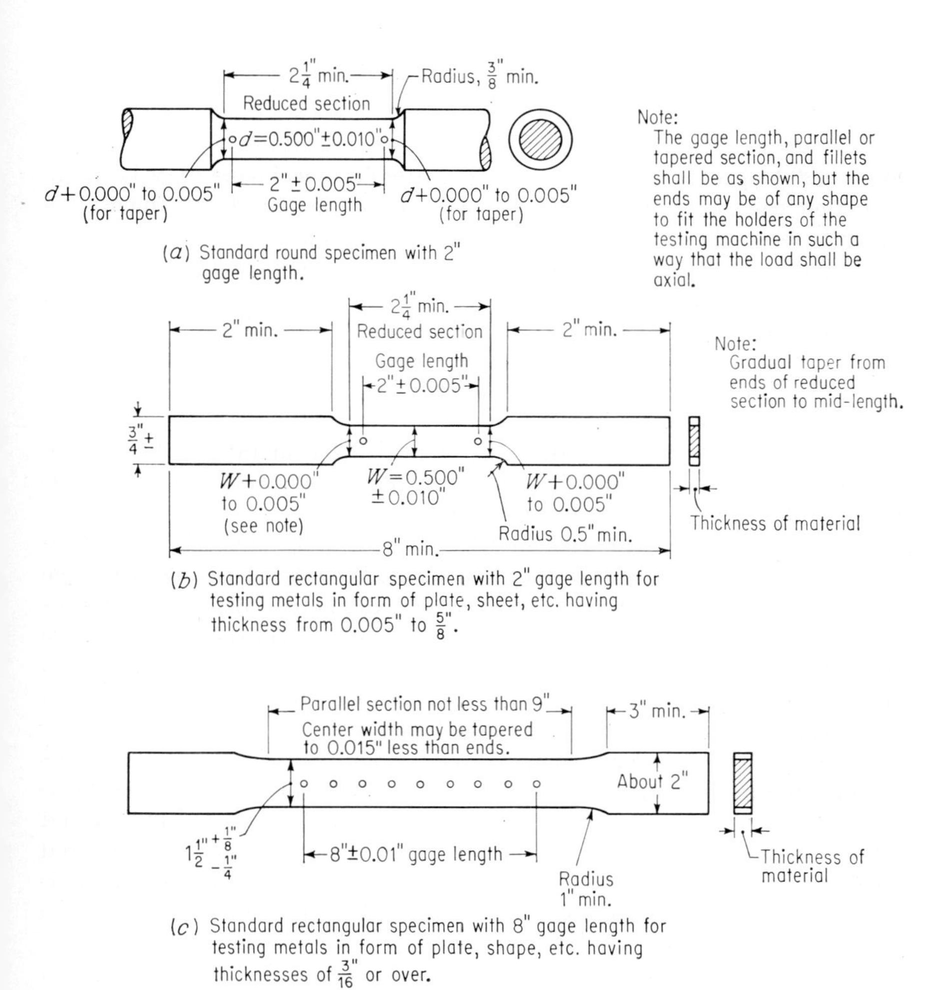

Fig. 5.3. **ASTM standardized forms of (ductile) metal tension-test specimens (ASTM E 8).**

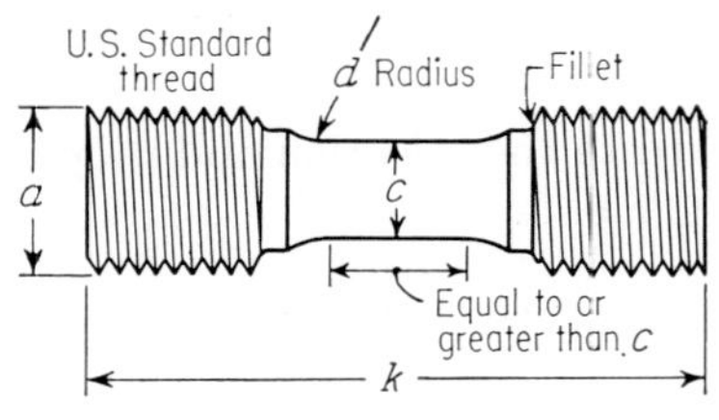

Dimensions of tension test specimens, in.

Dimension	Spec. *A*	Spec. *B*	Spec. *C*
a	$\frac{3}{4}$	$1\frac{1}{8}$	$1\frac{7}{8}$
c	0.500	0.750	1.25
d, min.	1.00	1.00	2.00
k, min.	3.75	4 00	$6\frac{3}{8}$

***Fig. 5.4.* ASTM Standard round tension-test specimen for cast iron (ASTM E 8, A 48).**

even 0.200-sq in. area of cross section. Smaller specimens may be used, provided the gage length is four times the diameter of specimen. If a taper is used, the difference in diameter at the ends and center of gage length should not exceed about 1 percent. Specimens from plate and flat stock are shown in Figs. 5.3*b* and 5.3*c*. Slight variations on these types of specimen may be found in various particular specifications.

The form of the ASTM Standard specimen for cast iron is shown in Fig. 5.4. Three sizes are used, the principal dimensions of which are shown on the figure.

The ASTM Standard specimen for die-cast metals is 0.25 in. in diameter and has a fillet of 3-in. radius and a gage length of 2 in. Otherwise it resembles the specimen shown in Fig. 5.3*a*.

Specimens from bar, rod, or wire stock usually have the full cross-sectional area of the product they represent. Where practicable, the gage length should be four times the specimen diameter, although for sizes ¼ in. and less a 10-in. gage length is often used. Tension tests of wire rope are made on lengths cut from commercial rope. The ends are held in special sockets filled with zinc that has been poured while molten around the splayed ends of the rope (see Fig. 5.14).

Small tubes (1 in. or less) are tested full size. Snug-fitting metal plugs are inserted in the ends far enough to permit the grips to clamp the specimens without causing collapse of the tube. The plugs should not extend into that part of the specimen upon which the elongation is measured (ASTM E 8). For large tubes that cannot be tested in full section, longitudinal specimens usually are cut, although transverse specimens are sometimes permitted (ASTM A 106).

The ASTM Standard tension-test specimen for wood is shown in Fig. 5.5.

The ASTM Standard specimen for tension tests of portland-cement mortars is shown in Fig. 5.6. The shape of a specimen that is used for

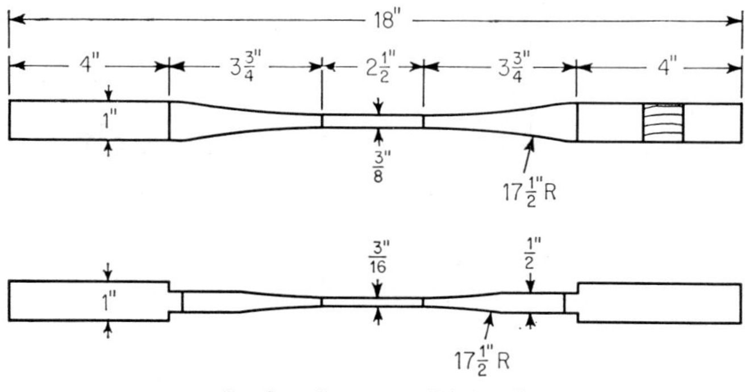

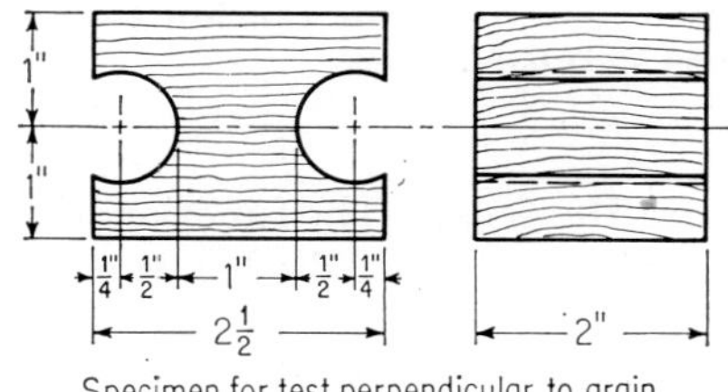

Fig. 5.5. **ASTM Standard tension-test specimens for wood (ASTM D 143).**

molded electrical insulation materials is shown in Fig. 5.7 (ASTM D 651).

Reference may be made to various other tension-test specimens that will not be discussed here, such as those for vulcanized rubber (ASTM D 412), hard rubber (ASTM D 530), and plastics (ASTM D 638).

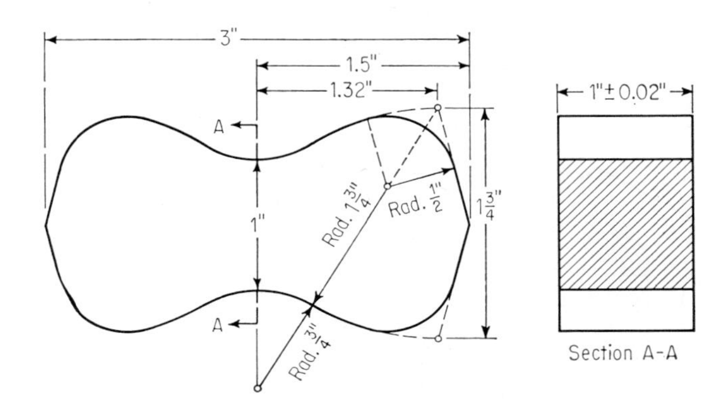

Fig. 5.6. **ASTM Standard tension specimen for portland-cement mortar (ASTM C 190).**

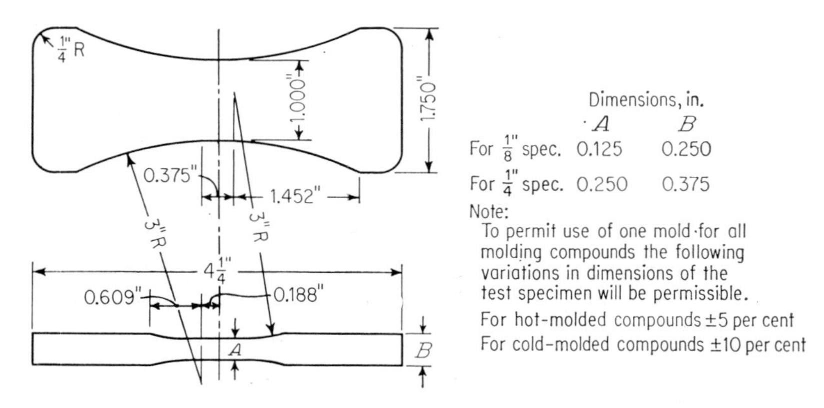

Fig. 5.7. **ASTM specimen for tension test of molded electrical insulating materials (ASTM D 651).**

5.4. Gripping Devices. The function of the gripping device or shackles is to transmit the load from the heads of the testing machine to the test specimen. The essential requirement of the gripping device is that the load be transmitted axially to the specimen; this implies that the centers of action of the grips be in alignment at the beginning and during the progress of a test and that no bending or twisting be introduced by action or failure of action of the grips. In addition, of course, the device should be adequately designed to carry the loads and should not loosen during a test.

Wedge grips, illustrated in Fig. 5.8, are a common type of holding device. They are satisfactory for commercial tests of ductile metal specimens of adequate length, because slight bending or twisting does not appear to affect the strength and elongation of ductile materials. No adjustment to prevent bending can take place with grips of this sort. Wedge-type grips are usually unsatisfactory for use with brittle materials,

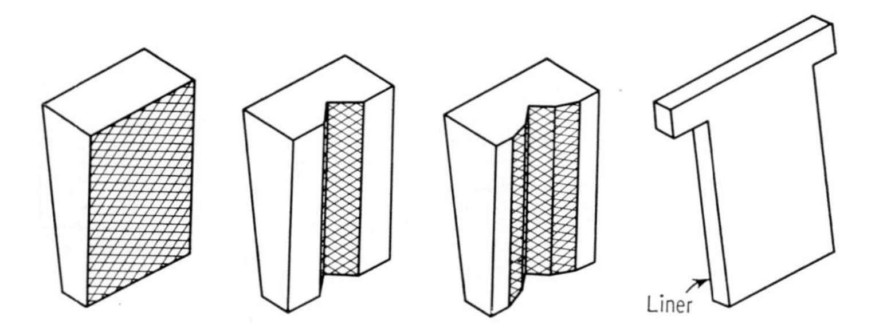

Fig. 5.8. **Wedge-grip units for tension tests of metals.**

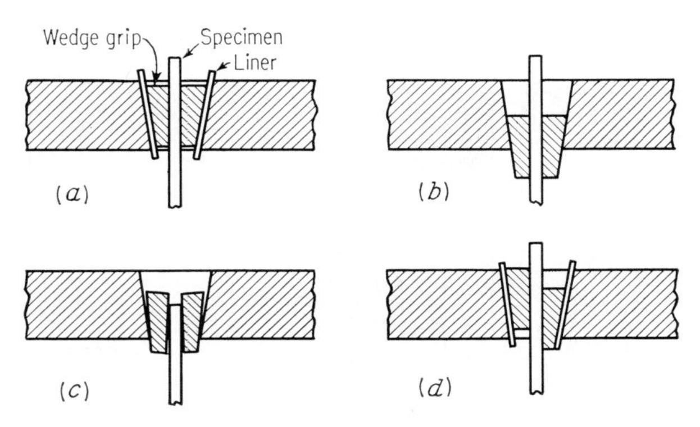

Fig. 5.9. **Correct** (*a*) **and faulty setting** (*b*, *c*, **and** *d*) **of wedge grips.**

because the crushing action of the wedges tends to cause failure at or near the grips. The faces of grips that contact the specimen are roughened or serrated to reduce slippage; for flat specimens the faces of the grips are flat, and for cylindrical specimens the grips have a V groove of suitable size. Adjustment is made by means of shims or liners so that the axis of the specimen coincides with the center line of the heads of the testing machine and so that the grips are properly located in the head. Correct and faulty settings of the grips are illustrated in Fig. 5.9.

Where assurance of more accurate alignment is necessary, which is highly important in tests of brittle materials, some type of universal joint is used in the holders at both ends; usually it is a spherically seated or pin-bearing arrangement (so-called "self-aligning" linkage). A schematic drawing of a device using spherically seated bearings at the heads of the testing machine is shown in Fig. 5.10 (ASTM E 8). The distance

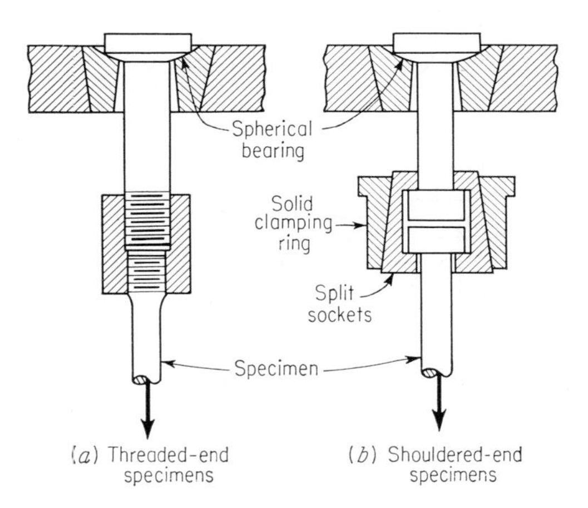

Fig. 5.10. **Spherically seated holders (ASTM E 8).**

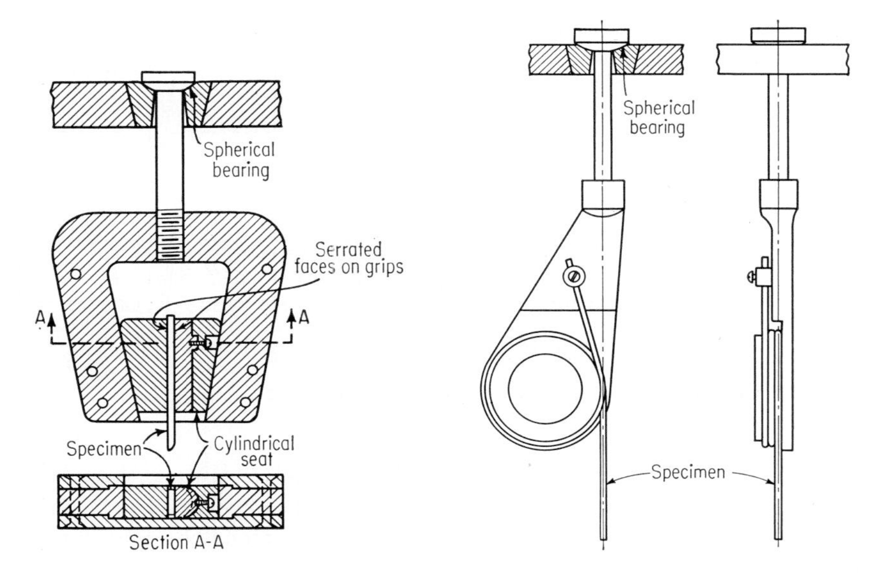

Fig. 5.11. **Templin grips.**

Fig. 5.12. **Snubbing device for testing wire.**

between spherical bearings should be as great as feasible. Such devices are not always entirely effective; obviously, spherical seats do not adjust themselves easily if not properly lubricated and may "freeze" at high loads, regardless of lubrication.

A device for adequately gripping thin sheet-metal specimens and wire (Templin grips) is illustrated in Fig. 5.11. A device for testing wire is shown in Fig. 5.12.

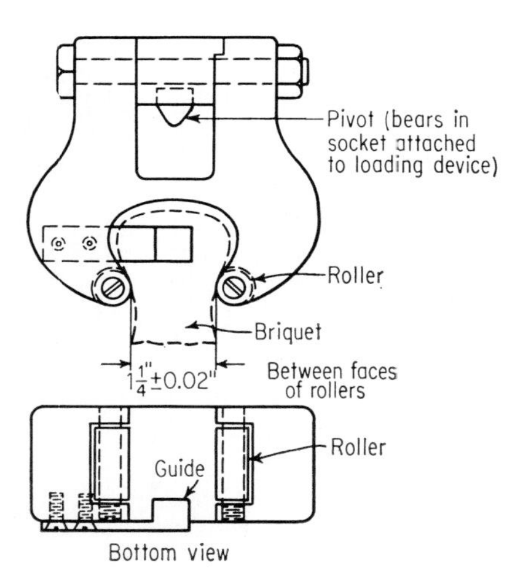

Fig. 5.13. **Briquet grip (ASTM C 190).**

Fig. 5.14. **Wire-rope socket.**

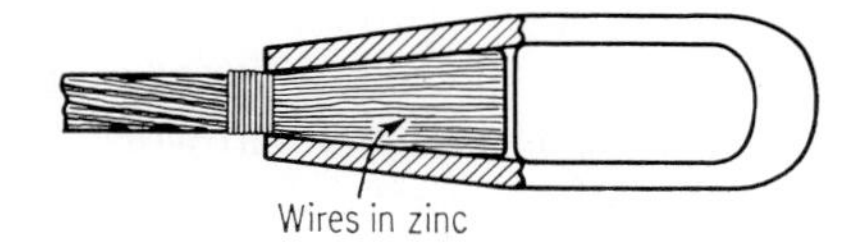

The type of grip used for tests of mortar briquets is shown in Fig. 5.13. A common type of wire-rope socket is shown in Fig. 5.14.

For testing prismatic specimens of concrete, rigid steel plates are bonded to the ends using an epoxy cement. Tensile loads are then applied to spherically seated axial steel rods connected to the end plates. Since the epoxy cement is stronger than the concrete, failure always occurs in the concrete.

Another type of test for determining the tensile strength of concrete is the splitting tension test which is covered by ASTM C 496-62*T*. As shown in Fig. 5.15, it uses a standard 6- by 12-in. cylinder which is loaded in compression along two axial lines 180° apart. Narrow strips of ⅛-in. plywood are used as a cushioning material along these load lines. The splitting tensile strength is computed from $\sigma = 2P/\pi ld$

where σ = splitting tensile strength, psi
P = maximum applied load, lb
l = length, in.
d = diameter, in.

This type of test is simpler than any axial tension test, yet the test results agree reasonably well (about 15 percent higher) with those from the more conventional type of test [506].

5.5. Conduct of Tests. In the commercial tension test of metals the properties usually determined are yield strength (yield point of ductile metals), tensile strength, ductility (elongation and reduction of area),

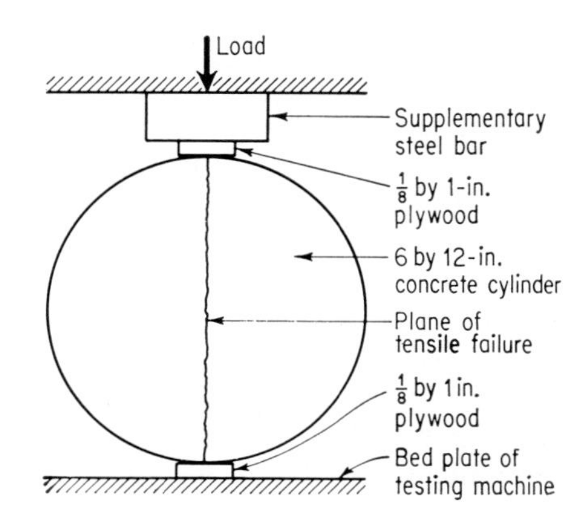

Fig. 5.15. **Method of loading concrete specimen for splitting tension test.**

and type of fracture. For brittle material, only the tensile strength and character of fracture are commonly determined. In more complete tests, as in much investigational work, determinations of stress-strain relations, modulus of elasticity, and other mechanical properties are included. See Chap. 2 for the basic procedures involved. The plan of the work and schedule of operations will obviously be adapted to the needs of the test.

Prior to applying load to a specimen, its dimensions are measured. Occasionally unit weight may be called for, requiring weight and volume determinations. Linear measurements are made with scales, calipers and scale, or micrometers, depending upon the dimension to be determined and the precision to be attained. In the simplest case, only the diameter or width and the thickness of the critical section are measured. Cross-sectional dimensions of metal specimens should ordinarily be read with a precision of about 0.5 percent. Except for small diameters and thin sheets, measurements to 0.001 in. meet this requirement. On cylindrical specimens, measurements should be made on at least two mutually perpendicular diameters.

If elongation measurements are to be made, the gage length is scribed or laid off. On ductile metal specimens of ordinary size, this is done with a center punch, but on thin sheets, or brittle material, fine scratches should be used. In any case, the marks should be very light so as not to damage the metal and thus to influence the break. Where much work is to be done, a double or multiple-pointed punch is sometimes used. It is convenient to lay round specimens in a V-shaped block while marking gage points. When an 8-in. gage length is used on steel specimens, marks are placed 1 in. apart.

Before using a testing machine for the first time, the operator should familiarize himself with the machine, its controls, its speeds, the action of the weighing mechanism, and the value of the graduations on the load indicator. Before a specimen is put in a machine, the weighing device of the machine should be checked for zero load indication and adjustments made if necessary.

When a specimen is placed in a machine, the gripping device should be checked to see that it functions properly. If stops or guards are used to prevent the grips from flying out of their sockets when sudden failure occurs, the stops should be fastened in place. The specimen should be so placed that it is convenient to make observations on the gage lines.

If an extensometer is to be used, the value of the divisions on the indicator and the multiplication ratio should be determined before the extensometer is placed on the specimen. It should be placed centrally on the specimen and properly aligned. When collar-type extensometers are used, the axis of the specimen and the axis of the extensometer should be made to coincide. After it is clamped in place, the spacer bar (if any) is removed and the adjustments are checked. Oftentimes a small initial load is placed on the specimen before the extensometer is set at zero reading.

The speed of testing should not be greater than that at which load and other readings can be made with the desired degree of accuracy, and if testing speed has an appreciable influence on the properties of the material, the rate of straining of the test piece should lie within definite limits, although studies have indicated they may be reasonably broad.

Methods of specifying testing speeds vary. A number of recommendations for testing speeds have been made and withdrawn from ASTM standards for metals. The remarks that follow here are intended only as general information until such time as generally acceptable bases for stating testing speeds become available.

Because of the former widespread use of screw-driven testing machines in this country, it has been customary in the past to indicate testing speed by the idling speed of the movable head of the machine. However, during the last decade or two there has been a marked increase in the use of hydraulic-type testing machines, in which the testing speed is controlled in terms of the rate of load application. It is desirable that alternative limits on crosshead speed and load rates be specified.

Various ASTM requirements on testing speeds are shown as a general guide in Table 5.1.

The speeds shown for metallic specimens are maximum values; slower speeds may be and often are used. For tests involving yield-point determinations by drop of beam, halt of load pointer, or by dividers, a rate of straining corresponding to a crosshead speed of about 0.05 in. per min probably represents about average practice, although in mill tests, higher speeds are not uncommon. Oftentimes the load is applied rapidly, at any convenient rate up to one-half the specified yield strength or yield point or up to one-quarter the specified tensile strength, whichever is the smaller. Above this point the load is applied at the specified rate.

Above the yield point of ductile metals, higher speeds are permitted because variation in speed does not appear to have so much effect upon the ultimate strength as upon the yield strength; the elongation, however, is sensitive to variation in speed at high rates of loading.

For tests involving extensometer measurements, either the load is applied in increments and the load and deformation are read at the end of each increment, or the load is applied continuously at a slow rate (generally at crosshead speeds ranging from about 0.01 to 0.05 in. per min) and load and deformation are observed simultaneously. The latter method is considered preferable.

Data are not available on which to base any simple general rule for transferring crosshead speeds to rates of load application, although a transference factor or "modulus" for a particular testing machine can be determined experimentally [531]. Within the elastic range, of course,

Table 5.1. **Various ASTM requirements on speed of testing in tension**

Material tested	ASTM Reference	Maximum speed of crosshead, in. per min		Load rate
		To yield	To ultimate	
Metallic materials......	E 8	....	0.5 per in. gage length (E 8 and A 370)	Max. of 100 ksi per min to yield
Steel products.........	A 370	0.062 per in. gage length		
Gray cast iron........	A 48	....	0.125 above 15 ksi	
		Specified rate of travel of power-driven grip, in. per min		
Plastics..............	D 638	0.05*	0.20–0.25*	
Hard rubber...........	D 530	.....		2.9–3.1 lb per sec.
Soft vulcanized rubber..	D 412	.....	20 ± 1	
Wood................	D 143			
Parallel to grain.....	.....	.....	0.05	
Perpendicular to grain............	.....	.....	0.10	
Cement mortar briquets	C 190	....		600 ± 25 lb per min

* Faster speeds permitted for some materials.

the load rate may be readily computed from the strain rate. A survey of practice made some years ago indicated that better than 50 percent of the laboratories involved used load rates within the limits 10 to 70 ksi per min [533]. A few laboratories used load rates as high as 1000 ksi per min for steel. A maximum load rate of 100 ksi per min has been suggested for yield-point determinations of metallic materials (ASTM E 8).

After the test specimen has failed, it is removed from the testing machine, and if elongation values are called for, the broken ends of a specimen are fitted together and the distance between gage points measured with a scale or dividers and scale to the nearest 0.01 in. The diameter

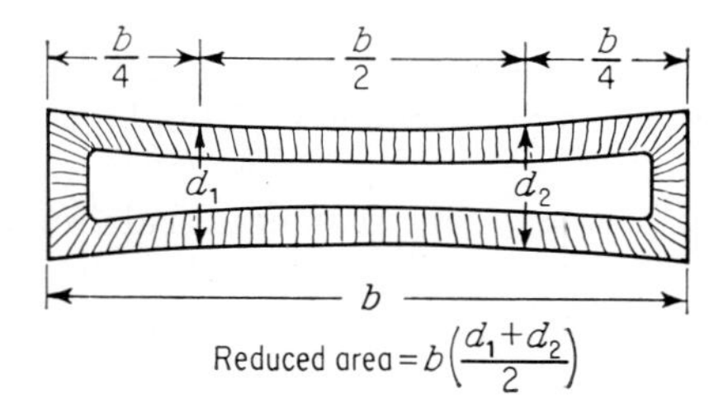

Fig. 5.16. **Determination of reduced section of rectangular test specimens.** (*From Withey and Aston, Ref.* 141.)

of the smallest section is calipered, preferably with a micrometer caliper equipped with a pointed spindle and anvil, for determining reduction in area. The same degree of precision should be employed as was used in measuring the original diameter. For symmetrical "cup-cone" fractures of rectangular specimens, Johnson recommends measurements for determining reduction of area as shown in Fig. 5.16 [141]. For irregular fractures, several measurements must be made, depending upon the nature of the break.

Directions for conducting a typical tension test are given in Part 2.

5.6. Observations of Test. Observations made during a test are recorded on some appropriate form, prepared before starting the test. Identification marks and similar pertinent information are noted. Original and final dimensions and critical loads are recorded as they are observed. If extensometer measurements are made manually, a log of the load and corresponding deformations are made. Some testing machines are equipped with an automatic attachment for drawing the stress-strain diagram. The character of the fracture and the presence of any defects are noted. At some place in the records, the conditions of test should be recorded, particularly the type of equipment used and speed of testing. Stresses, strength, percentage of elongation and reduction of area are computed on the basis of the original dimensions. A log sheet and a stress-strain diagram prepared therefrom are shown in Fig. 5.17. The log sheet contains most of the information pertinent to a tension test but is not intended to be complete since such items as the date and names of operator and recorder, which should be given, are not included.

The elongation is the increase in the gage length, expressed as a percentage of the original gage length. Both the percentage increase and the original gage length are reported. In ductile metals, if the break occurs near one end of the gage length, some of the effects of the localized "drawing out" or "necking down" will extend beyond the gage length. Hence, when the break occurs outside the middle third, specifications often call for a retest, although an approximate method for obtaining the elongation may be used as shown in Fig. 5.18.

The reduction of area is the difference between the area of the smallest cross section (at the break) and the original cross-sectional area, expressed as a percentage of the original cross-sectional area.

Tensile fractures may be classified as to form, texture, and color. Types of fracture as regards form are symmetrical: cup-cone, flat, and irregular or ragged; or asymmetrical: partial cup-cone, flat, and irregular or ragged. Various descriptions of texture are silky, fine grain, coarse grain or granular, fibrous or splintery, crystalline, glassy, and dull.

Certain materials are effectively identified by their fractures. Mild steel in the form of the standard cylindrical specimen usually has a cup-

TENSION TEST OF METAL

LOG SHEET

Material		Mild steel
Mark or number		A 618
Total length of specimen, in.		18.5
Length between shoulders, in.		11.2
Gage length, in.		8.00
Diameter of ends, in.		1.25
Diameter of reduced section, in.		1.001
Elongation in 8 inches, in.		2.50
Diameter of ruptured section, in.		0.613
Speed of machine, in. per min.	To yield str.	0.05
	After yield	0.2

Notes:

(1) 60,000-lb. Olsen machine. (No. 12)

(2) Federal dial extensometer with multiplier of 2 (No. 61)

(3) Three-quarter cup-cone fracture, fine grained in center, silky at edge

Elongation in each inch:
0.20, 0.22, 0.25, 0.35, 0.78, 0.27, 0.23, 0.20

Load, lb. (1)	Dial reading, in. (2)	Stress, psi	Strain, in./in.
3,410	0.002	4,330	0.000125
6,450	0.004	8,200	0.000250
9,160	0.006	11,640	0.000375
12,370	0.008	15,720	0.000500
14,830	0.010	18,860	0.000625
18,020	0.012	22,900	0.000750
20,780	0.014	26,400	0.000875
23,640	0.016	30,000	0.001000
26,370	0.018	33,500	0.001125
29,250	0.020	37,200	0.001250
31,600	0.022	40,200	0.001375
31,710	0.023	40,300	0.00144
31,520	0.024	40,000	0.00150
31,390	0.030	39,900	0.00188
31,100	0.040	39,500	0.00250
31,630	0.050	40,200	0.00312
31,650	0.075	40,200	0.00469
31,700	0.100	40,300	0.00625

Load, lb.	Scale reading, in.	Stress, psi	Strain, in./in.
31,800	0.10	40,400	0.0125
37,200	0.20	47,300	0.0250
41,400	0.30	52,600	0.0375
47,200	0.50	60,000	0.0625
50,200	0.70	63,800	0.0875
52,200	0.90	66,300	0.1125
53,100	1.10	67,500	0.1375
53,400	1.30	67,900	0.1625
53,500	1.50	68,000	0.1875
53,300	1.70	67,700	0.2125
53,000	1.90	67,300	0.2375
52,000	2.10	66,100	0.2625
38,800	2.50	49,300	0.3125
↖ Ruptured (3)			

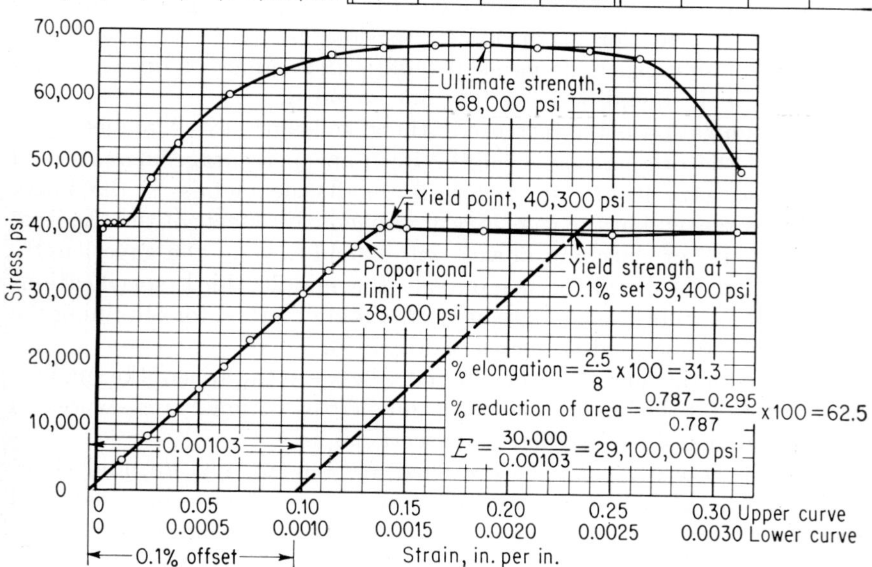

Fig. 5.17. **Log sheet and stress-strain diagram.**

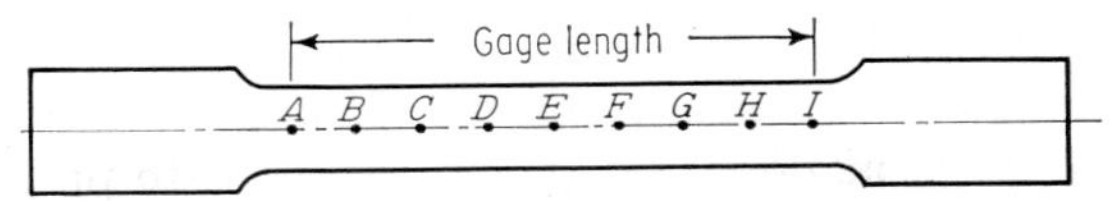

For breaks between C and G; Elongation = Final AI − Original AI
For breaks between A and C but nearer B than A;
Elongation = Final ($AC + 2CF$) − Original AI
For breaks within one half division of A; Elongation = Final $2AE$ − Original AI

***Fig. 5.18.* Determination of approximate elongation for breaks outside middle third of gage length.**

cone type of fracture of silky texture. Wrought iron has a ragged fibrous fracture, whereas the typical fracture of cast iron is gray, flat, and granular. An examination of the fracture may give a possible clue to low values of strength or ductility of the specimen. Nonaxial loading will cause asymmetrical types. Lack of symmetry may also be caused by nonhomogeneity of the material or a defect or flaw of some sort, such as segregation, a blowhole, or an inclusion of foreign matter, such as slag. On the fractured surface of material that has been cold-worked or has an internal stress condition due to certain heat-treatments, there is often the appearance of streaks or ridges radiating outward from some point near the center of the section; this is sometimes referred to as a "star fracture." A description of the fracture should be included in every test report, even though its value is incidental for normal fractures.

Illustrations of a number of typical fractures are shown in Fig. 5.19.

5.7. Effect of Important Variables. As has been repeatedly pointed out, the test conditions and the condition of the material at time of test have a very important influence on the results. Reports of investigations to determine such effects comprise a vast literature extending over many years. One object of many such investigations is to evaluate the effects of test conditions with a view to selecting a standard procedure that will

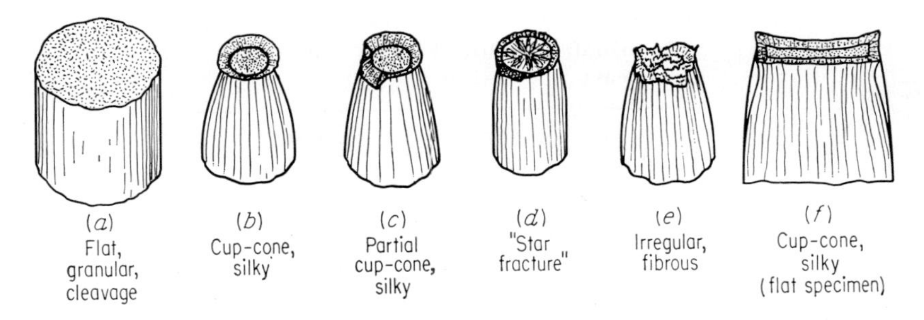

***Fig. 5.19.* Typical tensile fractures of metals.**

give results having a minimum of variability with reasonable fluctuation in test conditions; another object is to develop a basis for projecting the results of tests made under given conditions to probable behavior under some other conditions.

It is not possible here to discuss at length the effect of the numerous variables of testing. However, the general effect of a few of the more important variables will be pointed out in order that some appreciation of sources of error may be attained.

In general, with metallic materials, if the metal is of *uniform quality*, the size of geometrically similar specimens does not appear appreciably to affect the results of the tension test. Various investigations on structural steels have borne this out [511, 514]. However, it is important to remember that, in the course of making or processing parts or shapes, the quality of the metal often varies with the size of the piece being produced. Thus test results for specimens of different sizes may reflect the effect of massiveness on properties. In the case of hot-rolled steel, the ductility is affected to some extent by the work of rolling, although the yield and tensile strengths are but little affected. The strength of cold-drawn wire is markedly influenced by the drawing process. Because of the strain-hardening effect, there is a considerable increase in yield point and ultimate strength of cold-worked metals, but these changes are accompanied by a marked decrease in ductility. In the case of cast metals, the variation of strength with size of casting is marked, as illustrated by the results of certain tension tests of cast-iron bars shown in Table 5.2, but the differences represent largely actual differences in the properties of the specimens, as cast, rather than a real effect of size.

The total elongation of a ductile metal at the point of rupture is due to plastic elongation which is more or less uniformly distributed over the

Table 5.2. Tensile strength of various sizes of cast-iron bars*

Diameter, in.		Tensile strength, psi
Nominal, as cast	Reduced section	
0.25	0.1875	50,800
0.375	0.253	38,100
0.625	0.400	32,200
0.875	0.505	26,000
1.20	0.80	23,300

* H. L. Campbell, "Relation of Properties of Cast Iron to Thickness of Castings," *Proc. ASTM*, vol. 37, pt. II, 1937.

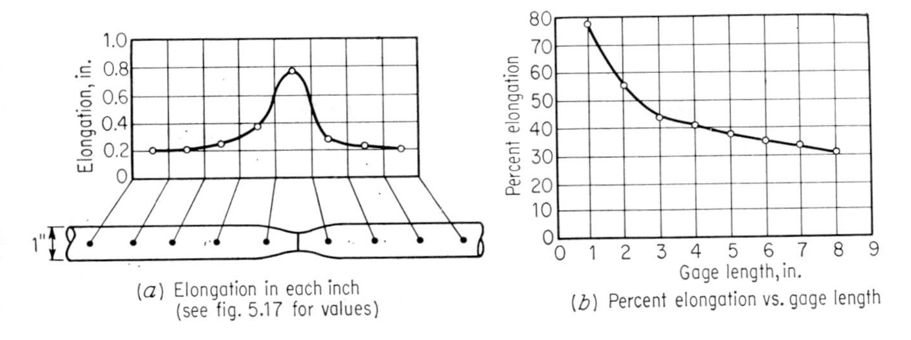

Fig. 5.20. **Effect of gage length on percentage of elongation.**

gage length, on which is superimposed a localized drawing out or extension of the necked section, which occurs just before rupture. The former is practically independent of the shape of cross section and gage length and is small compared with the final localized drawing out which is affected by the shape of the piece. The elongation in each inch of length up to 8 in. of a typical mild-steel tension specimen is illustrated in Fig. 5.20. The corresponding elongations over various gage lengths symmetrical about the break are also shown. The length affected by the final localized drawing out is of the order of two or three times the diameter of the specimen. It is apparent, then, why the diameter of piece and gage length (or ratio of diameter to gage length) must be fixed if comparable elongations are to be obtained and why specifications call for rejection of a test if the break is too near the ends.

The requirement of geometrical similarity of test pieces for comparable elongations was first stated by J. Barba in 1880 and is often referred to as Barba's law [510]. Numerous investigations since then have confirmed this general finding that when the gage length $L = k\sqrt{A}$, where A is the cross-sectional area and k is a constant for the type of specimen, the elongation is practically constant [511].

If the shoulders of a test bar are too close together, if the piece is notched or grooved transversely or contains holes, or if the sides of the specimen are curved, the strength and ductility of the piece may be appreciably affected. The severity of this effect depends upon the abruptness and relative magnitude of the change in section and the ductility of the material.

For a series of specimens of ductile metal which enlarge abruptly at the ends of the gage length, the effect of the length-diameter ratio (L/d) upon both the elongation and reduction of area is shown in Fig. 5.21a [142, p. 39]. For values of L/d greater than about 2, the reduction of area is independent of L/d, but for lower values it is reduced because the

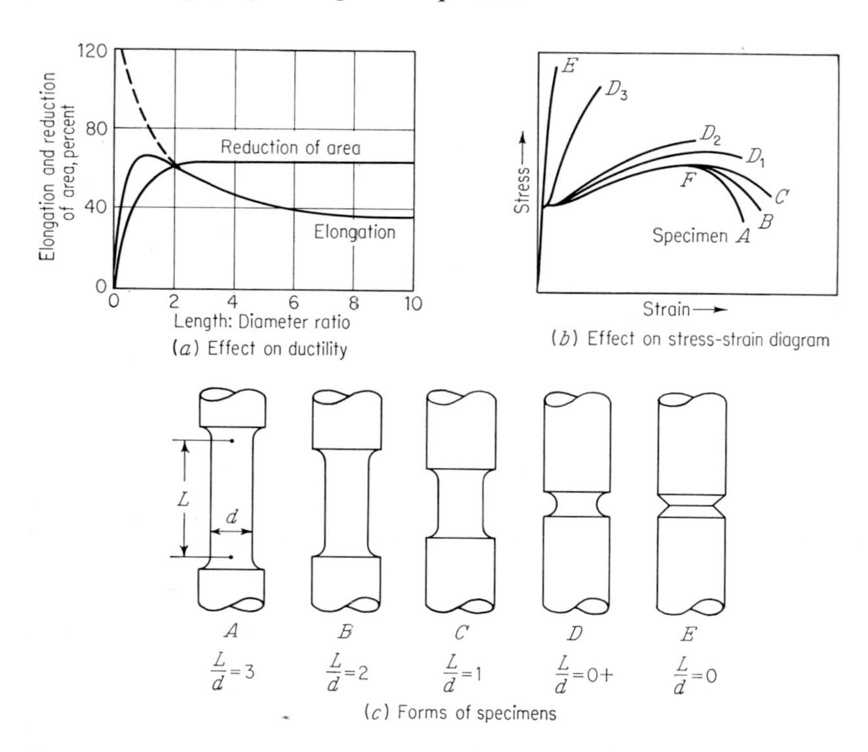

Fig. 5.21. **Variation in strength and ductility of steel with length of reduced section.** *(From Upton, Ref.* 142.)

enlarged ends provide lateral restraint against the reduction of area. In the extreme case it is reduced to zero. Also, the elongation curve does not continue sharply upward to the left as shown in Fig. 5.20*b* and by the dotted line in Fig. 5.21*a*, but it also turns downward to zero because of this same restraint. Figure 5.21*b* shows the stress-strain diagrams for these same ductile specimens. When L/d exceeds about 1, the only effect is upon the segment of the curve beyond the maximum strength at F. The shorter the length L, the longer is the segment of the curve beyond point F. When L is shortened so that the ends give support to the reduced section, the curve is raised and shortened as shown by curves D_1, D_2, and D_3. Finally the specimen E with zero length L gives a high strength with very little deformation.

The effect of notching is (1) to suppress the drawing out of the reduced section, owing to the support given by the mass of the adjacent larger sections, and (2) to cause stress concentration in the material at the base of the notch. The former tends to increase the apparent strength and reduce the apparent ductility of ductile materials. The latter tends to cause a reduction in apparent strength, and with brittle materials, which have little elongation up to rupture, this is very pronounced. This effect in a ductile material is shown by comparing the strength of a standard specimen with that of a grooved specimen of the same material, where the diameter at the base of the groove is the same as that of the standard specimen. For a certain high-carbon steel, the ultimate strength of the standard specimen was 102,000 psi, whereas that for a specimen with a $\frac{1}{32}$-in. (wide) square groove was 163,000 psi, both computed in the conventional manner [138]. In this latter specimen the usual drawing out and reduction in area were prevented, thus causing an increase in strength greater than the decreasing effect of high stress concentrations at the edge of the groove. In a brittle material, where the elongation is of little consequence, the groove would have caused a decrease in strength.

Owing to the effects of curvature of the side of the piece and to the closeness of the grips, the stress distribution across the net section of the standard mortar tension briquet is not uniform but varies as shown in Fig. 5.22. The ratio of maximum to average stress on the critical cross section has been estimated from photoelastic studies to be about 1.75 [520].

For further information regarding the effects of notches and the stress concentration caused by changes of section of various types, see Refs. 141, 521, and 522.

Eccentric loading produced by the gripping device causes nonuniform stress distribution in the test bar. An example of this is shown in Fig. 5.23, where the separate stress-strain curves for three gage lines 120° apart around a cylindrical test specimen are plotted. In the test for which the results are shown, the wedge grips at the ends of the specimen were out of alignment by 0.035 in., a relatively small amount. This caused certain parts of the specimen to reach the proportional limit before other parts and thus resulted in a proportional limit, as deter-

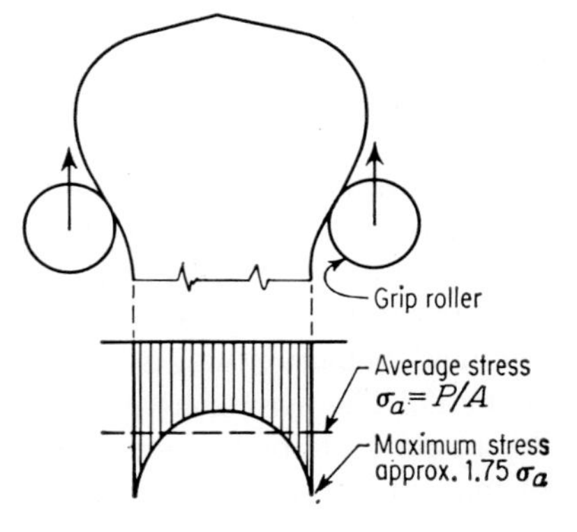

***Fig. 5.22.* General nature of stress distribution in test specimen having shape of ASTM Standard mortar briquet.**

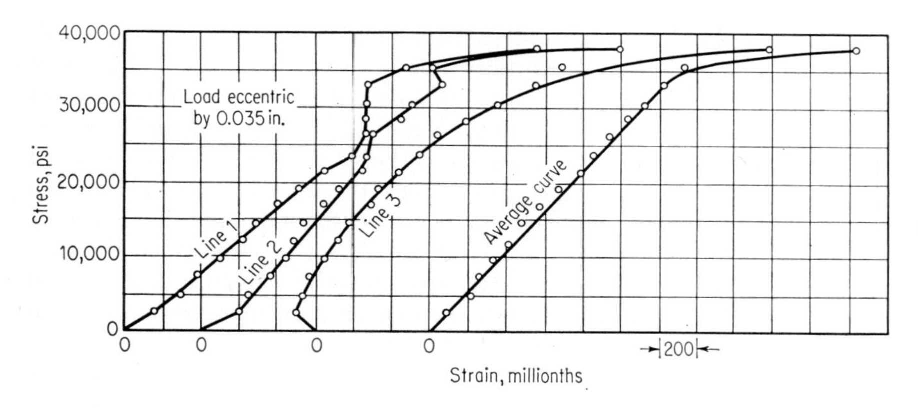

Fig. 5.23. **Effect of eccentric load on strains along three gage lines 120° apart around cylindrical specimen.**

mined from the average stress-strain diagram, somewhat lower than if the specimen were stressed uniformly. The average stress-strain diagram (which might also have been obtained with an averaging extensometer) gives no apparent indication of the effect of eccentric loading. In tests to determine elastic properties of materials, the effect of eccentric loadings is important. Under conditions of slightly eccentric loading, the averages of strain measurements taken on two opposite elements appear to give satisfactory values of the modulus of elasticity [540]. The strength of ductile materials does not appear to be greatly affected by slight eccentricities of load; the strength of brittle materials may be appreciably affected.

Over a wide range of speed, the rate of loading has an important effect upon the tensile properties of materials. Strengths tend to increase and ductility to decrease with increased speeds. For example, certain tests have indicated that with a speed ratio of about 14,000:1 the yield point of mild steel was increased about 30 percent [532]. In general, the change in strength or elongation appears to vary approximately as the logarithm of the speed [502, 530]. The effect appears to be more pronounced for materials having low melting points, such as lead, zinc, and plastics than for materials having high melting points, such as steel. With some materials, notably wood [141], but apparently also steel [538], the effect of very slowly applied loads (long-time tests) is to decrease the strength over that observed at normal testing speeds.

The term *viscoelasticity* is used to designate a time-dependent mechanical behavior which is a function of both elastic and viscous components. The strength-time data shown in Fig. 5.24 for rapidly loaded rectangular-shaped specimens having a 4-in. uniform length are typical of many plastic materials which exhibit this characteristic. The influence

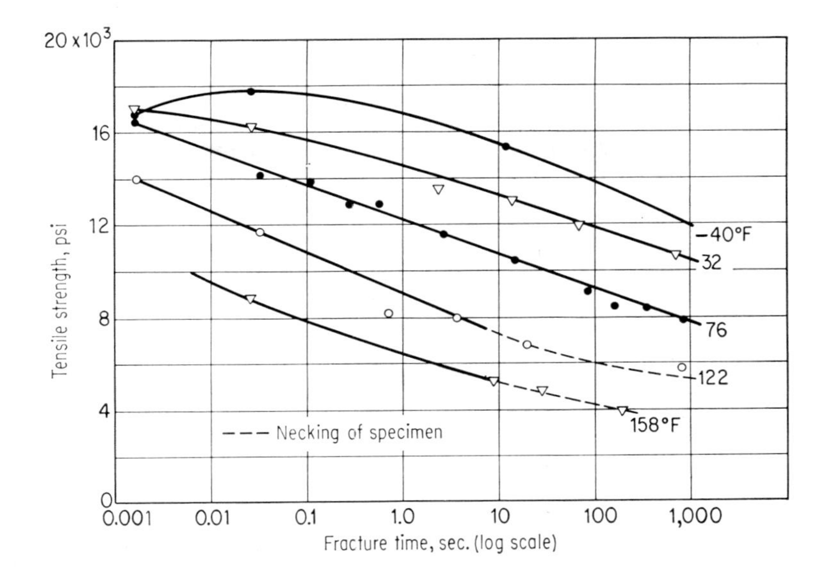

Fig. 5.24. **Strength-time data for polymethyl methacrylate [535].**

of temperature at time of test for such materials is also shown. In general, such materials exhibit a lowered modulus of elasticity at slower strain rates and also at higher temperatures.

Fortunately, recent investigations have shown that over the range of speed used with ordinary testing machines the effects of moderate variation in speed on the tensile properties of metals is fairly small, and rather wide tolerances can be permitted without introducing serious error in the results of tests for ductile metals [530]. For example, in tests of standard specimens of a structural steel it was found that an eightfold increase in rate of *strain* increased the yield point by about 4 percent, the tensile strength by about 2 percent, and decreased the elongation by about 5 percent. In the machine in which these tests were made, this change corresponded to a change in idling speed of the head from 0.05 to 0.40 in. per min. For further data, see Refs. 531 and 533. The effect of speed variations within the range of normal rates of loading on the strength of brittle materials such as cast iron appears to be small.

A number of factors affect the character of the fracture. As discussed in Arts. 3.2 to 3.6 inclusive, the two fundamental types of tensile fractures are sliding (or shear) and separation (or cleavage) [138, 307, 503, 505]. If resistance to sliding is the greater, the material will fail by separation, overcoming the cohesive force with very little plastic elongation, and the material is said to be brittle. If the strength depends upon its resistance to sliding, i.e., resistance to separation is the greater, there is considerable plastic elongation and reduction of area before fracture occurs; such behavior characterizes ductile materials. Both these properties are functions of the temperature and the rate at which the load is applied. Experiments show that the resistance to separation is affected less than the resistance to sliding. Some materials that show considerable ductility under slowly applied loads fail with little plastic elongation when the load is applied suddenly. Both the resistance to separation and the resistance to sliding increase with decreasing temperature, but the resistance to sliding increases much more markedly; thus it is possible to obtain separation fractures (brittle behavior) in plain unnotched steel bars cooled in liquid air. For an illuminating discussion of this, see Ref. 245, p. 79.

A state of triaxial stress may reduce ductility or cause a separation fracture to occur in ductile materials under tension. Since the maximum shear is a function of the difference between the maximum and minimum principal (direct) stresses, as the magnitudes of the three principal stresses approach each other, the maximum shear may become very small, even though the principal stresses are high. A state of triaxial stress may be induced by abrupt changes in section or the presence of irregularities in a piece of material subject to uniaxial load; thus it is possible to reduce ductility markedly and to cause separation fractures at moderately low temperatures in steel [505].

After a standard test specimen of ordinary steel has necked down (shearing or sliding action), the state of stress in the central portion of the necked section is no longer that of simple tension; radial as well as axial stresses act on the crystals that compose the material. The maximum principal stress may be several times the maximum shear stress, instead of having the 2 to 1 relation that existed between tension and shear before necking began. Thus, at moderately low temperatures, the type of failure may be mixed, i.e., a separation failure in the central portion of a cup-cone fracture and sliding along the cone at the edges. However, experimental evidence indicates that at room temperatures the mode of failure of mild steel at the bottom of the cup can be sliding, even though the bottom of the cup appears granular to the naked eye; this has been attributed to a higher shear stress near the center than near the edges, which also accounts for the observation that the first crack can begin at the *center* of the section [245, 505].

THE COMPRESSION TEST

5.8. General Remarks. It has been pointed out in Art. 5.1 that, in theory at least, the compression test is merely the opposite of the tension test with respect to direction or sense of the applied stress. General reasons for choice of one or the other type of test were stated. Also, a number of general principles and concepts were developed throughout the section on tension testing which apply equally well to compression testing. There are, however, several special limitations to the compression test to which attention should be directed:

1. The difficulty of applying a truly concentric or axial load.
2. The relatively unstable character of this type of loading as contrasted with tensile loading. There is always a tendency for bending stresses to be set up and for the effect of accidental irregularities in alignment within the specimen to be accentuated as loading proceeds.
3. Friction between the heads of the testing machine or bearing plates and the end surfaces of the specimen due to lateral expansion of the specimen. This may alter considerably the results that would be obtained if such a condition of test were not present.
4. The relatively larger cross-sectional areas of the compression-test specimen, in order to obtain a proper degree of stability of the piece. This results in the necessity for a relatively large-capacity testing machine or specimens so small and therefore so short that it is difficult to obtain from them strain measurements of suitable precision.

It is presumed that the simple compression characteristics of materials are desired and not the column action of structural members, so that attention is here confined to the short compression block.

5.9. Requirements for Compression Specimens. For uniform stressing of the compression specimen, a circular section is to be preferred over other shapes. The square or rectangular section is often used, however, and for manufactured pieces such as tile, it is not ordinarily feasible to cut out specimens to conform to any particular shape.

The selection of the ratio of length to diameter of a compression specimen appears to be more or less of a compromise between several undesirable conditions. As the length of the specimen is increased, there is increasing tendency toward bending of the piece, with consequent nonuniform distribution of stress over a right section. A height-diameter ratio of 10 is suggested as a practical upper limit. As the length of the specimen is decreased, the effect of the frictional restraint at the ends becomes relatively important; also, for lengths less than about 1.5 times the diameter, the diagonal planes along which failure would take place in a longer specimen intersect the base, with the result that the apparent strength is increased. A ratio of length to diameter of 2 or more is commonly employed, although the height-diameter ratio used varies for different materials. To accommodate a compressometer of desired precision, it is sometimes necessary to use a relatively long specimen.

The actual size depends upon the type of material, the type of measurements to be made, and the testing apparatus available. For homogeneous materials for which only ultimate strength is required, relatively small specimens may be used. The size of specimens of nonhomogeneous materials must be adjusted to the size of the component particles or aggregates.

The ends to which load is applied should be flat and perpendicular to the axis of the specimen or, in effect, made so by the use of caps and adjustable bearing devices.

Gage lengths for strain measurements should preferably be shorter than the specimen length by at least the diameter of the specimen.

General requirements with regard to the selection and preparation of specimens have been outlined in Chap. 2.

5.10. Standard Test Specimens. Specimens for compression tests of metallic materials recommended by the ASTM (ASTM E 9) are shown in Fig. 5.25. The short specimens are intended for use with bearing metals, the medium-length specimens for general use, and the long specimens for tests to determine modulus of elasticity. Specimens for compression tests of sheet metal must be loaded in a jig that provides lateral support against buckling without interfering with the axial deformations of the specimen. The details of such jigs and the corresponding specimens are covered by the ASTM (ASTM E 9).

For concrete, the standard specimens are cylinders of height twice the diameter. For concrete with aggregate of maximum size not greater than 2 in., the standard size of cylinder is 6 by 12 in.; for concretes containing aggregates of maximum size up to 2½ in., an 8 by 16-in. cylinder

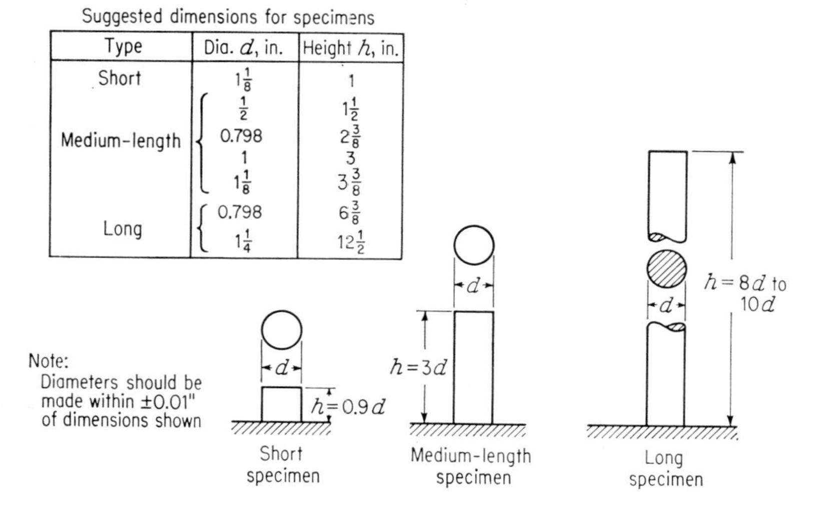

Suggested dimensions for specimens

Type	Dia. d, in.	Height h, in.
Short	$1\frac{1}{8}$	1
Medium-length	$\frac{1}{2}$	$1\frac{1}{2}$
	0.798	$2\frac{3}{8}$
	1	3
	$1\frac{1}{8}$	$3\frac{3}{8}$
Long	0.798	$6\frac{3}{8}$
	$1\frac{1}{4}$	$12\frac{1}{2}$

Fig. 5.25. **Test specimens for compression tests of metallic materials in other than sheet form (ASTM E 9).**

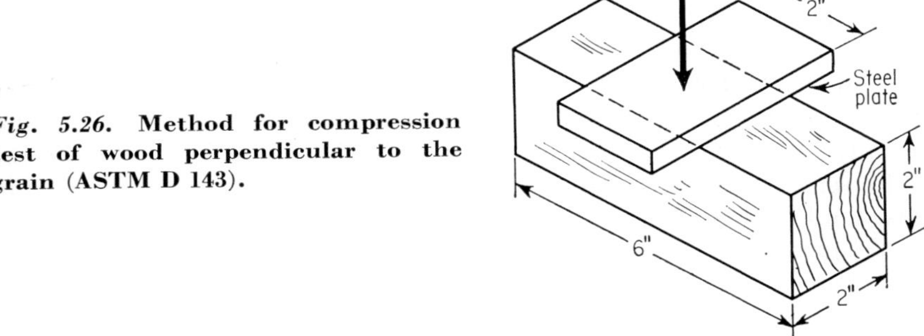

Fig. 5.26. **Method for compression test of wood perpendicular to the grain (ASTM D 143).**

is used (ASTM C 31). It is common practice in many laboratories to use 3 by 6-in. cylinders for concrete with aggregates up to ¾ in., and for tests of mass concrete with aggregates up to 6 in., 18 by 36-in. cylinders are used. Cubes are in use in England and in Europe; in England the 6-in. cube is a common size for ordinary concrete.

For mortars the 2 by 4-in. cylinder is often used, although the ASTM now specifies a 2-in. cube (ASTM C 109).

Specimens for compression tests of small, clear pieces of wood parallel to the grain are 2 by 2 by 8-in. rectangular prisms. Compression tests perpendicular to the grain are made on nominal 2 by 2 by 6-in. specimens, as shown in Fig. 5.26. The load is applied through a metal bearing plate 2 in. in width placed across the upper surface at equal distances from the ends and at right angles to the width (ASTM D 143).

The compressive strength of building brick is determined on a half brick, with approximately plane and parallel surfaces, tested flatwise (ASTM C 67).

Reference is made to the ASTM Specifications for type of specimens for compression tests of other materials such as drain tile (ASTM C 4), structural clay tile (ASTM C 112), sewer pipe (ASTM C 13, C 14), refractory brick (ASTM C 133), vulcanized rubber (ASTM D 395, D 575), molded insulating materials (ASTM D 48), timber in structural sizes (ASTM D 198), and building stone (ASTM C 170).

5.11. Bedments and Bearing Blocks. The ends of compression specimens should be plane or flat so as not to cause stress concentrations and should be perpendicular to the axis of the piece so as not to cause bending due to eccentric loading.

The end surfaces of metal specimens can be machined flat and at right angles to the axis. Wood test pieces can usually be trimmed so that these conditions are satisfied. For materials such as concrete, stone, and brick, however, a bedment, with or without the use of accompanying capping plates, is usually necessary. Materials commonly used for bedments are plaster of paris, Hydrostone (a high-strength gypsum compound), quick-setting cements, and sulfur compounds. In setting capping plates, precaution should be taken to ensure perpendicularity between the bearing surface and the axis of the specimen. A jig is sometimes used for the purpose. It is desirable that the capping material have a modulus of elasticity and strength at least equal to that of the material of the specimen. The cap should be as thin as practicable. If a capping compound contains water, it may affect the strength of absorbent materials like brick; so a coat of shellac or a sheet of waxed paper is placed on the ends of the specimen before capping. Loose materials, such as sand or small steel balls, have not proved successful for end bedments. Soft bedments such as rubber sheets and fiber boards should be avoided, since they tend to flow laterally under load and cause the specimen to split.

Plain bearing plates or capping plates should have machined, flat, parallel surfaces. The material of the bearing plate should be strong and hard relative to the test specimen. See ASTM C 39 and C 192 for typical detailed requirements.

Usually one end of the specimen should bear on a spherically seated block. Figure 5.27 shows satisfactory arrangements of the specimen and block. The object of the block is to overcome the effect of a small lack of parallelism between the head of the machine and the end face of the specimen, giving the specimen as even a distribution of initial load as possible. It is desirable that the spherically seated bearing block be at the upper end of the test specimen. In order that the resultant of the forces applied to the end of the specimen should not be eccentric with respect to the axis of the specimen, it is important that the center of the spherical surface of this block lie in the flat face that bears on the specimen and that the specimen itself be carefully centered with respect to the center of this spherical surface. Owing to increased frictional resistance as the load builds up, the

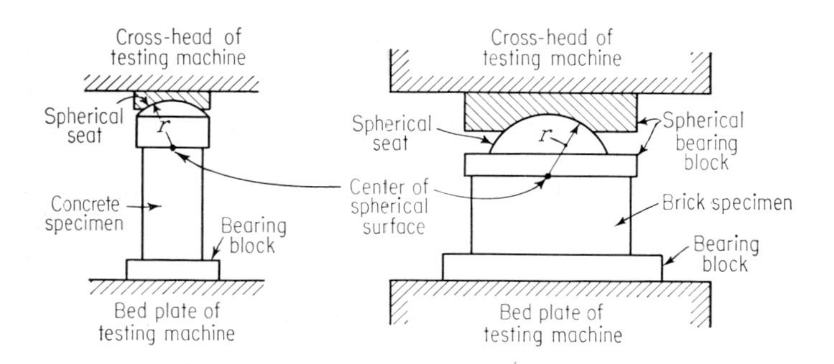

Fig. 5.27. **Spherical bearing blocks for compression tests.**

spherically seated bearing cannot be relied upon to adjust itself to bending action that may occur during the test. Under some conditions of test, the spherically seated bearing block may be omitted, whereas, under others, two such blocks may be required. The block should be of the same or slightly larger diameter than the specimen.

5.12. Conduct of Tests. In commercial tests, the only property ordinarily determined is the compressive strength. For brittle materials with which fracture occurs, the ultimate strength is definitely and easily determined. For materials where there is no unique phenomenon to mark ultimate strength, arbitrary limits of deformation are taken as the criteria of strength. See, for example, ASTM B 22 and D 575.

In tests to determine the yield strength of metals in compression, the usual criteria (described in Chap. 2) may be used.

The dimensions should be determined with appropriate precision. Recommended precisions for cross-sectional measurements for ordinary work are as follows: metals, to nearest 0.001 in.; concrete and wood, to nearest 0.01 in. On cylindrical specimens, measurements should be made on at least two mutually perpendicular diameters. If unit weights are required, specimens should ordinarily be weighed with a precision of about 0.5 percent.

Great care must be exercised to obtain accurate centering and alignment of specimen and bearing blocks in the testing machine. For careful work, effort should be made to have axes of the specimen and bearing blocks coincide with an axis through the centers of the head and base plate of the machines within 0.01 in. While the head of the machine is being lowered to contact with the spherical bearing block, it is desirable to rotate slightly by hand the upper part of the block in a horizontal plane so as to facilitate the seating of the block.

In testing metals the ends of the specimen and the faces of the bearing blocks should be cleaned with acetone or other suitable solvent immediately before testing to remove grease or oil which would influence the frictional restraint at the end surfaces (ASTM E 9).

For screw-gear machines, the speed of testing in compression is still commonly specified in terms of the idling speed of the movable crosshead. In Table 5.3 are recommended maximum crosshead speeds for use with standard specimens. In many instances slower speeds are desirable. For tests of concrete involving strain measurements, a crosshead speed of 0.01 or 0.02 in. per min is satisfactory. There is a growing tendency to use or specify rates of loading, as shown in Table 5.3. Procedures developed by the U.S. Bureau of Reclamation have specified 2000 psi per min (33 psi per sec) for strength tests of concrete. In many cases any convenient speed is permitted up to about one-half the maximum load, after

Table 5.3. **Various ASTM requirements on speed of testing in compression**

Material tested	Reference	Maximum idling speed of crosshead, in. per min	Load rate, psi per sec	Time to apply last half of load, sec
Metallic materials	E 9–33 T*			
1 to 3 in. in length	....	0.05		
3 in. and over in length	....	0.10		
Concrete	C 39	0.05	20–50	
Mortar	C 109	....	.....	20–80†
Wood	D 143			
Parallel to grain	....	0.024		
Perpendicular to grain	....	0.012		
Brick	C 67	....	.....	60–120
Clay tile	C 112	0.05		
Plastics	D 695	0.05 to yield point then 0.20 to 0.25		
	D 953	0.05		

* Requirements here shown withdrawn from current specifications.
† Entire loading time if ultimate load is less than 3000 lb.

which the specified rate (usually within stated tolerances) is required. For tests to determine the stress-strain relations in concrete, a load rate of about 600 to 1000 psi per min seems appropriate.

In a compression test, absolutely uniform stress distribution is practically never attained. In making precise stress-strain determinations with a view to finding the proportional limit, it is therefore desirable to measure strains along at least three gage lines 120° apart around a cylindrical piece. For ordinary determinations of modulus of elasticity, an averaging type of compressometer is usually sufficient.

Directions for conducting a typical compression test are given in Part 2.

5.13. Observations of Test. Identification, dimensions, critical loads, compressometer readings (if taken), type of failure, including sketches, etc., are recorded on a form appropriate to the type of test and extent of the data to be taken.

Brittle materials commonly rupture either along a diagonal plane, or with a cone- (cylindrical specimens) or a pyramidal- (square specimens) shaped fracture, sometimes called an *hourglass* fracture (see Fig. 5.28). Cast iron usually fails along an inclined plane, and concrete exhibits the cone type of fracture. Such fractures are essentially shear failures.

For a material whose resistance to failure is due to internal friction as

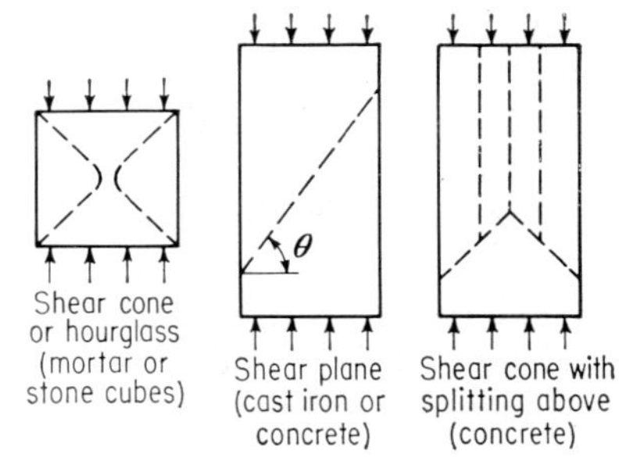

Fig. 5.28. **Types of failure of brittle materials under compressive loading.**

well as cohesion, and which behaves in accordance with the Mohr theory of rupture, the angle of rupture is not 45° (plane of maximum shear stress) but is a function of the angle of internal friction, ϕ. In Fig. 5.29 is shown, by means of the Mohr stress circle, the state of stress at failure in an element subjected to uniform principal stress in one direction only. From the representation of the angles of break on the Mohr circle diagram, it may be shown that $\alpha = 45° - \phi/2$ or $\theta = 45° + \phi/2$. (For an exposition of the Mohr circle and its use, see textbooks on strength of materials such as Refs. 138 and 594; see also Refs. 591, 592, and textbooks on soil mechanics [633] for discussions of failure of granular materials whose resistance is governed by internal friction.)

The behavior of such materials as cast iron, concrete, or ceramics does not conform exactly to that predicted by the Mohr theory of rupture, in part because their nonhomogeneous composition causes irregu-

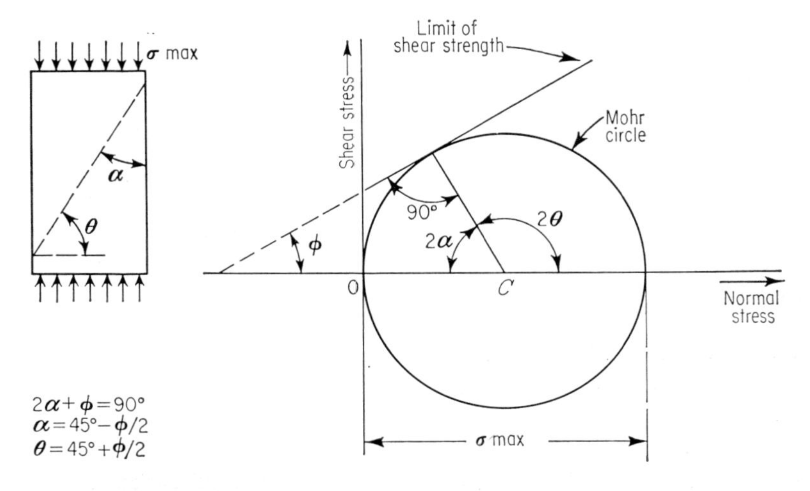

Fig. 5.29. **Relation between angle of rupture α and angle of internal friction ϕ.**

larity in the stress pattern. Further, the angle of rupture may be caused to deviate somewhat from the theoretical value owing to the complex stress condition induced in the end portions of compression specimens by restraint to lateral expansion under load caused by friction of the bearing plates on the end surfaces; this effect of lateral restraint of the ends becomes more pronounced in short specimens.

The observed values of θ for a number of materials including cast iron, sandstone, brick, and concrete vary roughly between 50 and 60° for specimens of sufficient length that normal failure surfaces can be developed [141], as shown in Fig. 5.28.

If the specimen is so short that a normal failure plane cannot develop within the length of the specimen, then the strength is appreciably increased, and other types of failure, such as crushing, may occur. With brittle materials in short specimens, when there is a combination of high compressive strength and unrestrained lateral expansion at the ends, the pieces often fail by separation into columnar fragments, giving what is known as a splitting failure or columnar fracture. Lateral flow of a bedment tends to produce a splitting failure.

Wood exhibits, under compressive loading, a behavior peculiar to itself. It is anything but an isotropic material, being composed of cells formed by organic growth which align themselves to form a series of tubes or columns in the direction of the grain. As a result of this structure, the elastic limit is relatively low, there is no definite yield point, and considerable set takes place before failure. These properties vary with the orientation of the load with respect to the direction of the grain. For loads normal to the grain, the load that causes lateral collapse of the tubes or fibers (crushing) is the significant load. For loads parallel to

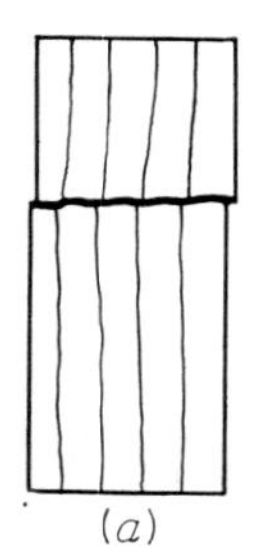

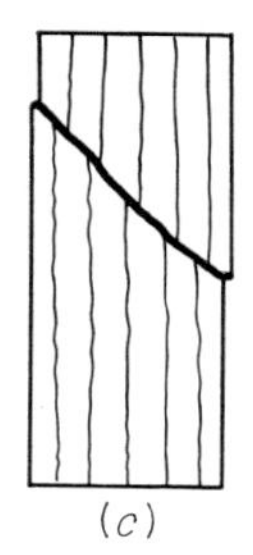

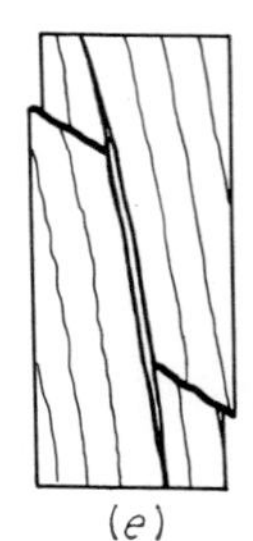

(*a*) Crushing (plane of rupture approximately horizontal).
(*b*) Wedge split (note direction of split: radial or tangential).
(*c*) Shearing (plane of rupture at acute angle with horizontal).
(*d*) Splitting.
(*e*) Shearing and splitting parallel to grain. (Usually occurs in cross-grained pieces).

***Fig. 5.30.* Types of failure in wood under compression parallel to the grain (ASTM D 143).**

the grain, not only is the "elastic" strength important but also the strength at rupture. Rupture often occurs because of collapse of the tubular fibers as columns. Various types of failure of wood loaded parallel with the grain are described by the sketches in Fig. 5.30.

Ductile and plastic materials with some tenacity bulge laterally and take on a barrel shape as they are compressed, provided, of course, that the specimen does not bend or buckle. Material of relatively low ductility and casehardened pieces develop surface cracks parallel to the loading axis as failure becomes pronounced.

5.14. Effect of Important Variables. The effect of size and shape of specimens of brittle materials on the compressive strength is illustrated by the results of an investigation of concrete summarized in Table 5.4. See also Ref. 563 for other information.

Table* 5.4. Effect of size and shape of specimen upon compressive strength of concrete

Type of specimen	Diameter, in.	Length, in.	Length-diameter ratio	Strength relative to 6 by 12-in. cylinder, percent
Cylinder	6	3	0.5	178
	6	6	1.0	115
	6	9	1.5	107
	6	12	2.0	100
	6	15	2.5	97
	6	18	3.0	95
	6	24	4.0	90
	8	16	2.0	96
Prism	6	12	2.0	93
	8	16	2.0	91
Cube	6	6	1.0	113
	8	8	1.0	115

* H. F. Gonnerman, "Effect of Size and Shape of Test Specimen on Compressive Strength of Concrete," *Proc. ASTM*, vol. 25, pt. II, 1925.

The ASTM gives correction factors to be applied to the strength of concrete specimens cored from concrete structures to obtain strengths equivalent to the standard cylinder having a length-diameter ratio of 2, as shown in Table 5.5.

Table* 5.5. Correction factors for concrete cylinders not of standard length

Length-diameter ratio	1.75	1.50	1.25	1.10	1.00	0.75	0.50
Multiplying factor	0.98	0.96	0.94	0.90	0.85	0.73	0.60

* Based on ASTM C 42.

The relative compressive strengths of large cylinders of concrete is illustrated in Table 5.6. These data are summarized from tests made by the U.S. Bureau of Reclamation [564].

Table 5.6.* Effect of size of test cylinder on the compressive strength of concrete

Cylinder size, in.	2 by 4	3 by 6	6 by 12	8 by 16	12 by 24	18 by 36	24 by 48	36 by 72
Relative strength. . . .	109	106	100	96	91	86	84	82

* R. F. Blanks, and C. C. McNamara, "Mass Concrete Tests in Large Cylinders," *Proc. ACI*, vol. 31, 1935, and discussion in vol. 32, 1936.

The end conditions at time of test, the capping method, and the end conditions before capping may have a pronounced effect upon the compressive strength of concrete test cylinders [565, 566]. Cylinders molded with machined plates so as to produce convex ends and tested without capping give pronounced reductions in strength even for a small amount of convexity. For a convexity of only 0.01 in. in a cylinder of 6-in. diameter, tests of 1:2 and 1:5 mixes have shown reductions in strength of about 35 and 20 percent, respectively [565]. This shows the importance

***Table 5.7.* Effect of capping materials and end conditions before capping on compressive strength of concrete cylinders[a]**

Type of capping material	3000-lb concrete				8000-lb concrete			
	Plane ends	Beveled ends[f]	Convex ends[g]	Concave ends[h]	Plane ends	Beveled ends[f]	Convex ends[g]	Concave ends[h]
Hydrostone[b].	100	99	102	99	100	94	97	93
Castite[c].	97	98	94	101	102	101	96	89
Shot[d].	88	90	93	75	89	90	90	56
Plaster of paris[e]. . .	97	85	88	88	87	66	54	34

Note. Cylinders with plane normal ends and Hydrostone caps taken as relative strength of 100.

[a] G. E. Troxell, "The Effect of Capping Methods and End Conditions before Capping upon the Compressive Strength of Concrete Test Cylinders," *Proc. ASTM*, vol. 41, 1941.

[b] A gypsum compound; 1-hr strength, 5000 psi; modulus, 1.6×10^6 psi.

[c] A sulfur-silica mixture; 1-day strength, 8500 psi; modulus, 2.2×10^6 psi.

[d] 1/16-in. steel shot. Oiled. Results with dry shot practically the same.

[e] A gypsum compound; 1-hr strength, 1700 psi; modulus, 0.5×10^8 psi.

[f] Plane but not perpendicular to axis. Slope of 3/16 in. in 3-in. diameter.

[g] Spherical bulge of 3/16 in.

[h] Spherical depression of 3/16 in.

of having plane ends on test specimens. Tests have also shown that the higher the compressive strength of the capping material, the higher the indicated strength of the concrete and the less the effect of irregular ends before capping on the indicated strength. With caps of plaster of paris or steel shot the indicated strength of normal concrete may be reduced as much as 10 percent even for flat-ended cylinders, but for irregular ends before capping, the strengths may be reduced as much as 25 percent. The results of tests showing the relative strengths obtained with several types of caps are summarized in Table 5.7.

The speed of testing has a definite effect on compressive strength, although the effect is usually fairly small over the ranges of speed used in ordinary testing. The results of tests on concrete indicate that the relation between strength and rate of loading is approximately logarithmic—the more rapid the rate, the higher the indicated strength [581, 582]. The strength of a specimen loaded, say, at 6000 psi per min would be about 15 percent greater than the strength of a specimen loaded at 100 psi per min. The modulus of elasticity also appears to increase with the loading rate, although most observers have attributed this effect to reduced creep during the test period.

For the effect of internal structure of the material on strength of various materials, see Ref. 141 in chapters on the properties of wood, stone, brick, concrete, iron, and steel; also Refs. 591 to 594.

For the effect of main and local buckling of elements on their compressive strength, see Arts. 3.8 to 3.10.

Chapter 6.

STATIC SHEAR AND BENDING TESTS

SHEAR TESTS

6.1. Behavior of Materials under Shearing Stress. In the following paragraphs are summarized some of the concepts of mechanics of materials pertinent to the problems of shear testing. For fuller discussion, particularly with reference to state of stress in a loaded body, texts on mechanics of materials should be consulted [e.g., 135, 138, 594]. While in the following discussion attention is focused on shear stresses and strains, it should be remembered that the physical bodies with which we actually deal are three dimensional, that any loading produces states of stress with various combinations of normal and shear stress on the planes through any given point, and that certain types of loading may have important three-dimensional effects.

A shearing stress is one that acts parallel to a plane, as distinguished from tensile and compressive stresses that act normal to a plane. Load-

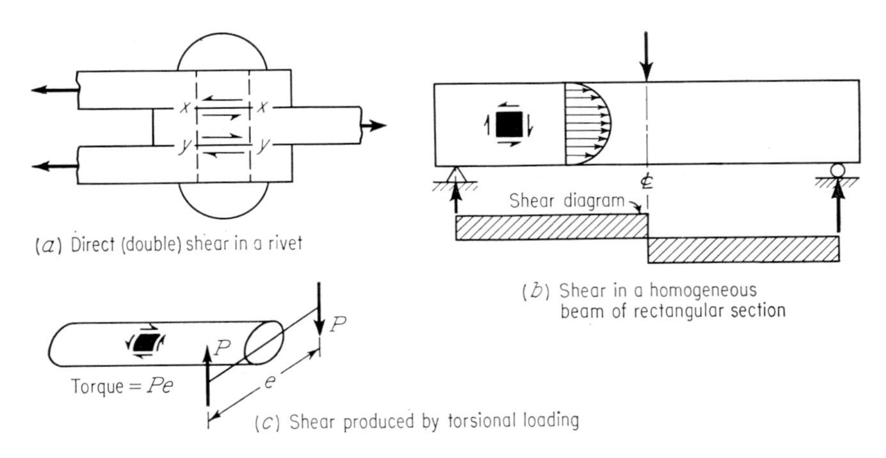

Fig. 6.1. **Loadings producing shear.**

ings that produce shear conditions of principal interest in materials testing are as follows:

1. The resultants of parallel but opposed forces act through the centroids of sections that are spaced "infinitesimal" distances apart. It is conceivable in such cases that the shearing stresses over the sections should be uniform and a state of pure direct shear would exist. This condition may be approached but is never realized practically. An approximation of this condition is the case of a rivet in shear as shown in Fig. 6.1*a*; here, for practical purposes, direct shear may be considered to exist within the rivet on planes xx and yy.

2. The applied opposed forces are parallel, act normal to a longitudinal axis of the body, but are spaced finite distances apart. Then, in addition to the shearing stresses produced, bending stresses are set up. In the case of a rectangular beam subjected to transverse loads (Fig. 6.1*b*), the shearing stresses on any cross section vary in intensity from zero at the upper and lower surfaces of the beam to a maximum at the neutral axis.

3. The applied forces are parallel and opposite but do not lie in a plane containing the longitudinal axis of the body; here a couple is set up which produces a twist about a longitudinal axis. This twisting action of one section of a body with respect to a contiguous section is termed *torsion*. Figure 6.1*c* represents a piece of shafting subjected to a torque. Torsional shearing stresses on circular cross sections vary from zero at the axis of twist to a maximum at the extreme fibers. If no bending is present, "pure shear" exists.

At any point in a stressed body, the shearing stresses τ in any two mutually perpendicular directions are equal in magnitude. If on some pair of planes at a point, only shear stresses act, the material at that point is said to be in "pure shear." These shears are greater than those on any other plane through the point. The pure-shear condition is illustrated by Fig. 6.2*a*, which represents an elementary block on which the stresses are uniformly distributed. On all planes inclined to the planes of maximum shear, normal tensile or compressive stresses act; and on mutually perpendicular planes at 45° with the planes of maximum shear, the tensile and compressive stresses are a maximum and the shear stress is zero. The maximum normal stresses are equal in magnitude to the maximum shearing stresses. Conversely, pure shear is induced by equal and opposite normal stresses, as shown in Fig. 6.2*b*. The secondary compression resulting from primary pure shear in thin plates may cause shear buckling, as discussed in Art. 3.12. The Mohr circle representation of the state of stress induced by pure shear is shown in Fig. 6.2*c*.

If a body is subjected to a tensile or compressive stress acting in only *one* direction, the shear stresses at 45° thereto are one-half the magnitude

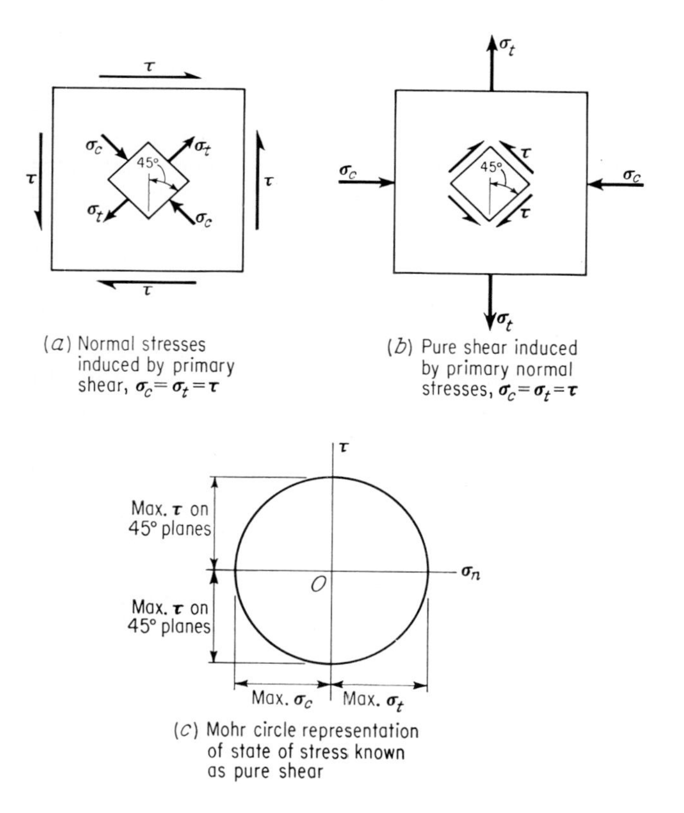

Fig. 6.2. **Relation between pure shear and normal stresses.**

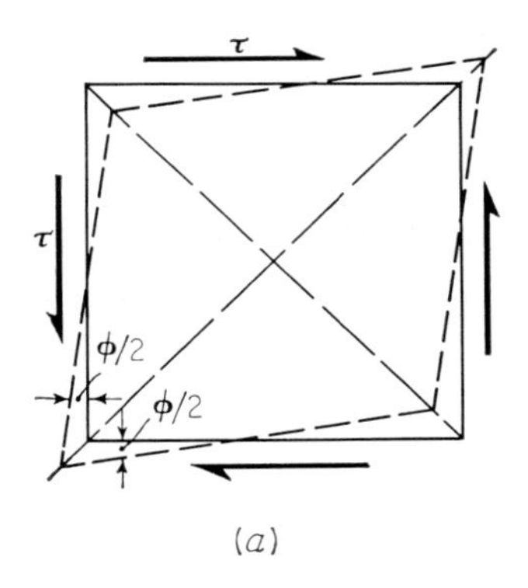

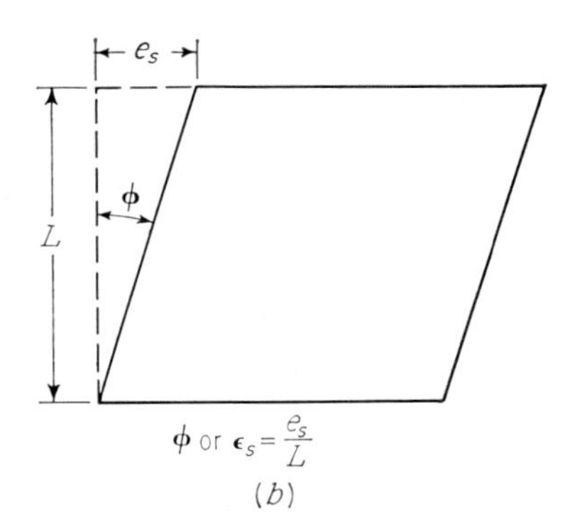

***Fig. 6.3.* Shearing strain.**

of the applied direct stress. In the general case, the maximum shear stresses are equal to one-half the difference between the maximum and minimum principal stresses and act on planes inclined at 45° with these stresses.

The strain that accompanies shear may be thought of as arising from the effort of thin parallel slices of a body to slide one over the other. Shearing strain or "detrusion" is a function of the change in angle between adjacent sides of an elementary block as it is distorted under shearing stresses, as illustrated in Fig. 6.3.*a* The total change in angle is more conveniently represented by a diagram such as Fig. 6.3*b*, in which it may be seen that the shearing strain is the tangent of the angular distortion. However, within the range of elastic strength of materials used for construction, the shearing strains are small; and the angle is commonly expressed in radians.

In so far as practical testing problems are concerned, the shearing stress-strain relations are of interest, principally in connection with torsional loading. In the common theory of torsion, it is assumed that plane sections remain plane after twisting. The circular section is the only one that conforms to this condition, and hence the simple theory of torsion does not apply satisfactorily to sections other than those of circular form. However, in practical calculations for noncircular sections, the results of the simple torsion theory are often used in conjunction with suitable correcting factors.

Various stress and strain relations for cylindrical pieces in torsion are stated below in terms of the following symbols:

T = torque or torsional moment
J = polar moment of inertia = $\pi r^4/2$ for a circle
ϕ = shearing strain
r = outside radius of a cylindrical test piece

r_1 = inside radius of a tubular test piece
L = distance between collars of strainometer
θ = angle of twist measured in distance L
τ = shearing stress at extreme fiber
E_s = modulus of rigidity or modulus of elasticity in shear

From geometrical relations between the various elements of a bar subjected to torsion (see Fig. 6.4),

$$\phi = \frac{r\theta}{L}$$

Within the elastic range all evidence indicates that the shearing strains are proportional to the distance from the axis of twist; and this relation appears also to hold approximately above the proportional limit.

Within the limit of proportionality,

$$\tau = \phi E_s = \frac{r\theta}{L} E_s$$

The stress varies linearly from zero at the axis of twist to a maximum at the extreme fiber and varies directly with the angle of twist (see Fig. 6.4*b*).

By summing up the stresses over a cross section, the relation between shearing stress on extreme fiber and the applied torque may be found: For solid sections:

$$\tau = \frac{Tr}{J} = \frac{2T}{\pi r^3}$$

For tubular sections:

$$\tau = \frac{2Tr}{\pi(r^4 - r_1^4)}$$

This relation is called *the torsion formula* and is applicable only to circular sections.

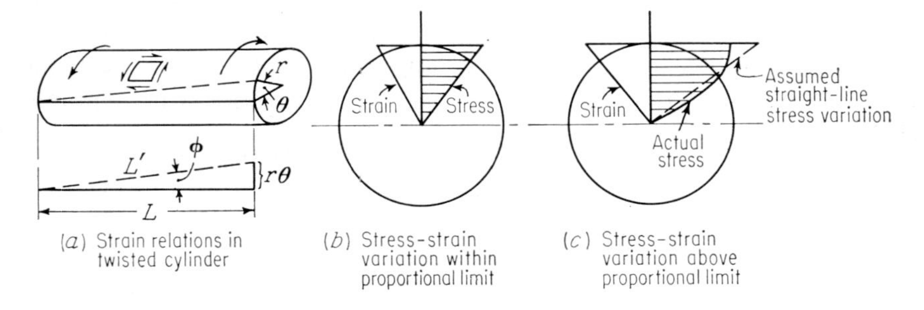

Fig. 6.4. **Strain and stress relationships in torsion.**

The expression for modulus of rigidity in terms of torque and angle of twist can be shown to be

$$E_s = \frac{TL}{J\theta} = \frac{2TL}{\pi r^4 \theta}$$

It can be shown theoretically that the modulus of rigidity or shearing modulus of elasticity for homogeneous isotropic materials is about 40 percent of the modulus of elasticity in tension [135]. This is closely borne out in the case of steel.

Ductility in a torsion test is determined by comparing the final fiber length L' (Fig. 6.4*a*) at rupture with the original fiber length or gage length L. The value of L' is computed knowing L and $r\theta$. The ductility is expressed as a percentage of elongation of the outer fiber and is equal to $[(L' - L)/L] \times 100$.

6.2. Failure under Shearing Stress. If the tensile strength of a material is less than its shearing strength, then failure under shear loading occurs by (tensile) separation along a plane making 45° with the plane of maximum shear. Under torsional loading, this results in a fracture with a helicoidal surface (see Fig. 6.9*b*). The ratio of shear strength to tensile strength appears to vary from perhaps 0.8 for ductile metals to values of about 1.1 or 1.3 for brittle ones such as cast iron. The shearing *elastic* strength of ductile and semiductile steels appears to be close to 0.6 of the tensile elastic strength [604].

In a solid cylindrical bar in torsion, the interior fibers are less highly stressed than the surface fibers. Consequently, when the surface fibers reach the proportional limit or yield point, they are, in a sense, supported by the interior fibers. Thus the effect of yielding of the surface fibers, during their early stage of plastic action, is masked by the resistance of the remainder of the section. It is not until considerable yielding has taken place that any noticeable effect is apparent with instruments ordinarily used for measuring angle of twist. This difficulty is overcome by the use of properly designed tubular test specimens which can be made to give more sensitive measures of the shearing elastic strength since all fibers are at about the same stress. However, if a thin sheet is subjected to shear or a thin tube to torsion, before the shear strength of the material is reached failure may occur by buckling due to the *compressive* stresses that act at 45° to the planes of maximum shear (see Fig. 6.9*c*). Thus, in tubular specimens for torsion tests, the relative thickness of the wall must be greater than some critical value if shear failure is to be assured.

In a bar subjected to torsional loading above the proportional limit, if straight-line variation of *strain* is assumed, the actual *stress* variation is something like that shown by the solid line in Fig. 6.4*c*. It is customary, however, for comparison purposes with similar materials, to compute a nominal extreme fiber stress at rupture by means of the torsion formula ($\tau = Tr/J$), which gives what is called the *modulus of rupture in torsion.* If proportionality between stress and strain were maintained up to the rupture point, the nominal straight-line stress distribution would be something like that shown by the dotted line in Fig. 6.4*c*. It can be seen that the modulus of rupture or maximum nominal stress is larger than the true maximum stress.

Upton shows a construction by means of which a corrected value of shearing stress may be determined from a diagram of nominal stress vs. strain [141, p. 23; 142, p. 52]. His reasoning indicates that the ratio of the stress at rupture to the calculated modulus of rupture must lie between $\frac{3}{4}$ and 1 and approaches $\frac{3}{4}$ as ductility increases. This value of $\frac{3}{4}$ is roughly corroborated by direct shear tests on steels.

For materials that fail by tension under torsional loading, the torsion modulus of rupture roughly approximates the breaking strength in tension, although it is always somewhat higher than the true tensile strength, since above the proportional limit the torsion formula gives stresses that are too high. For cast iron, the torsion modulus of rupture as determined from solid rods appears to be about 1.2 times the tensile strength [142].

In some materials (in particular, materials composed of granular elements, as concrete and soil) the resistance to rupture by shear is a function not only of the shearing strength of the material but also of the frictional resistance to sliding on the surface of rupture. For such materials it is necessary then to evaluate both these factors. A relation between them and total shearing resistance (Coulomb's law) is

$$\tau = c + \sigma_n \mu = c + \sigma_n \tan \phi$$

where τ = shearing resistance, psi
c = shear strength of material under no normal load, often referred to as cohesion, psi
σ_n = normal stress on plane of failure, psi
μ = coefficient of friction
ϕ = "angle of internal friction"

Reference is made to text on soil mechanics for detailed analysis of the problem of shear failure of granular materials [633, 634].

6.3. Scope and Applicability of Shear Tests. The types of shear tests in common use are the direct shear test and the torsion test. In certain instances, shear properties are evaluated by indirect methods (see Art. 6.7).

In the direct shear test, sometimes called the "transverse" shear test, it is usually the procedure to clamp or support a prism of the material so that bending stresses are minimized across the plane along which the shearing load is applied. Although the method suffices for an indication of what shearing resistance may be expected in rivets, crankpins, wooden blocks, etc., nevertheless, owing to bending or to friction between parts of the tool or to both, it gives but an approximation to the correct values of shearing strength. The results of such a test depend to a considerable degree on the hardness and the sharpness of the edges of the hardened plates that bear on the specimen. The transverse-shear test has the further limitation of being useless for the determination of the elastic strength or of the modulus of rigidity, owing to the impossibility of measuring strains.

The punching shear test is also a form of the direct shear test; its use is restricted to tests of flat stock, principally of metal. When a metal plate is punched, the punched area is removed by a slicing motion within a narrow ring of material adjacent to the cutting edge of the punch. The greater the clearance between the punch and the die, the greater the bending stresses accompanying the development of shear. Punching shear tests of brittle materials, such as concrete, cast iron, etc., usually give higher nominal strengths than do simple transverse-shear tests of the same materials, probably because of the greater friction and the smaller bending developed in the punching test. However, the results of punching shear tests are unsatisfactory as measures of shear strength and should be considered as giving simply a representation of overall load to cause punching.

For a more precise determination of shearing properties, the torsion test is made, employing either solid or hollow specimens of circular section. In such a test, the specimen can be of such length that a strainometer (called a troptometer in this type of test) can be attached to assist in determinations of proportional limit and yield strength in shear, shearing resilience, and stiffness (modulus of rigidity or modulus of elasticity in shear), the latter being obtained from the angle of twist and the applied torque. See Chap. 2 for a general discussion of these terms. The ultimate shearing strength or shearing modulus of rupture is usually obtained. The ductility of the material is determined from the amount of twist to rupture, toughness is represented by the amount of twist and the strength, and uniformity is indicated by the spacing, distribution, and appearance of the lines of twist. For accurate determination of the elastic strength, a tubular test specimen should be used.

The torsion test may be of especial use in the investigation of noncircular sections or of circular sections having various surface irregularities such as keyways and splines. In such cases, the object is to find the torsional resistance of the part, not the shearing strength of the material. Torsion tests are also used in investigations of the effect of various heat-treatment operations [611], particularly for parts subjected to treatments that tend to have a greater effect upon the metal near the surface than upon the interior of the member. For such studies the full-size part is subjected to test. For example, complete automobile- and truck-axle assemblies have been subjected to torsion tests.

The torsion method is inapplicable to determinations of the *shearing* strength of brittle materials, such as cast iron, since a specimen thereof would fail in diagonal tension before the shearing strength is reached, although the torsion test has been applied to cast iron [607, 608] and concrete [621] to determine other shearing properties or nominal over-all torsional resistance.

It is worthy of note that, in service, pieces that are subject to torque are usually machine parts and have to withstand impact loading and reversals of stress. In the interpretation of experimental investigations on such parts, these conditions should receive consideration.

Neither direct nor torsional shearing tests have been standardized except for direct shear tests of wood and building stone. They are rarely used as acceptance tests.

6.4. The Direct Shear Test. For the direct shear test of metals, a bar is usually sheared in some device that clamps a portion of the specimen while the remaining portion is subjected to load by means of suitable dies. In the Johnson type of shear tool, a bar of rectangular section, about 1 by 2 in., or a cylindrical rod of about 1 in. diameter is used. As shown in Fig. 6.5*a*, the specimen A is clamped to a base C. Force applied to the loading tool E ruptures the specimen in single shear. If the specimen

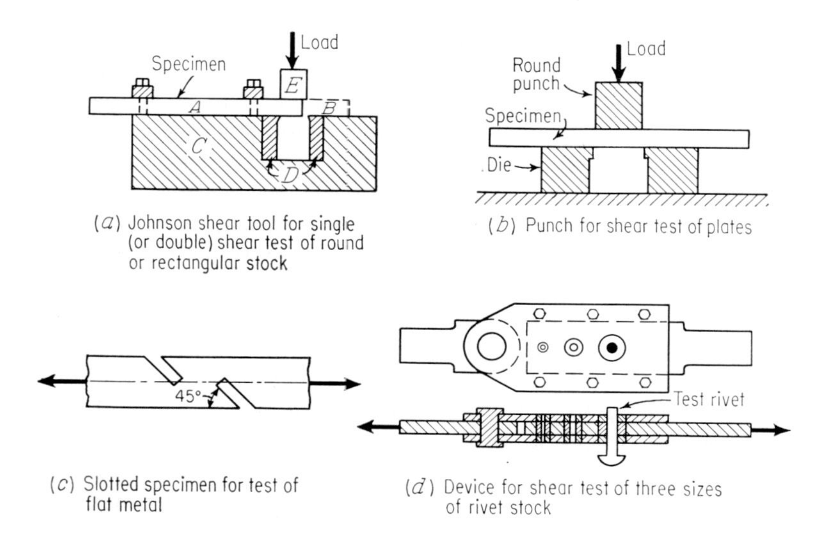

Fig. 6.5. **Methods of testing metals in direct shear.**

is extended to B and bridges the gap between the dies D, it is subjected to double shear. The dies and the loading tool are made of tempered tool steel ground to an edge. For metal plates, a round punching device is sometimes used, as illustrated schematically in Fig. 6.5*b*. In some tests of steels, a slotted specimen is used, as illustrated in Fig. 6.5*c*. For tests of rivet stock, a device illustrated in Fig. 6.5*d* has been used [613]. Direct shear tests are ordinarily made in compression- or tension-testing machines.

Directions for conducting a direct shear test of metal are given in Part 2.

For direct shear tests of brick, concrete, building stone, and cast iron, the Johnson type of shear tool has been employed, but use of this type of test of brittle materials has been largely abandoned.

For direct shear tests of wood, a special tool and specimen developed by the Forest Products Laboratory are used as shown in Fig. 6.6 (ASTM D 143). Failure tends to occur along the shear plane shown. The development of plywood for many uses has stimulated research in adhesives. For shear tests of glued joints, a specimen somewhat similar to that shown in Fig. 6.6 is glued along the dashed line and then loaded as indicated.

Direct shear tests of soils are made by use of a box that is in two parts so that one part may be moved with respect to the other, thus shearing the enclosed soil, as indicated in Fig. 6.7. To evaluate the internal friction, tests are being made at each of several "normal loads" [630, 631].

In the direct shear test, the testing device should hold the specimen firmly and preserve good alignment and the load should be applied evenly at right angles to the axis of the piece. In the single shear test, when equipment similar to that shown in Fig. 6.5*a* is used, the specimen must extend sufficiently beneath the loading tool E to avoid high bearing

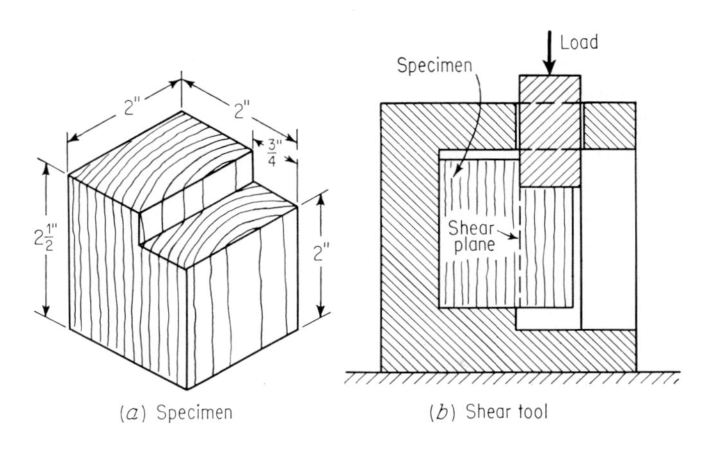

Fig. 6.6. **Method of testing wood in direct shear (ASTM D 143).**

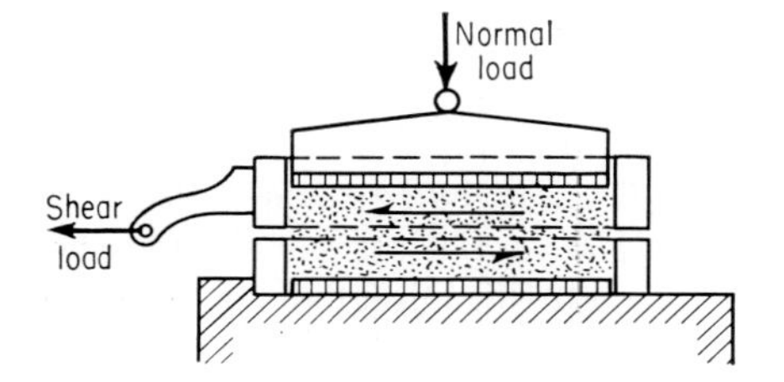

Fig. 6.7. Schematic arrangement of device for direct shear test of soils.

stresses. Likewise, in the double shear test, the specimen B must overlap the second die D sufficiently to avoid high bearing stresses. The speed of crosshead for applying the load should not exceed 0.05 in. per min for metals, stone, and concrete; for wood, the standard crosshead speed is 0.024 in. per min.

In the direct shear test, the only critical value that can be observed is the maximum load P. If A is the area subjected to the force, then the average shearing strength is taken simply as P/A. The shape and texture of the fractured surface should be reported.

Several series of tests of rivets and riveted connections show that the unit single shear strength of steel rivets is usually *greater* than the unit double shear strength. The difference may be as much as 20 percent, depending upon the material of the rivets and of the connected plates and upon the arrangement of the rivets [613].

Care should be taken to distinguish between pure shear failures and failures that may occur as a result of bending stresses or of diagonal tensile stresses.

6.5. The Torsion Test. The principal criteria in the selection of the torsion-test specimen appear to be that (1) the specimen should be of such size as to permit the desired strain measurements to be made with suitable accuracy, and (2) it should be of such proportions as to eliminate from that portion of the specimen on which measurements are made the effect of stresses due to gripping the ends. The ends should be such that they can be securely gripped without developing stresses sufficiently localized to cause failure in the grips. Ordinarily, the grips in the chucks of the machine are in the form of serrated blocks or cams, some types of which automatically tighten as the torque is applied. Care must be exercised in the gripping of the specimen that bending is not introduced. Centering points are usually provided in the chucks of the torsion machine for insertion into small centering holes in each end of the specimen; thus the specimen can be accurately centered in the machine.

As was mentioned in Art. 6.2, it is practically impossible to determine the proportional limit shearing strength of the extreme fibers of a solid torsion specimen. A thin tubular specimen is preferable for the determination of this property. Tubular specimens for ultimate shear-strength determinations should have short reduced sections with a ratio of length of reduced section to diameter (L/D) of about 0.5 and a diameter-thickness ratio (D/t) of about 10 to 12. For determinations of shearing yield strength and modulus of rigidity, a hollow specimen having a length of at least 10 diameters and a ratio of diameter to wall thickness of about 8 to 10, for its reduced section, is to be preferred [602, 143]. For larger ratios of diameter to thickness there is a tendency for failure to occur by buckling (see Fig. 6.9*c*), owing to inclined compressive stresses; this would appreciably affect the value determined for the yield strength. The actual dimensions of the specimen used are commonly chosen to suit both the size and type of testing machine available, as well as the product to be tested.

In making a torsion test on tubing, the ends usually must be plugged so that pressure from the jaws of the machine will not collapse the tubing. These plugs should not be so long that they extend within the test section. Occasionally it becomes necessary to test tubing that will not fit into the chucks of the torsion machine. In such cases, steel plug adapters may be welded, riveted, or screwed to the ends of the tube and the projecting ends of the adapters reduced in diameter to fit the chucks of the machine.

The torsion test of metals is carried out in a special testing machine designed for the purpose. An illustration of one type of torsion testing machine is shown in Fig. 6.8. A suitable driving mechanism turns a chuck with hardened serrated jaws, and the applied torque is transmitted through the test specimen to a similar chuck at the weighing head, which actuates some type of torque indicator. By one method, a lever system actuates a beam over which a movable poise can be made to travel; the beam is graduated in torque units, say inch-pounds. On some machines, the weighing system involves a pendulum connected to the pointer of a dial, which is graduated in inch-pounds; as torque is transmitted through the specimen, the pendulum rotates from its vertical equilibrium position until its static moment balances the applied torque. In other types of machines, a lever or arm attached to the chuck actuates (1) a hydraulic capsule connected to a load dial or (2) a series of levers connected to a small beam, the deflection of which is transmitted electronically to the load dial as for the machine shown in Fig. 6.8. To allow for longitudinal deformation of a specimen during test, as well as to accommodate specimens of various lengths, the part of the machine that carries the load-indicating mechanism is adjustable in position, often being mounted on rollers. This arrangement avoids the superposition of stress due to any axial loading upon the shear due to torsion. A capacity of 100,000 in.-lb is probably satisfactory for most ordinary testing of bar stock, although machines having a capacity of 2,000,000 in.-lb have been designed.

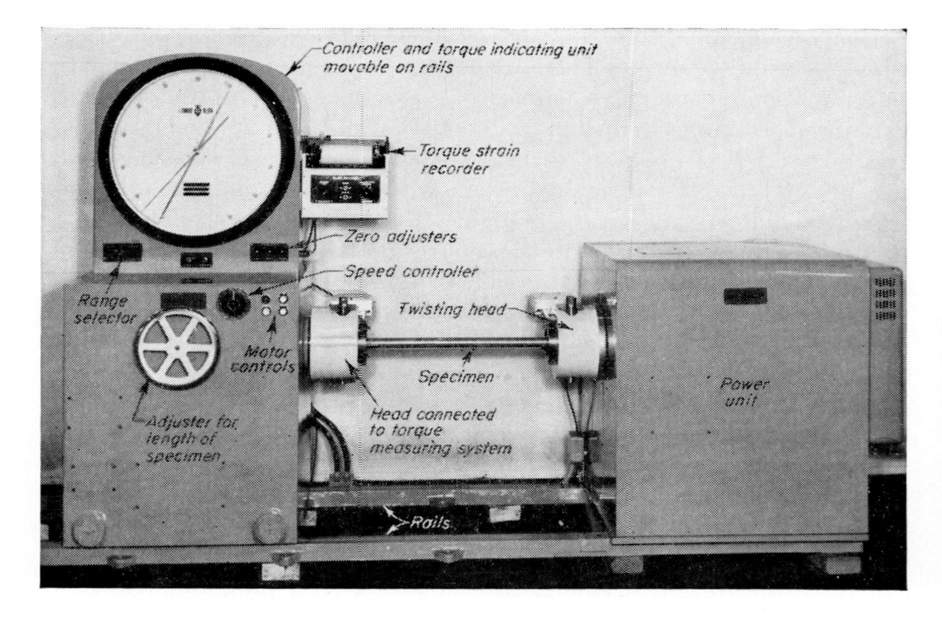

Fig. 6.8. **Torsion testing machine with electronic torque-bar load indicator.** (*Courtesy of Tinius Olsen Testing Machine Co.*)

Machines of low capacity, less than 10,000 in.-lb, are used for testing wire and light rod.

Various devices are used to measure the strain or angular twist in a torsion specimen. These torsion indicators or troptometers ordinarily consist of two collars secured to the specimen a given distance or gage length apart, with some means of measuring the relative angular displacement of the collars. In one type, a vernier attached to one collar moves around a graduated circle attached to the other collar; in another type, mirrors are attached to the collars, and observations are made with telescopes and scales. A very simple torsion meter consists of long radial arms attached to the collars, the arms being arranged to move around graduated arcs as the specimen twists. The precision of the first and third types may be approximately ± 0.0005 radian, but troptometers of the mirror type have been made to give a precision of ± 0.00005 radian.

The cross section of a torsion specimen should be measured to 1 part in 1000. Within the proportional limit of the material, the speed of the twisting head should not exceed about 0.01 rpm per inch of length of specimen, although the speed may be increased after the yield point is reached.

Directions for conducting a torsion test of metal are given in Part 2.

6.6. Observations of Test. The general types of observations and records of tests in direct shear and torsion are similar to those of tension and compression tests (see Arts. 5.6 and 5.13).

The shear fracture is quite distinct from either the tension or compression fracture; there is no localized reduction of area or elongation.

For materials that break in shear in the torsion test, the break in solid rods is plane and normal to the axis of the piece, as shown in Fig. 6.9*a*. For ductile steels, the fracture is usually silky in texture, and the axis about which the final twisting took place may usually be observed. Since the surfaces of the break may not be quite smooth, the outer portions, in moving past each other, act like cams, pushing the piece apart in the direction of its length. The center portions not yet broken by shear are possibly broken in tension by this cam action. In brasses, bronzes, and wrought iron, the material often breaks up into fibers, like a rope, before rupture.

The rupture of a material for which the tensile strength is less than the shearing strength occurs by separation in tension along a helicoidal surface, as mentioned in Art. 6.2. This type of break occurs when cast iron

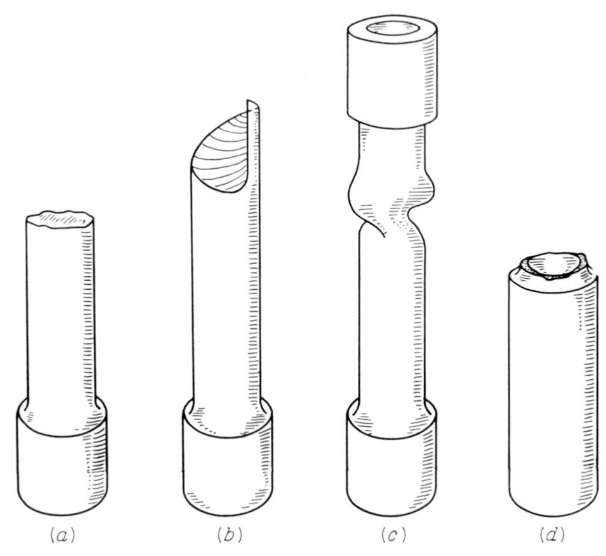

(*a*) Solid bar of ductile material. Fracture on plane right section.
(*b*) Solid bar of brittle material. Helicoidal fracture.
(*c*) Tubular specimen of ductile material. Failure by buckling.
(*d*) Tubular specimen of ductile material; short reduced section. Failure on plane right section.

***Fig. 6.9.* Types of failure in torsion.** (*Report of ASTM Committee E*-1, *Ref*. 601.)

or plain concrete is subjected to torsion. The outline of the fracture makes a complete revolution of the bar, the ends of the helix being joined by an approximate straight line, as illustrated in Fig. 6.9*b*. The helicoidal type of fracture may be easily obtained by breaking a piece of chalk in torsion with the fingers.

Thin-walled tubular specimens of ductile material having a reduced section of length greater than the diameter fail by buckling as shown in Fig. 6.9*c*, but those having a short reduced section fail in torsion on a right section, as shown in Fig. 6.9*d*.

6.7. Special and Indirect Determinations of Shear Properties. Helical springs tested in axial compression or tension constitute one type of shear test, since the stresses developed are largely those of torsional and direct shear but principally the former. The modulus of rigidity of the material can be determined from the applied loads and the deflections, whereas the shearing stresses developed are a direct function of the loads.

The torsional rigidity of small rods and wire is sometimes determined from observations of the period of the torsional vibrations when a specimen is hung vertically with a known mass at the lower end. The modulus of rigidity may also be determined by applying a known torque to the lower end and measuring the angle of twist; the torque may simply be applied through a drum on the lower end of the rod by means of a system of strings, pulleys, and weights.

The shearing resistance of soils is often determined by means of a triaxial-compression test. Tests of cylindrical samples are made in a closed container in which hydrostatic pressures can be applied as well as an axial load. The internal friction and (shearing) cohesion are evaluated from an analysis of the stress conditions at failure. For materials such as plain portland-cement concrete and bituminous mixtures, a triaxial-compression test appears to offer a satisfactory means for obtaining a measure of the shearing resistance. However, the conduct and interpretation of triaxial-compression tests are specialized procedures involving considerable skill and knowledge of the variables affecting the behavior of the material [632, 633].

In reinforced-concrete beams, failure may occur owing to cracking caused by diagonal tensile stresses. Because these diagonal tensile stresses are caused by shear, it is customary to design the web reinforcement on the basis of allowable shearing stresses. The allowable shearing stresses are obtained from a study of the failure of beams of various proportions. Limiting safe values are chosen from a general correlation of the results, rather than by a logical analysis. This is an example of a problem in which the results of practical observations of the over-all behavior of a complex member are a more significant criterion of design than what might be called the "fundamental" method of approach.

BENDING TESTS

6.8. Behavior of Materials Subject to Bending. If forces act on a piece of material in such a way that they tend to induce compressive stresses over one part of a cross section of the piece and tensile stresses over the remaining part, the piece is said to be in bending. The common illustration of bending action is a beam acted upon by transverse loads; bending can also be caused by moments or couples such, for example, as may result from eccentric loads parallel to the longitudinal axis of a piece.

In structures and machines in service, bending may be accompanied by direct stress, transverse shear, or torsional shear. For convenience, however, bending stresses may be considered separately, and in tests to determine the behavior of materials in bending, attention is usually confined to beams. In the following discussion, it is assumed that the loads are applied so that they act in a plane of symmetry, so that no twisting occurs, and so that deflections are parallel to the plane of the loads. It is also assumed that no longitudinal forces are induced by the loads or by the supports. For more complicated cases of bending, texts on mechanics of materials should be consulted [e.g., 138, 594].

Figure 6.10 illustrates a beam subjected to transverse loading. The bending effect at any section is expressed as the "bending moment" M which is the sum of the moments of all forces acting to the left (or to the right) of the section. The stresses induced by a bending moment may be termed *bending stresses.* For equilibrium, the resultant of the tensile forces T must always equal the resultant of the compressive forces C. The resultants of the bending stresses at any section form a couple that is equal in magnitude to the bending moment. When no stresses act other than the bending stresses, a condition of "pure bending" is said to exist.

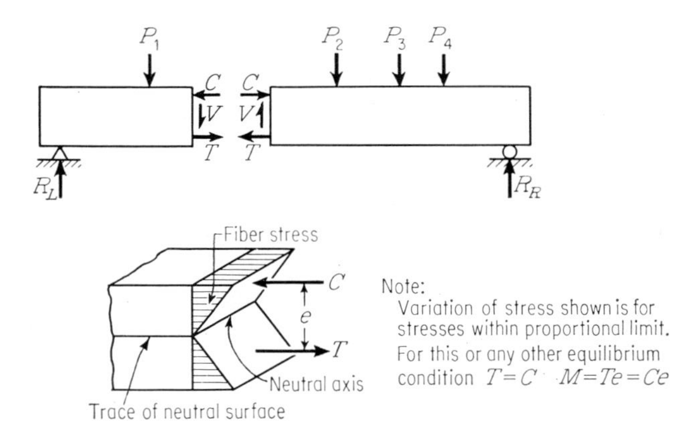

Fig. 6.10. **Bending of a beam.**

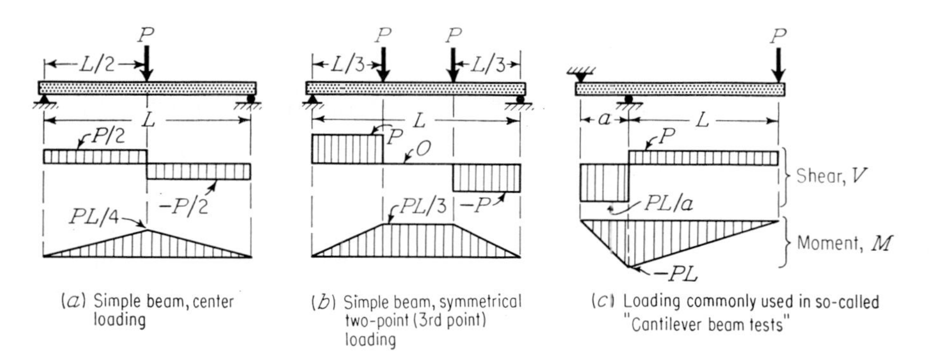

Fig. 6.11. **Shear and moment diagrams.**

Pure bending is developed under certain loading conditions; in the usual case, bending is accompanied by transverse shear. The resultant of the shearing stresses across a transverse section equals the total transverse shear V, which is computed as the algebraic sum of all transverse forces to the left (or to the right) of a section. Bending action in beams is often referred to as "flexure." As used in this text, the term *flexure* refers to bending tests of beams subject to transverse loading.

The variations in total transverse shear and in bending moment along a beam are commonly represented by shear and moment diagrams, which are illustrated for several cases of concentrated loading in Fig. 6.11. It should be noted that symmetrical two-point loading gives a condition of pure bending (constant moment) over the central portion of the span (see Fig. 6.11*b*).

In a cross section of a beam, the line along which the bending stresses are zero is called the *neutral axis*. The surface containing the neutral axes of consecutive sections is the *neutral surface*. On the compressive side of the beam the "fibers" of the beam shorten, and on the tensile side they stretch; thus the beam bends or deflects in a direction normal to the neutral surface, becoming concave on the compressive side (see Fig. 6.12).

It has been well established by many observations that in pure bending the *strains* are proportional to the distance from the neutral axis; and this appears to hold, at least with good approximation, in the range of inelastic action as well as within the range of elastic action. This is referred to as a condition of "plane bending," i.e., plane sections before bending remain plane sections after bending. The relative rotation of one cross section of an initially straight beam with respect to a reference cross section is illustrated by Fig. 6.12*a*. The elongation or shortening of the fibers in any given length of beam over which the moment is constant divided by that length gives unit fiber strain, as illustrated in Fig. 6.12*b*.

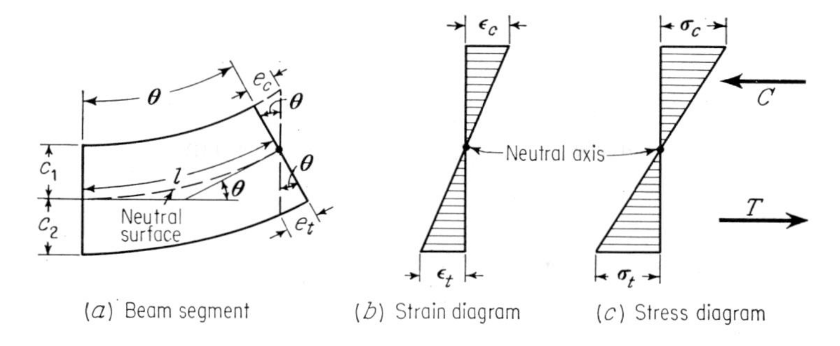

Fig. 6.12. **Fiber strains and stress due to bending within the proportional limit.**

If the stresses are proportional to the strains (say within the proportional limit), the stress variation across a section is linear, as shown in Fig. 6.12*a*.

By summing the moments of the stresses about the neutral axis, the resisting moment, within the proportional limit, may be found in terms of the extreme fiber stress:

$$M = \frac{\sigma I}{c} \qquad \text{the "flexure formula"}$$

where σ = stress on extreme fiber
c = distance from neutral axis to extreme fiber
I = moment of inertia of section about the neutral axis (I for a rectangular section is $bd^3/12$; for a circular section, $\pi d^4/64$; in which b = breadth and d = depth or diameter)

In terms of the extreme fiber strains, the moment may be stated as

$$M = \epsilon \frac{EI}{c}$$

where ϵ = extreme fiber strain per unit length of beam

For pure bending the moment may also be found from the change in slope of the beam:

$$M = EI\frac{\theta}{x} = \frac{EI}{\rho}$$

where θ = change in slope between two cross sections
x = distance between two cross sections
ρ = radius of curvature of neutral surface

The deflection of a beam is the displacement of a point on the neutral surface of a beam from its original position under the action of applied loads. Within the proportional limit, the deflection due to bending under a given type of loading may be computed from the modulus of elasticity of the material and the properties of the section. Transverse deflections for two common cases are

Center deflection of a "simple beam" (i.e., one freely supported at the ends) with concentrated load P at mid-span $= PL^3/48EI$

Center deflection of a simple beam with concentrated loads, *each* equal to P, at third points of span $= 23PL^3/648EI$

Deflection is a measure of the overall stiffness of a given beam and can be seen to be a function of the stiffness of the material and the proportions of the piece. Measurement of deflections serves as a means for determining the modulus of elasticity of the material in flexure. If the modulus of elasticity in tension and compression is not the same, the modulus of elasticity computed from a flexure test tends to be intermediate between those for tension and compression. Furthermore, if there are transverse shearing stresses, the modulus of elasticity in flexure tends to be slightly below that for axial stress, since the shearing strains tend to increase an observed deflection over that due to fiber stains alone, as discussed below.

The shearing stress at any point on a cross section of a beam loaded within the elastic range may be computed as follows:

$$\tau = \frac{V}{Ib} az$$

where V is the total transverse shear on the section
I is the moment of inertia of the section
a, b, and z are as shown in Fig. 6.13*a*

If the beam is of rectangular section, the shearing stress is a maximum at the neutral axis and varies parabolically from a maximum at the neutral axis to zero at the extreme fibers as indicated in Fig. 6.13*b*; the maximum shearing stress is $\frac{3}{2}\frac{V}{A}$. If shearing stresses act, a plane

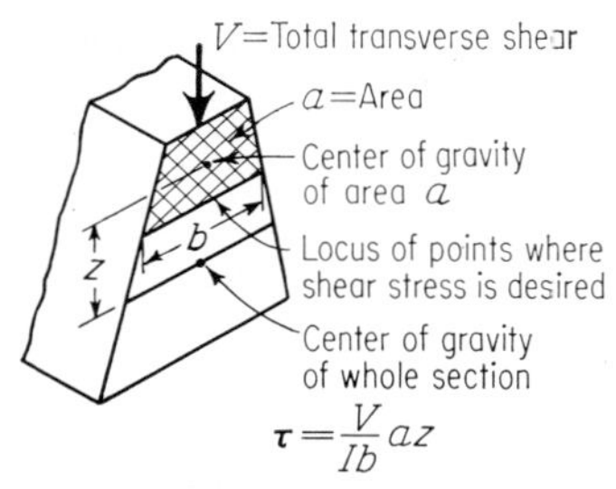

(*a*) Relation of shearing stress to transverse shear

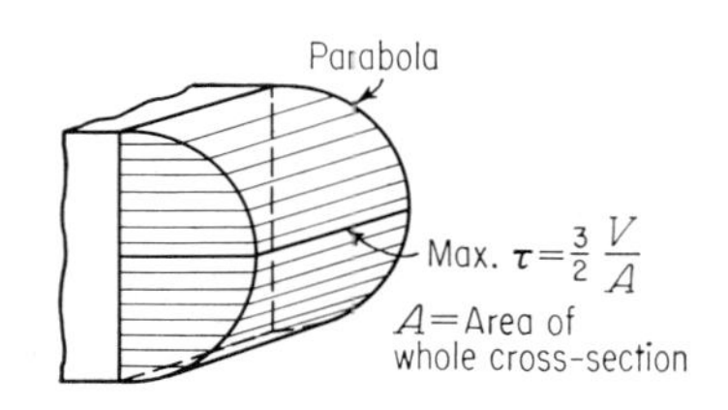

(*b*) Variation in shearing stress across a rectangular section

***Fig. 6.13.* Shear in a beam.**

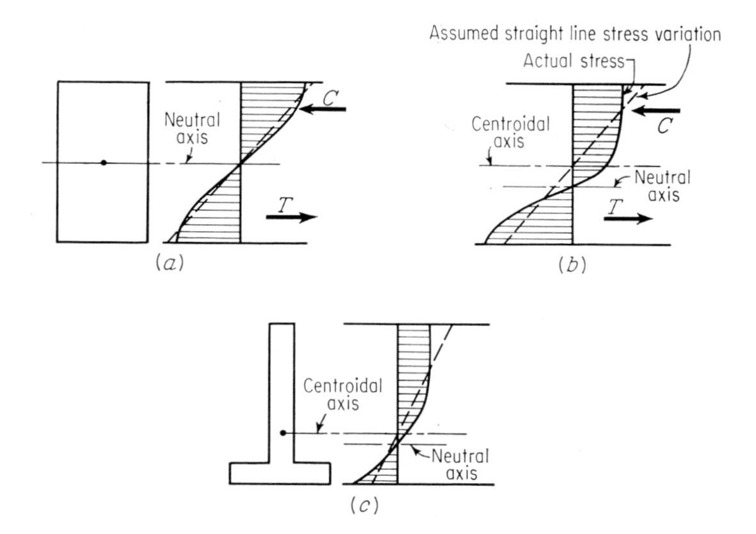

Fig. 6.14. **Fiber stress above the proportional limit.**

cross section does not remain plane under load. The deflections due to shear may be computed by summing up the shear strains in the various elements of a beam (see texts on mechanics of materials). In beams in which the ratio of length to depth is about 10 or more, the shearing deflections are sufficiently small compared with bending deflections that they can usually be neglected in practical testing.

Above the proportional limit, bending stresses do not vary linearly across a section, because stress is not proportional to strain. Illustrations of common cases are shown in Fig. 6.14. In Fig. 6.14*a*, the solid line shows the stress variation in a homogeneous beam of symmetrical section for a material that has the same stress-strain characteristics in both tension and compression. The stress distribution computed on the basis of the flexure formula $\sigma = Mc/I$ is shown by the dotted line. The extreme straight-line fiber stresses computed on the basis of this formula are seen to be greater than the true maximum fiber stresses.

If the material does not have the same stress-strain characteristics in tension as in compression, the neutral axis must shift toward the stiffer side of the beam in order to maintain equality of the resultants of the tensile and compressive forces, as shown in Fig. 6.14*b*. In this case the fiber stress computed by the flexure formula is less than the true fiber stress on the stiffer side of the beam and greater than the true fiber stress on the less stiff side.

If the beam has a cross section such as that shown in Fig. 6.14*c*, the stresses are lower on the side nearer the centroidal axis. This side of the beam is, in effect, stronger because of the concentration of material to resist stress. Yielding of the more highly stressed fibers on the opposite side then causes a shift of the neutral axis toward the stronger side, giving a stress distribution similar to that shown.

6.9. Failure in Bending. Failure may occur in beams owing to one of several causes, listed below. Although these modes of failure are stated primarily with reference to beams of ductile material, in their general aspects they may apply to any material.

1. The beam may fail by yielding of the extreme fibers. When the yield point is reached in the extreme fibers, the deflection of the beam increases more rapidly with respect to an increase of load; and if the beam is of a thick, stocky section or is firmly held so that it cannot twist or buckle, failure takes place by a gradual sagging which finally becomes so great that the usefulness of the beam as a supporting member is destroyed.

2. In a beam of long span, the compression fibers act somewhat as do the compression fibers of a column, and failure may take place by buckling. Buckling, which in general occurs in a sidewise direction, may be either the primary or the secondary cause of failure. In a beam in which excessive flexural stress is the primary cause of failure and in which the beam is not firmly held against sidewise buckling, overstress may be quickly followed by the collapse of the beam due to sidewise buckling, since the lateral stability of a beam is greatly lessened if its extreme fibers are stressed to the yield point. Sidewise buckling may be a primary cause of beam failure, in which case the computed fiber stress, in general, does not reach the yield-point strength of the material before buckling occurs. Buckling often limits the strength of narrow, deep beams, especially beams of I section or channel section with tension and compression flanges connected by a thin web. Whether it is a primary cause of failure or a final manner of failure, sidewise buckling results in a clearly marked and generally a sudden failure.

3. Failure in thin-webbed members, such as an I beam, may occur because of excessive shearing stresses in the web, or by buckling of the web under the diagonal compressive stresses which always accompany shearing stress. If the shearing stress in the web reaches a value as great as the yield-point strength of the material in shear, beam failure may be expected, and the manner of failure will probably be by some secondary buckling or twisting action. The inclined compressive stress always accompanying shear may reach so high a value that the buckling of the web of the beam is a primary cause of failure. Danger of web failure as a primary cause of beam failure exists, in general, only for short beams with thin webs.

4. In the parts of beams adjacent to bearing blocks that transmit concentrated loads or reactions to beams, high compressive stresses may

be set up, and in I beams or channel beams the local stress in that part of the web nearest a bearing block may become excessive. If this local stress exceeds the yield-point strength of the material at the junction of web and flange, the beam may fail primarily on account of the yielding of the overstressed part.

The failure of beams of brittle material such as cast iron and plain concrete always occurs by sudden rupture. Although, as failure is approached, the neutral axis shifts toward the compression face and thus tends to strengthen the beam, failure finally occurs in the tensile fibers because the tensile strength of these materials is only a fraction of the compressive strength. The ratio of tensile to compressive strength is about 25 percent for cast iron and about 10 percent for concrete.

The failure of reinforced-concrete beams may be the result of (1) so-called failure of the steel due to stresses above the yield point, resulting in vertical cracks on the tensile side of the beam; (2) failure of the concrete in compression at the outermost compressive fibers; and (3) failure of the concrete in diagonal tension, primarily due to excessive shearing stresses, resulting in the formation of cracks that slope downward toward the reactions, often becoming horizontal just above the main steel in simple span beams.

A wooden beam also may fail in a number of ways. (1) It may fail in direct compression at the concave compression surface. (2) It may break in tension on the convex tension surface. Since the tensile strength of wood parallel to the grain is usually greater than its compressive strength, the neutral axis shifts toward the tensile face so as to maintain equality of the tensile and compressive forces, as shown in Fig. 6.14*b*. Therefore, the first visible signs of failure may be in the tensile face even though the wood is stronger in tension than in compression. This type of failure occurs only in well-seasoned timbers, since green test pieces usually fail in compression before rupturing the tension fibers. (3) It may fail by lateral deflection of the compression fibers acting as a column. (4) It may fail in horizontal shear along the grain near the neutral axis. This type of failure is sudden and more common in well-seasoned timbers of structural sizes than in green timbers or in small beams. (5) It may fail in compression perpendicular to the grain at points of concentrated load.

Values of the proportional limit determined from beam tests are generally higher than values obtained from tension or compression tests because yielding of the extreme fibers is masked by the supporting effect of the less highly stressed fibers nearer the neutral axis.

For beams of brittle material, the nominal fiber stress at rupture as computed by the flexure formula (the "modulus of rupture" in bending) is usually appreciably greater than the true tensile strength of the material. The ratio of the modulus of rupture to true tensile strength is about

1.8 for cast iron and about 1.5 to 2 for concrete. The ratio of modulus of rupture to compressive strength is about 0.5 for cast iron, about 0.15 to 0.20 for concrete, and about 2 for wood (considering the compressive strength parallel to the grain).

As the load and deflection at rupture are functions of the cross-sectional dimensions of the specimen, the standard specifications for cast iron give correction factors to adjust values for specimens which are not exactly of standard size (ASTM A 48) (see also Part 2, Problem 10).

6.10. Scope and Applicability of Bending Tests. Most structures and machines have members whose primary function is to resist loads that cause bending. Examples are beams, hooks, plates, slabs, and columns under eccentric loads. The design of such structural members may be based upon tensile, compressive, and shearing properties appropriately used in various bending formulas. In many instances, however, bending formulas give results which only approximate the real conditions. While special analyses can often be made of stresses arising from unusual loading conditions and from local distortions and discontinuities, it is not always feasible to make such analyses, which may be very complicated. The bending test may serve then as a direct means of evaluating behavior under bending loads, particularly for determining the limits of structural stability of beams of various shapes and sizes.

Flexural tests on beams are usually made to determine strength and stiffness in bending; occasionally they are made to obtain a fairly complete picture of stress distribution in a flexural member. Beam tests also offer a means of determining the resilience and toughness of materials in bending.

Under the general designation of strength may be included the proportional limit, yield strength, and modulus of rupture. These properties may be determined with a view to establishing, with appropriate reduction factors, allowable bending stresses for use in design. The modulus of rupture also may be used simply as a criterion of quality in control tests.

The stiffness of a material may be determined from a bending test in which the load and deflection are observed. The modulus of elasticity for the material in flexure is computed by use of an appropriate deflection formula. The value of the modulus of elasticity may then be used to compute the elastic deflection of beams of the same material but of other size, shape, or loading, although some error may be involved owing to (1) ignoring shearing deflections, which are of importance in short, deep beams; (2) deviations from the straight-line relationship of stress and strain as expressed by Hooke's law; and (3) lack of uniformity of the material.

Because the loads required to cause failure may be relatively small and easily applied, bending tests can often be made with simple and inexpensive apparatus. Because the deflections in a bending test are many times the strains in a tension test, a reasonable determination of stiffness or resilience can be made with less sensitive and less expensive instruments than are required in a tension test. Thus the bending test is often used as a control test for brittle materials, notably cast iron and concrete. It is obviously unsuited for determining the ultimate strength of ductile materials.

For wire and sheet metals, a simple bend test is sometimes used as an arbitrary measure of relative flexibility. For ductile materials in the form of rods, such as reinforcement bars for concrete, a cold-bend test is used to determine whether or not the rod can be bent sharply without cracking and serves as an acceptance test with respect to this form of ductility.

6.11. Specimens for Flexure Tests. To determine the modulus of rupture for a given material, the beam under test must be so proportioned that it will not fail in shear or by lateral deflection before it reaches its ultimate flexural strength. To produce a flexural failure, the specimen must not be too short with respect to the beam depth, and conversely, if a shear failure is desired, the span length must not be too long. Values of $L = 6d$ to $L = 12d$ (the actual value depending upon the material, the shape of beam, and the type of loading), in which L = length and d = depth, serve as an approximate dividing line between the short, deep beams that fail in shear and the long, shallow beams that fail in the outer fibers.

Although beams of a variety of forms are used for special and research testing work, standardized specimens are used for routine and control testing of a number of common materials, such as cast iron, concrete, brick, and wood.

Test specimens of cast iron are cylindrical bars, cast separately, but under the same sand-mold conditions and from the same ladle as the castings they represent. The three common sizes of test bars are given in Table 6.1. They are tested as simple beams under center loading on spans dependent on the size of bar, also as given in Table 6.1.

Table* 6.1. Standard sizes of cast-iron test bars for flexure tests

Controlling section of casting, in.	Nominal dimensions, in.		Distance between supports, in.
	Diameter	Length	
0.50 and under	0.875	15	12
0.51 to 1.00	1.20	21	18
1.01 and over	2.00	27	24

* Based on ASTM A 48.

Current ASTM Specifications for testing plain concrete beams of rectangular section call for third-point loading (ASTM C 78) and for mid-span loading (ASTM C 293) on a simple span. The size of beam is not specified, but for aggregate up to 1½-in. maximum size, the cross section is usually 6 by 6 in.; for aggregate up to 2½-in. maximum size, 6 by 8-in. and 8 by 8-in. sections have been used. The beams are tested on a span of three times the depth of beam.

Building bricks are approximately 2 by 4 by 8 in. in size and are tested flatwise on a 7-in. span under center loading (ASTM C 67).

Standard test beams of small clear pieces of wood are 2 by 2 by 30 in. in size and tested on a 28-in. span under center loading (ASTM D 143). Timbers in structural sizes are often tested under third-point loading on a span of 15 ft (ASTM D 198); common sizes of large wood test beams are 16 ft long with nominal cross sections of 6 by 12 in. or 8 by 16 in.

Specimens of gypsum (plaster boards) are 12 in. wide by 16 in. long and are tested under center loading on a 14-in. span (ASTM C 26). Specimens of building stone are 2¼ by 4 by 8 in. and are tested flatwise under center loading on a 7-in. span (ASTM C 99). Specimens of slate are 1 by 1½ by 12 in. and are tested flatwise under center loading on a 10-in. span (ASTM C 120). Specimens of molded insulating materials are ½ by ¼ by 5 in. and are tested under center loading on a 4-in. span (ASTM D 48).

6.12. Apparatus for Flexure Tests. The principal requirements of the supporting and loading blocks for beam tests are as follows:

1. They should be of such shape that they permit use of a definite and known length of span.
2. The areas of contact with the material under test should be such that unduly high stress concentrations (which may cause localized crushing around the bearing areas) do not occur.
3. There should be provision for longitudinal adjustment of the position of the supports so that longitudinal restraint will not be developed as loading progresses.
4. There should be provision for some lateral rotational adjustment to accommodate beams having a slight twist from end to end, so that torsional stresses will not be induced.
5. The arrangement of parts should be stable under load.

A number of specifications describe in detail the type of support to be used with particular materials. The principal features of representative supporting arrangements are shown in Fig. 6.15.

Many flexure tests are conducted in universal testing machines, with the supports placed upon the platen or an extension thereof and with the loading block fastened to or placed under the movable head. However, for control tests of some materials (e.g., foundry tests of cast iron and

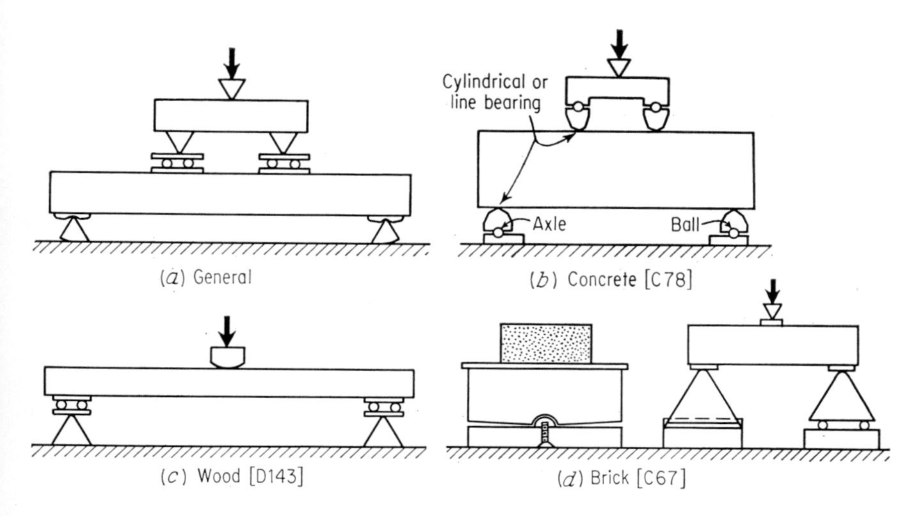

Fig. 6.15. **Principal features of supporting and loading devices for beam tests indicating provision for longitudinal and lateral rotational adjustment at supports.**

field tests of concrete) specially designed hand-operated machines are often employed.

Apparatus for measuring deflection should be so designed that crushing at the supports, settlement of the supports, and deformation of the supporting and loading blocks or of parts of the machine do not introduce serious errors into the results. One method of avoiding these sources of error is to measure deflections with reference to points on the neutral axis above the supports. Typical arrangements are shown in Figs. 6.16*a* and 6.16*b*. In general, deflections within the proportional limit should be read to at least $\frac{1}{100}$ of the deflection at the proportional limit; for greater deflections they should be read to at least $\frac{1}{100}$ of the deflection at rupture.

For determining fiber strains, a surface strainometer, SR-4 resistance-wire gages, or a portable strain gage may be used to measure strains along

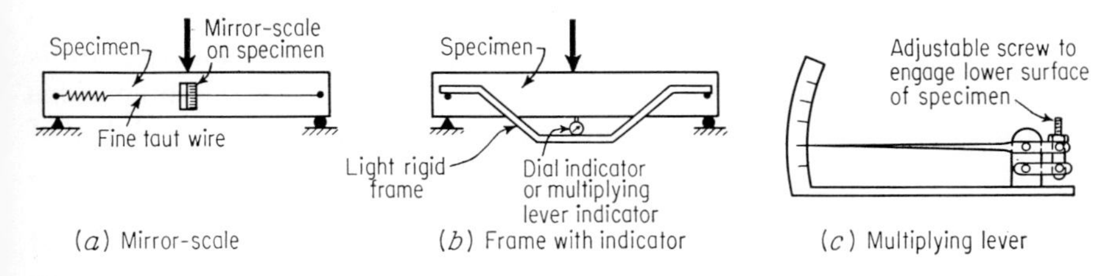

Fig. 6.16. **Deflection measuring devices (schematic).**

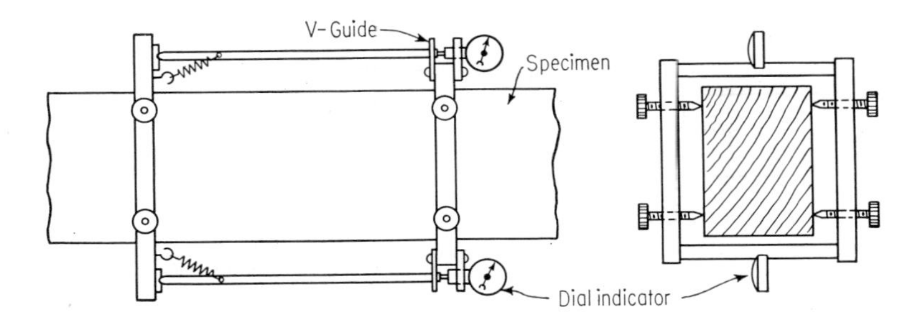

Fig. 6.17. **Strainometer for beam tests.**

desired gage lines, or a special device may be used to indicate the relative rotation of plane cross sections at some given distance apart, as shown in Fig. 6.17. The dial indicators of this device give the movement of the two collars with respect to one another to the nearest 0.0001 in. By drawing to scale a simple sketch of the strainometer, including the location of the extreme fibers of the beam and the change in dial readings, the position of the neutral axis and the strains in the extreme fibers of the beam can be read from the drawing or computed.

Deflectometers should be designed so that they will not be injured by failure of the test piece, or precautions should be taken to remove them well in advance of final rupture.

For tests requiring fiber-strain measurements the symmetrical two-point loading (Fig. 6.11*b*) is desirable in that the portion of the beam length between the loads is subject to constant moment and hence constant stress along any fiber. A reasonable gage length can thus be used for strain measurements. Further, in the section of constant moment, the shear is zero, so that the observed strains are due to bending stresses only. This type of loading also has an advantage over center loading in tests of brittle materials to determine the modulus of rupture, because with center loading the failure is forced to occur near the center, and defects in the material would probably remain undiscovered unless they happen to occur close to the mid-span.

6.13. Conduct of Flexure Tests of Beams. The conduct of routine flexure tests is usually simple. Ordinarily only the modulus of rupture is required; this is determined from the load at rupture and the dimensions of the piece (span and critical cross section). When the modulus of elasticity is required, a series of load-deflection observations are made, following the general principles laid down in Art. 2.5.

The dimensions of cast-iron specimens are measured to the nearest 0.001 in., of wood and concrete specimens to the nearest 0.01 in. The supporting and loading blocks are located with a reasonable degree of accuracy, say 0.2 percent of the span length. The assembly of supports and specimen should be placed centrally in the testing machine and should be checked to see that they are in proper alignment and can function as intended. Deflectometers and strainometers should be located carefully and checked to see that they operate satisfactorily and are set to operate over the range required.

For cast-iron flexure bars, the load should be applied at such a rate that fracture is produced in not less than 15 sec for the 0.875-in.-diameter bar, 20 sec for the 1.20-in. bar, and 40 sec for the 2.0-in. bar (ASTM A 48). Concrete beams may be loaded rapidly at any desired rate up to 50 percent of the breaking load, after which loads should be applied at a rate such that the extreme fibers are stressed at 150 psi or less per minute (ASTM C 78). For small clear pieces of wood the load should be applied throughout the test at a rate of movement of the movable crosshead of 0.1 in. per min (ASTM D 143); this corresponds in the standard test set-up for small clear specimens to a loading rate of about 400 lb per min. For larger clear timbers, 8 in. or less in depth, the load should be applied so as to produce a rate of strain in the extreme fibers of 0.0007 in. per in. per min. For clear timbers over 8 in. in depth, the rate of strain in the extreme fibers should be 0.0015 in. per in. per min; this corresponds in a beam 16 in. deep under third-point loading on a 15-ft span to a speed of movable crosshead of about 0.26 in. per min (ASTM D 198). The speed of the testing machine should not vary more than 25 percent from that specified for a given test in order to keep the variation in results due to this cause within 1 percent. For brick and molded insulating materials, the load rate should be not greater than 2000 lb per min or at a rate not to exceed that corresponding to a crosshead speed of 0.05 in. per min (ASTM C 67). For plastics, the load rate should be that which produces a rate of strain in the extreme fibers of 0.01 in. per in. per min (ASTM D 790). For building stone, the load rate should be not more than 1000 lb per min (ASTM C 99). For special purposes, other speeds are used, but in any case the speed should be no greater than that at which observations can be made accurately. The speed should be recorded as part of the data of the test.

Directions for conducting typical flexure tests of cast iron, wood, reinforced concrete, and of a steel I beam are given in Part 2.

6.14. Observations of Test. The general types of observation and record of tests in flexure are similar to those of tension and compression tests.

The conditions under which the modulus of rupture is determined (type of specimen, span length, type and rate of loading, etc.) should always be recorded, since these markedly affect the results.

In computing the modulus of elasticity from load-deflection data, the

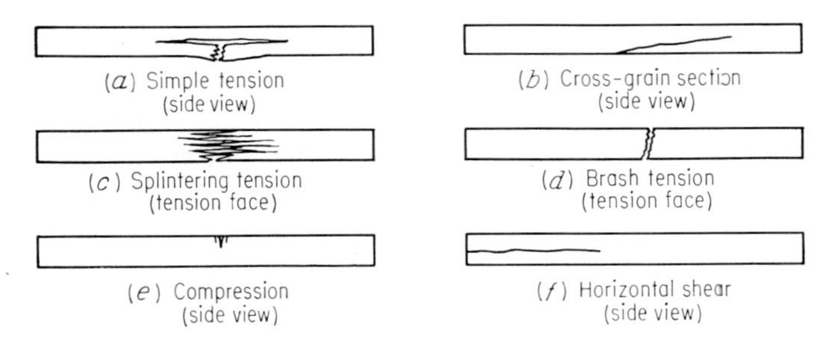

Fig. 6.18. **Various modes of failure of wood beams (ASTM D 143).**

simplest procedure is to plot a load-deflection diagram and from the slope of the diagram compute loads and deflection for substitution in the pertinent deflection formula.

The fracture of materials such as cast iron and concrete is definite, usually occurring on an approximately plane surface at a section of maximum moment. The texture of the fracture may be of significance and should be noted. The designations of various modes of failure of wood are indicated in Fig. 6.18. In tests of beams that do not fail by sudden rupture, indications of impending failure, such as cracks, localized yielding, buckling, etc., should be observed carefully. The causes of primary and secondary failure of structural beams should be noted (see Art. 6.9).

6.15. Effect of Important Variables in Flexure Tests. In flexure tests of brittle materials, some of the more important factors that affect the results are type and rate of loading, length of span, and cross-sectional dimensions of the beam.

The effect of type of loading is illustrated by the results of numerous tests of concrete, which tests have indicated the relative magnitudes of the modulus of rupture for three common types of loading to be as follows [668, 669, 670]:

1. In a simple span, the largest value of modulus of rupture is obtained from center loading. Values computed on the basis of the moment at the center of span tend to be somewhat greater (about 7 percent) than values computed on the basis of the moment at the section of break.

2. Cantilever loading tends to give slightly higher results than center loading on a simple span, although on the average the difference is not great.

3. Third-point loading on a simple span gives results invariably somewhat less than center loading (roughly 10 to 25 percent). It seems reasonable to suppose that since the strength of the material varies somewhat throughout the length of the beam, in the third-point loading, the weakest section (of those subjected to constant moment) is sought out.

These relations would probably hold, at least in principle, for other brittle materials. In general, the third-point loading method appears to give the most concordant results.

Tests of both cast iron and concrete have shown that, for beams of the same cross section, the shorter the span length, the greater the modulus of rupture [657, 668]. Typical results for cast-iron bars tested on various span lengths are shown in Table 6.2.

Table* 6.2. Effect of span length on modulus of rupture of 1¼-in.-diameter cast-iron bars

Length of span, in.	*Modulus of rupture, psi*
12	45,900
18	44,600
24	42,800

* C. D. Matthews, "Test of Cast Iron Arbitration Test Bars," *Proc. ASTM*, vol. 10, 1910.

The shorter the span, the less is the computed value of the modulus of elasticity of cast iron, although the difference is not over about 10 percent for length-diameter ratios ranging from 10 to 30.

The shape of the cross section of a beam may appreciably affect the resistance of the beam. Tests of cast-iron beams having a variety of shapes but of about the same cross-sectional area show that in general the modulus of rupture and the modulus of elasticity are lower for beams having a relatively larger proportion of the cross-sectional area concentrated near the extreme fibers, as is the case with an I section, although the breaking *loads* are considerably greater for such sections [141].

Tests of both cast iron and concrete indicate lower strengths for beams of larger cross-sectional dimensions [659, 668, 670]. These results are in line with the results of tension and compression tests on cast iron and concrete. Test results for wood show that both the strength and modulus of elasticity of large timbers are less than for small clear specimens on short spans; this is attributed to the presence of defects in members of structural size.

Speed of testing has the same general effect in the flexure test as in the tension and compression tests, i.e., the greater the speed, the higher the indicated strength. Tests on timber beams in which the time from zero load to failure varied from ½ sec to 5 hr indicate that it requires a tenfold increase in rate of loading to produce a 10 percent increase in bending strength. Timber beams under continuous loading for years will fail under loads one-half to three-quarters as great as required in the usual static bending test, which ruptures the test specimen in a few minutes [1330]. For concrete beams, increasing the rate of stressing from 20 to 1140 psi per min resulted in an increase of about 15 percent in the modulus of rupture [670].

As indicated previously (Art. 6.9), the greater the ratio of length to width of beam, the greater the tendency toward lateral buckling of the compression flange, unless the beam is laterally supported. Although the problem is not determinate, it would appear desirable to maintain a length-width ratio of something less than 15 in tests of beams to determine properties of a material.

6.16. "Bend" Tests for Metals. The "bend" tests (of which the "cold-bend" test is the most common) offer a simple, somewhat crude, but often satisfactory means of obtaining an index of ductility. Essentially the test consists in sharply bending a bar through a large angle and noting whether or not cracking occurs on the outer surface of the bent piece. Oftentimes the angle of bend at which cracking starts is determined. The severity of the test is generally varied by using different sizes of pins about which the bend is made.

Bend tests are sometimes made to check the ductility for particular types of service or to detect loss of ductility under certain types of treatment. Thus, cold-bend tests, which, as the name implies, are made by bending a metal at ordinary temperatures, may serve to detect too high a carbon or phosphorus content or improper rolling conditions in steel. Cold-bend tests are required in the specifications for many steels, particularly those in the form of rod and plate, e.g., bars for concrete reinforcement (ASTM A 15, A 16), rivet steel (ASTM A 141), structural steel

Table 6.3. **Bend-test requirements for concrete-reinforcement bars***

Bar designation number	Plain bars			Deformed bars		
	Structural grade	Intermediate grade	Hard grade	Structural grade	Intermediate grade	Hard grade
Under 6	180° $d = t$	180° $d = 2t$	180° $d = 4t$	180° $d = 2t$	90° $d = 3t$	90° $d = 4t$
6, 7, 8	180° $d = t$	90° $d = 2t$	90° $d = 4t$	180° $d = 3t$	90° $d = 4t$	90° $d = 5t$
9, 10, 11	180° $d = t$	90° $d = 2t$	90° $d = 4t$	180° $d = 4t$	90° $d = 5t$	90° $d = 6t$

Note. d = diameter of pin around which specimen is bent
t = diameter of specimen

* Based on ASTM A 15.

(ASTM A 7), steel plate for pressure vessels (ASTM A 285), etc. The bend test is also often used for testing the ductility of welds.

The specified angle of bend and size of pin around which the piece is to be bent without cracking depends upon the grade of metal and the type of service for which it is to be used. In the case of concrete-reinforcement bars, which must be bent cold on the job, the requirements are shown in Table 6.3. The requirements for structural steel are shown in Table 6.4, the specimen being bent through 180° in each case. It is required that a specimen of rivet rod stock be bent flat on itself.

A "hot-bend" test is sometimes made, for example, on wrought iron by heating it to a welding temperature (about 1800°F) and bending the heated piece on an anvil; the test serves to detect too high a sulfur content. The "quench-bend" test is sometimes used in connection with rivet steels for boilers and is made by heating, quenching, and then bending; the test in this case is used to detect too high a carbon content.

Table 6.4.* Bend-test requirements for structural steel

Thickness of material, in.	*Ratio of pin diameter to thickness of specimen*
3/4 and under	1/2
Over 3/4 to 1	1
Over 1 to 1 1/2	1 1/2
Over 1 1/2 to 2	2 1/2
Over 2	3

* Based on ASTM A 7.

A "nick-bend" test is made when it is desired to make a rapid examination of a metal for coarse crystalline structure or for the occurrence of internal defects. Sometimes the specimen is nicked with a cold chisel, clamped in a vise, and bent with a hammer. In more carefully made tests, the nick or groove may be made by a hack saw or in a milling machine, and after a slight bend is started with a hammer, it is completed by axial loading in a testing machine.

Similar to the nick-bend tests are tests in which a hole is made in the specimen by punching or drilling. The effect of such operations upon ductility is then qualitatively determined by bending the metal at the restricted section.

Although rough qualitative tests on the job are often made by use of a hammer and vise or anvil, in the laboratory some special apparatus or procedure is usually employed. The essential features of two types of cold-bend apparatus are shown schematically in Fig. 6.19. In the Olsen machine the angle of bend may be measured. In the ASTM test for

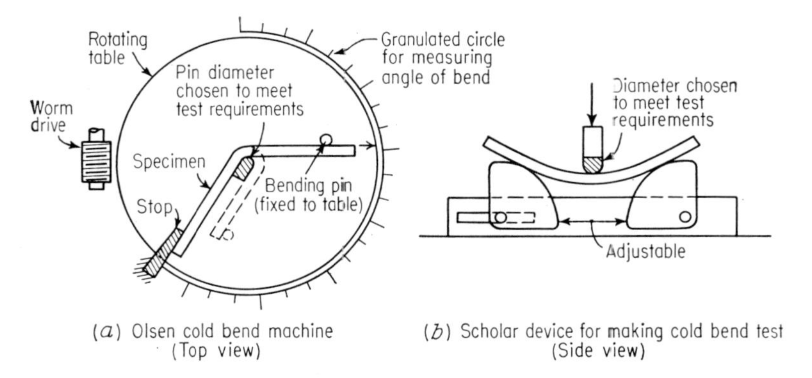

Fig. 6.19. **Cold-bend test apparatus (diagrammatic).**

ductility of welds, the bend is started by hammering or by transverse loading and is completed by loading the ends of the specimen as a strut in a vise or press or testing machine until failure occurs at the outside fibers (ASTM E 16).

Fiber-strain measurements are sometimes made in connection with cold-bend tests. What is to be the convex surface of the bend specimen is marked by scribe marks at intervals over a distance of several inches. The elongation of the outer fiber is then determined through the use of a flexible tape. For welds, a gage length of about ⅛ in. less than the width of the face of the weld is used (ASTM E 16).

From a consideration of the bending action it may be seen that the elongation of the outer fiber varies directly as the thickness of the specimen and inversely as the radius of curvature. It is for this reason that metals of various thicknesses are bent around pins of different diameters.

6.17. Stiffness in Flexure. In bend tests of some materials, such as wire [686] (ASTM F 113) and plastics (ASTM D 747), the ASTM specifies that the bending moment as well as the angle of bend be observed. As the observed angle has both elastic and plastic components, a true elastic modulus cannot be calculated from the test data. However, an apparent value is obtained and is defined for purposes of the test as the stiffness of the material in flexure. It is determined from the equation

$$E = \frac{ML}{3I\phi}$$

where E = stiffness in flexure, psi
M = bending moment, in.-lb
L = span length, in.
I = moment of inertia of specimen, in.^{4}
ϕ = angular deflection, radians

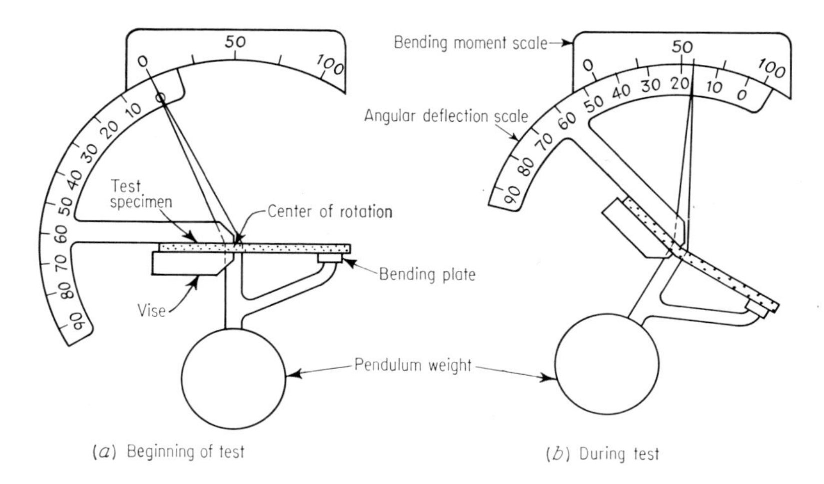

Fig. 6.20. **Mechanical system of stiffness tester (ASTM D 747).** *(From LaTour and Sutton, Ref.* 687.)

Figure 6.20 shows schematically a good arrangement for obtaining simultaneous values of moment and angle of bend of the specimen. During the test the vise holding the specimen is rotated by a motor and the bending moment is measured by the pendulum system.

6.18. Miscellaneous Bending Tests. For particular uses a number of special bending tests have been developed, such as those for investigating the ductility of thin sheet metals [685], flexing tests for ply separation and cracking of rubber products (ASTM D 430), and the "flexivity" of thermostat metals (ASTM B 106).

Chapter 7.

HARDNESS TESTS

7.1. Hardness. The general concept of hardness as a quality of matter having to do with solidity and firmness of outline is easily comprehended, but no single measure of hardness, universally applicable to all materials, has yet been devised. The fundamental "physics" of hardness is not yet clearly understood.

A number of different arbitrary "definitions" of hardness form the basis for the various hardness tests now in use. Some of these definitions are

1. Resistance to permanent indentation under static or dynamic loads—indentation hardness
2. Energy absorption under impact loads—rebound hardness
3. Resistance to scratching—scratch hardness
4. Resistance to abrasion—wear hardness
5. Resistance to cutting or drilling—machinability

Such definitions generally develop with the necessity for some way of expressing quantitatively performance requirements under differing conditions of service. In spite of their apparent divergence in meaning, the method of test implied by each definition has a certain useful field of application.

Although all the hardness measures are, no doubt, functions of interatomic forces, the various hardness tests do not bring these fundamental forces into play in the same way or to the same extent; thus no method of measuring hardness uniquely indicates any other single mechanical property. Although some hardness tests seem to be more closely associated than others with tensile strength, some appear to be more closely related to resilience, or to ductility, etc. In view of this situation, it is obvious that a given type of test is of practical use only for comparing the relative hardnesses of similar materials on a stated basis. The results of ball-indentation tests on steel, for example, have no meaning when compared with results of such tests when performed on rubber but serve nicely to evaluate the effectiveness of a series of heat treatments on a given steel or even to classify steels of various compositions.

Detailed discussions of the property of hardness and its measurement may be found in Refs. 701 and 702.

7.2. Scope and Applicability of Hardness Tests. Hardness tests have a wide field of use, although as commercial tests they are perhaps more commonly applied to metals than to any other class of material. The results of a hardness test may be utilized as follows:

1. Similar materials may be graded according to hardness, and a particular grade, as indicated by a hardness test, may be specified for some one type of service. The degree of hardness chosen depends, however, upon previous experience with materials under the given service and not upon any intrinsic significance of the hardness numbers. It should be observed that a hardness number cannot be utilized directly in design or analysis as can tensile strength, for example.
2. The quality level of materials or products may be checked or controlled by hardness tests. They may be applied to determine the uniformity of samples of a metal or the uniformity of results of some treatment such as forming, alloying, heat treatment, or casehardening.
3. By establishing a correlation between hardness and some other desired property, e.g., tensile strength, simple hardness tests may serve to control the uniformity of the tensile strength and to indicate rapidly whether more complete tests are warranted. It should be noted, however, that correlations apply only over a range of materials on which tests have previously been made; extrapolation from empirical relations should rarely be made and then only with great caution.

Most of the operations called hardness tests can be classified as shown in Table 7.1.

Table 7.1. **Classification of hardness tests**

Active element or tool	Line of action of load applicator	Fixed load; variable indentation or attrition		Fixed indentation or attrition; variable load	
		Static	Dynamic	Static	Dynamic
Two specimens, one pressed against the other	Normal to specimen	Réaumur (1722)			
Tool of material harder than specimen	Normal to surface of specimen	Brinell (1900) Rockwell (1920) Vickers (1925) Knoop (Tukon) (1939)	Shore scleroscope (1906) Ballentine Cloudburst Schmidt Various abrasion tests—e.g., sandblast type	Monotron Wood-hardness tool	
	Parallel to surface of specimen	Marten sclerometer (1889) Bierbaum sclerometer Herbert pendulum (1923) Machinability (cutting, drilling) tests Various wear or abrasion tests		Allcut and Turner sclerometer (1887) Various qualitative scratch-hardness tests—Mohs (1822)	

The fundamental idea that hardness is measured by the resistance to indentation is the basis for a variety of instruments. The indenter, either a ball or a plain or truncated cone or pyramid, is usually made of hard steel or diamond and ordinarily is used under a static load. Either the load that would produce a given depth of indentation or the indentation produced under a given load could be measured. The variable would be a function of the hardness. A similar choice would exist in the case of dynamic or impact loads. With most of the dynamic machines, however, the force, whether developed by a drop or a spring load, is of fixed magnitude, as with the scleroscope, for which the height of rebound of the indenter is taken as a measure of the hardness.

Indentation and rebound-type tests, because of their simplicity, have become one of the important quality-control tests for metals. Because they are relatively inexpensive, require relatively little experience for their conduct, and are nondestructive, some hardness tests may be employed for a "100 percent inspection" of finished parts.

Probably the most commonly used hardness tests for metals in this country are the Brinell and Rockwell tests. However, the increasing use of very hard steels and hardened-steel surfaces has brought into use a number of other tests, such as those made with the Shore scleroscope, Vickers, Monotron, Rockwell-superficial and Herbert machines. Also, the need for determining the hardness of very thin materials, very small-size parts, and the hardness gradients over very small distances has led to the development of the so-called "microhardness" tests, such as that using the Knoop indenter.

A static ball-indentation test has been standardized for use with wood, although the test is not in common use. A dynamic indentation test using the Schmidt hammer (see Art. 10.31) is made to determine the hardness (and probable compressive strength) of concrete in place. The Rockwell L-scale test is applied to hard rubber (ASTM D 530), and a special ball-indentation test is used for soft rubber (ASTM D 314). The Durometer (Art. 7.14) is used for rubber and plastics.

Abrasion or wear tests have found their principal use in connection with paving materials, and a number of such tests have been standardized. Wear or abrasion tests have also been proposed and used experimentally for tests of metals and concrete-floor surfaces. In the case of concrete aggregates and brick, the sample (composed of a number of pieces) is tumbled in a drum—the "rattler" test (ASTM C 131, C 7). Sandblast tests as well as rubbing or tumbling tests are in the class of abrasion tests.

For determining the machinability of metals, various special tests have been proposed. The hardness reported as the depth of hole made by a special drill in a given time while running at a constant speed and pressure is sometimes called the "Bauer drill test" after the originator. It has been described as testing cutting hardness or machinability. There are other tests described as grooving and cutting tests, but all have a limited field of application.

For a qualitative classification of materials over a wide range, perhaps the most applicable type of test is the scratch test in which an arbitrary scale is set up in terms of several common materials, each of which will just scratch the material of next lower hardness number. In familiar form, this is the mineralogist's or Mohs' scale. To take into account the recently developed abrasive materials of extreme hardness, a modification of this scale has been proposed (see Art. 7.19).

STATIC INDENTATION HARDNESS TESTS

7.3. The Brinell Test. The Brinell test consists in pressing a hardened steel ball into a test specimen. In accordance with the ASTM Specifications (ASTM E 10), the provisions of which are followed herein, it is customary to use a 10-mm ball and a load of 3000 kg for hard metals, 1500 kg for metals of intermediate hardness, and 500 kg (or even as low as 100 kg) for soft materials, as shown in Table 7.2.

Table 7.2. **Hardness range for standard Brinell loads***

Ball diameter, mm	Load, kg	Recommended range of Brinell hardness
10	3000	96 to 600
10	1500	48 to 300
10	500	16 to 100

* From ASTM E 10.

Various types of machines for making the Brinell test are available. They may differ as to (1) method of applying load: oil pressure, gear-driven screw, weights with lever; (2) method of operation: hand, motive power; (3) method of measuring load: piston with weights, bourdon gage, dynamometer, weights with lever; and (4) size: large (usual laboratory size), small (portable). The Brinell test may be made in a small universal testing machine by the use of a suitable adapter for holding the ball, as well as by the use of special machines designed for the purpose. For tests of thin sheet-metal products such as cartridges and cartridge cases, a small hand-plier device using a $3/64$-in. ball and a 22-lb spring pressure has been employed.

The principal features of a typical hydraulically operated Brinell testing machine are illustrated in Figs. 7.1 and 7.2. The specimen is placed on the anvil and raised to contact with the ball. Load is applied by pumping oil into the main cylinder, which forces the main piston or plunger downward and presses the ball into the specimen. The plunger has a ground fit so that frictional effects are usually negligible. The bourdon gage is used only to give a rough indication of the load. When the desired

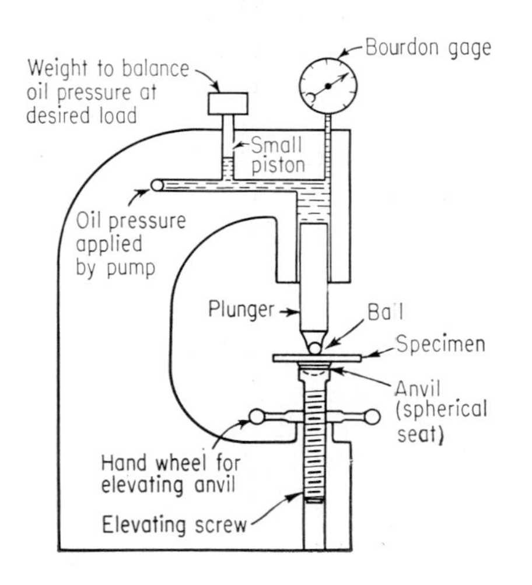

Fig. 7.1. **Features of hydraulic-type Brinell machine.**

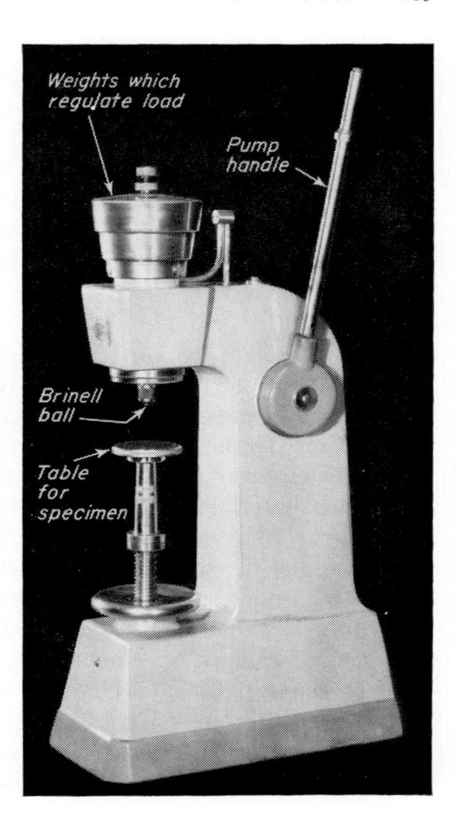

Fig. 7.2. **Hydraulic Brinell machine.** *(Courtesy of Riehle Testing Machine Division.)*

load is applied, the balance weight on top of the machine is lifted by action of the small piston; this ensures that an overload is not applied to the ball.

It is required that the ball be within 0.01 mm of the nominal 10-mm diameter. This requirement is necessary in order to obtain concordant results with different machines. When used on very hard steels, it is required that the ball should not show a permanent change in diameter of more than 0.01 mm. For this reason, carboloy (tungsten carbide) balls are often used for testing the harder steels.

In the standard test, the diameter of the indentation is measured by use of a micrometer microscope, having a transparent engraved scale in the field of view. The scale has divisions corresponding to 0.1 mm, and the measurements are made, by estimation, to at least 0.02 mm. The diameter is taken as the average of two readings taken at right angles to each other. Sometimes the depth of indentation is measured by means of a dial indicator fastened to the plunger and actuated by a gimbal ring held in contact with the surface of the specimen.

Although the Brinell test is a simple one to make, several precautions are necessary in order to obtain good results. It is not adapted to testing extremely hard materials, because the ball itself deforms too much, nor is it satisfactory for testing thin pieces such as razor blades, because the usual indentation may be greater than the thickness of the piece. It is not adapted to testing casehardened surfaces, because the depth of indentation may be greater than the thickness of the case and because the yielding of the soft core invalidates the results; also, for such surfaces, the indentation is almost invariably surrounded by a crack that may cause fatigue failure if the part is used in service. Obviously the Brinell test should not be used for parts the marring of the surface of which impairs their value.

7.4. Brinell Procedure. To make a test, the surface of the specimen should be flat and reasonably well polished; otherwise difficulty will be experienced in making an accurate determination of the diameter of the indentation. If the specimen is prepared from rough stock, the surface should be dressed with a file and then polished with fine emery cloth. For some materials, the edge of the indentation is very poorly defined, even when the surface finish is good. To increase the sharpness of definition of the edge of the indentation, the ASTM suggests the use of a movable lamp for illumination, placing it so that the contrast of light and shade will bring first one edge of the indentation, then the other, into sharp definition. For some specimens, the indentation may be made more distinct by using balls lightly etched with nitric acid or by the use of some pigment, such as prussian blue, on the ball.

In the standard test, the full load is applied for a minimum of 15 sec for ferrous metals and 30 sec for softer metals, after which interval the load is released and the diameter of the indentation is measured to the nearest 0.02 mm with the microscope. Often, however, a 30-sec interval is used for ferrous metals and a 60-sec interval for other metals. In rapidly made control tests, a time interval of less than standard is sometimes used.

The material of the specimen is permanently deformed for an appreciable distance below the surface of the indentation. If this deformation extends to the lower or opposite surface, the size of the indentation may be greater for some materials and less for others than for a thicker specimen of the same material. From tests at the National Bureau of Standards on a large variety of materials, it was noted that under each indentation made, where the thickness was less than the "critical" value, a spot of altered surface was visible on the underside of the specimen [721]. The ASTM specifies that no marking shall appear on the side of the piece opposite the indentation and also requires the thickness of the specimen to be at least 10 times the depth of indentation. To satisfy this requirement the minimum hardness for a given thickness of a specimen should be as shown in Table 7.3.

If an indentation is made too near the edge of the specimen, it may be both too large and unsymmetrical. If made too close to a previous one, it may be too large owing to lack of sufficient supporting material or too

Table 7.3. **Thickness of Brinell specimen***

Thickness of specimen, in.	Minimum Brinell hardness for which a Brinell test may safely be made		
	500-kg load	1500-kg load	3000-kg load
1/16	100	301	602
1/8	50	150	301
3/16	33	100	201
1/4	25	75	150
5/16	20	60	120
3/8	17	50	100

* From ASTM E 10.

small owing to work-hardening of the material by the first indentation. However, tests have shown that the errors may be neglected if the distance of the center of the indentation from the edge of the specimen or from the center of adjacent indentations is equal to or greater than 2½ times the diameter of the indentation.

If the compressive properties of a flat specimen are not uniform, owing perhaps to direction of rolling or to cooling stresses, a noncircular indentation will result. In this case, the average Brinell hardness of the material may be obtained if the diameter is taken as the average in four directions, roughly 45° apart.

It is frequently necessary in practice to measure the Brinell number on a curved surface rather than a plane one, but an indentation on a curved surface of a specimen having uniform properties will not have a circular boundary unless the curvature is constant in all directions as in the case of a sphere. Provided the radius of the specimen is not less than 1 in., the diameter of the impression may be taken as the average of the maximum and minimum diameters. For smaller radii a flat spot may be prepared on the surface of the specimen.

The Brinell hardness number is nominally the pressure per unit area, in kilograms per square millimeter, of the indentation that remains after the load is removed. It is obtained by dividing the applied load by the area of the surface of the indentation, which is assumed to be spherical. If P is the applied load (in kilograms), D is the diameter of the steel ball (in millimeters), and d is the diameter of the indentation (in millimeters), then

$$\text{Brinell hardness number (Bhn)} = \frac{\text{load on ball}}{\text{indented area}}$$

$$= \frac{P}{\dfrac{\pi D}{2}\left(D - \sqrt{D^2 - d^2}\right)}$$

In practice the Brinell numbers corresponding to a given observed diameter of indentation are taken from tables, such as given in ASTM Specification E 10 (see Table 7.7).

The hardness numbers obtained for ordinary steels (with 3000-kg load) range from about 100 to 500; the medium-carbon structural steels have hardness numbers of the order of 130 to 160; for very hard special steels the hardness numbers may be as high as 800 or 900, but the Brinell test itself is not recommended for materials having a Bhn over 630.

For a standardized test such as this, the projected circular area computed from the diameter of indentation might just as well have been used. In fact, it seems just as logical as the assumed spherical area of the indentation as first used by Brinell. However, a change appears undesirable now because of wide familiarity with and use of the standard Brinell numbers.

The surface of the indentation is not truly spherical because the ball undergoes some deformation under load and because there is some recovery of the test piece when the load is removed. Thus the indentations made by different sized balls and different loads are not geometrically similar. However, for testing small or thin specimens, it is sometimes necessary to make Brinell hardness tests with a ball less than 10 mm in diameter. Such tests (which are not to be regarded as standard Brinell tests) will approximate the standard tests more closely if the relation between applied load P, in kilograms, and diameter of ball D, in millimeters, is the same as in the standard tests

where $P/D^2 = 30$ for 3000-kg load and 10-mm ball
$= 15$ for 1500-kg load and 10-mm ball
$= 5$ for 500-kg load and 10-mm ball

As the indentations made by different sized balls and different loads are not geometrically similar, it is essential that the ball size and the load be reported with the hardness number whenever the 3000-kg load and the 10-mm ball have not been used.

When the depth of indentation is to be measured, the observation is made just after the load is released. A hardness number is computed from the depth of indentation by use of the following equation:

$$\text{Brinell hardness number} = \frac{\text{applied load}}{\text{indented area}} = \frac{P}{\pi D t}$$

in which $t =$ depth of indentation, mm
$D =$ diameter of ball, mm

However, the observed depth of indentation t_1 (usually determined from the relative motion of the ball plunger and the specimen) and the actual depth t corresponding to the diameter of indentation d do not agree, owing to the possible formation of an encircling ridge (Fig. 7.3*a*) or depression (Fig. 7.3*b*). Soft materials such as copper and mild steel show the former

Fig. 7.3. Cross sections of indentations in Brinell test.

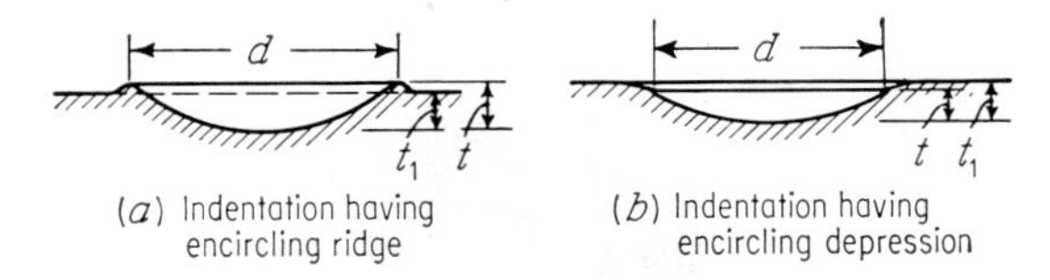

effect, and hard materials such as manganese steel and some bronzes develop a depression. Although the observed depth of indentation t_1 appears to be a logical basis of comparison, it is not the basis of the Brinell hardness number according to definition. However, it is possible for routine control work to establish by test the relation between depth and diameter of indentation for each kind of material in different stages of hardness. These relations may be plotted on a graph or recorded in tabular form so that rapid routine tests in mass production may be made to yield standard Brinell values.

7.5. Effect of Important Variables. A rapid rate of applying the load affects the diameter of the Brinell indentation in two ways. (1) The effect of inertia of the piston and weights and the friction of the plunger cause a momentary rise of the load above 3000 kg and consequently enlarge the indentation. (2) A rapid rate of penetration allows less time for the plastic flow of the material, resulting in a decreased size of the indentation.

Tests have shown that the error due to inertia of the plunger can be, and often is, very much greater than that due to an insufficient period of sustaining the load to permit plastic flow to occur. This inertia effect is of importance not only when the load is first applied but also when, during the maintenance of the required load, the downward drift of the weights is reversed into an upward motion by a stroke of the pump in order to maintain the weights in floating equilibrium. By exercising care in operating the pump, this inertia effect may be reduced to a negligible factor.

From tests at the National Bureau of Standards on 29 steels and nonferrous metals, it was found that the flow for most materials is quite rapid during the first 30 sec under required load; it is much less rapid in the interval from 30 to 120 sec [721]. For most metals the Brinell number varies less than 1 percent for loading intervals between 30 and 120 sec.

The error in the Brinell number is less than 1 percent as long as the error in diameter does not exceed 0.01 mm. Errors in reading the diameter of the indentation may be ascribed to two causes: (1) to an error in reading the microscope and (2) to indefiniteness of the boundary of the indentation. The error in reading a modern Brinell microscope should not exceed 0.02 mm, provided it is in proper adjustment as determined by comparison with the calibrated scale furnished with the microscope. The indefiniteness of the boundary of the indentation may cause considerable

***Fig. 7.4.* Rockwell hardness tester.** *(Courtesy of Wilson Mechanical Instrument Co.)*

uncertainty in the magnitude of the diameter; some of the precautions that may be taken to obtain a clear indentation have been outlined in Art. 7.4.

The variations from standard size and shape of modern well-made balls are usually too small to introduce appreciable errors. However, the flattening of the ball, particularly when the hardness of the specimen approaches that of the ball, may lead to serious errors. For testing materials with Brinell hardness numbers greater than about 400, the ball should be frequently checked for distortion, and for testing materials having hardness numbers over 450, balls of harder material than steel should be used. Carbide balls may be used for values to 630, which is the maximum hardness for which the Brinell test should be used. In quoting Brinell numbers greater than 200, the material of the ball should be stated, since the results depend to some extent upon the material of which the ball is made.

7.6. Calibration of Brinell Apparatus. The load-measuring device may be calibrated (1) by the use of weights and proving levers, (2) by an elastic calibration device, or (3) by making a series of indentations on specimens of different degrees of hardness and comparing with a second series of indentations made by the use of any standardized testing machine and standard test ball. According to the ASTM Standards, a Brinell machine is acceptable for use over a loading range within which its load-measuring device is correct within 1 percent.

The size and uniformity of the ball are checked by measurements with a micrometer caliper of suitable accuracy.

The Brinell microscope is checked by comparing its readings with a standardized scale. The error of reading throughout the range should not exceed 0.02 mm.

7.7. The Rockwell Test. The Rockwell test is similar to the Brinell test in that the hardness number found is a function of the degree of indentation of the test piece by action of an indenter under a given static load [723]. Various loads and indenters are used, depending on the conditions of test. It differs from the Brinell test in that the indenters and the loads are smaller, and hence the resulting indentation is smaller and shallower. It is applicable to the testing of materials having hardnesses beyond the scope of the Brinell test, and it is faster because it gives arbitrary direct readings. It is widely used in industrial work. The procedure outlined herein follows that now standardized by the ASTM (ASTM E 18).

The test is conducted in a specially designed machine that applies load through a system of weights and levers. A photograph of one model is shown in Fig. 7.4. The indenter or "penetrator" may be either a steel ball or a diamond cone with a somewhat rounded point. The hardness value, as read from a specially graduated dial indicator, is an arbitrary number that is related inversely to the depth of indentation.

In the operation of the machine, which is explained diagrammatically in Fig. 7.5, a minor load of 10 kg is first applied, which causes an initial indentation that sets the indenter on the material and holds it in position. The dial is set at the "set" mark on the scale, and the major load is applied. This major load is customarily 60 or 100 kg when a steel ball is used as an indenter, though other loads may be used when found necessary, and it is usually 150 kg when the diamond cone is employed. The ball indenter is normally $\frac{1}{16}$ in. in diameter, but others of larger diameter such as $\frac{1}{8}$, $\frac{1}{4}$, or $\frac{1}{2}$ in. may be employed for soft materials. After the major load is applied and removed, the hardness reading is taken from the dial while the minor load is still in position.

There is no Rockwell hardness value designated by a number alone because it is necessary to indicate which indenter and load have been employed in making the test. Therefore, a prefix letter, as shown in the first column of Table 7.4, is employed to designate the test conditions.

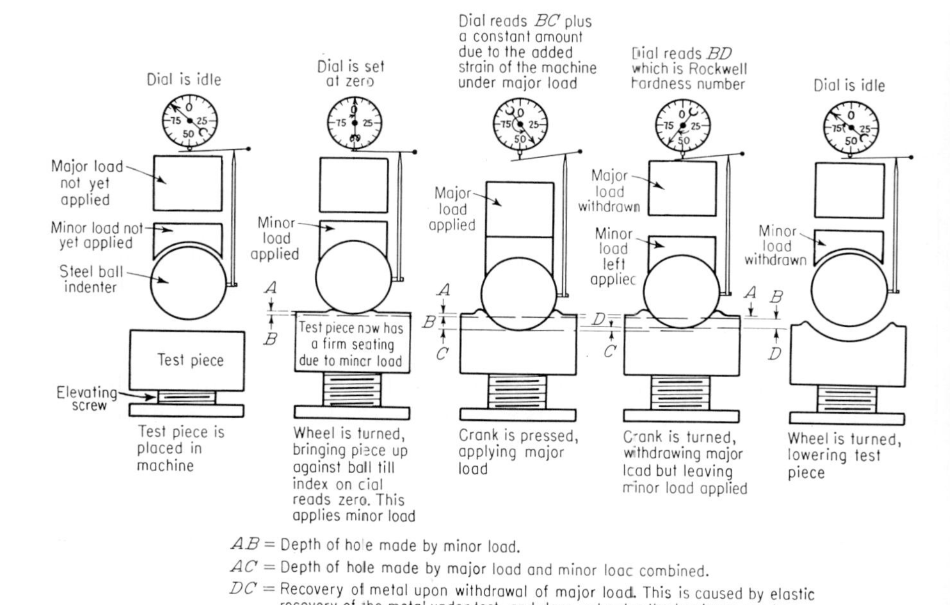

Fig. 7.5. **Procedure in using Rockwell hardness tester.**

The dial of the machine has two sets of figures, one red and the other black, which differ by 30 hardness numbers. The dial was designed in this way to accommodate the B and C scales, which were the first ones standardized and are the most widely used. Two advantages were gained by this shifting of the zero points: (1) negative numbers were avoided on soft material such as brass, when tested on the B scale; and (2) this established B 100 as the upper practical limit of hardness that might be tested with the 100-kg major load and the $\frac{1}{16}$-in. ball without deforming the ball.

In the designation of scales (Table 7.4), it will be noted that red figures are used for readings obtained with ball indenters regardless of size of ball or magnitude of major load and that black figures are used only for the diamond cone.

The B scale is for testing materials of medium hardness such as low- and medium-carbon steels in the annealed condition. The working range of this scale is from 0 to 100. If the ball indenter is used to test material harder than about B 100, there is danger that it will be flattened. Furthermore, because of its shape, the ball is not so sensitive as the rounded conical indenter to differences in hardness of hard specimens. If the $\frac{1}{16}$-in. ball is used on material softer than B 0, there is danger that the

Table* 7.4. Rockwell hardness scales and prefix letters

Scale symbol and prefix letter	Indenter	Major load, kg	Dial numerals	Typical applications of scales
	Group one†			
B†	1/16-in. ball	100	Red	Copper alloys, soft steels, aluminum alloys, malleable iron
C†	Diamond cone	150	Black	Steel, hard cast iron, pearlitic malleable iron, deep case-hardened steel
	Group two			
A	Diamond cone	60	Black	Cemented carbides, thin steel, shallow case-hardened steel
D	Diamond cone	100	Black	Thin steel, medium case-hardened steel
E	1/8-in. ball	100	Red	Cast iron, aluminum and magnesium alloys, bearing metals
F	1/16-in. ball	60	Red	Annealed copper alloys, thin soft sheet metals
G	1/16-in. ball	150	Red	Phosphor bronze, beryllium copper, malleable iron
H	1/8-in. ball	60	Red	Aluminum, lead, zinc
K	1/8-in. ball	150	Red	Bearing metals and other very soft or thin materials. Use smallest ball and heaviest load that does not give anvil effort.
	Group three			
L	1/4-in. ball	60	Red	
M	1/4-in. ball	100	Red	
P	1/4-in. ball	150	Red	
R	1/2-in. ball	60	Red	
S	1/2-in. ball	100	Red	
V	1/2-in. ball	150	Red	

* Based on ASTM E 18.
† Commonly used scales and indenters.

cap of the indenter that holds the ball in place will make contact with the specimen or that the weight arm will descend too far and rest on its stop pin. Below B 0, the $\frac{1}{16}$-in. ball, owing to its shape, becomes supersensitive and the readings are erratic.

The C scale is the one most commonly used for materials harder than B 100. The hardest steels run about C 70. The useful range of this scale is from C 20 upward. Any inaccuracies that occur in grinding the diamond cone to its proper shape have a proportionately greater effect on small indentations, and it should therefore not be used below this lower value. Also, on soft materials, the spherical apex (0.2-mm radius) of the cone is driven into the material a considerable distance by the minor load, and unless the speed with which the indenter makes contact against the work and the time interval between applying minor and major loads are standardized, there will be considerable variation in the readings from these causes. These effects are negligible when using the cone on harder material.

In general, a scale should be selected to employ the smallest ball that can properly be used, because of the loss of sensitivity as the size of indenter increases. An exception to this is when soft nonhomogeneous material is to be tested, in which case it may be preferable to use a larger ball that makes an indentation of greater area, thus obtaining more of an average hardness.

The Rockwell scales are divided into 100 divisions, and each division or point of hardness is the equivalent of 0.002 mm in indentation; thus the difference in indentation between dial readings of B 53 and B 56 is 3×0.002, or 0.006 mm. Since the scales are reversed, the number is higher the harder the material, as shown by the following expressions which define the Rockwell B and C numbers:

$$\text{Rockwell B number} = 130 - \frac{\text{depth of penetration (mm)}}{0.002}$$

$$\text{Rockwell C number} = 100 - \frac{\text{depth of penetration (mm)}}{0.002}$$

7.8. Rockwell Procedure. Because of the smallness of the indentation and because of the way it is measured, there are some differences in selecting and preparing test pieces for the Rockwell test as compared with the Brinell test. Certain precautions are necessary, of which the following are the more important.

The test surface should be flat and free from scale, oxide films, pits, and foreign material that may affect the results. A pitted surface may give erratic readings, owing to some indentations being near the edge of a depression. This permits a free flow of metal around the indenting tool and results in a low reading. Oiled surfaces generally give slightly lower readings than dry ones because of the reduced friction under the indenter.

The bottom surface should be free from scale, dirt, or other foreign materials that might crush or flow under the test pressure and so affect the results.

The thickness of the piece tested should be such that no bulge or other marking appears on the surface of the piece opposite the indentation, since in such cases the depth of the indentation is noticeably affected by the supporting anvil. For very hard materials, the thickness may be as little as about 0.01 in. [731]. Selector charts for the proper scale to be used with thin sheets of various ranges of hardness are shown in ASTM E 18.

All hardness tests should be made on a single thickness of the material, regardless of the thickness of the piece. The use of more than one piece of thin material to give adequate thickness does not yield the same result as for a solid piece of the same thickness as the combined pieces. Relative movement takes place on the surfaces between the various pieces, and lack of flatness of the separate pieces will lead to compression of the pile under the major load.

The hardness number determined by indenting a curved surface is in error because of the shape of the surface. In industrial applications, this problem is often encountered, commonly in the case of shafts. If it is feasible, a small flat spot may be filed on the rod before making the indentation. However, corrections to be added to observed Rockwell values on cylindrical specimens having diameters from $\frac{1}{4}$ to $1\frac{1}{2}$ in. are given in ASTM E 18.

The steps in applying load and reading the hardness number have been described in Art. 7.7 and Fig. 7.5.

If the table on which the Rockwell hardness tester is mounted is subject to vibration, the hardness numbers will be too low, since the indenter will sink farther into the material than when such vibrations are absent. If, when the operating handle is being returned to its normal position, the latch operates with such a snap as noticeably to change the position of the dial pointer, felt or rubber washers should be placed under the trip mechanism in order to cushion this blow. If this snap is severe, a difference in reading of several hardness numbers may result.

Specimens should be prepared with care. If curved plates are tested, the concave side should face the indenter. If such specimens are reversed, an error will be introduced owing to the flattening of the piece on the anvil. Specimens that have sufficient overhang so that they do not balance themselves on the anvil should be properly supported. To prevent injury to the anvil and indenter, they should not be brought into contact without a test specimen between them.

The speed and time of application of the major load should be established and accurately adhered to and reported when comparing results. The dashpot should be adjusted so that the operating handle completes its travel in 4 to 5 sec with no specimen in the machine and with the

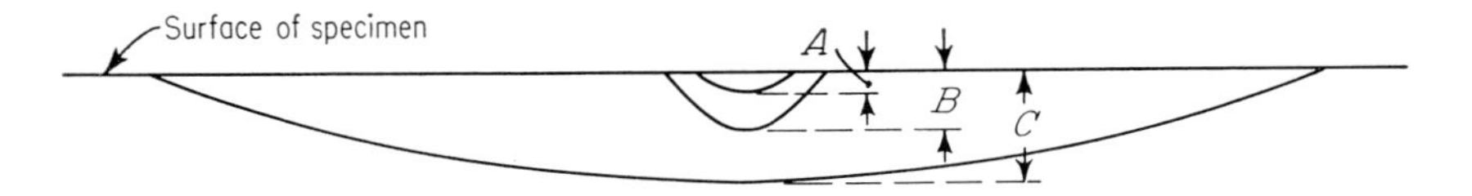

Fig. 7.6. **Comparative impressions in steel (Rockwell C 39) using Brinell, common Rockwell, and Rockwell superficial testers. (*A*) Superficial Rockwell, N diamond cone, 30-kg load, 0.0018 in.; (*B*) common Rockwell, C diamond cone, 150-kg load, 0.0052 in.; (*C*) Brinnell, 10-mm ball, 3,000-kg load, 0.010 in. All impressions magnified about 30 times.** (*From Lysaght, Ref.* 703.)

machine set up to apply a major load of 100 kg. An interval of full application of the major load of not more than 2 sec is specified. For soft metals, plastic flow may cause variations as high as 10 hardness numbers. With such materials, the operating lever should be brought back the moment it is seen that the major load is fully applied.

It is advisable to check the ball indenters regularly to see that they do not become flattened and the diamond cone to see that it has not become blunted or chipped.

7.9. Calibration of Rockwell Hardness Tester. The accuracy of Rockwell hardness testers is checked by the use of special test blocks that are available for all ranges of hardness. If the error of the tester is more than ± 2 hardness numbers, it should be reconditioned and brought into proper adjustment.

7.10. Rockwell Superficial-hardness Tester. This tester is a special-purpose machine, intended exclusively for hardness tests where only very shallow indentation is possible and where it is desired to know the hardness of the specimen close to the surface. It was designed particularly for testing nitrided steel, safety-razor blades, lightly carburized work, and brass, bronze, and steel sheet. (See also Art. 7.13 for "microhardness" tests.) The relative sizes of the indentations made by the Brinnell, ordinary Rockwell, and Rockwell superficial hardness testers are illustrated in Fig. 7.6.

The "superficial" tester operates on the same principle as the regular Rockwell tester but employs lighter minor and major loads and has a more sensitive depth-measuring system. Instead of the 10-kg minor load and the 60-, 100-, or 150-kg major loads of the regular Rockwell, the superficial tester applies a minor load of 3 kg and major loads of 15, 30, or 45 kg. One point of hardness on the superficial machine corresponds to a difference in depth of indentation of 0.001 mm.

Since the diamond cone in these superficial machines is intended especially for use on "nitrided" work and the $\frac{1}{16}$-in. steel ball for testing "thin" sheet, the letters N and T have been selected for these two scale

Table 7.5. **Rockwell superficial-hardness scales***

Major load, kg	Scale symbols				
	N scale, diamond cone	T scale, 1/16-in. ball	W scale, 1/8-in. ball	X scale, 1/4-in. ball	Y scale, 1/2-in. ball
15	15 N	15 T	15 W	15 X	15 Y
30	30 N	30 T	30 W	30 X	30 Y
45	45 N	45 T	45 W	45 X	45 Y

* Based on ASTM E 18.

designations. The W, X, and Y scales are used for very soft materials. Although these machines have but one set of dial graduations, scale symbols must be used as given in Table 7.5 to indicate the indenter and major load used.

7.11. Vickers Hardness Tester. This machine is somewhat similar to the Brinell in that an indentation is made and the hardness number is determined from the ratio P/A of the load P in kilograms to the surface area A of the indentation in square millimeters. The indenter is a square-based diamond pyramid in which the angle between the opposite faces is 136° (ASTM E 92). The load may be varied from 5 to 120 kg in increments of 5 kg.

In conducting a test, the specimen is placed on the anvil and raised by a screw until it is close to the point of the indenter. By tripping the starting lever, a 20:1 ratio loading beam is unlocked and the load slowly applied to the indenter and then released. Operation of a foot lever resets the machine. After the anvil is lowered, a microscope is swung over the specimen and the diagonal of the square indentation measured to 0.001 mm. The machine is also arranged to make tests with 1- and 2-mm ball indenters.

One advantage of the Vickers machine claimed by some operators is in the measurement of the indentation: a much more accurate reading can be made of the diagonal of a square than can be made of the diameter of a circle where the measurement must be made between two tangents to the circle. It is a fairly rapid method and can be used on metal as thin as 0.006 in. It is said to be accurate for hardnesses as high as 1300 (about 850 Brinell) and to indicate the friability of nitrided-steel cases. The hardness so determined seems to be a good criterion of the wearing qualities of nitrided steel.

7.12. Monotron Hardness Tester. This device also operates on the indentation principle; however, it is essentially a constant-depth indicator since the hardness is arbitrarily taken as the pressure in kilograms per square millimeter necessary to give a fixed indentation of 0.0018 in. This corresponds to a depth of indentation of 6 percent of the diameter of the ¾-mm spherical-tipped diamond indenter and yields an indentation 0.36 mm in diameter. The load is applied by a hand lever that makes it difficult to control precisely. The load and depth are read from separate dials. Since the dial for reading the depth is rather insensitive, it is difficult to produce indentations of exactly the same size. The depth is measured under load and from the *original* surface of the specimen; it is the depth of the "unrecovered" indentation. The machine is well adapted to determining the hardness of thin materials or casehardened surfaces. It is usable over the entire range of hardness of metals and is quite rapid in its operation.

7.13. Microhardness Testers. Because of a real need for a device that would determine the hardness of a material over a very small area and

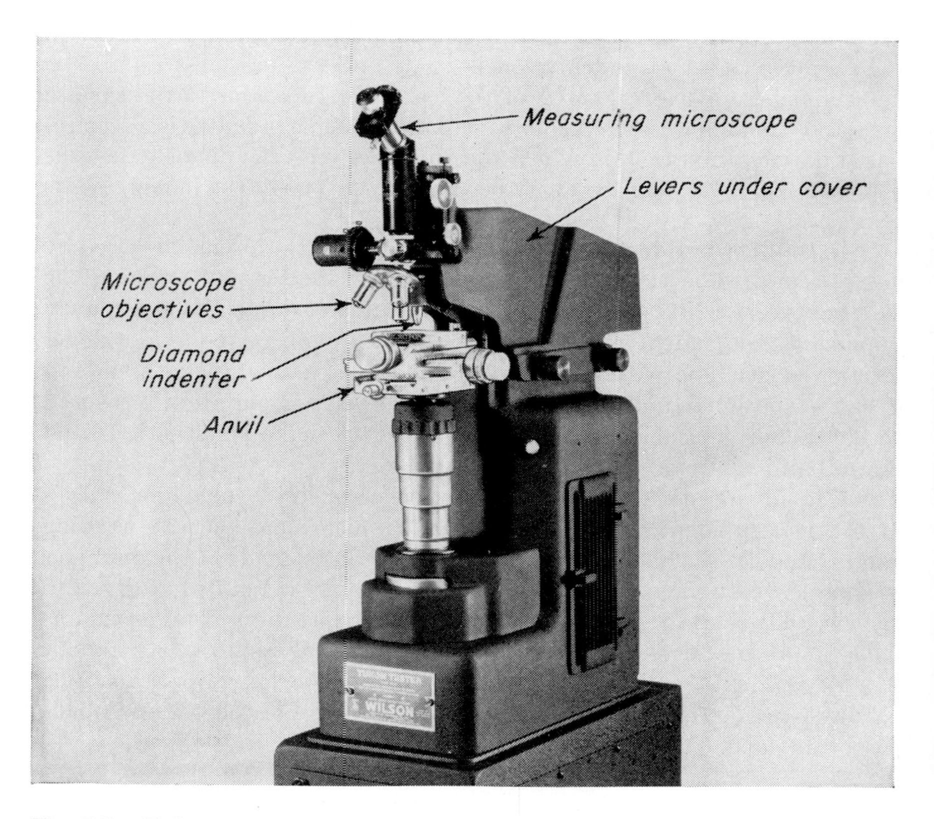

Fig. 7.7. **Tukon tester with Knoop Indenter.**

that would produce a small indentation, the National Bureau of Standards developed the Knoop indenter [740]. It is made of diamond and is ground so that it produces a diamond-shaped indentation, the ratio of the long to the short diagonal being about 7:1.

The Tukon tester (see Fig. 7.7) with which the Knoop indenter is used can apply loads of 25 to 3600 g. It is fully automatic in making the indentation. The operator selects the position for test under high microscopic magnification, places the selected area under the indenter, and finally relocates the specimen under the microscope for reading the length of the diagonal of the impression from which the Knoop hardness number is calculated. This number is the ratio of the applied load (in kilograms) to the unrecovered projected area (in square millimeters). Tables have been prepared to simplify obtaining the hardness number.

The Tukon-Knoop device, or a somewhat similar Wilson-Knoop device, is useful for hardness tests of small parts such as those in watches, thin materials, small wires, tips of cutting tools, single crystals or constituents of alloyed metals, and surface layers and for exploring variations in hardness of small areas such as over the thickness of thin sheets or adjacent to a critical surface.

Although the Tukon tester is normally supplied with the Knoop indenter, it can easily be adapted to the 136° Vickers diamond-pyramid indenter. The results of a survey of hardness across the thickness of a thin sheet, using the Tukon-Vickers combination, are shown in Fig. 7.8.

Another type of microhardness tester is the Eberbach, which uses a spring-loaded Vickers indenter and an electronic device to indicate when the full indenting load is applied to the specimen.

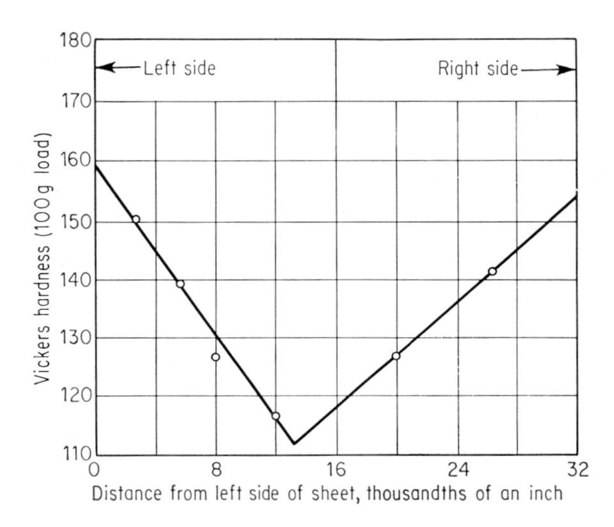

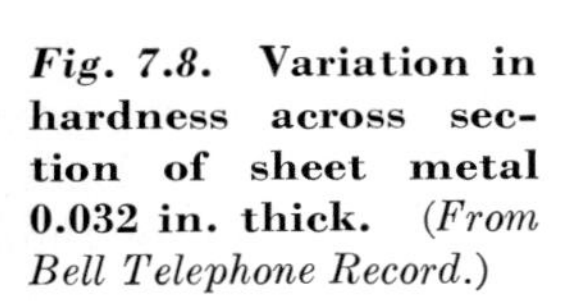
Fig. 7.8. **Variation in hardness across section of sheet metal 0.032 in. thick.** (*From Bell Telephone Record.*)

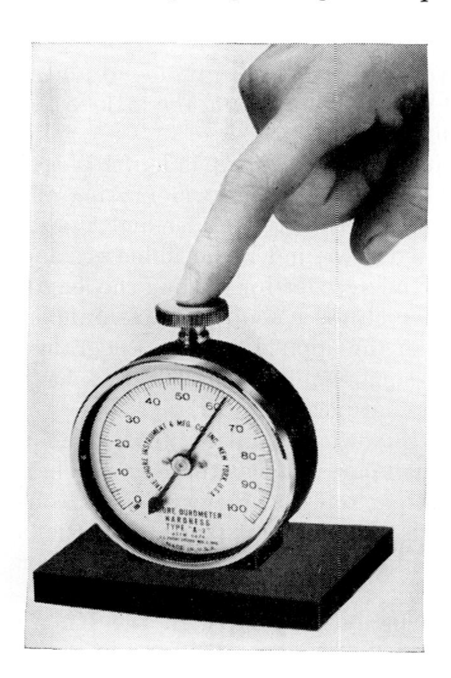

***Fig. 7.9.* Type A-2 Durometer for measuring hardness of rubber-like materials.** (*Courtesy of Shore Instrument and Manufacturing Co.*)

7.14. Durometer Hardness Tester. Durometers are of various types depending upon the material to be tested. The type A-2 shown in Fig. 7.9 is used for soft rubber and nonrigid plastics, while the type D is used for harder rubber and plastics (ASTM D 676, D 1484, and D 1706). All Durometers are quite similar, differing primarily in the sharpness of the point of the conical steel indenter and the magnitude of the load applied to the indenter by a calibrated spring, the type D having the sharpest and more heavily loaded indenter. The Durometer hardness is a measure of the depth of indentation; it varies from 100 at zero indentation to 0 at an indentation of 0.100 in. and is automatically indicated on a scale. The load acting on the indenter usually varies inversely with the depth of penetration, being a maximum at zero penetration and reducing to practically zero load at a penetration of 0.100 in., although some Durometers use a weight to apply a constant load. Test specimens should be at least ¼ in. thick and no determinations should be made within ½ in. of the edge. Results obtained on one type of Durometer cannot be correlated with those obtained on another type.

7.15. Hardness Test of Wood. The only standardized hardness test for wood is of the indentation type (ASTM D 143). The hardness is determined by measuring the load required to embed a 0.444-in. steel ball to one-half its diameter into the wood. It is of value for comparative purposes only. The approximate range in hardness of air-dry wood is from about 400 lb for poplar to 4000 lb for persimmon. The hardness of Douglas fir is about 900 lb.

DYNAMIC-HARDNESS TESTS

7.16. Dynamic Hardness. Most dynamic-hardness tests are virtually indentation tests, irrespective of the manner of loading. As in most dynamic testing, since the methods of calculating the energy absorbed by the specimen are questionable, a test procedure is fixed and the results are therefore arbitrary. In order to secure comparable results, specified equipment and procedure must be employed.

Perhaps the first dynamic-hardness tests were those of Rodman, who experimented with a pyramidal punch in 1861. Later investigations using a hammer with a spherical end verified Rodman's tests and showed that the work of the falling hammer is proportional to the volume of the indentation. The hardness is expressed as the work required to produce unit volume of indentation. This method would be useful in determining the hardness of metals at high temperatures since the hammer does not stay in contact with the specimen long enough to be affected by the heat.

There are a number of foreign machines that use a dynamic load. In some, the hardness numbers are calculated by dividing the net energy of the blow by the volume of indentation. Among the more important are the Pellin hardness tester, in which the indentation is produced by a falling rod of known weight having at the lower end a steel ball 2.5 mm in diameter; the Whitworth auto punch, which is a hand Brinell machine actuated by the release of a spring in the handle that supplies a standard striking energy to the ball indenter in the bottom of the punch (the diameter of indentation is measured as in the Brinell test); the Waldo hardness tester, which uses a conical-pointed steel indenter weighing 0.1 lb and dropped from a height of 12 in. (hardness is based on the diameter of indentation); the Duroskop tester, which depends on the rebound of a pendulum hammer; and Avery's modification of the Izod impact machine for dynamic-hardness tests.

At the present time, the Shore "scleroscope" is probably the most widely used device of the dynamic type. The hardness measured by this instrument is often referred to as "rebound hardness."

7.17. The Shore Scleroscope. Scleroscope hardness is expressed by a number given by the height of rebound of a small pointed hammer after falling within a glass tube from a height of 10 in. against the surface of the specimen. The standard hammer is approximately $\frac{1}{4}$ in. in diameter, $\frac{3}{4}$ in. long, and weighs $\frac{1}{12}$ oz, with a diamond striking tip rounded to a 0.01-in. radius.

The indications obtained by the use of this instrument depend upon the resilience of the hammer as well as that of the material tested, but the permanent deformation of the material is also an important factor. When the hammer falls onto a soft surface, it penetrates that surface to some extent before rebounding and produces a minute indentation. In so doing, part of the energy of fall is absorbed, and the energy available for rebound is comparatively small. If the hammer is dropped on a hard surface, the size of the indentation is much smaller, so that less energy is absorbed in making it. The rebound of the hammer in this case is therefore much higher than before. The height of rebound, rather than the volume, diameter, or depth of indentation, any one of which might logically have been used, is taken as the measure of hardness.

The scale is graduated into 140 divisions, a rebound of 100 being equivalent to the hardness of martensitic high-carbon steel. For this material, the area of contact between the hammer and specimen is only about 0.0004 sq in. and the stress developed is over 400,000 psi. Notwithstanding the lightness of the hammer, the forces exerted in all cases are sufficient to overcome the surface resistance of the hardest materials used in engineering practice; however, the indentations produced are so minute that they do not seriously impair a finished surface.

It should be noted that the scleroscope hardness numbers are arbitrary and that they are comparable only when determined on similar material. Obviously there would be little relationship between the scleroscope hardness numbers of two such dissimilar materials as rubber and steel.

There are two types of scleroscopes, one a direct-reading or visual type, as shown in Fig. 7.10*a*, in which the height of rebound must be caught by eye, and the other an improved dial-recording instrument, as shown in Fig. 7.10*b*, in which the dial hand remains at the height of rebound until reset. Both instruments are portable, and tests can be made quite rapidly with them. In the direct-reading type, the instrument is provided with an ingenious automatic head by means of which the hammer is lifted and released by air pressure from a rubber bulb.

A magnifier hammer is available for use on soft materials. This has a larger point area than the standard hammer and gives higher readings, thus magnifying small but significant variations of hardness.

The minimum thickness of the specimen which can be tested depends upon its hardness. For hard steel, as in safety-razor blades, the thickness should be at least 0.006 in.; for cold-rolled unannealed brass and steel it should be 0.010 in.; and for annealed sheets 0.015 in. The scleroscope is very useful for testing the hardness of casehardened surfaces provided they are at least $\frac{1}{64}$ in. thick.

Various precautions must be observed if reliable results are to be obtained. The surface of the specimen should be flat, smooth, and free

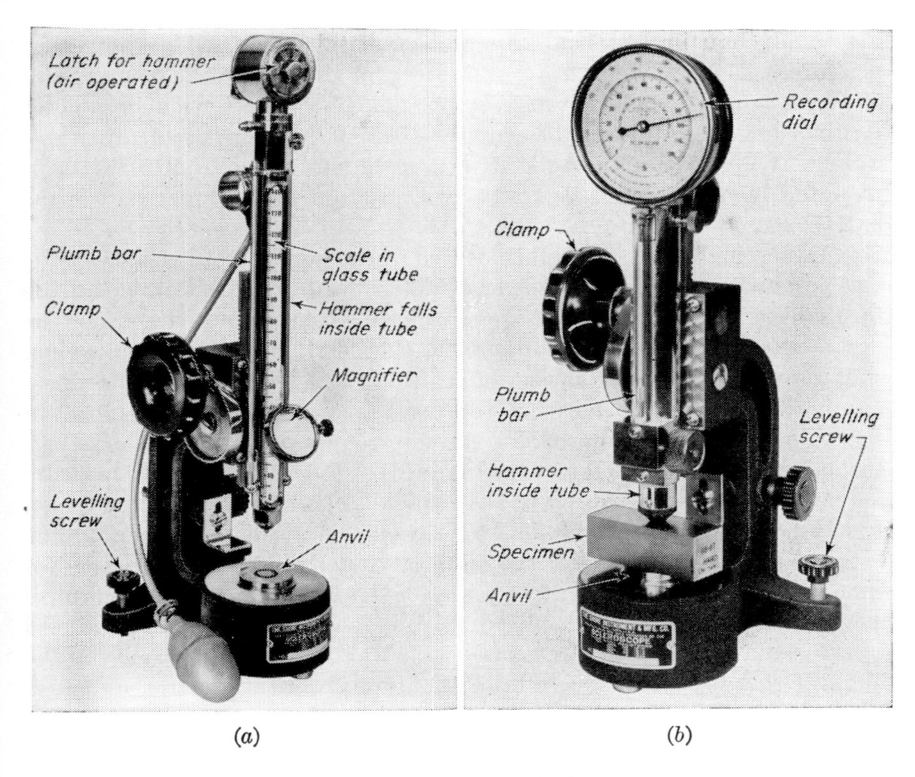

(*a*) (*b*)

Fig. 7.10. **(*a*) Direct-reading scleroscope and (*b*) dial-recording scleroscope.**

from oil or other materials. The glass tube through which the hammer falls must be carefully plumbed to avoid rubbing of the hammer on its inner surface. All specimens should be securely clamped to the anvil to avoid inertia effects.

When making a hardness determination, several readings should be taken to get a fair average, but the hammer should not be dropped more than once on any one spot because of the possible effect of strain hardening.

The condition of the diamond point is very important and should be checked frequently by using hardened-steel reference blocks. The indicated hardness should not vary more than ± 5 percent from that stamped on the blocks; larger variations are indicative of possible cracking or chipping of the diamond.

7.18. The Herbert Pendulum Device. The Herbert hardness tester operates on a principle different from that of any other type of device. The test may be considered in the class of dynamic tests, although it does not involve an impact load. An arched metal frame, which acts as a pendulum, is supported on a 1-mm steel (or diamond) ball. The center of gravity of the pendulum may be brought to a predetermined distance (ordinarily 0.1 mm) below the center of the ball by means of an adjustable weight. A curved bubble tube with a scale graduated from 0 to 100 is mounted on the top of the frame, with the 50 mark directly over the ball [752].

In making a test, the ball is located at the point on the surface of the test piece where the hardness is to be determined. The pendulum is made to oscillate, and various measures of hardness may be obtained. In the "scale test," the angular oscillation of the pendulum during one swing, starting with the bubble at zero, gives what is called the "scale hardness number"; it is said to measure the resistance of the metal to flow. In the "time test," the number of seconds required for 10 single swings through a small angle is the "time hardness number"; it is said to measure indentation hardness, similar to that which is observed in the Brinell test. If the pendulum is moved first to the extreme right and then to the extreme left, the hardness of the material under the ball is changed. If a time test is made after each two passes of the ball as just described, until a condition of maximum time hardness is reached, this maximum induced time hardness is taken as a measure of the work-hardening capacity of the metal. Other procedures have been set up to give other related measures of hardness [752]. The device appears to have many possibilities in the study of work-hardening problems that arise in connection with the machining of metals. It also affords an easy means of measuring hardness at elevated temperatures while the specimen is in a special furnace at a controlled temperature.

SCRATCH- AND WEAR-HARDNESS TESTS

7.19. Scratch Hardness. A convenient and definite hardness scale, like a temperature scale, is lacking because materials are not available that have invariable hardness to serve as calibration points analogous to the various boiling and freezing points that are standards of comparison for thermometers and pyrometers.

An approach to this is the arbitrary mineralogical scale of hardness in which a mineral will scratch other minerals that are lower on the scale (smaller hardness number) and will in turn be scratched by minerals higher on the scale. The well-known scale used by mineralogists is Mohs' scale, shown in Table 7.6. With the development in recent years of extremely hard abrasives has come the need for more adequately distinguishing between materials in the range of hardness between that of quartz and that of diamond. An extension of Mohs' scale devised for this purpose is also shown in Table 7.6.

Table 7.6. **Scratch hardness—mineralogical basis**

Mohs' scale		Extension of Mohs' scale*		Metal equivalent
Hardness No.	Reference mineral	Hardness No.	Reference mineral	
1	Talc	1	Talc	
2	Gypsum	2	Gypsum	
3	Calcite	3	Calcite	
4	Fluorite	4	Fluorite	
5	Apatite	5	Apatite	
6	Feldspar (orthoclase)	6	Orthoclase	
		7	Vitreous pure silica	
7	Quartz	8	Quartz	Stellite
8	Topaz	9	Topaz	
		10	Garnet	
		11	Fused zirconia	Tantalum carbide
9	Sapphire or corundum	12	Fused alumina	Tungsten carbide
		13	Silicon carbide	
		14	Boron carbide	
10	Diamond	15	Diamond	

* R. R. Ridgway, A. H. Ballard, and B. L. Bailey, "Hardness Value for Electrochemical Products," *Trans. Electrochem. Soc.*, vol. 43.

7.20. Sclerometers. In an attempt to obtain a quantitative measure of hardness on the scratch principle, a number of tests have been proposed in which there is measured either the pressure required to make a given scratch or the size of scratch produced by a stylus drawn across the surface under a fixed load. The device for making such a test is often referred to as a "sclerometer." Although sclerometer tests are simple in principle and have in the past been regarded with considerable interest, they are difficult to standardize and interpret and have not come into general use except in the Bierbaum scratch hardness test of plastics (ASTM D 1526). In this method an accurately ground diamond point, shaped in the form of the corner of a cube and carrying a load of 3 g, is moved laterally by means of a worm gear causing the point to cut a groove in the surface of the test specimen. The Bierbaum scratch hardness equals the load on the diamond point, in kilograms, divided by the square of the width of the scratch, in millimeters.

A good discussion of scratch-hardness tests is given in Ref. 702.

7.21. The File Test. A widely used test that is of the scratch type is the file test. In many modern plants it is used as a qualitative or inspection test for hardened steel. In addition to the standard and more accurate methods of testing used on representative samples from each lot, each piece in the lot may be gone over with a file, and by making one pass over the surface with a file of appropriate hardness the operator is able to cull out unsatisfactory pieces. Obviously, it is impossible to make fine distinctions in hardness, but at the critical hardness the file will slide over the surfaces of acceptable pieces and bite into pieces that are too soft. Obviously, the method depends upon the file and the way it is used, but despite the large personal element in the test it is widely used and is a satisfactory, speedy test in the hands of a skilled operator. An improvement on this general idea is a proposed file scratch test using a series of needle-pointed files having Rockwell hardnesses ranging from C 25 to C 65 [750]. Such a test would extend the usefulness of the file test.

7.22. Wearing Hardness. Wearing or abrasion tests have been applied to metals, both dry and lubricated, and to other materials, but with the exception of the test of stone and paving bricks, they have not been generally adopted.

In the standard test of stone, the Deval abrasion test, a charge of 50 specimens is "tumbled" in a cylinder, and with a standardized machine and test procedure the percentage of loss in weight is determined; sometimes a so-called "French coefficient" of wear is computed by dividing 400 by the wear in grams per kilogram of rock used (ASTM D 2).

The standard abrasion test for coarse aggregate for concrete, the Los Angeles rattler test, is made by tumbling a charge of the aggregate together with a charge of steel balls in a drum for a specified period and determining the percentage of wear (ASTM C 131). A similar rattler test is made to determine the wear resistance of brick (ASTM C 7).

HARDNESS CORRELATIONS

7.23. Relations between Various Systems of Hardness Numbers. No precise relationship exists between the several types of hardness number. However, approximate relationships have been determined by tests of the same material using the various devices. Since these relationships vary with the different materials and with the mechanical and heat treatment given them, too much reliance upon them must be avoided. Table 7.7 presents comparative values for steel as determined by the more common types of equipment. Additional relations covering alloys are shown in ASTM E 140.

7.24. Relation of Hardness to Tensile Strength. It has been aptly stated by Williams that

> In the solid state the cohesive and adhesive forces are so strong that the atoms retain fixed positions relative to each other. This produces an aggregate that has and retains definite form. Resistance to change in form, i.e., to change of

Table 7.7. **Approximate hardness relations for steel***

Brinell, 3,000 kg			**Vickers diamond pyramid**	**Rockwell, using cone**				**Scleroscope**	**Mohs**	**Tensile strength, 1000 psi**
Diameter, mm	**Standard ball**	**Tungsten carbide ball**		**C 150 kg**	**D 100 kg**	**A 60 kg**	**30 N**			
2.35	...	682	737	61.7	72.0	82.2	79.0	84		
2.40	...	653	697	60.0	70.7	81.2	77.5	81		
2.45	...	627	667	58.7	69.7	80.5	76.3	79	8.0	323
2.50	...	601	640	57.3	68.7	79.8	75.1	77	...	309
2.55	...	578	615	56.0	67.7	79.1	73.9	75	...	297
2.60	...	555	591	54.7	66.7	78.4	72.7	73	7.5	285
2.65	...	534	569	53.5	65.8	77.8	71.6	71	...	274
2.70	...	514	547	52.1	64.7	76.9	70.3	70	...	263
2.75	495	...	539	51.6	64.3	76.7	69.9	...	...	259
	...	495	528	51.0	63.8	76.3	69.4	68	...	253
2.80	477	...	516	50.3	63.2	75.9	68.7	...	...	247
	...	477	508	49.6	62.7	75.6	68.2	66	...	243
2.85	461	...	495	48.8	61.9	75.1	67.4	...	...	237
	...	461	491	48.5	61.7	74.9	67.2	65	...	235
2.90	444	...	474	47.2	61.0	74.3	66.0	...	7.0	226
	...	444	472	47.1	60.8	74.2	65.8	63	...	225
2.95	429	429	455	45.7	59.7	73.4	64.6	61	...	217
3.00	415	415	440	44.5	58.8	72.8	63.5	59	...	210
3.05	401	401	425	43.1	57.8	72.0	62.3	58	...	202
3.10	388	388	410	41.8	56.8	71.4	61.1	56	...	195
3.15	375	375	396	40.4	55.7	70.6	59.9	54	6.5	188
3.20	363	363	383	39.1	54.6	70.0	58.7	52	...	182
3.25	352	352	372	37.9	53.8	69.3	57.6	51	...	176
3.30	315	341	360	36.6	52.8	68.7	56.4	50	...	170
3.35	331	331	350	35.5	51.9	68.1	55.4	48	...	166
3.40	341	321	339	34.3	51.0	67.5	54.3	47	...	160
3.45	311	311	328	33.1	50.0	66.9	53.3	46	...	155
3.50	302	302	319	32.1	49.3	66.3	52.2	45	6.0	150
3.55	293	293	309	30.9	48.3	65.7	51.2	43	...	145
3.60	285	285	301	29.9	47.6	65.3	50.3	42	...	141
3.65	277	277	292	28.8	46.7	64.6	49.3	41	...	137
3.70	269	269	284	27.6	45.9	64.1	48.3	40	...	133
3.75	262	262	276	26.6	45.0	63.6	47.3	39	...	129
3.80	255	255	269	25.4	44.2	63.0	46.2	38	...	126
3.85	248	248	261	24.2	43.2	62.5	45.1	37	5.5	122
3.90	241	241	253	22.8	42.0	61.8	43.9	36	...	118
3.95	235	235	247	21.7	41.4	61.4	42.9	35	...	115
4.00	229	229	241	20.5	40.5	60.8	41.9	34	...	111

Table 7.7. **Approximate hardness relations for steel** (*Continued*)*

Brinell, 3000 kg		Rockwell						Scleroscope	Mohs	Tensile strength, 1000 psi
		Cone								
Diameter, mm	Standard ball	D 100 kg	A 60 kg	B 1/16-in. ball	E 1/8-in. ball	30 N	30 T			
4.05	223	40	60	97	...	41	80.5	33	...	108
4.10	217	39	60	96	...	40	80.0	32	...	105
4.15	212	38	59	95	...	39	79.0	31	...	102
4.20	207	37	59	94	...	38	78.5	31	...	100
4.25	202	37	58	93	110	37	78.0	30	...	98
4.30	197	36	58	92	110	36	77.5	29	...	96
4.35	192	35	57	91	109	35	77.0	28	5.0	94
4.40	187	34	57	90	109	34	76.0	28	...	92
4.45	183	34	56	89	109	33	75.5	27	...	90
4.50	179	33	56	88	108	32	75.0	27	...	88
4.55	174	33	55	87	108	31	74.5	26	...	86
4.60	170	32	55	86	107	30	74.0	26	...	84
4.65	166	32	54	85	107	30	73.5	25	...	82
4.70	163	31	53	84	106	29	73.0	25	...	81
4.75	159	31	53	83	106	28	72.0	24	...	79
4.80	156	30	52	82	105	27	71.5	24	...	77
4.85	153	...	...	81	105	...	71.0	23	...	76
4.90	149	...	...	80	104	...	70.0	23	4.5	75
4.95	146	...	...	79	104	...	69.5	22	...	74
5.00	143	...	...	78	103	...	69.0	22	...	72
5.05	140	...	...	76	103	...	68.0	21	...	71
5.10	137	...	...	75	102	...	67.0	21	...	70
5.15	134	...	...	74	102	...	66.0	21	...	68
5.20	131	...	...	73	101	...	65.0	20	...	66
5.25	128	...	...	71	100	...	64.0	...	...	65
5.30	126	...	...	70	100	...	63.5	...	...	64
5.35	124	...	...	69	99	...	62.5	...	...	63
5.40	121	...	...	68	98	...	62	...	...	62
5.45	118	...	...	67	97	...	61	...	...	61
5.50	116	...	...	65	96	...	60	...	...	60
5.55	114	...	...	64	95	...	59	...	...	59
5.60	112	...	...	63	95	...	58	...	...	58
5.65	109		...	62	94	...	58	...	...	56
5.70	107	...	...	60	93	...	57	...	...	55
5.75	105	...	...	58	92	...	55	...	...	54
5.80	103	...	...	57	91	...	54	...	...	53

* Based on *Metals Handbook*—1961, American Society for Metals, Cleveland, Ohio, 1961. See ASTM E 140 for additional relations.

relative positions of the atoms composing a body, may be defined as the rigidity of that material. The stress necessary to produce permanent deformation of the structure of the solid is most intimately allied to the property of hardness, for, in the measurement of hardness by a penetration method, it will be necessary to vary permanently these fixed positions that the atoms bear with respect to each other [702].

No correlation exists between any indentation hardness and the yield strength determined in a tension test since the amount of inelastic strain involved in the hardness test is much greater than in the test for yield strength. However, because of the greater similarity in inelastic strain involved in the test for ultimate tensile strength and indentation hardness, empirical relations have been developed between these two properties, at least in the case of certain steels. For example, in the commonly used carbon and alloy steels, the tensile strength in pounds per square inch is approximately 510 times the Brinell hardness number. The approximate tensile strength of steel for the several types and degrees of hardness is indicated in the last column of Table 7.7.

Chapter 8.

IMPACT TESTS

8.1. Dynamic Loading. Although many structures are at some time subjected to dynamic loads, many machines and machine parts are commonly subject to such loads. To estimate the safe performance of structures and machines, or their parts, under dynamic loading involves not only analyses to determine the general response of the structure or machine but also consideration of the properties of the component materials under such loading. The behavior of materials under dynamic loading may sometimes differ markedly from their behavior under static or slowly applied loads.

An important type of dynamic loading is that in which the load is applied suddenly, as from the impact of a moving mass. This chapter is concerned with some of the aspects of the behavior of materials under such impact loads. The behavior of materials under rapidly fluctuating loads involves another phenomenon, known as fatigue, which is discussed in Chap. 9.

As the velocity of a striking body is changed, there must occur a transfer of energy; work is done on the parts receiving the blow. The mechanics of impact involve not only the question of stresses induced but also a consideration of energy transfer and of energy absorption and dissipation.

The energy of a blow may be absorbed in a number of ways: through elastic deformation of the members or parts of a system, through plastic deformations in the parts, through hysteresis effects in the parts, through frictional action between parts, and through effects of inertia of moving parts. The effect of an impact load in producing stress depends upon the extent to which the energy is expended in causing deformation. In dealing with problems involving impact loading, the predominant way in which the load is to be resisted obviously determines the type of information that is needed.

In the design of many types of structures and machines that must take impact loading, the aim is to provide for the absorption of as much energy as possible through *elastic* action and then to rely upon some kind of damping to dissipate it. In such structures the resilience (i.e., the elastic energy capacity) of the material is a significant property, and resilience data derived from static loading may be adequate.

Satisfactory performance of certain types of machine parts, such as parts of percussion drilling equipment, parts of automotive engines and transmissions, parts of railroad equipment, track and buffer devices, depends upon the toughness of the parts under shock loadings. Although a direct approach to this problem would seem to be the use of tests that involve impact loads, the solution to the problem is not simple. Without doubt the results of impact tests have contributed indirectly to the improved design of certain types of parts, but in general such tests, to date at least, have proved to be of limited significance in producing basic design data.

In most tests to determine the energy-absorption characteristics of materials under impact loads, the object is to utilize the energy of the blow to cause *rupture* of the test piece. There is thus a distinction to be made between problems that largely involve elastic energy absorption and problems to which data on energy capacity at rupture are pertinent. This difference contributes to a basic limitation to the general applicability of the results of the ordinary impact test.

8.2. Some Aspects of Impact Testing. In some experimental studies of the properties of materials under impact loading, detailed determinations have been made of stress-strain-time relationships. Generally, such studies of dynamic properties require very special and complicated experimental procedures which are not adapted to everyday testing, although notable contributions have been made to the understanding not only of the impact problem but of the mechanics of plastic strain and fracture. The development in recent years of electric strain gages, electronic measuring equipment, and high-speed oscillographs has greatly extended the scope of research into the behavior of materials under very rapidly applied loads.

In some tests there are made limited measurements of strain or deflection under impact loading, but in the most commonly used impact tests, the so-called "notched-bar" tests, only the energy to produce rupture is determined. In these tests one objective is to obtain a relative measure of the tendency to exhibit brittleness with a decrease in temperature, especially as affected by the presence of minor constituents or small variations in composition or structure of a particular metal or other material.

For many uses, knowledge of the over-all or macroscopic behavior of a material provides adequate information. It is being found, however, that the failure of materials under some conditions can be adequately explained only in terms of the behavior of their microstructure. An understanding of the behavior of metals in the form of notched bars under impact loading requires a knowledge of the mechanism of fracture as influenced by a polycrystalline structure [814].

8.3. Behavior of Materials under Impact Loading. The property of a material relating to the work required to cause rupture has been designated as "toughness" (Art. 2.15). Toughness depends fundamentally upon strength and ductility and would appear to be independent of the type of loading. It is a fact, however, that the rate at which the energy is absorbed may markedly affect the behavior of a material, and thus different measures of toughness may be obtained from impact loadings than from static loadings.

All materials do not respond in the same way to variations in speed of load application; some materials display what is termed "velocity sensitivity" to a much more marked extent than others. Among striking examples of materials that display radically different behavior under slow- and high-speed loadings are ordinary glass, which is punctured with a fairly clean hole by a high-speed bullet but shatters under slowly applied point loading, and sealing wax, a stick of which breaks as if it were brittle under a sharp blow but slowly sags plastically under its own weight if supported as a beam.

Over the range of low- and medium-carbon steels, the *relative* toughness of a series of steels determined from impact and static tests of *plain (unnotched) tension* specimens appears to be more or less the same, although the *actual* work required to cause rupture under impact loadings (where the striking velocities are less than some critical value, as is the case with the usual impact tests) runs probably 25 percent greater than the work as obtained from the usual static stress-strain diagram. But the toughness as determined by impact loading is not necessarily greater than that determined by static loading; in the case of chrome-nickel steel, for example, the impact toughness is *less* than the static toughness [803, pp. 101 and 155]. However, it has been pointed out by Mann [820] that, for velocities obtained with the ordinary impact machines, good correlation between (unnotched) tension test results under impact and static loading is obtained if the area under the true stress–conventional strain diagram is used to calculate the energy to rupture.

With a given material the toughness does not vary greatly over a considerable range in striking velocity, but above some critical speed (different for various materials) the energy required to rupture a material appears to decrease rapidly with increase in speed [821, 825]. This critical velocity is found to be associated with the rate of propagation of plastic strain and is affected by the length of the piece subjected to impact loading [823, 825].

In addition to the velocity effect, the form of a piece may have a marked effect upon its capacity to resist impact loads. At ordinary temperatures a plain bar of ductile metal will not fracture under an impact load in flexure. In order to induce fracture to take place under a single blow, test specimens of a ductile material are notched. The use of a notch causes high localized stress concentrations, restricts the drawing-out action (i.e., artificially tends to reduce ductility), causes most of the energy of rupture to be absorbed in a localized region of the piece, and tends to induce a brittle type of fracture. The tendency of a ductile material to act like a brittle material when broken in the form of a notched specimen is sometimes referred to as "notch sensitivity." Materials that have practically identical properties in static tension tests, or even in impact tension tests when unnotched, sometimes show marked differences in notch sensitivity. It has been appropriately suggested that impact testing and notched-bar testing really belong in different categories [803, p. 141].

From various illustrations cited above, it appears that all materials do not respond in the same way to impact loadings. An analogous situation was pointed out in connection with the results of hardness tests.

8.4. Scope and Applicability of Impact Tests. In research on the behavior of materials under dynamic loading, many devices and techniques have been used, and many more will probably be developed in attempts to learn the detailed mechanism of deformation and fracture as affected by the many variables of composition, temperature, velocity of loading, and geometry of specimen. Reference is made herein to some of the results of this kind of research, in order to explain some of the behavior of materials under test. However, the remainder of this chapter will be devoted primarily to the conduct and use of the more commonly used and more-or-less standardized impact tests.

The ideal impact test would be one in which all the energy of a blow is transmitted to the test specimen. Actually this ideal is never realized;

some energy is always lost through friction, through deformation of the supports and of the striking mass, and through vibration of various parts of the testing machine. In some tests, it is impossible to obtain a truly accurate measure of the energy absorbed by a specimen. Further, the particular values obtained from an impact test depend very much upon the form of specimen used. These facts require close attention to standardization of details in any given type of test if concordant results are to be obtained, and usually preclude direct comparisons of results from various different types of impact tests. Each type of impact test has its own specialized field of use, and its applicability depends largely upon satisfactory correlation with performance under service conditions. In this connection, it may also be observed that the applicability of a test may not necessarily be confined to materials for use in parts that are to be subject to impact.

Committee A-3 on steel, of the ASTM, describes the impact tests of steel for production purposes as follows [865]:

An impact test is a dynamic test in which a selected specimen, machined or surface-ground and usually notched, is struck and broken by a single blow in a specially designed testing machine and the energy absorbed in breaking the specimen is measured. The energy values determined are qualitative comparisons on a selected specimen and cannot be converted into energy figures that would serve for engineering design calculations. The notch behavior indicated in an individual test applies only to the specimen size, notch geometry, and testing conditions involved, and cannot be applied to other sizes of specimens and conditions. Minimum impact requirements are generally specified only for quenched and tempered, normalized and tempered or normalized materials, as provided in the appropriate product specifications.

In making an impact test, the load may be applied in flexure, tension, compression, or torsion. Flexural loading is the most common; tensile loading is less common; compressive and torsional loadings are used only in special instances. The impact blow may be delivered through the use of a dropping weight, a swinging pendulum, or a rotating flywheel. Some tests are made so as to rupture the test piece by a single blow; others employ repeated blows. In some tests of the latter type, the repeated blow is of constant magnitude; in others, the "increment-drop" tests, the height of drop of the weight is increased gradually until rupture is induced. In Table 8.1, several of the various impact tests are grouped in accordance with these classifications.

Perhaps the most commonly used impact tests for steels in this country are the Charpy and the Izod tests, both of which employ the pendulum principle. Ordinarily these tests are made on small notched specimens broken in flexure. In the Charpy test, the specimen is supported as a simple beam, and in the Izod test it is supported as a cantilever. In such tests a large part of the energy absorbed is taken up in a region immedi-

Table 8.1. **Classification and summary of impact tests**

Means of applying blow	Type of loading	Single-blow tests				Repeated-blow tests							
						Blows of constant magnitude				Blows of increasing magnitude ("increment-drop tests")			
		Machine	Maximum capacity, ft-lb	Maximum striking velocity, fps	Ref.*	Machine	Maximum capacity, ft-lb	Maximum striking velocity, fps	Ref.*	Machine	Maximum capacity, ft-lb	Maximum striking velocity, fps	Ref.*
Dropping weight	Flexure	Hatt-Turner Fremont	3200 440	21 29	D 143 O	Krupp-Stanton	0.48	2.5	101, 870	AREA† Hatt-Turner Army Ordnance (cast iron)	50,000 3,200	40 21 ...	4 O, 801 871
	Tension	Olsen Guillotine Calif. Inst. Tech.	3500	21	O 826 823								
	Compression	Olsen Calif. Inst. Tech.	3500	21	O 827					Page	13	14	D 3
Swinging pendulum	Flexure	Charpy Izod Russell (cast iron) Oxford	2–240 2–260 500	11–17 11–17 11	O, R, W O, R, W 870 803	Heisler	50	11	O	Heisler	50	11	O
	Tension	Modified Charpy or Izod	250	11–17	O, R								
	Shear	McAdam	400	16	O								
Rotating flywheel	Flexure	Guillery	430	29	101								
	Tension	Mann-Haskell Calif. Inst. Tech.	....	1000	821 823								
	Torsion	Carpenter	>138	....	W, 850								

Note. A number of other impact machines, principally variations on the types listed above, are made. See makers' catalogs.
* Reference in which description may be found. Single letters refer to makers' catalogs: O = Olsen, R = Riehle, W = Wiedemann-Baldwin.
† American Railway Engineering Association.

ately adjacent to the notch, and a brittle type of fracture is often induced. It should be observed that these tests do not, and are not intended to, simulate shock loading in service; they simply give the resistance of a particular notched metal specimen to fracture under a particular type of blow. It has been found that the results indicate differences in condition of a metal that are not indicated by other tests. The results appear to be particularly sensitive to variations in the structure of steel as affected by heat treatment; by certain minor changes in composition which tend to cause "embrittlement," such as variations in the sulfur or phosphorus content; and by various alloying elements. Also, these tests, when made on specimens at low temperatures, have proved useful in indicating whether or not adequate toughness is maintained at those temperatures. While Charpy or Izod tests may not directly predict the ductile or brittle behavior of steel as used in large structural units, they find use as acceptance tests or tests of identity for different lots of the same steel or in choosing between different steels, when correlation with reliable service behavior has been established. As shown in Art. 8.11, impact-test results of many steels are very sensitive to temperature changes within the normal atmospheric range. For this reason the actual temperature of test may be quite important and should be reported in conjunction with all test results; preferably the transition temperature range for the given steel should be determined.

Procedures for the Charpy and Izod tests as applied to metals have been standardized (ASTM E 23), and formal specification of impact-strength limits has been made in the case of materials for a number of products such as airplane-engine parts, transmission gears, parts for tractor belts, turbine blading, many types of forgings, and pipe and steel plate for low-temperature service.

The Charpy and Izod tests have been adapted to molded electrical insulating materials, and the procedure for testing such materials is standardized (ASTM D 256). This procedure is also applied to molded plastics.

For wood, the Hatt-Turner test—a flexural-impact test of the increment-drop type—is used, and the procedure is standardized (ASTM D 143). In this test the height of drop at which failure occurs is taken as a measure of toughness, but from the data of the test the modulus of elasticity, the proportional limit, and the average elastic resilience may also be found.

Impact tests of metals in tension have been made largely for experimental purposes. They have been made as single-blow tests in a drop-weight machine, in a pendulum machine when suitably modified to accommodate a tension specimen, and in a flywheel machine. The impact-tension test affords opportunity to study the behavior of ductile materials under impact loading without the complications introduced by the use of a groove or notch, although notched tensile specimens have also been used in some tests. The flywheel type of machine, in particular, is capable of providing very high impact velocities; studies concerned with the "transition" or "critical" velocity have been made with this type of machine [821, 823, 825].

For testing tool steels, a torsion-impact test, the Carpenter test, has come into some prominence [850–852]. The test appears to offer a method for investigating and controlling optimum heat-treatment conditions of products such as drills, taps, and rock-drill parts.

Cast iron is not often used in parts that must have high shock resistance, and it is felt by some that the results of the static-flexure test give most of the information needed for estimating the relative energy capacity of cast irons [870, 871]. However, a number of investigations on impact resistance of cast iron have been made, and a variety of procedures have been employed, including single-blow pendulum tests, increment-drop tests, and repeated-blow tests. A repeated-blow method and both Charpy and Izod test procedures for cast iron have been standardized in ASTM A 327. All these test methods employ flexural loading; impact-tension tests of cast iron do not appear to give reliable results. Unnotched specimens are standard, since they seem to be more satisfactory than notched specimens. Cast iron is not as "notch-sensitive" as steel; this may be due to the notchlike effect of the graphite flakes in cast iron, which is not greatly increased by an additional effect of an external notch.

A compression test for toughness of rock for road-building purposes is made by an increment-drop test (ASTM D 3). Any form of machine meeting the requirements of the standardized test procedure may be used. The Page impact machine has been employed for this purpose.

Impact loading is used in tests for the acceptance of a number of metal products such as rails and axles. The American Railway Engineering Association specification for steel rails calls for an increment-drop test (ASTM A 1) by use of a special machine, the details of which are also specified [4]. Likewise, for axles a repeated-drop test is specified in ASTM A 383. Tests of this kind have significance in relation to performance in service.

8.5. General Features of Impact Machines. The effect of a blow depends in general upon the mass of the parts receiving the blow as well as upon the energy and mass of the striking body. Items that require standardization are foundation, anvil, specimen supports, specimen, striking mass and its velocity.

The principal features of a *single-blow pendulum* impact machine are (1) a moving mass whose kinetic energy is great enough to cause rupture of the test specimen placed in its path, (2) an anvil and a support on which the specimen is placed to receive the blow, and (3) a means for measuring the residual energy of the moving mass after the specimen has been broken.

The kinetic energy is determined from and controlled by the mass of the pendulum and the height of free fall, measured with respect to the center of mass (energy = WH, where W is the weight of the mass and H is the vertical distance through which the mass falls). The pendulum should be supported so that it falls in a vertical plane without possibility of lateral play or lateral restraint, and the bearings should be such that friction is small. The pendulum should be sturdy enough so that excessive vibrations do not cause serious variations in the results. The release mechanism should not influence the free-fall movement of the pendulum by causing any binding, accelerating, or vibrating effects.

The anvil should be heavy enough in relation to the energy of the blow so that an undue amount of energy is not lost by deformation or vibration. The device for supporting the specimen should be such that the specimen is held accurately in position prior to the instant of impact.

The striking edge of the pendulum should coincide with a vertical line through the center of the rotation when the pendulum is hanging unrestrained. The line of action of the reactive force between the specimen and the pendulum should pass through the center of percussion at the instant of impact. It is considered desirable that the center of percussion be as close as possible to the striking edge.

To indicate the swing of the pendulum of the Charpy- and Izod-type machines after the specimen has been broken, an arm attached to the pendulum moves a "friction pointer" over an arc graduated in degrees or in foot-pounds. The friction pointer, the axis of rotation of which coincides with that of the pendulum, is simply an arm that can rotate on a pin bearing of such tightness that the pointer is prevented from changing position under its own weight. The bearing pressure should be adjusted to a minimum that will prevent overcarrying or dropping down of the pointer. At the beginning of each test, this pointer is positioned to make contact with the pendulum and to indicate the proper reading when the latter is hanging vertically.

In the Oxford machine the anvil is mounted on a pendulum that can swing independently of the striking pendulum and thus measure the energy transmitted to the anvil [803]. The energy relations are obtained from the motions of both pendulums.

In *drop-weight* machines the principal features are the moving mass of known kinetic energy, but not necessarily of such magnitude that rupture is caused by a single blow, and an anvil. Most drop-weight machines do not have a device for measuring the residual kinetic energy of the weight, hammer, or "tup" after it has ruptured the specimen (one exception is the Fremont machine). In an increment-drop test, in which the height of drop is increased gradually, an approximate measure of the least energy load required to cause rupture is obtained. In some machines the variation in velocity of the tup before and after impact may be obtained from autographic time-displacement data, from which the energy relationships may be calculated; however, measurements of this sort are very difficult to make accurately. In repeated-blow machines there must be some provision for keeping flexure specimens from being displaced without restraining or fixing the ends.

In drop-weight machines it is necessary that the axes of the tup and the guides be vertical and in alignment and that the supports and anvil be so placed that the blow can be delivered squarely to the specimen. Friction in the guides should be minimized by keeping them free from grease or rust; they may be lubricated by powdered graphite.

The general requirements for *flywheel* machines are similar to those for pendulum machines, although the mechanical details are, of course, different. The Guillery machine has a fixed anvil, and the energy of the blow is determined from the change in velocity of rotation of the flywheel, before and after impact [101]. In the Mann-Haskell machine, the anvil is carried on a pendulum, the displacement of which is determined in order to obtain the energy of rupture [821]. In both these machines the blow is delivered through the use of a retractible striker arm which is carried flush with the flywheel until the wheel is brought up to speed. By means of a tripping device the striker can be released, and it is then forced to its outstanding position by centrifugal action.

For the purpose in hand, detailed discussion of impact-test procedures is confined to the Charpy and Izod tests of metals and the Hatt-Turner test of wood. These may be considered fairly representative of most ordinary impact-test work. The details of other types of tests are available in various technical publications such as those noted in Art. 8.4.

8.6. The Charpy Test for Metals and Plastics. The Charpy type of machine is available in a variety of sizes. A usual size is one having a capacity of about 220 ft-lb for metals and 4 ft-lb for plastics. A common design of machine is shown in Figs. 8.1 and 8.2. The following abbreviated description of the test is based on the requirements of ASTM E 23.

The pendulum consists of a relatively light although rigid rod or I section on the end of which is a heavy disk. The pendulum is suspended from a short shaft that rotates in ball bearings and swings midway between two rigid upright standards, near the base of which are the specimen supports or anvils. The knife or striking edge is slightly rounded, as shown in Fig. 8.3*d*, and should be so aligned as to make contact with the specimen over its full depth at the instant of impact.

The standard flexure test specimen is a piece 10 by 10 by 55 mm notched as shown in Fig. 8.3*a* (ASTM E 23). Other sizes are used in special cases. In many commercial specifications a keyhole notch or a

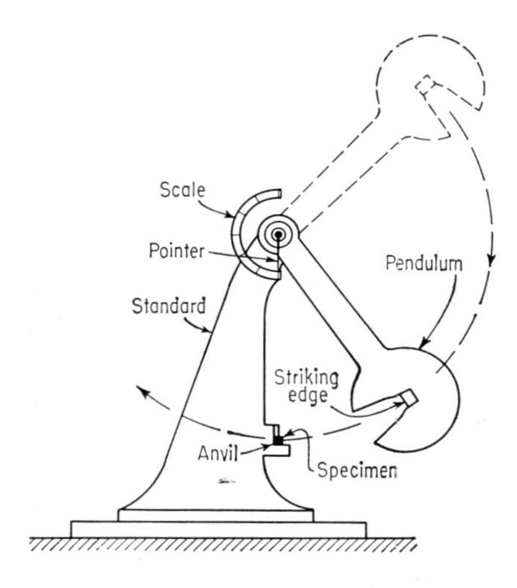

Fig. 8.1. **Diagram of Charpy impact machine.**

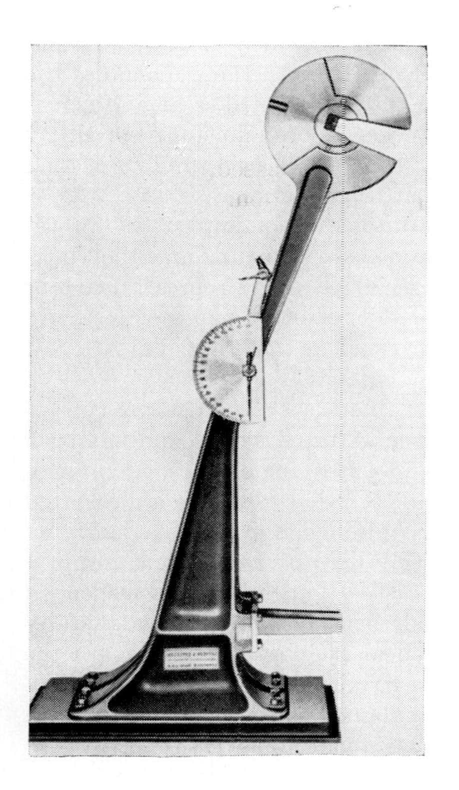

Fig. 8.2. **Charpy impact machine.**

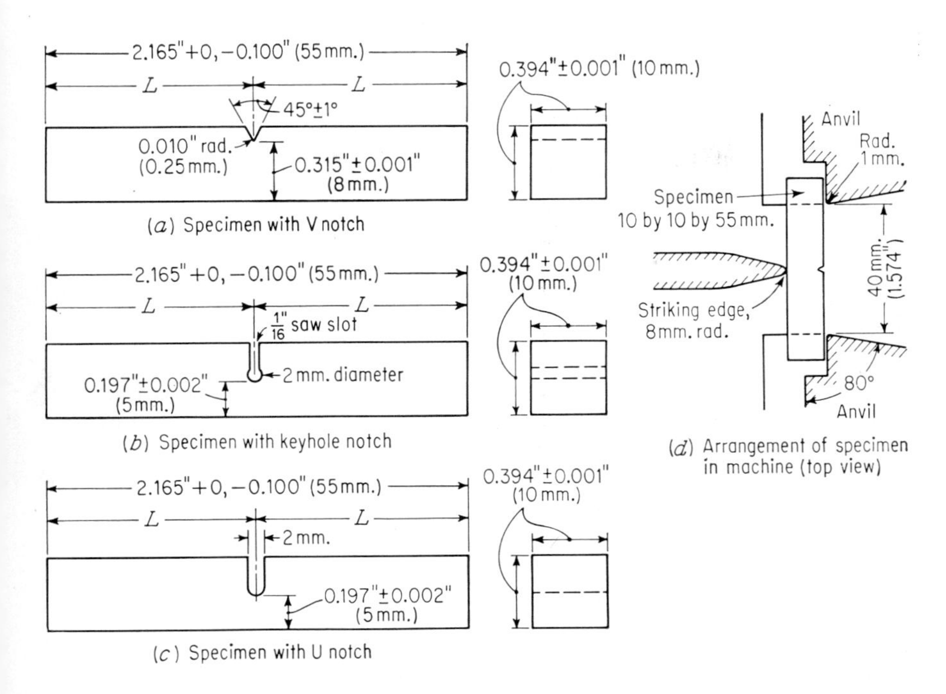

Fig. 8.3. **Charpy specimens.**

U notch is required, as shown in Fig. 8.3*b* and *c*. The specimen, which is loaded as a simple beam, is placed horizontally between the two anvils, as shown in Fig. 8.3*d*, so that the knife strikes opposite the notch at the mid-span. The pendulum is raised to its topmost position and held by a catch adjusted to give a constant height of fall for all tests. It is then released and allowed to fall and rupture the specimen.

In its upward swing the pendulum carries the friction pointer over a semicircular scale graduated in degrees or in foot-pounds. The energy required to rupture the specimen is a function of the angle of rise, as will be shown later. The weight of the pendulum, the position of the center of gravity, and the height of fall counted from the center of gravity are all determined experimentally.

The machine should be so constructed that the space between the anvils does not decrease in width in the direction of motion of the pendulum. In a standard machine the width should increase, as shown in Fig. 8.3*d*, to prevent drag between the specimen and the anvils.

For impact-tension tests a specimen is secured to the back edge of the pendulum. As the pendulum falls, a hammer block secured to the outstanding end of the specimen strikes against two extended anvils, the specimen being ruptured as the pendulum passes between the two anvils. Tension specimens may be plain or with a circumferential notch. One type of plain specimen has a diameter of 6 mm; a corresponding notched specimen has a diameter of 10 mm except at the root of the notch which is 1 mm wide and 2 mm deep, giving a net diameter of 6 mm as for the first type. The tension test has not been standardized and is not used to any great extent in commercial practice.

8.7. Charpy Tests at Low Temperatures. Tests to determine the impact resistance of metals at low temperatures are commonly conducted by immersing the specimens in some cool liquid in a widemouthed vacuum jar, with at least 1 in. of liquid above and below the specimens. For temperatures from ambient to −109°F, this liquid is usually alcohol or acetone, cooled to the desired temperature by the addition of small lumps of dry ice. For lower temperatures the cooling agent is usually liquid nitrogen (−319°F), and the liquid for immersion is usually alcohol to −190°F, isopentane to −250°F, and the liquid nitrogen itself to −319°F.

Thermometers suitable for determining the temperature of the coolant are the mercurial type to −38°F, alcohol or bimetallic types to −150°F, and copper-constantan thermocouples or pentane-type thermometers for lower temperatures.

Specimens should be held at temperature for 15 min and the bath temperature should be held constant within +0, −3°F during the last 5 min before testing. The test should be completed within 5 sec after removing the specimen from the coolant. As shown in Fig. 8.4, at a test

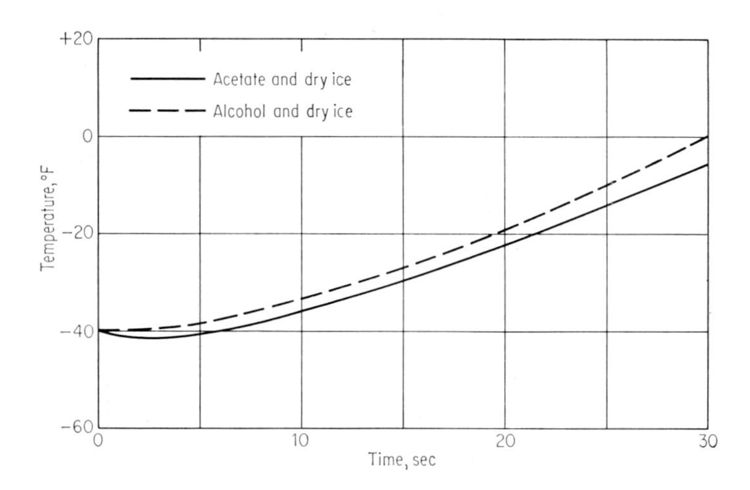

Fig. 8.4. **Temperature change in Charpy specimens when removed from bath at −40°F.** (*From Driscoll, Ref.* 860.)

temperature of $-40°F$, no appreciable temperature changes occur within 5 sec when alcohol is used.

8.8. The Izod Machine. The Izod impact testing machine is commonly made with a 120-ft-lb capacity, although other capacities are available. The pendulum consists of a hammer mounted at the end of a relatively light member, the upper end of which is mounted on ball bearings in uprights that are bolted down to a cast-iron bedplate. An illustration is shown in Fig. 8.5. The pendulum strikes against the specimen, which is clamped to act as a vertical cantilever 10 by 10 mm in section and 75 mm long having a standard 45° notch 2 mm deep. The mounting of the specimen and the relative position of the striking edge are shown in Fig. 8.6. The angle rise of the pendulum after rupture of the specimen or the energy to rupture the specimen is indicated on a graduated scale by a friction pointer. Some experimenters prefer the Charpy to the Izod ma-

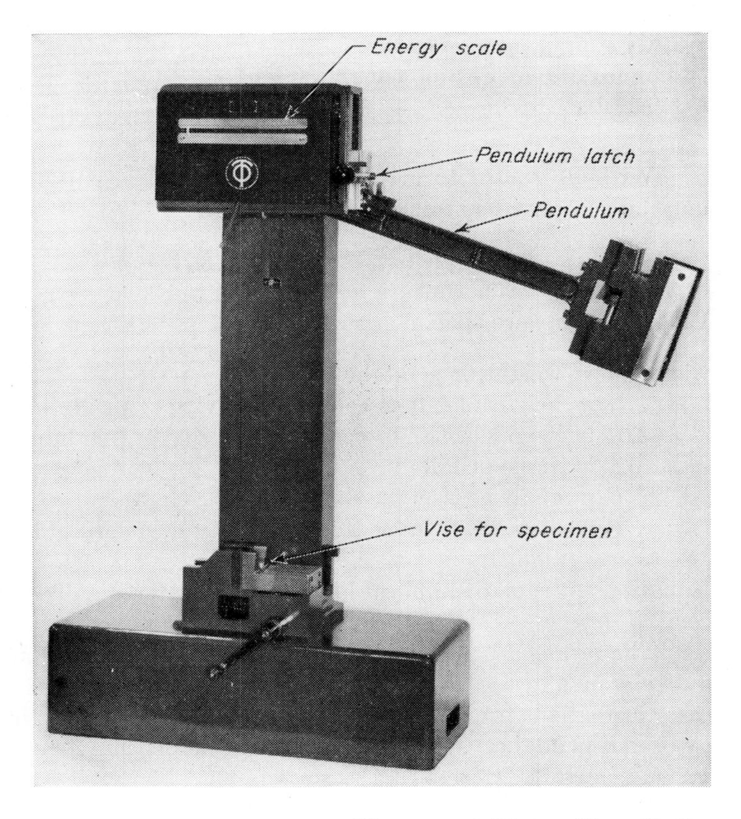

Fig. 8.5. Izod impact-testing machine. *(Courtesy of Tinius Olsen Testing Machine Co.)*

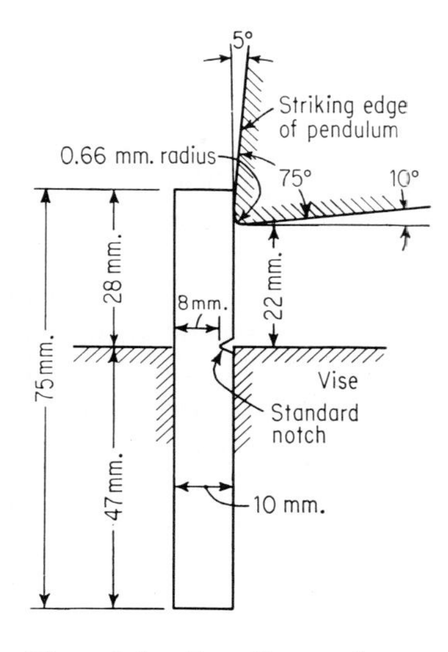

Fig. 8.6. **Cantilever beam specimen and mounting for the Izod test.**

chine, since in the former the test piece is not stressed in the region of the notch by the grip of a vise.

8.9. Calculation of Energy Relations. The impact strength or energy absorbed in breaking the specimen is equal to the difference between the energy in the pendulum before and after impact. The difference in energy is a function of the decrease in rotational velocity and can be computed from the weight and the height of fall of the pendulum before impact and the height of rise afterward. For accurate results (within the limitations of the test) a correction may be necessary for loss due to air drag, for the energy absorbed by friction in the machine bearing and by the indicator arm, and for the energy used in moving the broken test piece.

Without regard to losses, the energy used in rupturing a specimen may be computed as follows (see Fig. 8.7):

$$\text{Initial energy} = WH = WR(1 - \cos A) \quad (1)$$
$$\text{Energy after rupture} = WH' = WR(1 - \cos B) \quad (2)$$
$$\text{Energy to rupture specimen} = W(H - H') = WR(\cos B - \cos A) \quad (3)$$

where W = weight of pendulum
H = height of fall of center of gravity of pendulum
H' = height of rise of center of gravity of pendulum
A = angle of fall
B = angle of rise
R = distance from center of gravity of pendulum to axis of rotation O

Equation (3) is made applicable to angles greater than 90° by noting that $\cos (90 + \theta) = -\sin \theta$.

The results are usually given in foot-pounds or kilogram-meters without reference to the volume of metal involved.

Corrections for the energy losses due to bearing friction in the indicator and pendulum and to air drag on the latter may be made as follows:

1. Without placing a specimen on the anvil and with the indicator set at zero, release the pendulum in the normal manner. Record the resulting indicator reading as angle B_1.

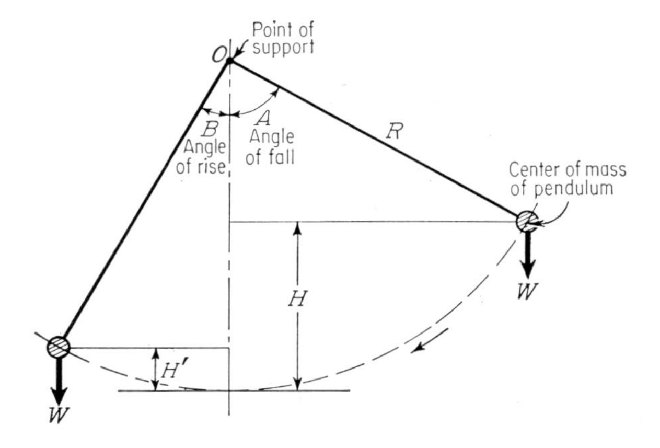

***Fig. 8.7.* Space relations for pendulum machine.**

2. Without resetting the indicator, again release the pendulum in the normal manner. Since it usually requires some small amount of energy to move the indicator forward, this second swing should move the pointer higher because the pendulum is not retarded by the pointer except at the very end of the swing. Repeat this process until the indicator is not moved by the pendulum. Record the final indicator reading as angle B_2.

3. With the indicator at angle B_2 again release the pendulum in the normal manner and allow it to make 10 swings (five forward and five back); then move back the indicator just enough so that on the eleventh swing the pendulum will move it forward not more than 1°. Do not touch the pendulum during these 11 swings. Record the final reading of the indicator as angle B_3.

The energy required to move the indicator and that lost in air drag and bearing friction of the pendulum may be assumed to be distributed uniformly over their ranges of action. The computations for a typical determination follow:

Let $B_1 = 160.1°$, $B_2 = 160.2°$, $B_3 = 156.6°$

The energy required to move the indicator through an angle of 160.2° is represented by an angle $B_2 - B_1 = 0.1°$

The average angle of rise between readings B_2 and B_3 is $(B_2 + B_3)/2$; hence a complete average swing, down and up, is approximately

$$(B_2 + B_3)/2 \times 2 = 316.8°$$

The energy lost in air drag and pendulum friction during one average forward swing is represented by an angle

$$(B_2 - B_3)/10 = (160.2 - 156.6)/10 = 0.36°$$

The energy lost in this manner during either a downward or an upward swing of 160.2° is represented by an angle

$160.2/316.8 \times 0.36 = 0.2°$

Effective angle of fall [A in Eq. (3)] is $160.2° + 0.2° = 160.4°$
Total energy lost in air drag and bearing friction of the pendulum and in friction of the indicator during an upward swing of 160.2° is represented by an angle of $0.2° + 0.1° = 0.3°$
Corrected angle of rise after rupture of specimen [B in Eq. (3)] is

$$\text{Observed angle} + 0.3 \times \frac{\text{observed angle}}{160.2}$$

The small amount of energy lost in imparting motion to the broken specimen is usually not taken into account in commercial testing. However, by assuming that the broken specimen leaves the anvils at the speed of the swinging pendulum, its magnitude can be calculated from the formula

$$e = wr(1 - \cos B) \tag{4}$$

where e = energy used in moving broken specimen
w = weight of specimen
r = distance of specimen from rotational axis of pendulum
B = angle of rise of pendulum after rupture of specimen

For machines reading directly in foot-pounds, a corresponding procedure may be followed to determine the above corrections.

In view of the status of impact testing of metals at the present time, it is recommended that the report of a test should include the following (ASTM E 23):

1. Type, model, and capacity of machine used
2. Type and size of specimen used
3. Maximum linear velocity in feet per second of the hammer
4. Energy loss due to friction, foot-pounds
5. Energy of blow with which specimen was struck, foot-pounds
6. Energy absorbed by specimen in breaking, foot-pounds
7. Temperature of specimen
8. Appearance of fractured surface
9. Number of specimens failing to break

8.10. The Hatt-Turner Test of Wood. The Hatt-Turner impact machine is used chiefly for flexure-impact tests of wood in which the height of drop is increased by increments until failure occurs (ASTM D 143).

A tup weighing 50 lb is held by an electromagnet which is raised by a motor. The tup drops between vertical guide columns when the magnet

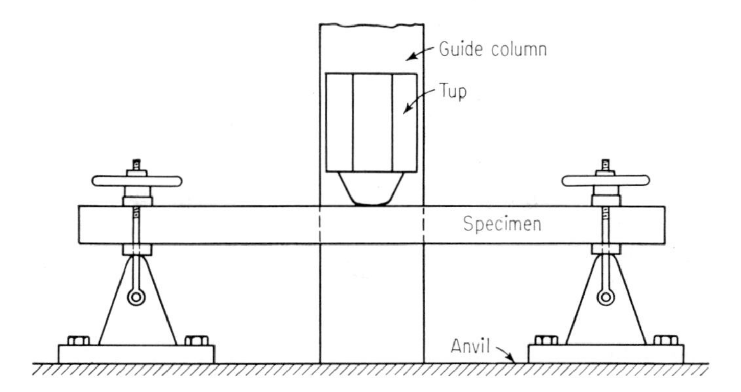

Fig. 8.8. **Method of supporting specimen in Hatt-Turner test.**

circuit is opened by a relay, this relay being actuated by an electric contact between the magnet and a movable arm that can be adjusted to any desired position along a vertical scale on one guide column.

The machine base carries the guide columns for the tup and suitable supports for the specimen. A hollow drum, free to rotate about its vertical axis, is also mounted on the base and near the guide columns. A pencil carried by the tup is pressed by a spring against a piece of paper secured to the surface of the drum and thus gives a graphic record from which there can be computed the height of drop of the tup and the corresponding deflection of the specimen.

The specimen is a clear piece of wood having nominal dimensions 2 by 2 by 30 in. The actual dimensions are recorded for each test. The piece is simply supported on a span of 28 in. so that the tup drops exactly at the center of the span. The tup has a curved striking face, and the supports are arranged as shown in Fig. 8.8.

The first drop is 1 in., and succeeding drops are increased by 1-in. increments until a height of 10 in. is reached, after which 2-in. increments are used until either complete rupture or a 6-in. deflection is reached.

Data from the drum records may be used to compute the proportional limit, modulus of elasticity, and resilience as explained in ASTM D 143.

8.11. Effect of Important Variables. The results of impact tests of metals made with various machines differ, owing to (1) variation in amounts of energy transformed at impact into vibrations of parts of the machines, (2) variations in striking velocity of the hammers, and (3) size and form of the specimen.

Provided the same form of notch is used, the results from various ordinary designs of Charpy and Izod machines are fairly comparable, although the Charpy results tend to be somewhat higher than the Izod results. The tougher the material, the greater the spread appears to be [101].

Over the range in velocity developed in most ordinary machines (about 10 to 18 ft per sec), the velocity does not appear appreciably to affect the results. However, experiments conducted with a machine capable of developing velocities up to 1000 fps indicate that, above some critical velocity, impact resistance appears to decrease markedly. The magnitude of the critical velocity and the rate of decline in impact resistance with increase in velocity differ for different metals. In general, the critical velocity is much less for annealed steels than for the same steels in a hardened condition [821, 823, 825]. The range of velocities developed in the usual Charpy and Izod tests appears to be well below the critical velocity for ordinary carbon steels; some alloy steels appear to have critical velocities near, if not within, the range of velocity of the ordinary Charpy and Izod tests [803, p. 101].

In some cases it is not possible to obtain a specimen of standard width from the stock available. Decreasing either the width or the depth of the specimen decreases the volume of metal subject to distortion, and thereby tends to decrease the energy absorption when breaking the specimen. However, any decrease in size also tends to decrease the degree of restraint and by reducing the tendency to cause brittle fracture, may increase the amount of energy absorbed. This is particularly true where a standard specimen shows a brittle fracture, and a narrower specimen may require more energy for rupture than one of standard width. Actual tests have shown that the Charpy values for specimens at room temperature, having widths of one-fourth and two-thirds that of the standard specimen, are roughly proportional to the width of the subsize specimen, but at low temperatures the narrow specimens of some steels may show up to three times the total energy resistance of a standard specimen [843].

Test results for mild steel as presented in Table 8.2 for a notch depth of 5 mm and a root radius of 0.67 mm show that the angle of notch does not appreciably affect results until it has exceeded 60°. One ASTM standard form of notch for impact tests of metals is a 45° V (see Fig. 8.3*a*).

The sharpness of the root of the notch may have an appreciable influence upon the energy of rupture of the test piece [101, 845]. As shown in Table 8.3, the energy of rupture decreases as the sharpness of the notch increases owing to the increase in stress concentration. It has been shown that the sharper the notch, the greater the difference between test results for brittle and tough materials, and that the tougher the material, the less the effect of radius of the root of the notch [101]. Since a dead sharp notch is difficult to produce, a 0.25-mm radius has been adopted as standard for the V notch.

The use of a shallow notch, in place of a deep notch, gives a greater spread of impact values for tough and brittle metals, and, further, the

Table 8.2. **Effect of angle of notch on energy of rupture of mild steel***

Angle of notch, °	Sketch of specimen	Charpy impact value, ft-lb
0	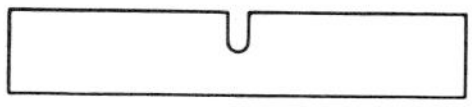	22.1
30	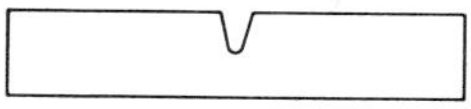	24.4
60	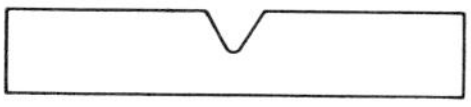	23.1
90		25.9
120	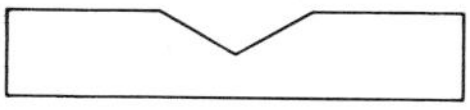	41.8
150	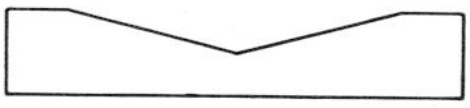	66.2
180	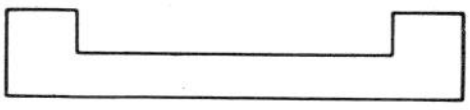	63.1

* J. J. Thomas, "The Charpy Impact Test on Heat-treated Steels," *Proc. ASTM*, vol. 15, pt. II, 1915.

Table 8.3. **Effect of root radius of 45° V notch on energy of rupture of 0.65 percent carbon steel***

Root radius of notch, 2 mm deep, mm	*Charpy impact value, ft-lb*
Sharp	4.0
0.17	6.9
0.34	8.3
0.68	13.7

* R. G. Batson, and J. H. Hyde, *Mechanical Testing*, Vol. I: *Testing of Materials of Construction*, Dutton, New York (Chapman & Hall, London), 1922.

shallow notch appears to be more sensitive to differences in either composition or temperature (see Fig. 8.11). For this reason, in Charpy impact testing some prefer the 2-mm V notch instead of the keyhole notch having a 5-mm depth, although the latter is commonly used for tests at low temperatures.

In contrast to the relatively small effect of temperature on the static strength and ductility of metals, at least within the atmospheric range, temperature has a very marked effect on the impact resistance of notched bars. Figure 8.9 illustrates in very generalized form the nature of the

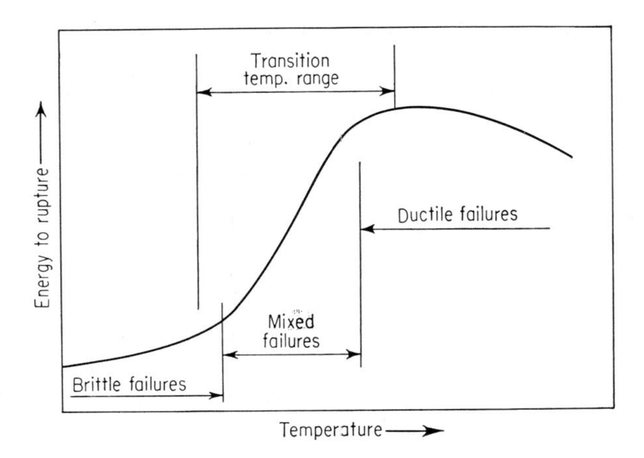

Fig. 8.9. **General nature of variation with temperature of energy to rupture in impact tests of metals.**

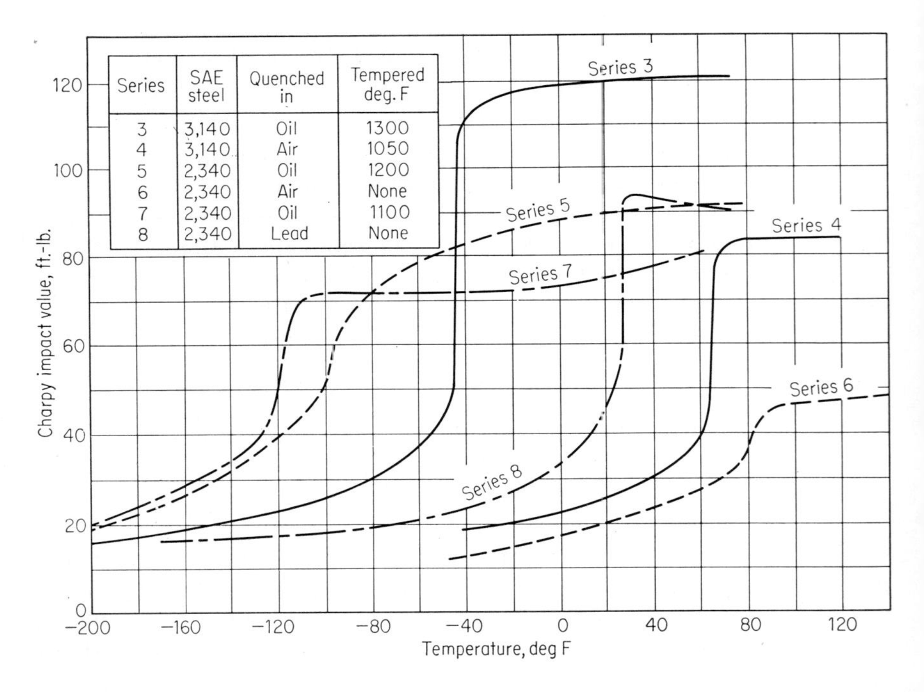

Series	SAE steel	Quenched in	Tempered deg. F
3	3,140	Oil	1300
4	3,140	Air	1050
5	2,340	Oil	1200
6	2,340	Air	None
7	2,340	Oil	1100
8	2,340	Lead	None

Fig. 8.10. **Variation in transition-temperature range for steel in the Charpy test.** (*From Dolan and Yen, Ref.* 814.)

variation of the energy to produce rupture in the impact test over a considerable range in temperature. For a particular metal and type of test, below some critical temperature the failures are brittle, with low energy absorption. Above some critical temperature the failures are ductile, with energy absorption that may be many times that in the brittle-fracture range. Between these temperatures is what has been termed the "transition-temperature range," where the character of the fracture may be mixed. This transition-temperature range may be very short or abrupt for some steels (see Fig. 8.10) or may extend over some appreciable range for others (see Fig. 8.11) [811, 812, 845]. A significant fact to be noted is that in or near this critical-temperature range, a variation in testing temperature of only a few degrees may cause very appreciable differences in impact resistance. Above the transition-temperature range, the impact resistance decreases more or less slowly with increase in temperature until a temperature of perhaps 1200°F is reached.

With the standard V notch the critical range for many steels appears to occur between the freezing point and room temperature; in some metals it may extend to temperatures well below the freezing point. The critical range appears to be higher with more highly "notch-sensitive" steels and with test pieces having sharper or deeper notches.

Coarse grain size, strain hardening, and certain embrittling elements tend to raise the transition range of temperature, whereas fine grain size, ductilizing and refining heat treatments, and the addition of certain alloying elements tend to enhance notch toughness, even at fairly low temperatures [803]. Grain size resulting from the work done in forming the steel and from the heat-treatment operations is an important factor in determining impact resistance. Steels having a fine-grained structure may be expected to show superior impact values, especially at low temperatures. In Fig. 8.11 is shown the effect of temperatures as low as −200°F, and the type of notch, on the Charpy impact resistance of a

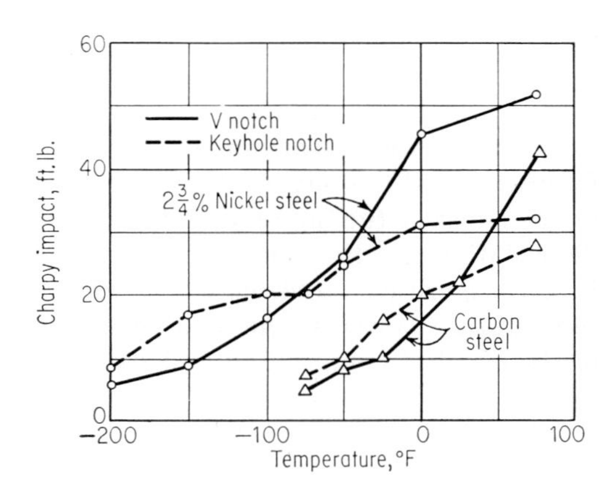

Fig. 8.11. **Influence of low temperatures and type of notch on impact strength of two normalized and drawn cast steels, both containing 0.18 percent carbon.** *(From Armstrong and Gagnebin, Ref.* 832.)

carbon steel and a nickel steel of about the same carbon content. The general beneficial effect of nickel on impact resistance at low temperatures is apparent.

8.12. Factors Affecting Failure. Two modes of fracture may occur in a metallic or crystalline material—a separation fracture and a shear or sliding fracture (see Art. 3.2–3.5 incl.). In the separation fracture a cleavage may occur within the crystals along critical planes of the crystal lattices, or it may occur by separation along crystal boundaries or through intercrystalline material. The shear type of fracture results from extreme detrusion; it may take place across crystals or in the intercrystalline material.

The final fracture may occur after little or much plastic deformation. When little or practically no plastic deformation occurs before rupture, the failure is said to be "brittle." It is generally considered that a brittle failure is accompanied by a separation fracture (although this is not necessarily the case). In a brittle failure with a separation fracture, the fractured surface usually has a granular or crystalline appearance; the energy absorbed is relatively small because little work has been done in producing plastic strain.

When considerable plastic strain takes place before rupture occurs, the failure is said to be "ductile." While the fracture which terminates in a ductile failure is usually a shear fracture, this is not always necessarily the case. The fractured surface following ductile action and shear fracture often has a fine, silky appearance; the energy absorbed in a ductile failure is relatively high, sometimes very high, because of plastic strain that takes place throughout the specimen.

Factors which inhibit plastic flow and, hence, tend to cause a brittle type of fracture in otherwise ductile metals are (1) a state of stress which holds the shearing stresses to a small magnitude relative to the tensile stresses, (2) a localization of the deformation by the presence of discontinuities or notches, (3) a very rapid application of stress (or high rate of strain), (4) lowered temperatures, and (5) certain types of structure or composition.

In the notched-bar impact test, the transition from ductile to brittle failure is usually brought into the atmospheric temperature range because of the effect of the notch and of the rapid strain rate in reducing plastic deformation. As the condition of the material at the base of the notch becomes a critical factor in determining the type of failure, the transition temperature as well as the relative amount of energy absorbed are indices of the effect of composition and structure on toughness or brittleness

Chapter 9.

FATIGUE TESTS AND TESTS OF METALS AT LOW AND HIGH TEMPERATURES

FATIGUE TESTS

9.1. Repeated Loadings. Most structural assemblages are subject to variation in applied loads, causing fluctuations in stresses in the parts. If the fluctuating stresses are of sufficient magnitude, even though the maximum applied stress may be considerably less than the static strength of the material, failure may occur when the stress is repeated a sufficient number of times. A failure induced in this manner is called a "fatigue failure."

Two methods of designating the nature of stress variations are (1) a statement of the numerically maximum stress, together with the ratio of the minimum to the maximum stress, called the *range ratio;* and (2) a statement of the mean value of the fluctuating stress, together with the alternating stress that must be superimposed on the mean stress to produce the given variation in stress conditions. A classification of types of "repeated" stresses is given in Table 9.1. In addition to designating the degree of stress *variation*, the *kind* of stress (tension, compression, or shear) must also be stated for complete definition of a stress condition. The stresses may be caused by axial, shearing, torsional, or flexural loadings or by combinations of them. For determinations of the fatigue characteristics of metals, one of the most commonly used types of repeated loading is completely reversed bending.

In only certain types of structures does the question of fatigue require consideration. In general, the fluctuations in the stresses in bridges and

Table* 9.1. Classification of types of repeated stresses

In stating numerical values of stresses, the kind of stress should always be designated as tension, compression, or shear. The kind of loading should also be designated as axial, torsional, direct shear, or bending.

Type of stress variation		Range-ratio nomenclature		Mean-stress nomenclature	
Description	**Diagram**	**Maximum stress**	**Range ratio**	**Mean stress**	**Alternating stress**
Steady stress, σ_1	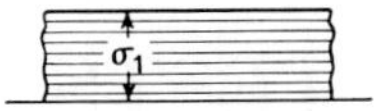 σ_1	σ_1	$\frac{\sigma_1}{\sigma_1} = 1.0$	σ_1	0
Pulsating stress, between σ_1 and σ_2	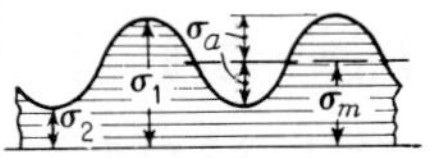 σ_a, σ_1, σ_m, σ_2	σ_1	$0 < \frac{\sigma_2}{\sigma_1} < 1$	σ_m	$\pm\sigma_a$
Pulsating stress, between σ_1 and 0	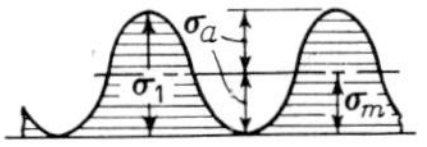 σ_a, σ_1, σ_m	σ_1	$\frac{0}{\sigma_1} = 0$	σ_m	$\pm\sigma_a$
Partly reversed between σ_1 and $(-)\sigma_2$, where $\sigma_2 < \sigma_1$ and of opposite sign	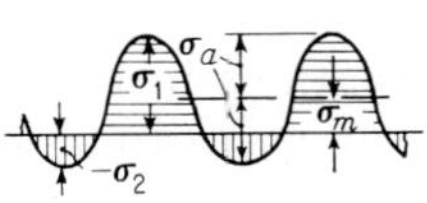 σ_a, σ_1, σ_m, $-\sigma_2$	σ_1	$-1 < \frac{-\sigma_2}{\sigma_1} < 0$	σ_m	$\pm\sigma_a$
Completely reversed stress, between σ_1 and σ_2, where $\sigma_2 = \sigma_1$	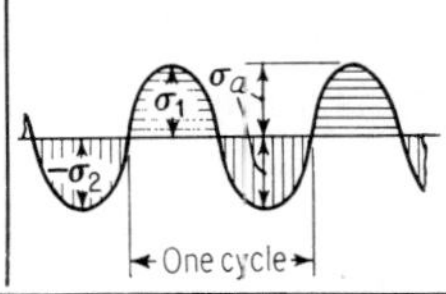 σ_a, σ_1, $-\sigma_2$, One cycle	σ_1	$\frac{-\sigma_2}{\sigma_1} = -1.0$	0	$\pm\sigma_a = \sigma_1$

Note. $\sigma_m = (\sigma_1 + \sigma_2)/2$ and $\sigma_a = (\sigma_1 - \sigma_2)/2$, with due regard to signs.

* "Report of the ASTM Research Committee on Fatigue of Metals," *Proc. ASTM*, vol. 37, pt. I, 1937.

buildings (except for certain slender elements that may be subject to vibration) are not large enough nor do they occur often enough to produce failure. It has been estimated that the stresses in the members of an ordinary railway bridge are repeated less than 2 million times in a period of 50 years.

In some instances, notably in rapidly moving machines and in parts subject to severe vibrations, appreciable stress fluctuations may occur, running to billions of repetitions during the useful life of the machine or structure. The crankshaft of a piston-type airplane motor is subjected to about 20 million reversals of stress in less than 200 hr of flying. Furthermore, the stresses are relatively high, since the motor is constantly operated at nearly maximum power and the size of the shaft is kept as small as possible to reduce its weight. The stresses in the shaft of a steam turbine, if operated continuously for 10 years, would be reversed about 16 billion times, whereas the stresses in the blades would be reversed about 250 billion times. Fatigue must be considered in the design of many parts subject to cycles of stress such as motor shafts, bolts, springs, gear teeth, turbine blades, airplane and automobile parts, steam- and gas-engine parts, railway rails, wire rope, car axles, and many machine parts subjected to cyclic loading.

The stress at which a metal fails by fatigue is herein termed the *fatigue strength*. It has been found that for most materials there is a limiting stress below which a load may be repeatedly applied an indefinitely large number of times without causing failure. This limiting stress is called the *endurance limit*. The magnitude of the endurance limit depends upon the kind of stress variation to which the material is subjected. Unless qualified, the endurance limit is usually understood to be that for completely reversed bending. For most constructional materials, the endurance limit in completely reversed bending varies between about 0.2 and 0.6 of the static strength, although for a given kind of material the ratio of endurance limit to static strength, called the *endurance ratio*, will range between narrower limits.

Most elements of a structure or machine that are subjected to repeated stresses are made of metal—principally steel—so that this discussion is concerned primarily with the fatigue testing of metallic materials. References on fatigue of wood and concrete are listed in Appendix H. Flexural-fatigue tests of plastics (ASTM D 671) and compressive-fatigue tests of vulcanized rubber (ASTM D 623) are covered by ASTM standards.

9.2. Nature of Fatigue in Metals. Commercial metals are composed of aggregations of small crystals with random orientations. The crystals themselves are usually nonisotropic. Experiments indicate that some crystals in a stressed piece of metal reach their limit of elastic action sooner than others, owing, no doubt, to their unfavorable orientation, which permits slip to occur. Also, the distribution of stress from crystal to crystal within a piece of stressed metal is probably nonuniform, and when a piece is subjected to cyclic stress variation, the constituent particles tend to move slightly with respect to one another. This movement finally weakens some minute element to such an extent that it ruptures. In the zone of failure a stress concentration develops, and with successive repetitions of stress the fracture spreads from this nucleus across the entire section. For this reason fatigue failures are often referred to as "progressive fractures."

High localized stress is also developed at abrupt changes in cross section, at the base of surface scratches, at the root of a screw thread, at the edge of small inclusions of foreign substances, and at a minute blowhole or similar internal defect. These are typical conditions which accentuate the susceptibility to failure by fatigue. It should be noted too that failure of machine parts is often the result of a combination of fatigue and damage by occasional overstrain.

The relative movement of the elements of minute steel crystals, when under stress, was first observed by Ewing and Rosenhain in 1899 [906]. The movement became evident as parallel lines, called "slip lines," across the face of individual crystal grains as they were viewed under the microscope when illuminated by oblique lighting. Four years later, Ewing and Humphrey observed that the slip lines developed in steel by subjecting it to repeated cycles of stress would develop into microscopic cracks which in turn spread and cause failure of the piece [240].

Fatigue failures occur suddenly without any appreciable deformation, and the fracture is coarsely crystalline as in the case of a static failure of cast iron or brittle steel. The appearance of the fracture and the characteristic suddenness with which failure occurs gave rise to a theory of cold crystallization of metal—there are still those who refer to metal that has failed by fatigue as having become "crystallized." Many experiments have shown that this idea is incorrect, although it is fostered by the fact that sudden failures display the crystalline structure of a metal, which, in the case of gradual failure as in a static test, is disguised by the distortion of the piece owing to its ductility.

Metals crystallize when they solidify from the liquid state, and there is no change in their inherent crystallinity due to the action of repeated stresses. The large facets in a fatigue failure are formed while the deterioration is in progress by a continuous cleavage plane extending through two or more adjacent crystals, which tends to exaggerate the apparent size of the crystals on the surface of the break. However, a microscopic examination of the metal behind a fatigue fracture shows no change in its structure or increase in size of the individual crystals. The idea of "crystallization" undoubtedly arose from the fact that many parts, ruptured under repeated loading, showed a coarsely crystalline fracture that may have been due to overheating, defective chemical composition, or some maltreatment in fabrication. The parts broke in many cases because these defects made them particularly weak in resisting repeated stresses.

9.3. Scope and Applicability of Fatigue Tests. Numerous tests of various types have been successfully employed in developing a fairly adequate fund of data on the endurance limits of many metals. On the whole, the data appear to be reasonably reliable and seem to form a satisfactory basis for design. It is worthy of note, however, that no fatigue tests for metals have been standardized by the ASTM, although certain types of machines, specimens, and procedures are in common use.

One of the simplest and probably the most widely used type of test for determining the endurance limit of a *material* employs completely reversed flexural loading on rotating-beam specimens (see Art. 9.4), the maximum stress being computed by the use of the simple flexure formula. When carefully prepared smoothly finished specimens without abrupt changes in cross section are used, fairly concordant results are obtained. When grooved or notched specimens are used, the endurance limits have been found to be functions of the true maximum concentrated stresses developed. In fact, repeated-load tests have been used for evaluating stress-concentration factors.

Although the effect of range in stress on the endurance limit has not been fully explored, sufficient information is available for what seems to be a fair (though rough) estimate of the endurance limit of metals under types of loading other than those giving complete reversal of stress. Data are also available that make possible an estimate of fatigue strengths under axial and under torsional loadings. For each of a number of given types of materials these show some correlation with static ultimate strengths.

Owing to uncertainties in the stress analysis it is difficult to apply the results of fatigue tests on small specimens to the design of complicated parts and built-up units. To obtain direct information on the behavior of such parts or units under repeated loadings, fatigue tests have been made on a number of kinds of full-sized specimens, notably axles and riveted, bolted, and welded joints [910, 920–923]. A fatigue test is generally unsuitable for an inspection test, or quality-control test, owing to the time and effort required to collect the data.

9.4. Machines for Fatigue Tests of Metals. Machines for making fatigue tests under cycles of repeated or reversed stress may be classified according to the type of stress produced:

1. Machines for cycles of axial stress (tension, compression)
2. Machines for cycles of flexural stress
3. Machines for cycles of torsional shearing stress
4. Universal machines for axial, flexural, or torsional shearing stresses or combinations thereof

All repeated-stress testing machines must be provided with a means for applying load to a specimen and with a means for measuring the load. Also, there must be provided a counter for recording the number of cycles applied and some device that, when the specimen breaks, automatically disengages the counter. Frequently the disengaging device is also designed to stop the testing machine itself.

Machines for applying flexural stresses are the most common, and they have been used in the majority of investigations. The popularity of this type of machine is due to its simplicity of operation, the accuracy with which it can be calibrated, and the fact that it produces a common condition of stress. Flexural-testing machines are generally of the rotating-beam type, one example of which is shown in Fig. 9.1. For this machine a specimen (Fig. 9.2) is held at its ends in special holders and loaded through two bearings equidistant from the center of the span. Equal loads on these two bearings are applied by means of weights that produce a uniform bending moment in the specimen between the two loaded bearings, as shown in Fig. 6.11*b*. To apply cycles of stress the specimen is rotated by a motor; since the upper fibers of the rotated beam are always in compression while the lower fibers are in tension, it is apparent that a complete cycle of reversed stress in all fibers of the beam is produced during each revolution.

Two types of testing machines are in use in which a cantilever specimen carries a load at its free end. In one type, the specimen is rotated while a gravity load is applied to the free end. In another, a bearing at the free end carries the load from a compression spring acting in a plane normal to the longitudinal axis of the specimen; thus the latter is deflected, and as the spring rotates around it, cycles of reversed stress are produced.

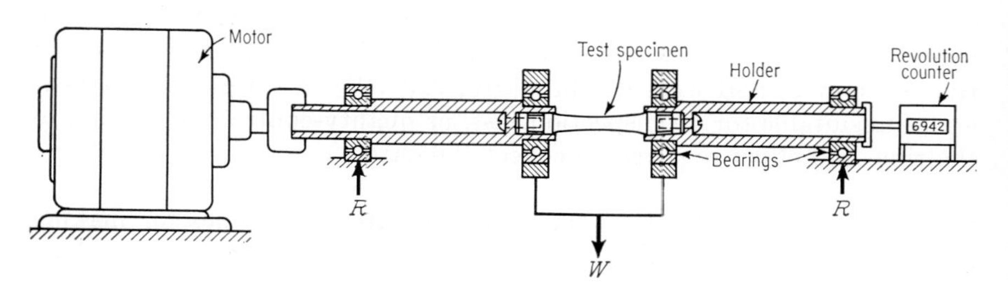

Fig. 9.1. **One type of rotating-beam reversed-stress testing machine.**

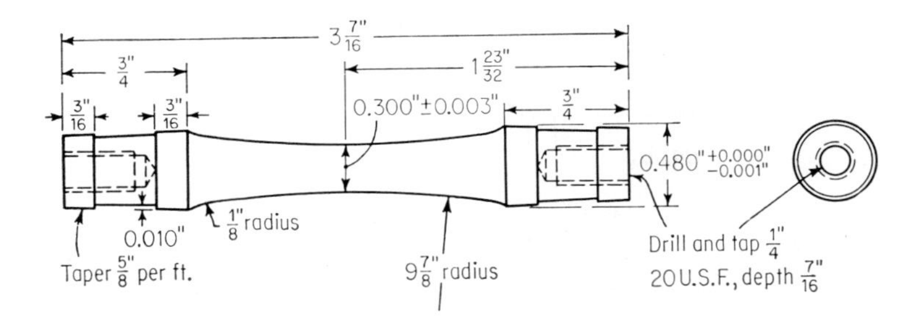

Fig. 9.2. **Specimen for rotating-beam test.** (*From Metals Handbook, American Society for Metals, Cleveland, Ohio,* 1948.)

Several types of fatigue-testing machines are universal in application. They have one stationary head or fixed platen and one vibratory platen, with provision for connecting various fixtures with attached specimens between them. The vibratory platen exerts a controlled motion or force on the specimen: if exerted axially, tensile and compressive stresses will be developed. By use of various fixtures, torsional or flexural stresses can be developed. Four basic methods are used on universal machines to generate and control the force.

In the Haigh machine, which is of the a-c magnet type, the specimen is subjected to cycles of axial stress by being attached to an armature that moves rapidly back and forth between two electromagnets energized by two-phase alternating current, one phase being connected to each magnet. Thus, when axially loaded, the specimen is alternately stretched and compressed by the action of the magnets [901].

Another type of machine uses a variable-throw crank, as shown in Fig. 9.3, to generate and control the force acting on the specimen. For any one test the throw is constant and the load on the specimen is measured by the deflection of the transmission beam. Static preload is applied by moving the fixed platen with either a hydraulic or screw mechanism. Some machines have special controls that sense the load and automatically maintain constant preload and vibratory forces to compensate for creep and slippage. The flexplates, shown in Fig. 9.3, eliminate all transverse motion of the vibratory platen and ensure a sinusoidal motion in one plane only.

A third type of universal fatigue-testing machine uses a hydraulic control and power system to move a piston through a fixed-amplitude stroke. The main loading piston is controlled by a small actuating piston which is operated by an adjustable-throw crank.

The fourth type of universal machine, which is shown schematically in Fig. 9.4, uses a constant force instead of a motion of constant amplitude. The load on the specimen is produced by a mechanical oscillator which

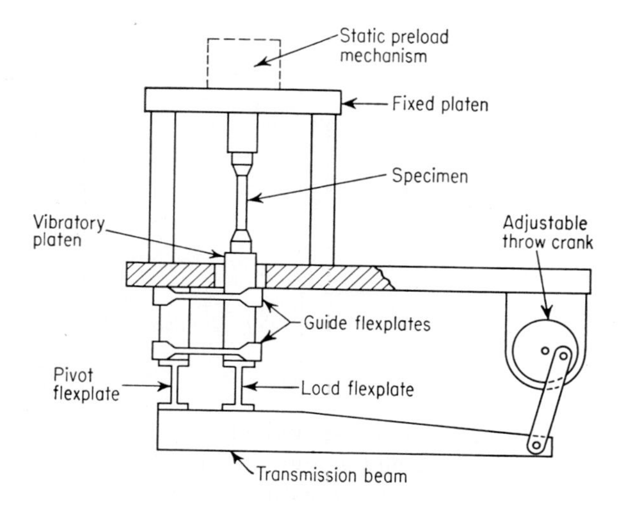

Fig. 9.3. **Constant-amplitude type of universal fatigue-testing machine set up for a tension-compression test.** (*From Breunich, Ref.* 914.)

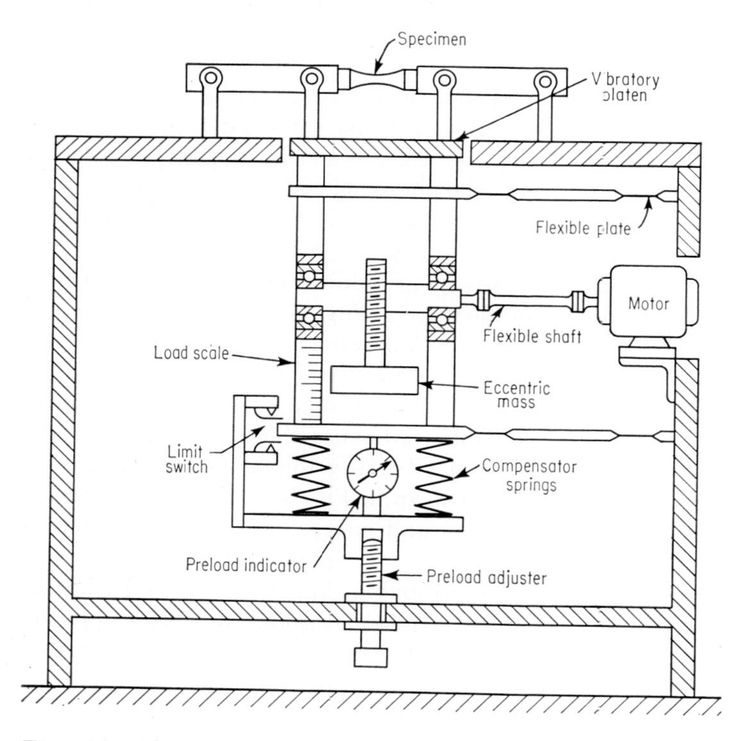

Fig. 9.4. **Sonntag inertia-type constant-force universal fatigue-testing-machine setup for a flexure test.** (*From Breunich, Ref.* 914.)

is, essentially, an eccentrically mounted mass rotated by a synchronous motor. The unbalance of this mass can be changed from zero to a maximum, and its position is calibrated to indicate the dynamic load in pounds. By adjusting the preload springs, any desired ratio of maximum to minimum stress can be produced.

The horizontal component of the centrifugal force is absorbed by restraining members, leaving a sinusoidal force in a vertical direction which is transmitted to the test piece.

9.5. General Procedure. To determine the endurance limit of a metal, it is necessary to prepare a number of similar specimens that are representative of the material. The first specimen is tested at a relatively high stress so that failure will occur at a small number of applications of stress. Succeeding specimens are then tested, each one at a lower stress. The number of repetitions required to produce failure increases as the stress decreases. Specimens stressed below the endurance limit will not rupture,

The results of fatigue tests are commonly plotted on diagrams in which values of stress are plotted as ordinates and values of number of

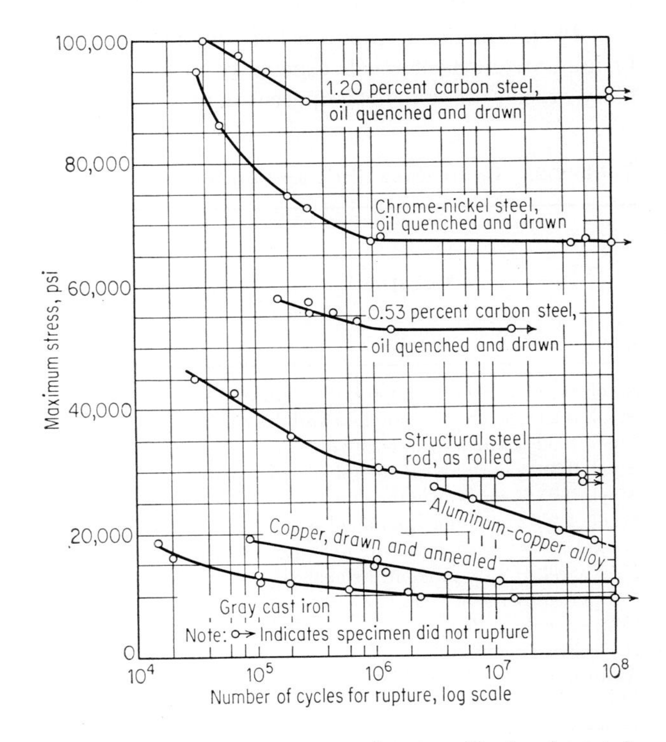

***Fig. 9.5.* Typical *SN* diagrams for determining endurance limit of metals under reversed flexural stress.** (*From Moore, Ref.* 202.)

cycles of stress for fracture are plotted as abscissas. Such diagrams are called SN diagrams (S for stress, N for number of cycles). In general, the SN diagrams are drawn using semilogarithmic plotting as shown in Fig. 9.5, which presents the results for various typical materials. For all ferrous metals tested, and for most nonferrous metals, the SN diagrams become horizontal, as nearly as can be determined, for values of N ranging from 1,000,000 to 50,000,000 cycles, thus indicating a well-defined endurance limit. The SN diagrams for duralumin and monel metal do not indicate well-defined endurance limits.

9.6. Effect of Important Variables. The fatigue strength of metals varies with composition, grain structure, heat-treatment, and mechanical working. The endurance limits as well as the static strengths of a few representative metals are shown in Table 9.2.

Various alloying elements increase both the static strength and the fatigue strength of steel. No alloying element is known that distinctly improves the fatigue strength without also increasing the static strength. Alloying elements (other than carbon) do not affect fatigue strength as much as does the microstructure of the steel [934].

Proper heat-treatment is beneficial to both the static strength and the fatigue strength of steel, especially in the case of high-carbon steel and alloy steels, for which the percentage of increase may be as high as 200.

Table* 9.2. Endurance limit and endurance ratio of various metals

Metal	Static tensile strength, psi	Endurance limit in reversed flexure, psi	Endurance ratio
Steel, 0.18% carbon, hot-rolled	62,700	30,900	0.49
Steel, 0.24% carbon, quenched and drawn	67,500	29,500	0.44
Steel, 0.32% carbon, hot-rolled	65,700	31,300	0.48
Steel, 0.38% carbon, quenched and drawn	91,500	33,500	0.37
Steel, 0.93% carbon, annealed	84,100	30,500	0.36
Steel, 1.02% carbon, quenched	200,400	105,000	0.51
Nickel steel, SAE 2341, quenched	282,000	112,000	0.40
Cast steel, 0.25% carbon, as cast	67,200	27,000	0.40
Copper, annealed	32,400	10,000	0.31
Copper, cold-rolled	52,000	16,000	0.31
70-30 brass, cold-rolled	73,200	17,500	0.24
Aluminum alloy 2024, T36	72,000	18,000	0.25
Magnesium alloy AZ63A	40,000	11,000	0.27

* Adapted from H. F. Moore and J. B. Kommers, *Fatigue of Metals*, McGraw-Hill, New York, 1927.

However, as in the case of composition of the steel, no heat-treatment is distinctly beneficial solely to the fatigue strength. In general, heat-treatment and chemical composition are closely related, inasmuch as the use of certain alloying elements is largely responsible for the depth of penetration and degree of heat-treatment possible. If a machine part is loaded only in flexure or torsion, then surface-hardening treatments may improve fatigue resistance, but if a machine part is to be used in axial loading, then the uniformity of properties throughout the whole cross section is important, and a heat-treated alloy steel that has good depth-hardening properties is superior to one that has not. Specimens refrigerated before tempering have higher fatigue strength than unrefrigerated specimens since refrigeration tends to reduce the amount of retained austenite [934]. Some heat-treatments may increase the static yield strength but decrease the fatigue strength because of minute quenching cracks or decarburization of the surface.

Cold-working of ductile steel increases its fatigue strength to about the same degree as it increases the static strength. For nonferrous metals the percentage of increase of fatigue strength due to cold-working is sometimes less than the percentage of increase in tensile strength.

It is difficult to carry out repeated stress tests of specimens under cycles of alternating direct tension and compression owing to the possibility that any slight eccentricity of load may cause serious flexural stresses and that high localized stresses are likely to occur at the shoulders of axially loaded specimens. These stress concentrations in items subjected to repeated stresses are of considerable importance even for ductile materials, although they have very little effect on the static tensile strength. In general, carefully performed tests have shown that the endurance limit for cycles of alternating direct tension and compression is practically the same as the endurance limit for cycles of reversed flexural stress.

The endurance limit for shearing stresses is usually determined from tests in repeated or reversed torsion. Most of these determinations have been made on carbon and alloy steels. For tests of carbon steels the ratio of the endurance limit in reversed torsion to the endurance limit in reversed flexure has been found to vary from about 0.48 to 0.64, with an average of 0.55. For alloy steels the ratio varies from 0.44 to 0.71, with an average of 0.58. The average ratio for a few nonferrous metals is 0.52 [901].

For specimens of metal subjected to repeated stresses involving a range of stress less than complete reversal, the smaller the range of stress, the higher the endurance limit. The limiting value is, of course, the static strength. The general nature of the variation of strength with range in stress is shown in Fig. 9.6. Three methods of representing fatigue data involving the variable of range in stress are shown. Figure 9.6*a* shows the Goodman-Johnson type of diagram in which the minimum stress is plotted to give a straight line (the horizontal scale being without significance), and the endurance limit corresponding to any minimum stress is plotted vertically above, giving the upper curved line. The range of stress is represented by the vertical ordinate between the lower and upper solid lines, whereas the mean stress is represented by the curved dashed line. Thus, for any minimum stress AC, the endurance limit is BC, the mean stress is DC, and the range of stress is AB.

Figure 9.6*b*, the Schenck-Peterson diagram, is drawn in much the same manner except that the curved line representing mean stresses is drawn as a straight line at an angle of 45° with the horizontal axis. This makes the minimum stress line a curve and permits the horizontal axis to represent the mean stresses to the same scale as on the vertical axis.

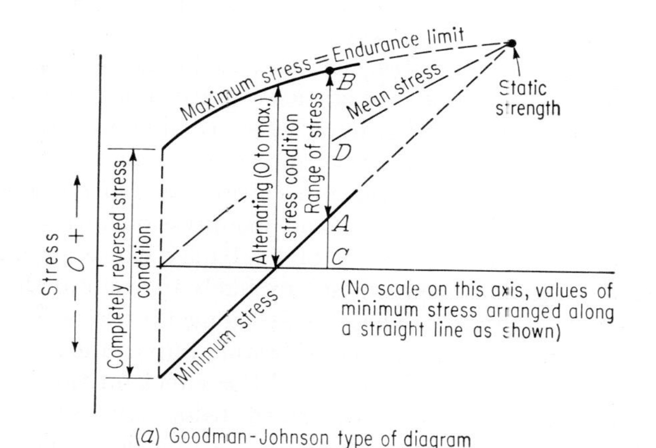

(*a*) Goodman-Johnson type of diagram

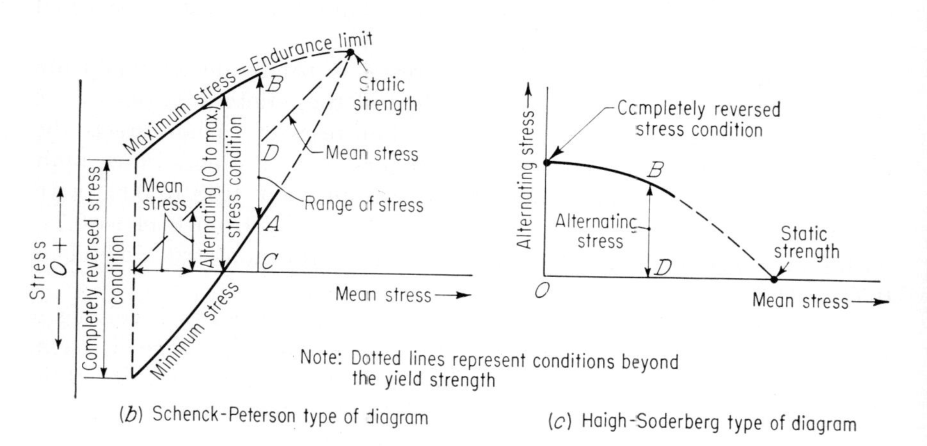

(*b*) Schenck-Peterson type of diagram

(*c*) Haigh-Soderberg type of diagram

Fig. 9.6. **General variation of strength with range of stress.**

Figure 9.6*c*, the Haigh-Soderberg diagram, represents the upper half of Fig. 9.6*b*, but with the 45° line representing mean stresses turned to the horizontal position. In this diagram ordinates to the curve represent the maximum value of the alternating stress BD, which can be applied simultaneously with a mean stress OD without causing failure by fatigue. For the purpose of estimating values of fatigue strength for use in design, a number of formulas, based on idealizations of the data, have been derived [901, 912].

The repeated loading of metals to stresses below the endurance limit appears to produce a mild cold-working of the material resulting in a slight increase in its original or normal endurance limit, but the strengthening varies widely for different metals; for progressively increasing cyclic understress, the increase in the endurance limit may be considerable. On the other hand, many preliminary cycles of stress above the normal endurance limit appear to start the formation of microscopic cracks that cause a reduction of the endurance limit; the higher the preliminary stress and the greater the number of preliminary cycles of stress, the greater the damage done and the lower the resulting modified endurance limit. Also, the static load capacity will be decreased due to the high stress developed at the root of the microscopic fatigue crack. Some materials appear to be damaged more than others by preliminary overstressing. Since many machines or structures are likely to be subjected to temporary overloads, the possible effect of such overloads on the endurance limit should be considered [909].

In machine design, although there may be no intention of using allowable stresses above the endurance limit, inability to analyze the stress condition accurately often results in higher stresses than those calculated. Many repetitions of this overstress may cause the part to fail. However, if during the life of a structure an occasional overload occurs owing to carelessness in operation or other unforeseen circumstance, such overstressing, unless it is excessive, will probably not materially affect the ability of the structure to carry its normal working loads.

The surface finish of a test specimen has a definite effect on the fatigue strength; the rougher the finish, the lower the endurance limit. In fact, a very rough surface finish, particularly if the scratches or machine marks are normal to the applied force, may lower the endurance limit of a metal by as much as 15 to 20 percent in comparison with the endurance limit for a polished surface. However, a part which is machine polished may fail more rapidly than one with a good tool finish, because of the tension set up in the skin by the heat of the polishing.

Shot blasting the surface of unpolished machined specimens of some kinds of steel has a beneficial effect on the endurance limit, in comparison with the result for polished but not shot-blasted specimens, although the tensile strength is not correspondingly benefited [930]. For wire specimens 0.162 in. in diameter which were not machined or polished, the endurance limits were as follows:

Untreated	74,000 psi
Shot blasted	96,000 psi
Shot blasted and tempered at 400°F	101,000 psi

There does not appear to be any advantage, as regards endurance limits, in shot blasting surfaces where high residual stresses due to quenching and insufficient drawing are present.

Abrupt changes in cross section definitely lower the nominal fatigue strength owing to the high concentration of stress at such transitions. The results of tests by Moore and Kommers, presented in Table 9.3, are typical of those obtained by other investigators. They show that a sharp V notch may reduce the nominal endurance limit of a test specimen in reversed bending by about 65 percent, even though the net cross-sectional area remains constant. In actual machine parts, subject to cycles of reversed stress, any abrupt change in section due to holes, grooves, notches, screw threads, and shoulders must be given special consideration. However, the effect of such stress raisers is not so serious as would appear from the results of computations made on the basis of the theory of elasticity or photoelastic analysis.

Some of the alloy steels, such as chrome nickel, seem to be more sensitive to stress-concentration effects, a property that has been called the *tenderness* of a material. Cast iron is considerably less sensitive to imposed stress concentrations in fatigue tests than are steels, the reason apparently being that the inherent irregularities in the microstructure of cast iron produce considerable stress concentration, so that the effect of an imposed discontinuity is somewhat masked.

Table* 9.3. Effect of shape of test specimen on nominal endurance limit

Specimen diameter, in.		Means of diameter reduction	Reduction in nominal endurance limit, percent
At ends	At center		
0.40	0.275	Groove with 10-in. radius	0
0.40	0.275	Groove with 1-in. radius	5
0.40	0.275	Groove with ¼-in. radius	10
0.40	0.275	Shoulder with small fillet	25
0.40	0.275	Square shoulder	50
0.40	0.275	90° V notch	65

* Adapted from H. F. Moore and J. B. Kommers, *Fatigue of Metals*, McGraw-Hill, New York, 1927; and H. F. Moore, T. M. Jasper, and J. B. Kommers, "An Investigation of the Fatigue of Metals," *Univ. Ill. Eng. Expt. Sta. Bulls*. 124 (1921), 136 (1923), 142 (1924).

Chrome plating decreases the fatigue strength, but nickle plating will either increase or decrease the strength depending on the bath used.

The influence of the frequency of stress reversal on the endurance limit has been under discussion for some time. It now seems clear that in the range of ordinary testing speeds of say 2000 to 15,000 cycles per min there is little if any variation in the endurance limits obtained, although there appear to be inconsistencies in results when speeds as high as 30,000 cycles per min are used [903, 904, 913, 926, 927].

9.7. Correlation with Other Properties. Correlation between the endurance limit and one of the more readily determinable mechanical properties is desirable because of the expense involved and the time required to determine the endurance limit by a long-time test. However, there is no direct relationship between the endurance limit and any of the other physical properties that will apply to all metals. Elastic strength, ultimate strength, and ductility seem to influence the fatigue strength. Furthermore, fatigue failure is a progressive fracture of a metal. This may account for a closer correlation between fatigue strength and tensile strength than there is between fatigue strength and any other single physical property.

The results of many tests indicate that the endurance limit under reversed flexure for rolled or forged steel and iron is usually between 45 and 55 percent of the tensile strength, averaging about 50 percent, although in some cases it may be as low as 35 or as high as 60 percent. For steel and iron castings this endurance ratio averages about 40 percent. For nonferrous metals the endurance ratio varies over a wide range, from less than 25 percent for cold-drawn copper and certain aluminum-copper alloys to 50 percent for annealed bronze. For representative values from various individual tests, see Table 9.2.

The Brinell hardness test seems to furnish a fair index of the endurance limit for rolled and forged steel. For such materials the endurance limit under reversed flexure is about 250 times the Brinell hardness number. This relationship fails to hold when the Brinell number is in excess of 400.

9.8. Corrosion Fatigue. Simultaneous corrosion and repeated stress is called *corrosion fatigue*. Cycles of repeated stress and simultaneous corrosion by so mild an agent as fresh water may reduce the fatigue strength of a metal to a value below one-half its value for fatigue tests in air. Corrosion-resistant metals seem to be somewhat superior in their resistance to corrosion fatigue. The actual surface corrosion of a specimen that fails by corrosion fatigue seems to consist in the mechanical breaking down by repeated stress of the more or less protective film that is formed as a result of reaction between the metal and the corroding agent, and consequent progressive spread of corrosion pits followed by the formation of a spreading fatigue crack. The time of exposure to the corroding agent plays an important part in corrosion fatigue because deeper pitting and consequent higher stress concentrations result from longer periods of exposure of the metal to corrosive agents while subject to repeated stress. Certain inhibitors, notably sodium chromate, have been effective in reducing corrosion fatigue.

LOW- AND HIGH-TEMPERATURE TESTS

9.9. Tests of Metals at Low Temperatures. Many materials are used at low temperatures. Such applications are made in refrigeration equipment, many chemical operations, and in both machines and structures located in areas where cold weather occurs. Most metals show an increase in tensile strength at lower temperatures, but their yield strengths are not always affected to the same degree, as shown in Fig. 9.7. In general, ductility decreases with lower temperatures so that their toughness is materially reduced under impact loadings, as shown in Fig. 8.9.

9.10. Long-time High-temperature Loadings. Many types of engineering machines and structures are subjected to stress while at relatively

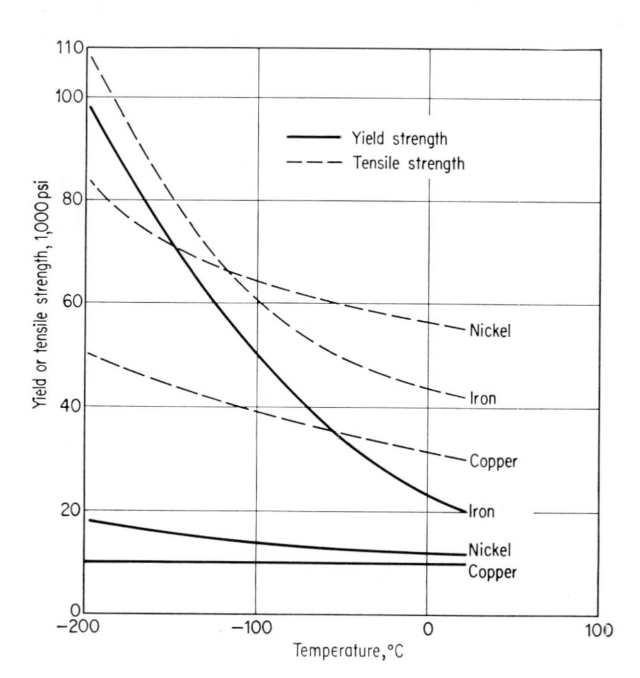

Fig. 9.7. **Variation of yield and tensile strengths of iron, copper, and nickel with temperature.** (*From Amer. Soc. Metals Symposium, Ref.* 946.)

high temperatures, a few common examples being internal-combustion and jet engines, high-pressure boilers and steam turbines, and cracking stills such as those used in the chemical and petroleum industries. Some types of equipment are operated for considerable periods of time at temperatures as high as 1200°F and sometimes at higher temperatures.

At high temperatures many materials act somewhat like a very viscous liquid so that a stress often much less than that required to cause failure in a few minutes may cause failure if sufficient time is allowed. The physical process that brings about failure is a comparatively slow but progressively increasing strain called *creep*. At any temperature, creep may be rapid or slow; its rate decreases rapidly as the stress is lowered. Although at some relatively low stress the possibility of rupture due to creep within a long period may be negligible, such a stress may nevertheless cause undesirable structural *distortion* in the course of time. Examples of the effects of such excessive distortion are the loosening of flanged joints caused by creep in the connecting bolts, and the undesirable changes in clearance of steam turbine blading.

Experience has shown that certain materials can be successfully used at high temperatures provided the stresses in them are kept sufficiently low or that they are replaced before too long a period of service. In many actual applications absence of appreciable distortion has been demonstrated. This has led to the desire to raise still further either or both allowable stresses and temperatures in the design of new units and to search for new materials capable of withstanding the more severe conditions. Progress in the production of new materials has been rapid; although some of these materials have given satisfactory service, others have proved inadequate, thus necessitating frequent replacements. The need for adequate high-temperature tests is obvious. Some of these tests are conducted over long periods of time to determine the resulting strain for a given temperature and stress. In other tests the time-for-rupture is determined at a given temperature and stress. Both types of test are covered in ASTM E 139.

9.11. Nature of Creep. Most structural metals and alloys possess elastic properties at room temperature. As illustrated in Fig. 9.8, if a stress σ below the proportional limit is applied, an elastic strain OA will occur immediately upon application of the load, and regardless of the duration of application the strain will remain constant. The strain characteristics under this condition will be represented by the curve OAB.

At elevated temperatures a strain OC occurs upon the application of the same stress σ, OC being greater than OA owing partly to the lower value of the modulus of elasticity at the higher temperature. The strain OC may be entirely elastic, or elastic plus plastic, depending upon the material, the temperature, and the stress. Under suitable stress-temperature conditions, however, the strain increases as the time of load applications is extended, the strain following the curve *CDEF*. This continued increase in strain with time, while under a constant stress, is defined as creep.

As indicated in Fig. 9.8, the creep sequence consists of three stages. In the first stage, creep continues at a decreasing rate that becomes approximately constant during the second stage; the beginning of the third stage is marked by a rapid increase in creep rate which continues until fracture occurs. In the majority of tests conducted the stress and temperature conditions are selected so as to preclude the development of the third stage. Although too broad generalizations are not permissible, it is generally found that the duration of each stage tends to vary inversely with the stress.

The characteristic failure of metals at low temperatures is by fracture through the crystals themselves, whereas failure at high temperatures occurs at crystal boundaries. The temperature at which the manner of fracture changes from intra- to intercrystalline is called the *equicohesive* temperature. This temperature is not accurately known; for ordinary mild steel it is about 850°F, but for certain alloy steels it is somewhat higher.

The mechanism of strain also appears to change as the testing temperature is increased. At temperatures below the equicohesive temperature, strain apparently occurs largely as an elastic intracrystalline movement so that the metal exhibits elastic properties. At temperatures above the equicohesive temperature, however, it is believed that strain may

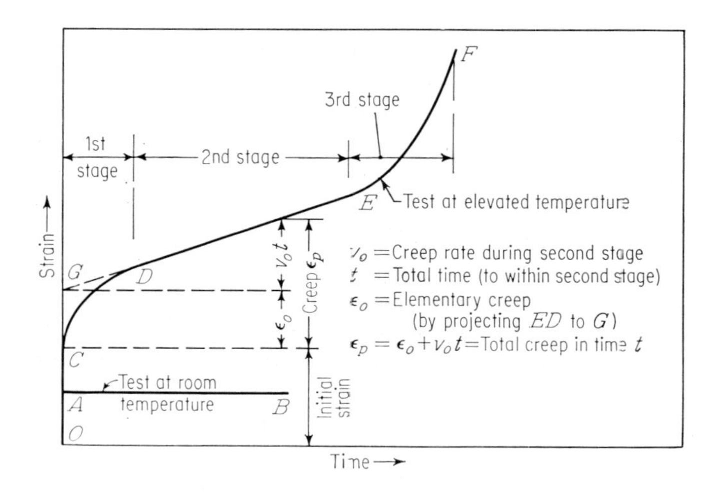

Fig. 9.8. **Strain-time curves at room and at elevated temperatures.**

occur through quasiviscous intercrystalline movement. However, the characteristics that render a material resistant to extension under high-temperature loading are not completely known.

The magnitude of the creep resulting from the application of a given stress over a given time period appears to depend upon the opposing effects of yielding of the material and strain hardening caused by such yielding. At or below the equicohesive temperature, strain hardening tends to predominate, and continuous measurable creep will not occur unless the stresses are sufficiently great to overcome the resistance caused by strain hardening. If strain hardening predominates, the diagram representing the second stage of Fig. 9.8 becomes a horizontal line. For example, the yielding of steel above the proportional limit at room temperature is a creep phenomenon, but measurable creep soon stops as a result of strain hardening. At temperatures above the equicohesive temperature, however, the yielding rate exceeds the strain-hardening rate, and creep will proceed even under low stresses. Creep could probably be detected at these higher temperatures even under very low stress if sufficiently sensitive apparatus were available.

If a specimen undergoing creep is unloaded, some of the strain is recovered, as illustrated in Fig. 9.9, but an appreciable plastic strain has become permanent, its amount depending upon the material and the test conditions of time of loading, temperature, and stress.

A measurement closely related to creep is that of *stress relaxation*. The specimen is first subjected to a fixed strain at an initial stress σ_o, and the load required to maintain this strain is observed progressively with time. A typical relaxation curve for a carbon steel is shown in Fig. 9.10. The importance of this stress is very evident. Various studies have been made to permit the prediction of relaxation from creep data [954].

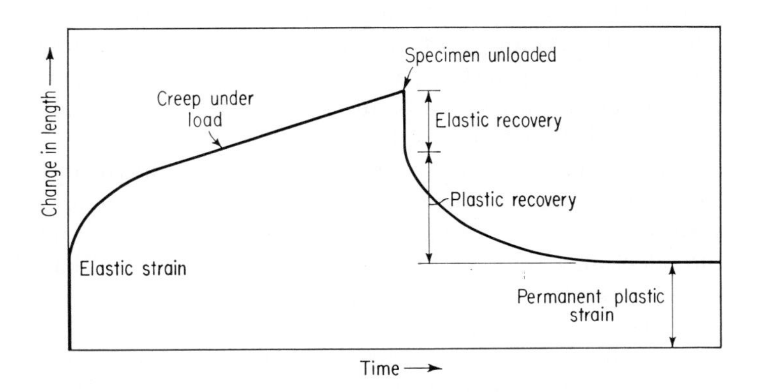

Fig. 9.9. **Recovery of strain after unloading creep specimen.**

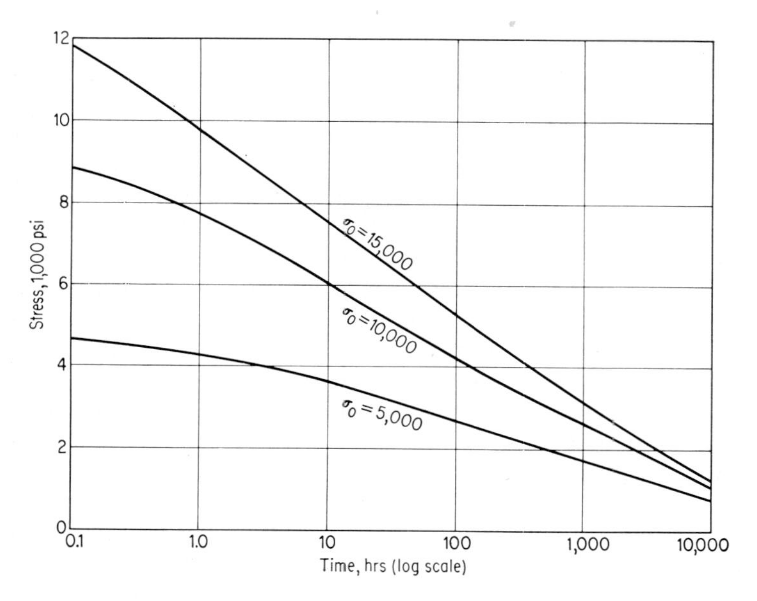

Fig. 9.10. **Relaxation of 0.35 percent carbon steel at 850°F.** (*From Roberts, Ref.* 954.)

9.12. Scope and Applicability of Creep Tests. Creep tests at high temperatures appear to be the only satisfactory guide to the performance of metals for high-temperature service, although creep testing as a scientific technique should be considered to be still in the developmental stage. Although the general phenomena associated with creep are fairly well known, there is much detailed information yet to be obtained in order to complete a well-rounded picture. Creep tests are inherently long-time tests, but the test periods may nevertheless be short in comparison with periods of high-temperature service in actual structures, so that extrapolation of creep-test data must be made with judgment. Creep tests require too much time to be used as acceptance tests. However, they are the basis of data to be used in design.

Extrapolation of creep data from tests that have not been carried into the second stage (uniform creep rate, see Fig. 9.8) tends to give excessive values of predicted creep, although this procedure admittedly results in the selection of more conservative allowable stresses than those based on tests conducted over longer periods. On the other hand, extrapolation from data of tests that have been carried well into the second stage does not, in itself, give assurance that under very long-time service the metal will not enter the third or failure stage, although the use of a small permissible creep (commonly 1 percent) ordinarily gives adequate protection against such an occurrence.

In view of the importance of having some common basis for making creep tests, the Joint ASTM-ASME Research Committee on Effect of Temperature on Properties of Metals has prepared a standard test procedure (ASTM E 139). In general, the method involves maintaining a standard test specimen at a constant temperature under a fixed tensile load and observing from time to time the strains that occur. Such a test may cover several months or even years, although some tests cover perhaps only about two or three months. Desirable test periods depend upon the reasonable life expected from the material in service. Test periods of less than 1 percent of the expected life are not considered to give significant results. Tests extending to 10 percent of the expected life are preferred. The test may be conducted on individual specimens at each of several loads and several temperatures. The results for a given material are often summarized in the form of composite diagrams, from which can be read the limiting stress for a given percentage of creep under stated temperatures and periods of time (see Art. 9.15).

For use under moderately high temperature service but below the equicohesive temperature, or for materials highly resistant to creep, it may be that strength is of more significance than creep alone. In such cases, *short-time* elevated-temperature tests, involving the determination of yield or tensile strengths, are pertinent. A standardized procedure is available (ASTM E 21). It should be obvious that, if the yield strength of a steel, as determined by a short-time test, is less than the creep limit for any given temperature, the yield strength is used in establishing the allowable stress.

9.13. Apparatus for Creep Tests. The determination of the creep characteristics of metals at high temperatures requires the use of three pieces of major equipment: (1) an electric furnace with suitable temperature-controlling apparatus, (2) an extensometer, and (3) a loading device.

Various types of electric furnace are used for creep tests, the one shown in Fig. 9.11 being typical of many installations for one specimen although some furnaces are arranged to accommodate as many as 12 or more specimens. A winding of nickel-chromium wire is spaced around a tube that is usually of fused silica or alundum. The winding can be held in position by a heat-resisting cement or may be laid in grooves in the tube. A large-diameter tube affords greater uniformity of temperature from center to outside of the test piece, and a long tube helps in reducing the temperature gradient between the center of the test length and the pull rods. A fair thickness of heat-insulating material is advisable for economy of heat.

It is common practice to space the turns along the tube in such a manner that more heat is supplied to the pull rods attached to the specimen than to the specimen itself in order to reduce temperature gradients

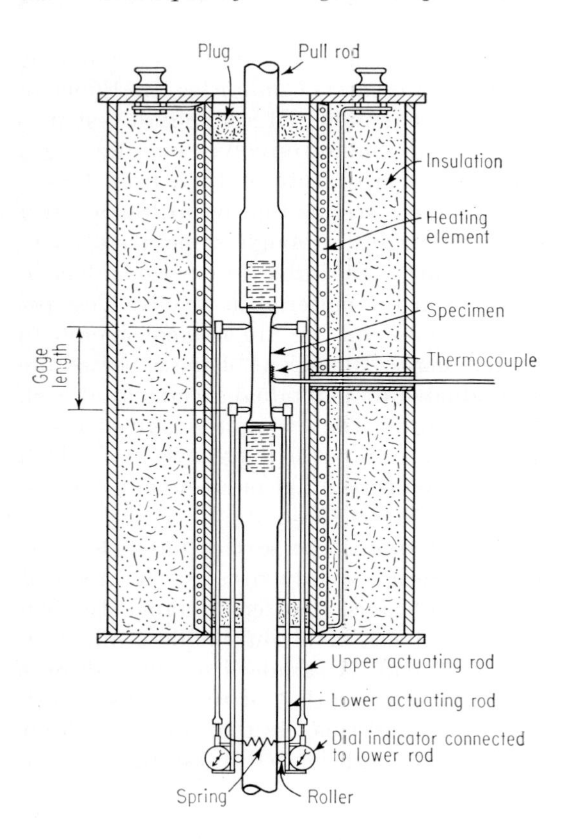

Fig. 9.11. **Electric furnace and extensometer for creep testing.**

along the test piece. Recognizing the undesirability of nonuniform temperatures, the ASTM specifies that the maximum variation of temperature over the gage length should not exceed ±3°F from the average test temperature for temperatures up to and including 1800°F, and ±5°F from the average test temperature for higher temperatures. Also, temperature variation of the specimen from the indicated nominal test temperature should not exceed ±3°F. The extent of the variation in temperature must be stated in reporting each test.

The ends of the furnace should be closed to prevent free circulation of air that would produce undesirable temperature variations and might cause oxidation of the surface of the specimen. In some cases an inert gas such as nitrogen is used inside the furnace.

Temperature is measured with a thermocouple whose junction is in direct contact with the specimen. Ordinarily, asbestos is used to protect the junction from direct radiation from the walls of the furnace tube.

An automatic controller, connected to the thermocouple, is used to ensure a constant temperature in the furnace. Relatively small fluctuations of temperature will not only introduce thermal strains to cause errors in the observed creep, but, near some critical high temperature, small changes in temperature cause large changes in the creep characteristics of the specimen. Therefore, the extent of the fluctuations of furnace temperature should be stated in reporting the tests. Since fluctuations of the room temperature affect the extensometer readings as well as control of the furnace temperature, thermostatic control of the room temperature within close limits is desirable.

Three types of extensometer are in common use to measure the strains in the specimen. One type makes use of two telescopes that sight through windows in the furnace on gage marks provided on the test piece. The strains are measured by means of a filar micrometer attached to one of the telescopes.

The time required for setting of a telescope and reading of a micrometer prevents rapid readings of the elastic strain. Since the early stages of creep occur many times as fast as the later stages, delay in obtaining measurements of elastic strain tends to give too high a value of the *initial* strain and too low a creep value.

The other two types of extensometer, mirror and dial indicator, employ two pairs of actuating rods that are clamped to the specimen at the ends of the gage length shown in Fig. 9.11. These rods should have good strength and show no appreciable scaling at high temperatures; a chrome-nickel alloy is generally used. In the simple type shown in Fig. 9.11, the actuating rods are connected to dial indicators that give direct readings of the strains. The ease and rapidity with which readings may be made with this latter type have a distinct advantage in that it permits quick determinations of the initial elastic strain without introducing errors caused by the very early but appreciable plastic strains. Thus elastic and plastic strains are more clearly defined. Readings should be taken on opposite sides of the specimen.

The load is applied to a specimen by means of direct weights or by a simple multiplying lever and weights as shown in Fig. 9.12. The applied load should be measured to within 1 percent. The outer ends of the pull rods are connected to the frame and loading lever through spherically seated nuts or other devices to permit freedom of rotation in two directions at right angles, thus ensuring axial loading of the test piece.

9.14. Test Specimens and Procedure. For round specimens, diameters of 0.505, 0.320, 0.252, or 0.100 in., and a gage length of 4 diameters, are recommended, but the 0.505-in. diameter is preferred. The surface should be finished smooth and free from tool marks and scratches.

In making tests at a given temperature, the unloaded specimen is first heated to the required temperature. When the temperature of the specimen is steady, the gage length is observed and the predetermined load

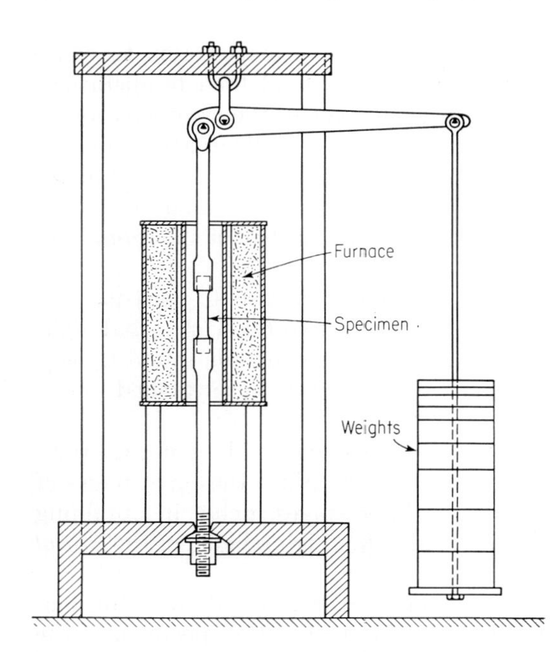

Fig. 9.12. **Loading apparatus for creep testing.** *(From Moore and Moore, Ref. 133.)*

applied quickly without shock. The resulting instantaneous extension is largely an elastic strain. Measurements of the subsequent creep are observed at sufficiently frequent intervals to define the strain-time curve; daily or weekly observations usually suffice. At the conclusion of the test there should be at least 50 observations of specimen temperature, the average of which should be reported as the actual test temperature.

9.15. Creep Data and Their Interpretation. In the design of members subjected to high temperatures, the allowable stress for a given operating temperature is defined by a permissible plastic strain in a specified time of service. Since these service periods may run into several years, it is usually necessary to extrapolate data obtained from relatively short-time tests. In some cases creep strains have been estimated on the assumption that creep proceeds at a constant rate for periods many times greater than the actual periods of observation, even though the test may not have been conducted as far as the second stage; that is done by extending a tangent to the end of the creep curve at some point in the first stage. Although this practice has proved helpful, it is only an expedient and an approximation. The creep test should be conducted long enough to establish the minimum uniform creep rate v_o of the second stage (Fig. 9.8). Then for a greater time t than is covered by the test, the total creep or plastic strain ϵ_p can be determined by the equation $\epsilon_p = \epsilon_o + v_o t$ (see Fig. 9.8).

An example of the reduction of typical creep data is shown in Fig. 9.13. Figure 9.13*a* shows creep curves plotted directly from experimental data for a temperature of 800°F. Extrapolated creep curves for various stresses are shown in Fig. 9.13*b*. Inasmuch as this scheme of evaluation is invalid for strains extending beyond the inflection toward the third phase of creep, it is recommended that extrapolations based upon it be carried no farther than the equivalent of 1 percent elongation. To obtain an allowable stress (a stress producing a specified creep in a stated time) a graph is first made between stress and percentage of creep for various

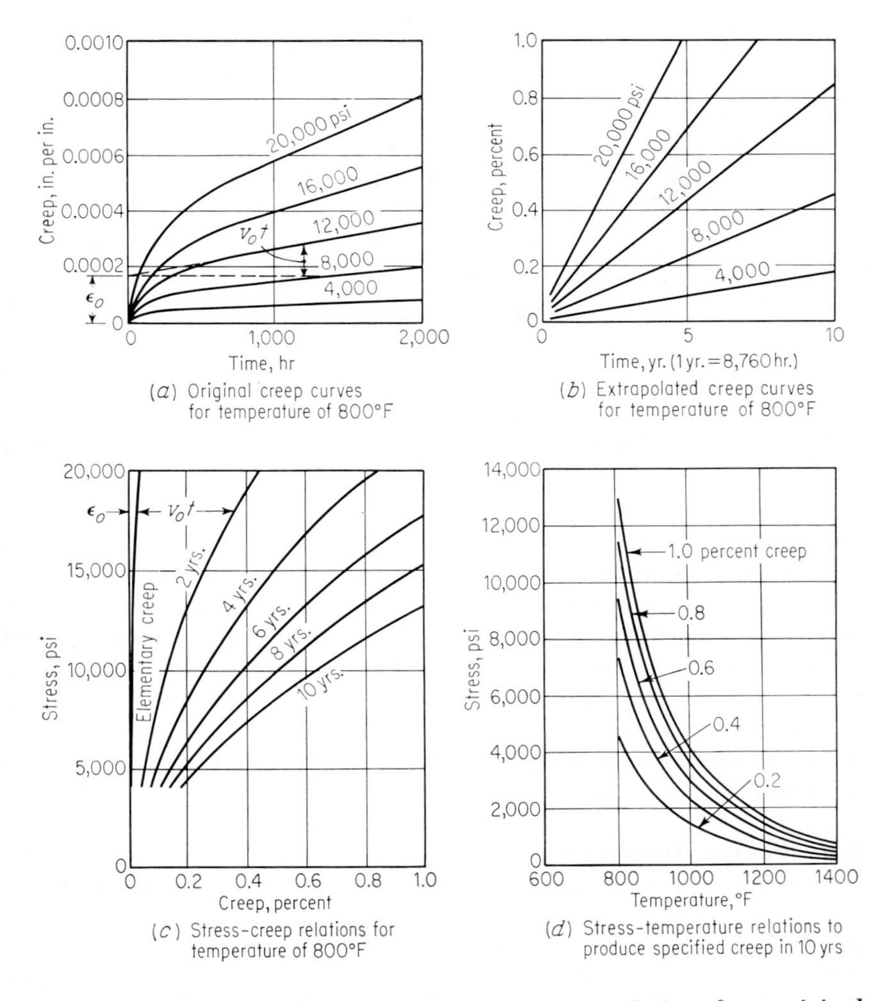

Fig. 9.13. **Development of stress-strain-temperature relations from original creep curves.** (*Based on McVetty, Ref.* 961.)

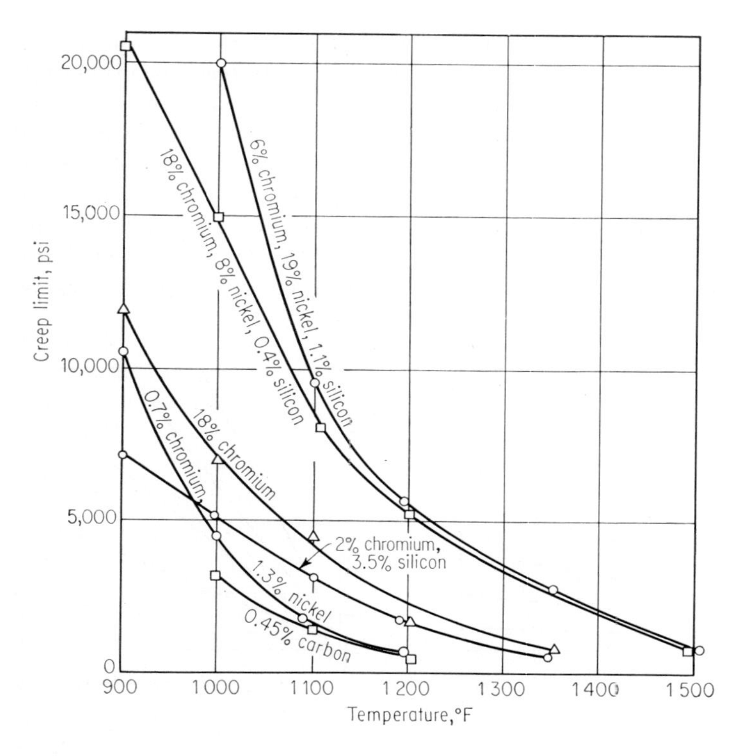

Fig. 9.14. **Creep limits [stresses producing 1 percent creep in 100,000 hr (11.4 years)] for various steels.** *(From Norton, Ref.* 952.)

periods of time (Fig. 9.13*c*); such relations are obtained directly from Fig. 9.13*b*. Thus, for a creep of 1 percent in 10 years, the allowable stress for a temperature of 800°F is 13,000 psi. For a temperature of 800°F these stresses are then plotted as ordinates in Fig. 9.13*d*. The stresses for other temperatures are also determined and the relations between stress and temperature for various creep values plotted as shown in Fig. 9.13*d*.

The limiting creep stresses for various types of steel at various operating temperatures for a period of 100,000 hr with 1 percent elongation are shown in Fig. 9.14.

9.16. Factors Influencing Creep. The influence of various factors on the creep of steel is summarized from Refs. 953 and 971 as follows:

The desired operating temperature determines the most suitable composition. At temperatures below the lowest temperature of recrystallization the creep resistance may be increased either by certain elements that largely enter into solid solution in the ferrite, such as nickel, cobalt and manganese, or by the carbide-forming elements, such as chromium, molybdenum, tungsten, and vanadium. At temperatures above the lowest temperature of recrystallization, however, the carbide-forming elements are the most effective in increasing the creep strength. Small additions of titanium and columbium to chromium-nickel stainless steels appreciably reduce their creep characteristics over a considerable range of high temperatures.

Creep resistance is influenced by heat-treatment. At temperatures of 1000°F or greater, the maximum resistance is usually produced by normalizing, provided the drawing temperature is about 200°F above the test temperature; the lowest resistance is produced by quenching and drawing. The creep resistance of the steel in an annealed condition is between those for the other two treatments.

Creep resistance is influenced by the manufacturing process. Results available indicate the electric-arc-furnace steel to be superior to the open-hearth product, whereas the induction furnace material is superior to the electric-arc-furnace steel.

Creep resistance is also influenced by the grain size of the steel. At temperatures below the lowest temperature of recrystallization a fine-grained steel possesses the greater resistance, whereas at temperatures above that point, a coarse-grained structure is superior.

9.17. Short-time Tests at High Temperatures. The loading of the components of rockets and guided missiles and missile power plants may occur within a few seconds, and the temperatures developed in some of the parts are very high. Obtaining satisfactory information for design of such parts has required a new set of high-temperature-testing techniques, since the test temperatures are usually in ranges where the metallurgical structure of the material is unstable and the creep rate comparatively high. Therefore, the new test procedure takes cognizance of the following factors in determining the strength of a material [998]:

1. Loss of strength at the elevated temperature
2. Creep rate of the material during the test
3. Effect of strain rate on the resultant strength level
4. Effect of aging or other metallurgical changes during the test

The *high-heating-rate test* for the evaluation of materials for missiles or similar service is performed by loading a special test specimen with weights, either directly or by means of levers, and heating the specimen, acting as a resistance unit, with an electric current until it fractures. The temperature and elongation are recorded by automatic equipment. The temperature-elongation curve shown in Fig. 9.15 resembles an ordinary stress-strain curve; the increase in temperature causes yielding and rupture to develop, whereas in the common tensile test the increase in stress causes failure. In Fig. 9.15, *OA* represents the initial strain due to loading.

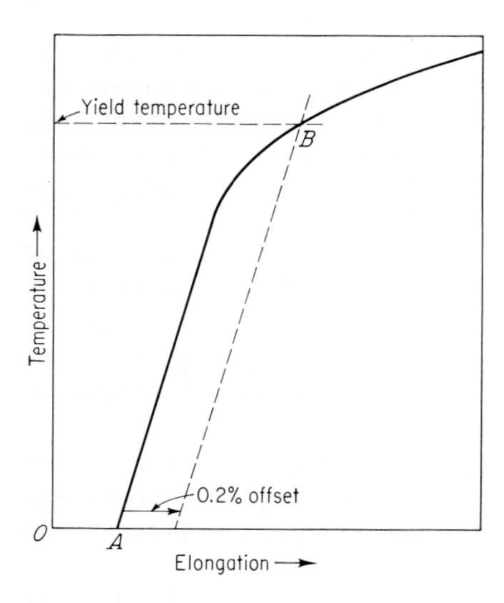

Fig. 9.15. **Method of determining yield temperature for a given stress and heating rate.**

During the early part of the test, as the temperature increases, the thermal expansions and decrease in the modulus of elasticity cause elastic elongations to occur. As for determinations of yield strength in an ordinary test, a line at 0.2 percent offset is drawn and extended to cross the plastic portion of the curve at *B*. This point determines the *yield temperature* for the given material, stress and heating rate.

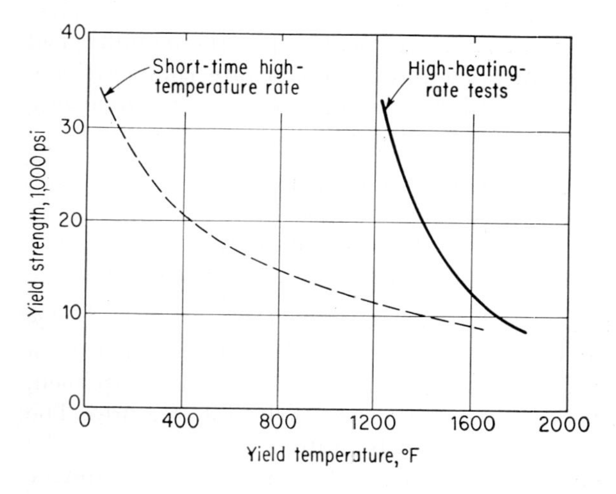

Fig. 9.16. **Comparison of short-time high-temperature and high-heating-rate yield strengths for annealed 18-8 stainless steel. Heating rate was 400°F per second.** (*From Amer. Soc. Metals, Ref.* 978.)

The results from a series of tests at a given heating rate but at different stresses are shown by the solid line in Fig. 9.16. The broken line shows the results from conventional short-time high-temperature tests in which the material is held for a while (or "soaked") at a given temperature without load and then is loaded to rupture over a period of time, usually several minutes in duration. The latter results show yielding at a lower temperature due to the effects of creep, the slower strain rate, and the metallurgical changes which occur during the test.

Chapter 10.

NONDESTRUCTIVE TESTING AND EXPERIMENTAL STRESS ANALYSIS

10.1. Need for Nondestructive Tests. Often it is desirable to know the characteristic properties of a product without subjecting it to destructive tests. With the exception of some hardness tests and proof loadings, the methods of testing discussed in the previous chapters will not permit the attainment of this objective, since most of the procedures, instead of using a finished product, use specially prepared specimens and test them to either partial or complete destruction.

The use of a particular material in a structure is ordinarily dependent upon its possessing certain characteristic properties which various destructive tests on similar parts have shown the material to possess normally. However, there is no assurance that the part used in the structure is the equal of those tested, in so far as mechanical properties and freedom from defects are concerned. If each part can be examined using some method by means of which its properties will become known within reasonable limits, or with reasonable assurance that any defects of consequence will be detected, it is much better than if only a few of a group be examined, even by a test so searching as to be destructive of the part.

Various types of defects are inherent in metallic products. A few of the more common ones are seams, subsurface flaws, and cracks resulting from quenching, embrittlement, and fatigue. Seam is a general term including strings of nonmetallic inclusions and elongated cracks, as well as defects caused by overfill or underfill in rolling operations; such defects run in the direction of rolling. Subsurface flaws form a large and important class of defects including seams, porosity, and inclusions. Quenching cracks are frequently developed during heat-treating operations, their occurrence being accentuated by sudden changes in section. Embrittlement cracks, as well as cracks produced by combined chemical action and mechanical stress, commonly occur in boilers, tanks, and digesters. As stated in Chap. 9, fatigue cracks result from many repetitions of stress; they are frequently due to stress concentrations at changes of section, grooves from machining operations, seams, quenching cracks, and forging laps.

Various methods of nondestructive testing used to determine certain properties or the presence of any defects that affect the structural action of the object will be presented in this chapter.

10.2. Types of Nondestructive Tests. Nondestructive tests may be divided into two general groups. In the first group are those tests used to *locate defects.* It includes various simple methods of examination such as visual inspection of the surface as well as the interior by the use of drilled holes, tests involving the application of penetrants to locate surface cracks, the examination of welded joints by use of a stethoscope to detect changes in sounds as caused by hidden flaws, and highly technical methods involving radiographic, magnetic, electrical, and ultrasonic techniques. Radiographic tests include those which make use of the radiation of short electromagnetic waves such as X rays and gamma rays. With the aid of such test methods it is possible to inspect the interior of opaque objects, of appreciable thickness, to ascertain their general homogeneity. Those tests which use the magnetic characteristics of the materials inspected include *electromagnetic analysis,* which depends upon changes in the electromagnetic characteristics of the material from point to point to detect any corresponding change in its mechanical or structural characteristics; and *magnetic particle tests,* which are used to locate defects by noting irregularities in the magnetic flux of the object. The class of tests that depend upon observations of *electrical characteristics* includes the Sperry detector for locating flaws in rails while in place and in tube and bar stock. *Ultrasonic tests* depend upon reflected sound effects to locate internal defects.

The second group of nondestructive tests includes those used for determining *dimensional, physical, or mechanical characteristics* of a material or part. In this group are tests for the thickness of paint or nickel coatings on metallic bases, the thickness of materials from only one surface, the determination of moisture content of wood by electrical means, certain hardness tests, proof tests of various kinds, surface roughness tests, and methods employing forced mechanical vibrations to determine the changes in natural frequency of the system due to changes in the properties of the material. One type of vibration test uses the sonic analyzer for determining the natural frequency, from which the modulus of elasticity can be computed. The simple determination of this latter property is of interest, since for some materials it is related, in general, to the quality of the material.

METHODS OF EXAMINATION FOR DEFECTS

10.3. Visual Examination. Visual inspection of the object should never be omitted whenever it is necessary to detect the presence of possible surface defects. Although it may appear unnecessary to list this as a test method, there has been a tendency to overlook the advantages to be gained by a careful visual inspection using low-power magnifying glasses as well as microscopes, if necessary. Microscopes equipped with photographic attachments are often used to get permanent records of defects, questionable areas, and variations in structure. The checking of dimensions by use of scales, tapes, micrometers, or special gages may also be considered as a type of visual inspection.

10.4. Penetrant Tests. All liquid-penetrant processes are nondestructive testing methods for detecting discontinuities that are open to the surface. They can be effectively used not only in the inspection of ferrous metals but are especially useful for nonferrous metal products and on nonporous, nonmetallic materials such as ceramics, plastics, and glass because magnetic-particle methods are not applicable. Surface discontinuities such as cracks, seams, laps, laminations, or lack of bond are indicated by these methods. They are applicable to in-process, final, and maintenance inspection. The various methods of liquid-penetrant inspection are covered in ASTM E 165.

The liquids used enter small openings such as cracks or porosities by capillary action. The rate and extent of this action are dependent upon such properties as surface tension, cohesion, adhesion, and viscosity. They are also influenced by factors such as the condition of the surface of the material and the interior of the discontinuity.

For the liquid to penetrate effectively, the surface of the material must be thoroughly cleaned of all material that would obstruct the entrance of the liquid into the defect. After cleaning, the liquid penetrant is applied evenly over the surface and allowed to remain long enough to permit penetration into possible discontinuities. The liquid is then completely removed from the surface and either a wet or a dry developer applied. The liquid that has penetrated the discontinuities will then bleed out onto the surface, and the developer will help delineate them. This will show the location and general nature and magnitude of any discontinuities present. To hasten this action the part may be struck sharply to produce vibrations to force the liquid out of the defect.

The oil-whiting test is one of the older and cruder penetrant tests used for the detection of cracks too small to be noticed in a visual inspection. In this method the piece is covered with a penetrating oil, such as kerosene, then rubbed dry and coated with dry whiting. In a short time the oil that has seeped into any cracks will be partially absorbed by the whiting, producing plainly visible discolored streaks delineating the cracks.

Fluorescent-penetrant inspection makes use of a penetrant known as "Zyglo" that fluoresces brilliantly under so-called "black light" having a wavelength of 3650 A, which is between the visible and ultraviolet in the spectrum. The coating of developer used with it is not fluorescent but is dark when viewed under black light. It acts to subdue background fluorescence on parts and causes any defects to show up more distinctly under black light. Pores show as glowing spots, cracks show as fluorescent lines, and where a large discontinuity has trapped a quantity of penetrant, the indications spread on the surface.

Some nonfluorescent penetrants can be easily seen in daylight or with visible light. They are usually deep red in color so that indications at defects produce a definite red color as contrasted to the white background of the developer. Depth of surface discontinuities may be correlated with the richness of color and speed of bleed-out.

Partek or filtered-particle inspection depends on the unequal absorption into a porous surface of a liquid containing fine particles in suspension. This preferential absorption causes the fine particles in the solution to be filtered out and concentrated directly over the crack, producing a visual indication.

Statiflux is a test method for locating cracks in nonconducting materials such as plastics, ceramics, and glass. The object is first covered with a special penetrant which conditions the defects, the surface is dried, and then a cloud of fine electrically charged particles is blown over the surface, causing a build-up of powder at the defect. The penetrant is not required when the nonconductor is backed by a conductor, as in enamelware.

10.5. Hammer Test. When a solid homogeneous object is struck with a hammer, it emits a clear ringing sound, whereas a defective (cracked) object has a well-known dull sound. This fact has long been the basis for one of the oldest nondestructive tests. Although it is highly satisfactory for simple forms, experience has shown that complicated shapes modify the sounds and tend to confuse the inspector.

Radiographic Examinations

10.6. Radiant Energy and Radiography. X rays and the still shorter gamma rays are electromagnetic types of waves used in industry to penetrate opaque materials and obtain a permanent record of the result on sensitized film. These short wavelengths of radiant energy have become a useful tool for the inspection of the interior of metals and other materials. When these rays pass through material of nonuniform structure containing defects such as cavities or cracks or portions of variable density, the rays passing through the less dense parts of the object are absorbed to a smaller extent than the rays passing through the adjacent sound material. Upon development of a light-sensitive film placed on the far side of the object exposed to short-wave radiation, there results a picture of light and dark areas, the latter representing parts of the material having a lower density. This film is called an exograph when produced by X rays and a gammagraph when produced by gamma rays; both types of film are termed *radiographs.* To be successful, a radiograph must show the size and shape of any significant defect or nonhomogeneity. The general principle involved in the production of a radiograph by the use of X rays and gamma rays as sources of radiant energy is

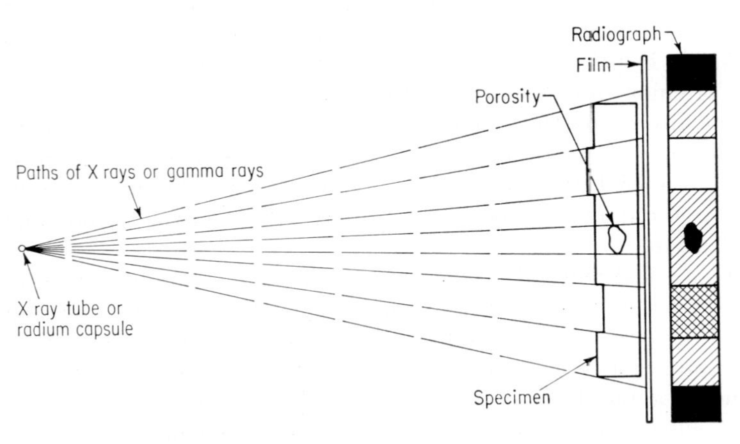

Fig. 10.1. **Production of a radiograph.**

illustrated in Fig. 10.1. The recommended practice for radiographic testing is covered in ASTM E 94.

10.7. Practical Applications. The two most common applications of industrial radiography are in the examination of welded products and castings; installations for the former are the most numerous. Installations in foundries are generally used to develop a proper foundry technique and less frequently for routine inspection.

The value of radiographic inspection of welded joints in showing the presence of injurious defects is well recognized. Welds in pressure vessels of all types, penstocks on most government power projects, and certain modern European rigid frame bridges are given a 100 percent inspection.

X rays have been used to determine the condition of wood poles near the ground line. These studies have disclosed that the amount of rot can be determined for dry poles [1011]. It would appear that the same method could be used in the inspection of structural timber in place, to detect the presence and extent of fungus growths.

10.8. Comparison of X- and Gamma-ray Radiography. In general, the use of X rays, even at 2000 kv, is limited to about a 9-in. thickness of steel, whereas gamma rays may be used for thicknesses up to about 10 in.

X rays are better than gamma rays for the detection of small defects in sections less than about 2 in. thick, but the two methods have equal sensitivity for sections about 2 to 4 in. thick.

The X-ray method is much more rapid than the gamma-ray method, requiring seconds or minutes instead of hours, although in the case of annular objects this difference may be partly offset when using gamma rays by placing films around the circumference, with the radium in the center, and exposing the whole circumference at one time.

Owing to less scatter, gamma rays appear to be more satisfactory than X rays for examining objects of varying thickness. However, for a uniform thickness of metal and under proper operating conditions, X rays appear to give clearer negatives than gamma rays.

Satisfactory portable X-ray equipment has been developed for shop use and even for limited types of field examination. However, the extreme portability and convenience of use of radium for certain field inspections have much in its favor.

The initial cost of 200 mg of radium is about the same as for a 300-kv X-ray unit. However, the former reaches its half strength in about 1580 years, whereas the latter units have become obsolete in about 10 years. Nevertheless the overall costs of the two methods compare favorably.

10.9. X-ray Equipment. X-ray tubes in common use are usually of the Coolidge type, as shown in Fig. 10.2. The X radiation is produced when

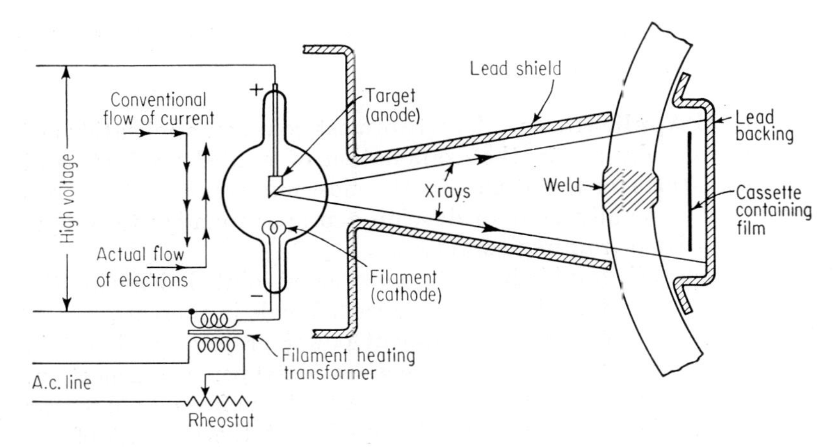

***Fig. 10.2.* Arrangement of equipment for radiographing a welded joint.** (*From St. John and Isenburger, Ref.* 1005.)

electrons, traveling at very high velocities, are suddenly stopped by impact with the tungsten anode. The intensity or quantity of the radiation depends upon the rate of liberation of electrons from the cathode, and this determines the exposure time necessary to produce a satisfactory radiograph. It is governed by the temperature of the incandescent filament and can be regulated by a rheostat in a separate low-voltage circuit. This intensity is directly proportional to the amperage of the high-voltage current that flows through the tube. The wavelength or quality of the radiation determines its penetrating power. This property depends upon the velocity with which the electrons are driven from the cathode to the target and is controlled by an adjustment of the high voltage supplied to the tube; the higher the voltage, the shorter the wave-length of the radiation and the greater the penetrating power. For industrial radiography the current supplied to the X-ray tube may be about 1 to 10 ma, and the voltage may range from 50 to 2000 kv.

The voltage applied to the tube and the current passing through it can be varied independently. Therefore, one tube may be adjusted over a wide range of operating characteristics and can be used for making radiographs through thin material of little density or through steel 5 in. thick.

In industrial applications the tendency is toward oil-immersed portable units because they are more adaptable to a variety of conditions. These units are on trunnions mounted on a truck so that the aperture of the tube can turn through 360°. A large lead-lined tubular shield is placed between the object and the window of the tube case to reduce the amount of radiation escaping to the neighborhood. A typical installation of this type is shown in Fig. 10.3. As accessories to such an installation, some device is often provided for easy handling of the objects to be

Fig. 10.3. **200,000-volt industrial X-ray unit. All X-ray controls are centered in a panel which is installed in the lead-lined operator's control booth (right).** *(Courtesy of General Electric X-ray Corp.)*

inspected. This may consist of a turntable, which serves well for large castings, or it may involve four rollers for rotating a large pressure vessel on its longitudinal axis.

10.10. Gamma-ray Equipment. Radium and its salts decompose at a constant rate, giving off gamma rays that are of much shorter wavelength and more penetrating than ordinary X rays. It is estimated that the wavelengths of gamma rays (about 0.02 Angstrom) correspond to the average for X rays excited by about 750 kv.

Radium sulfate, sealed in small silver capsules and enclosed in an outer duralumin container to facilitate ease in handling and protection against loss, is the form commonly used. For commercial radiography, 100 to 300 mg is usually placed in a unit, the intensity of the radiant energy varying directly with the amount used.

The apparatus necessary for gamma-ray radiography is very simple. The container of radioactive material is supported rigidly in front of the specimen to be inspected, and films in suitable holders are fastened to the back of the specimen. Cobalt 60, an isotope produced by neutron irradiation, can be used in place of radium, and since it is much cheaper it may eventually replace the latter.

10.11. The Radiograph and the Effect of Important Variables. A radiograph is simply a record of the differences in intensity of the radiation that penetrates through the various parts of the test object to produce on sensitized film a shadow picture called the image. The intensity of the emergent radiation and of the resultant image is a function of the distance between the source of the ray and the film, being inversely proportional to the square of this distance. It also depends upon the amount of radiation absorbed, which varies with the thickness and the material of the object examined; the absorption increases very rapidly with the thickness according to an exponential equation and increases with the density of the material. The absorption is also a function of the wavelength of the radiation: the longer the wavelength, the greater the absorption.

Several of these variable factors can be adjusted by the operator so that the radiation image has sufficient intensity to be registered adequately within a reasonable time on the film; this may take from a few seconds to several hours. To assist the operator in selecting the proper conditions, exposure charts are constructed for specific conditions of use.

Special film is used in radiograph work consisting of a cellulose acetate base coated on both sides with a silver bromide emulsion specially prepared for X-ray service. The double coating gives about double the contrast possible with a single coating and serves to reduce exposure times. Emulsions that are very sensitive to differences in intensity of the radiation will give the greatest contrast and differentiation of detail and will reveal very small defects in certain regions but will not show proper exposure for objects having an appreciable range of thickness. For objects that vary in thickness over a region to be covered by one film, it is best to use emulsions that have a fair degree of latitude (range of useful film intensities) even at the expense of some contrast. For the detection of small defects it may be preferable to use a film of high contrast and to give different exposure times to the portions of different thicknesses.

Exposure times can be reduced by the use of intensifying screens in intimate contact with each sensitive surface of the film. The screens for X-ray work are fluorescent, glowing principally in the blue region of the spectrum to which a film emulsion is highly sensitive. They are usually made of thin layers of calcium tungstate on a cardboard backing. There are appreciable differences between the calcium tungstate screens commercially available, the principal variables being the thickness of the active layer and the size of the grain. Thick screens are more active, but they also absorb more of the rays; so they should be used only for higher voltages. Larger grains of calcium tungstate tend to reduce the necessary exposure time, but they also tend to make the radiograph appear granular and less distinct. Ordinarily it is necessary to sacrifice some speed to obtain good definition in the radiograph; this is accomplished by using a screen with fine grains. Intensifying screens of thin lead foil are used with X rays at high voltages and for all gamma-ray radiography. They are less efficient in intensifying the radiographic image, but they produce finer grained and hence more sharply defined radiographs. The use of higher voltages and compensating filters of lead foil tends to reduce the need for fluorescent-type screens even on thin objects.

Radiography of objects of variable thickness requires special attention so that parts of the radiograph will not be over- or underexposed. Circular parts may be enclosed in pads of the same material so that the X rays will pass through a constant thickness. For irregular objects the best method for producing a sufficiently uniform effective thickness is to immerse them in a solution of a lead salt or the salt of some other heavy metal. Owing to the lower absorption of gamma rays, it is easier to obtain suitable radiographs of specimens of irregular thickness with them than with X rays. By the use of gamma rays, steel sections may vary as much as 2 or 3 in. in thickness and be radiographed successfully on the same film.

10.12. Requirements for a Good Radiograph. For a radiograph to have value, it must show all *significant* defects present and their size. This requires that small differences in density or effective thickness of the object produce sufficiently large differences in intensity of the emergent radiation to be clearly recorded on the film. Furthermore, there must not occur any disturbing influence to distort or fog the shadow picture produced.

To interpret a radiograph accurately, it is necessary to know the size of the smallest defect that it can be depended on to show. The size of the smallest discontinuity detectable on the radiograph depends upon the definition or sharpness of the record and the contrast between adjoining areas.

The size of the ray source is important in determining sharpness, because, as shown in Fig. 10.4, the smaller the source, the larger the intense umbral shadow u and the narrower the penumbral ring (p minus u). Also, it is evident from the figure that the closer the flaw is to the film, the sharper the shadow. Both of these conditions tend to facilitate the detection of the flaw. The same result is obtained by using a large distance D between the ray source and the flaw; the distance should be made as large as possible without unduly increasing the time of exposure. In general, the ratio D/t should never be less than 7; higher values up to

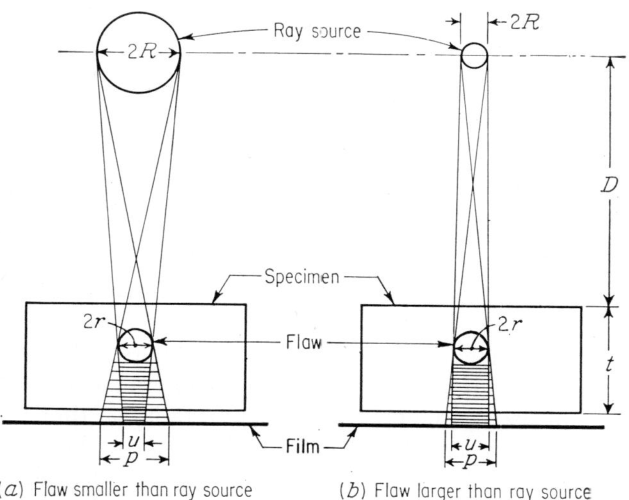

Fig. 10.4. Influence of size of source on size and sharpness of image. (*From Lester, Ref.* 1012.)

about 30 are preferable. Modern X-ray tubes are made so that the focal spot is small enough to produce sharp images at reasonable distances D, say, 30 in. To secure clear images in gamma-ray work the distance D should be governed by the amount of radium used.

One condition tending to fog the radiograph and to make it appear less sharp is the scattering of the X rays in the interior of the object. This causes any particular point P on the film to receive radiation from a number of different directions, as is shown in Fig. 10.5, so that the image formed at any point on the film does not result from the rays passing through a particular part of the specimen but results from the composite effect of rays passing through various portions of the specimen. Internal scattering explains the fact that a defect near the top of the specimen is less sharp than a similar defect close to the bottom.

The effect of scattering may be eliminated by the use of a Bucky grid between the specimen and the film. This device is a grid of lead strips

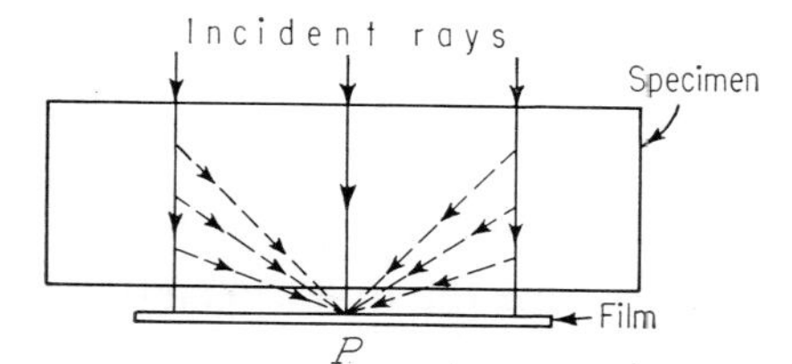

Fig. 10.5. Scattered radiation blurs the X-ray image. Scattered rays falling on point P shown as dotted lines. The direct ray is shown as a solid line. (*From ASTM Symposium on Radiography, Ref.* 1007.)

with slots of such shape that they permit the direct rays to strike the film but absorb the scattered radiation that comes in at an angle. The grid is slowly moved during the exposure so that it makes no record upon the film. The disadvantage of the increased time of exposure required with the grid is offset by the improvement in the results. The use of a grid is unnecessary with gamma rays, since they are of short wavelength and do not scatter much. This is also true when using the short wavelengths produced by higher voltages so that it is possible to eliminate scattered radiation by stepping up the voltage and then reducing the intensity of the rays by using metal filters at the tube port or between the object and the film.

Contrast in the radiograph is a function of the difference in intensity of the emergent radiation from point to point. The differences in intensity due to differences in thickness of the object vary with the tube voltage and the material of which the object is made. With low voltages, small differences in thickness produce larger differences in intensity than is the case with high voltages, so that it is common practice to use as low a voltage as will produce the necessary background density on the film in a reasonable length of time. For objects of variable thickness, it is necessary to use a voltage high enough for the thicker sections, in which case the contrast would be comparatively low. Thus, a compromise must be made between contrast and the range of thickness to be covered in one exposure.

Proper orientation of thin flat defects with respect to the path of the penetrating rays is essential. The optimum position is one parallel to the direction of the rays.

10.13. The Penetrameter. It is common practice in radiography to employ a thickness gage or "penetrameter" to assist in determining the magnitude of the smallest detectable defect. The ASME Code penetrameter shown in Fig. 10.6 has become more or less standard. It consists of a piece of material approximately the same as that being X rayed, but having a thickness of not more than 2 percent of the thickness of the part being radiographed and having three drilled holes of diameters 2, 3, and 4 times the thickness of the penetrameter. These penetrameters are placed on the tube side of the specimen, one showing on each end of the film where the angularity of the rays is greatest. The outline of the penetrameter and all three holes must show clearly to indicate a satisfactory technique.

Provided attention is given to the recording of a clear radiograph resulting in a sharp image of a penetrameter having holes of as small an area and depth as the imperfections to be detected, then it can be assumed that all corresponding flaws in the object will be detectable on the radiograph. Tests have shown that with X rays a sensitivity of 1 percent of the

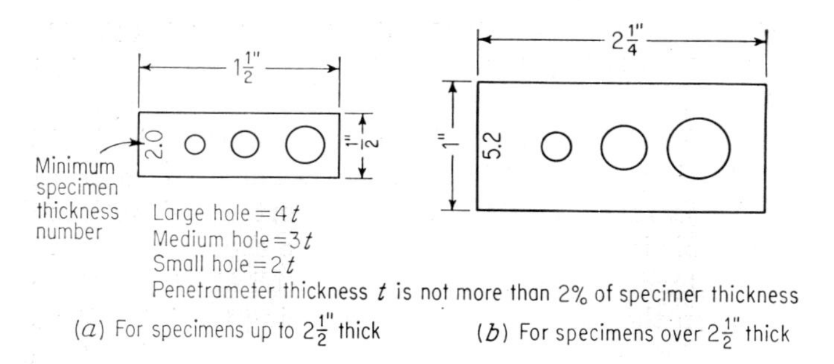

Fig. 10.6. **The ASME Boiler Code penetrameter.** (*From ASME Boiler and Pressure Vessel Code, Ref.* 1015.)

thickness can be obtained by careful procedure, provided a sufficiently large defective area is involved; a sensitivity of 2 percent is easily obtained in routine work. Using gamma rays with about a 2½ to 6-in. thickness of iron, the sensitivity obtainable is about 1.5 percent, but it decreases rapidly to about 5 percent for ½-in. thickness.

10.14. Interpretation of a Radiograph. The darker portions of the radiograph (negative) indicate the less dense parts of the object, whereas the lighter portions indicate the more opaque parts; the reverse is true for a print from the negative. This nonuniformity may be caused by (1) differences in thickness due to the form of the object, (2) differences in thickness due to voids, or (3) differences in density of the component parts, such as when metal parts are embedded in a material such as a plastic or when slag inclusions occur in welds.

Industrial radiographic standards for steel castings are covered in ASTM E 71.

Common defects and their characteristic appearance on the negatives of castings are as follows [143]:

1. Gas cavities and blowholes are indicated by well-defined circular dark areas.
2. Shrinkage porosity appears as a fibrous irregular dark region having an indistinct outline.
3. Cracks appear as darkened areas of variable width.
4. Sand inclusions are represented by gray or black spots of an uneven or granular texture with indistinct boundaries.
5. Inclusions in steel castings appear as dark areas of definite outline. In light alloys the inclusions may be more dense than the base metal and thus cause light areas.

When the specimen being examined has a rough surface, its radiograph should be studied so as not to confuse any dark or light areas, which

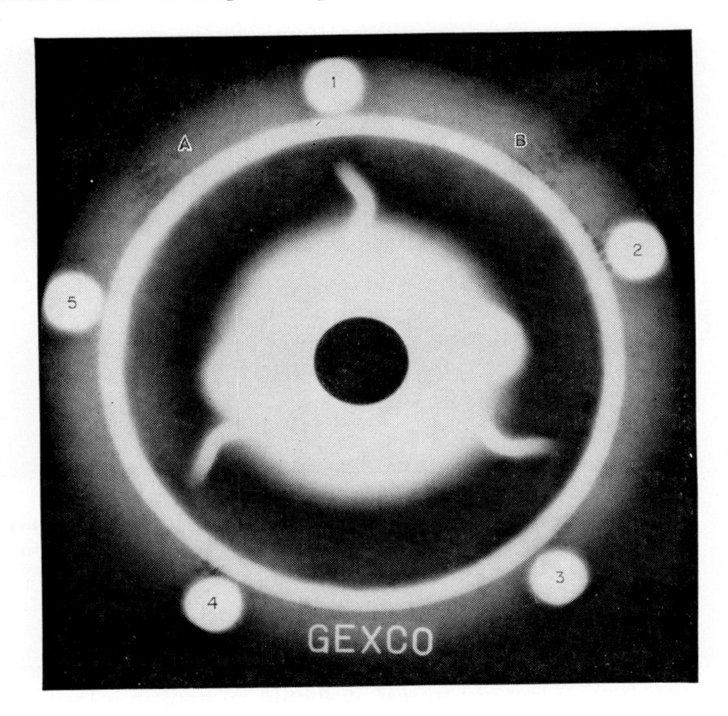

Fig. 10.7. **Radiograph of a steel casting which shows the shrinkage areas under the gates *A* and *B* and bosses 2, 3, and 4.** *(Courtesy of General Electric X-ray Corp.)*

result from surface roughness, with similar markings that might indicate internal irregularities.

A radiograph of a steel casting is shown in Fig. 10.7. The darkened areas that indicate shrinkage under the gates at *A* and *B* and adjacent to the bosses 2, 3, and 4 are plainly visible in the negative but are difficult to reproduce clearly.

In steel welds the most common defects are slag inclusions, porosity, cracks, and incomplete fusion. Since these flaws are less dense than steel, they produce darkened areas on the negative. Incomplete fusion produces a dark line parallel to the joint; the other imperfections appear as described for castings. Figure 10.8 shows a radiograph of a welded joint; note the image of the penetrameter in the lower right corner.

10.15. Safety Precautions. X rays should be shielded so that the operators or other persons nearby are not exposed to them. Ordinarily, this is accomplished by the use of sheet lead, as shown in Fig. 10.2. The use of such shields in front of and behind the film permits an operator to remain inside a vessel being radiographed.

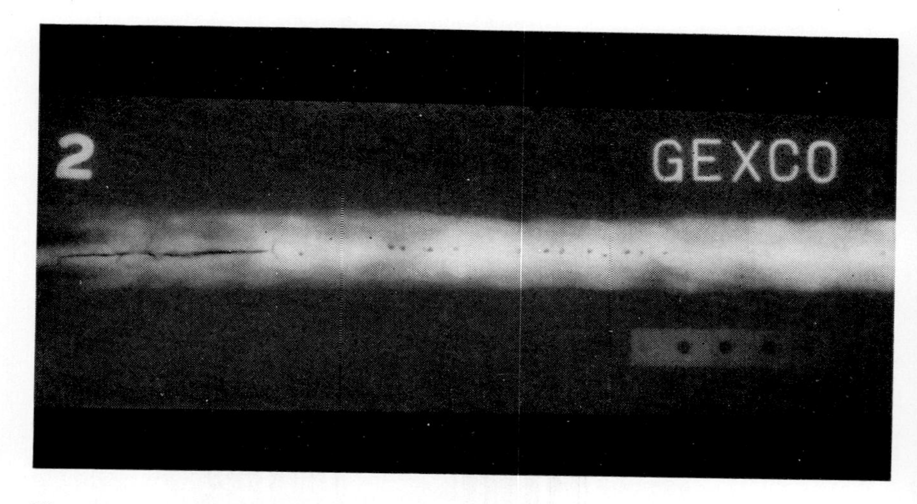

Fig. 10.8. **Radiograph of a ¾-in. weld which reveals a large crack in the internal weld metal (left) and a profusion of gas pockets through the center line of the weld.** (*Courtesy of General Electric X-ray Corp.*)

Radium must be handled with extreme care to prevent injury to the operators. When not in use, it should be kept in a lead container having a minimum wall thickness of 6 in. In transit, a portable lead case having a 2-in. wall thickness may be used. Upon removal of the container from the lead case, everyone should remain at least 10 ft from it, unless shielded by thick metal. It will be necessary to violate this precaution when placing the container for a test; even then, however, the operator should never touch the container, even with gloves on, but may handle it with long tongs or string. He should not remain close to it any longer than is absolutely necessary.

Excessive exposure to the radiation may cause a decrease in the number of white blood corpuscles (leukopenia), anemia, and low blood pressure, but severe exposure will even result in skin burns that may prove fatal. To detect undue exposure, operators should keep close watch on a piece of sensitized film, carried on their person, for fogging. Perceptible fogging within a 2-week period indicates that more shielding should be provided. Occasional blood-corpuscle counts are advisable.

10.16. Xeroradiography. Xeroradiography (pronounced "zero radiography") is an advanced technique in X ray, employing a reusable dry plate to record the image. Xeroradiography uses static electricity, much as a comb does in picking up a piece of paper. The static electricity arranges a fine powder on a specially coated aluminum plate and produces an image similar to that on conventional X-ray film. Using this process, an X-ray image is available for viewing within 45 sec, whereas conventional X-ray film requires 1 hr to develop and dry for viewing.

This method uses a conventional X-ray tube and requires no new X-ray equipment. Need for a darkroom is eliminated, and a permanent record of the image on the plate may be made in a few seconds by conventional photographic methods.

Methods of Magnetic Analysis

10.17. General Principles. It has been recognized for some time that the magnetic characteristics of a material are related to its composition or structure and its mechanical properties. It appears possible, at least theoretically, to locate differences and discontinuities of structure and variations in the dimensions and various properties of a given material by making certain magnetic measurements on the material and comparing them with corresponding measurements on some standard or reference material. Certain practical limitations are involved, however, owing to the fact that the magnetic properties may be subject to irregular changes that do not correlate well with changes that occur in the mechanical properties of the material. In addition to variables that influence the mechanical properties, a magnetic test is sensitive to internal strains and differences in temperature in the material, which exert little effect on the mechanical properties. The effect of variations due to internal strains is most noticeable when high frequencies or low magnetizing forces are employed. These conditions make it most difficult to determine definite quantitative relationships and to make satisfactory practical applications of the principles.

Although numerous factors such as variations in chemical composition, structure, internal strains, temperature, and dimensions are known to influence the magnetic measurements, sufficient progress has been made with the method to devise magnetic tests for a variety of commercial applications that are in daily use. In these applications the difference in magnetic properties between the object tested and a piece known to have the requisite mechanical properties and freedom from defects is determined. The magnitude of the difference that can be tolerated is determined by experiment. In these industrial applications, only one magnetic property is ordinarily used in a given type of test for evaluating the comparative quality of a material. This property may be simply the magnetic permeability, some quantity such as residual induction or coercive force derived from the hysteresis loop, or the waveform of an alternating induced voltage as determined by the shape of the hysteresis loop.

Although, in general, d-c methods have considerable value for certain commercial applications, it has been found that a-c methods have many advantages. For example, observations that depend upon the shape of the hysteresis loop developed by the use of alternating current are often found to be a better indication of structural characteristics than simple permeability as determined by the use of direct current. In the a-c methods the sample under observation forms the core of a transformer. The various characteristics of the core cause shifts in the phase angle between the current and the voltage of the induced current and give rise to harmonics in the a-c wave. By measuring the shifts or the harmonics, differences in magnetic characteristics of the cores (samples) can be distinguished. Alternating-current methods are most suitable for material of uniform cross section.

Although the relationship between magnetic and mechanical properties is not so direct and simple as was formerly surmised, the results obtained with modern equipment appear to justify confidence in the further development of magnetic analysis.

10.18. Magnetic Analysis of Steel Bars and Tubing. Equipment is available which permits continuous magnetic analysis of bar stock and pipe ¼ to 3½ in. in outside diameter [1026]. It includes a test-coil assembly, an indicator with control cabinet, and a feeding mechanism. The test-coil assembly consists of a heavy primary coil and a secondary coil. The material to be inspected is passed through the secondary coil, located coaxially within the primary coil, and is energized by alternating current. The secondary coil contains a number of windings connected to the test circuits, and thus it acts as a pickup or detector.

One of three available circuits is used as a flaw detector. Its use is based upon the magnetic-flux leakage method and depends upon the action of special detector coils in connection with a compensating network and high-gain amplifier. The output of the amplifier is measured by means of a microammeter and controls the flashing of a light. Employing an average speed of about 120 fpm, short defects may not cause impulses sufficient for a noticeable deflection of the microammeter, although they are strong enough to flash the light. Since this method is based upon a comparison of one section of the test bar with another section of the same bar, it is limited to indicating the beginning and end of a defect only. For this reason, continuous defects (seams) even of considerable depth are sometimes not detectable.

The other two circuits are of the wave-analyzer type. These permit the investigation of variations of amplitude at certain arbitrarily fixed phase points of the curves representing the two electromotive forces induced in the two separate detector windings. These circuits are used for the detection of flaws as well as the investigation of composition variations. Although less sensitive than the main flaw-detector circuit, they outline the full length of detectable defects. Using alternating current, the depth of penetration of this method is limited. This precludes the possibility of locating internal defects and restricts magnetic inspection to the detection of surface flaws.

In commercial work one bar is selected as the standard of comparison, preferably after a thorough check as to correct composition, dimension, heat-treatment, and freedom from flaws. Passing this bar through the coil, the test circuits are individually and successively compensated to show zero deflection for the desired degree of sensitivity and selectivity. The operator then passes the bars in succession through the coil, watching the resultant meter deflections, as well as light flashes. The types of product particularly suited to magnetic inspection are hot-rolled and cold-drawn straight bar stock and several types of pipe and tubing.

Sucker rods used in drilling oil wells have been inspected by magnetic methods for the presence of fatigue cracks. In this application the sucker rod is passed through a suitably energized test coil as it is withdrawn from the oil well. A pickup coil is connected to a recorder to make a graph of the magnetic characteristics. An analysis of these graphs permits the detection of parts of rods having abnormal magnetic characteristics. Past records appear to show that such characteristics indicate rods that are susceptible to fatigue failures.

The Magnetic-particle Method

10.19. Principles Involved. The magnetic-particle method of inspection is a procedure used to determine the presence of defects at or near the surface of ferromagnetic objects. It is based on the principle that, if an object is magnetized, irregularities in the material, such as cracks or nonmetallic inclusions, which are at an angle to the magnetic lines of force, cause an abrupt change in the path of a magnetic flux flowing through the piece normal to the irregularity, resulting in a local flux leakage field and interference with the magnetic lines of force. This interference is detected by the application of a fine powder of magnetic material, which tends to pile up and bridge over such discontinuities. Under favorable conditions, a surface crack is indicated by a line of the fine particles following the outline of the crack, and a subsurface defect by a fuzzy collection of the fine particles on the surface near the defect. A more complete statement of the procedure is presented in Ref. 1030, from which much of this material is taken. Also see ASTM E 109 and E 138.

10.20. Scope and Applicability. Various types of defects are detectable by the magnetic particle method. A few of the more common ones are cracks caused by quenching, fatigue, and embrittlement, seams, and subsurface flaws. Typical quenching crack indications are shown in Fig. 10.9. An example of fatigue cracks in an airplane gear that were invisible to the eye but detected by the magnetic particle method is shown in Fig. 10.10.

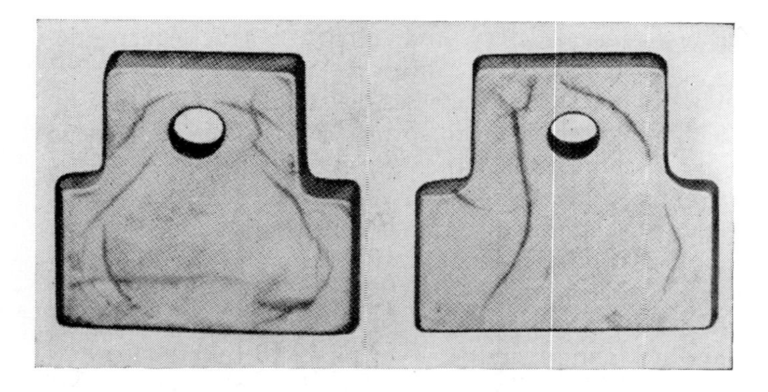

Fig. 10.9. **Quenching crack indications as shown by the magnetic particle method.** (*From Doane, Ref.* 1030.)

Subsurface defects can be located only when relatively close to the surface, but usually the desirability of locating them increases as they approach the surface.

The materials to be examined must necessarily be capable of being magnetized to an appreciable degree. Alloys that are only very faintly magnetic do not produce very satisfactory results, and the method is not applicable to austenitic steels, such as the 18-8 chrome-nickel type.

The orientation of the defect with respect to the direction of the flux lines will influence its detection. It will be located most readily when its axis is normal to the flux lines; it may go undetected until the part is magnetized so that this condition is obtained.

Fig. 10.10. **Fatigue cracks in airplane gear detected by the magnetic particle method.** (*From Doane, Ref.* 1030.)

The character of any porosity is of importance in determining the degree of success with which it may be located. If it is composed of numerous small scattered cavities, chances for detection are unfavorable; whereas the type that contains a few larger cavities, resembling blowholes, is more favorable for detection.

In welded joints the surface cracking that cannot be seen by the eye constitutes a certain percentage of the defects that occur, but by far the largest number of damaging defects, such as porosity and lack of fusion, are found to lie beneath the surface where they are difficult to detect. Also, there is frequently a change in permeability at the junction between weld metal and parent metal that is sufficiently abrupt to cause a line of powder adherence, but does not signify lack of fusion. In some cases it is difficult to distinguish this line from an indication that does signify poor fusion.

The sensitivity of the magnetic-particle method requires that its application and interpretation be supervised by personnel with experience in engineering and metallurgy. A conscientious and intelligent operator can soon become adept at identifying the type of defect, especially after routine inspection of a number of similar parts, followed by sectioning and etching of several parts at the location of various typical powder patterns, but familiarity with the stress analysis of the part and the effect of defects on the strength of the materials is a prerequisite if unnecessary and costly rejections are to be avoided.

The method has found application, for inspection purposes, in the manufacture and also the overhaul of certain parts for airplanes, automobiles and trucks, railroad equipment, and steam turbines.

10.21. Magnetizing Methods. Various magnetizing methods may be used for practically any steel part, but some of these methods will produce more satisfactory results than others. One basis of classification divides them into *residual* and *continuous* methods. In the first method the residual magnetism in the part is relied on when the magnetic powder is applied. In the continuous method the current inducing the magnetic flux in the part to be inspected is allowed to flow while the powder is applied.

For ordinary steels the continuous method is considerably more sensitive than the residual method, since the strength of the magnetic leakage field at the defect is proportional to the intensity of the generating flux.

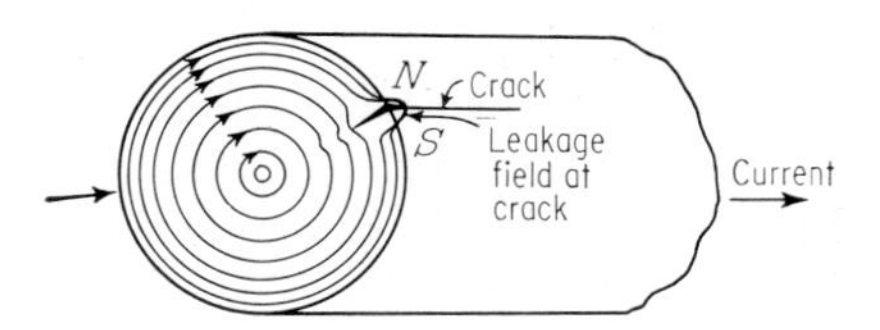

Fig. 10.11. **Circular field produced by longitudinal currents.** (*From McCune, Ref.* 1032.)

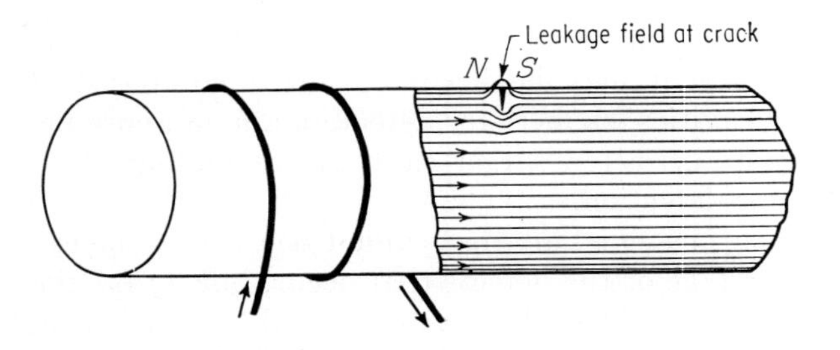

Fig. 10.12. **Longitudinal field produced by circular current.** (*From McCune, Ref.* 1032.)

For steels having low magnetic retentivity, as in the case of soft steels, only the continuous method can be used.

Another basis of classifying the methods of magnetizing objects for magnaflux inspection depends on the character of the induced magnetic field; circular fields produce circular magnetization, and solenoid fields produce longitudinal magnetization. These two types of magnetization are illustrated in Figs. 10.11 and 10.12.

Circular magnetization can be produced by passing current through the part itself or, if it is hollow, by passing current through a central conductor such as a copper bar or cable. In both cases there is little appreciable external field except that which might be generated by a defect at an angle to the field.

Longitudinal magnetization is produced by passing current through a solenoid coil or several turns of conductor surrounding the object, the latter serving as the core of the solenoid. Cable wrappings are commonly used on large objects such as tanks, boilers, large castings, large crankshafts, and similar parts.

A third basis of classifying the methods used in magnetizing depends on the kind of current, direct or alternating, which is used. Direct current is often used, since it appears to permit the detection of defects lying more deeply in the section, but in many cases either alternating or direct current can be used so that the character of the available power supply may be the deciding factor.

10.22. Magnetizing Equipment. For d-c magnetization, 6-volt automobile-type storage batteries are used effectively, since they will furnish high currents at low voltages for brief intervals of time. Motor generator sets of the type used in welding are also occasionally used as sources of high-amperage low-voltage direct currents.

A common type of magnetizing equipment is shown in Fig. 10.13. Parts to be circularly magnetized rest on two Bakelite supports between two copper contact plates, the head supporting the right-hand contact

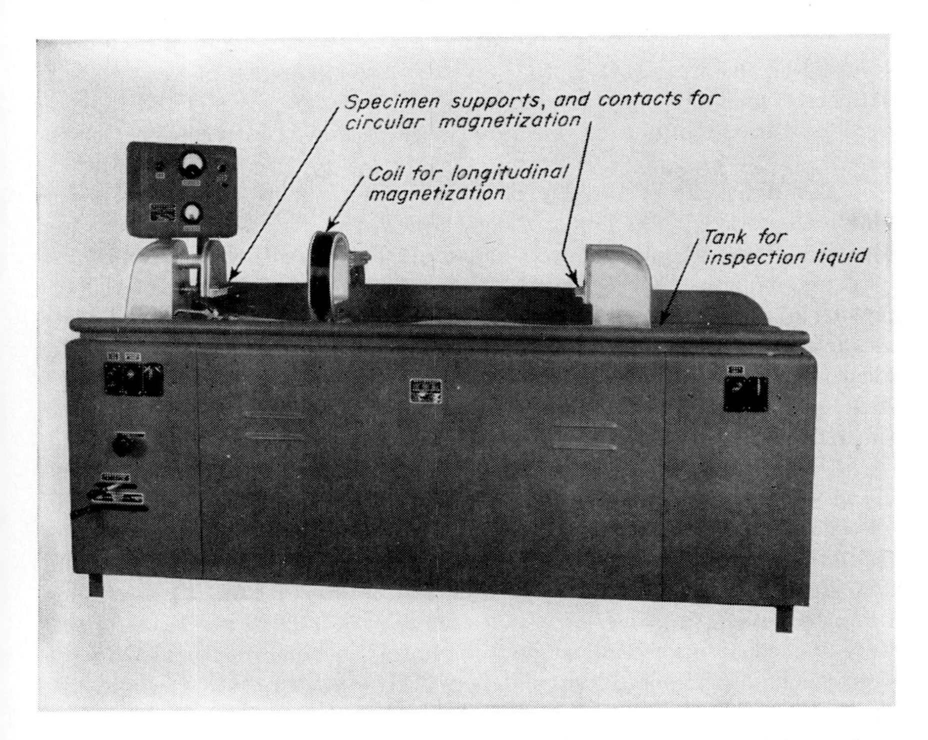

Fig. 10.13. **Direct-current magnetizing unit for magnetic particle testing.** (*From Doane, Ref.* 1030.)

plate being movable by hand or by motor. A low-voltage current, usually 4000 to 10,000 amp, is passed through the part; duration of flow of the current is controlled by a time-delay relay.

For longitudinal magnetization a movable coil that slides along the bed formed by the guide rails is moved into position along and around the part. An angular vector field can be produced in the part by impressing on it a longitudinal and a circular field simultaneously.

10.23. The Inspection Medium. The magnetic powder, the base of which is generally iron or black magnetic iron oxide, is ground to pass a 100-mesh sieve. The individual particles are elongated, rather than globular, to obtain a better polarization, and those of metallic iron are coated to prevent oxidation and sticking. The powder may be applied dry or wet, and is available in both black and red colors. A color is selected that shows up to best advantage on the parts inspected. A more recent development called "Magnaglo" uses magnetic particles prepared with a fluorescent coating, inspection being carried out under ultraviolet light so that every crack is marked by a glowing indication.

For the dry method the powder is applied in the form of a cloud or spray. For the wet method the powder is suspended in a low-viscosity noncorrosive fluid such as kerosene. Either the magnetized part is immersed in the liquid, or the liquid may be flowed or sprayed over the part.

An advantage of the dry powders is the fact that they are not so messy to work with as is oil. Where large areas are to be inspected and recovery of the oil-paste suspension would be difficult, the dry method is used extensively. From a magnetic standpoint an advantage of the dry powder over the wet is that it is better for locating near-surface defects, as shown in Fig. 10.14. The wet method is often considered superior for locating minute surface defects, because at present the available paste is finer than the available powder. The ability to reach all surfaces of the part being inspected, including vertical surfaces and the underside of horizontal surfaces, by hosing or by immersion, is a considerable advantage in favor of the wet method.

The results of an experiment made to show the effect of various test procedures on the indications obtained are shown in Fig. 10.14 [1030]. The specimen was a 1½-in. length of 4-in.-diameter SAE 1020 seamless tubing with a ¾-in. wall. Nine holes made with a 0.028-in.-diameter drill were spaced at intervals around the ring at various distances from the surface. The "threshold" current is the lowest current that would produce a distinctly noticeable powder pattern on the ring. Referring to the figure, it will be noticed that, as the depth of defect increases, the amount of alternating current required for threshold indications increases considerably faster than does the amount of direct current needed. Also, if the curves should be extended below the point where they intersect, it can be

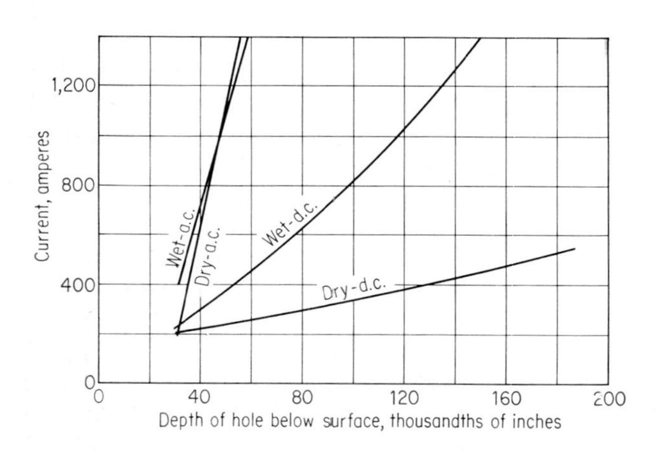

Fig. 10.14. **Threshold indications of near surface cavities.** (*From Doane, Ref.* 1030.)

concluded that, for defects at or extremely near the surface, alternating current at a given amperage may be more revealing than direct current. The curves also show that there is very little difference between the sensitivity of the wet and dry method when alternating current is used, whereas using direct current the dry method gives satisfactory indications at considerably lower current values than for the wet method.

10.24. Preparation of Surfaces. Reasonably smooth surfaces of uniform color offer the most favorable conditions for formation and examination of the powder pattern. When the dry method is used, grease and soft loose rust should be removed. Paint and metallic plating have the effect of converting surface flaws into subsurface flaws. In considering whether or not paint is to be removed, the relative thickness of the paint layer and the size of the smallest flaws, which are being sought by the inspection, must be considered. Rough surfaces, such as that of a rough weld, interfere with the powder pattern; such surfaces should be ground reasonably smooth.

10.25. Demagnetization. The placing of a magnetized part in service after a magnetic-particle inspection may result in the attraction of steel particles to areas of local polarity, and if two moving parts are in contact, undue wear may occur. In airplanes it is important that there be as little external field as possible in the vicinity of a ship's compass and that this field undergo as little variation as possible once compensation has been made. However, for many types of service, demagnetization of parts is not necessary. If demagnetization is necessary, the most practical and convenient method for accomplishing it is to subject the piece to the action of a magnetic field that is continually reversing in direction and, at the same time, gradually decreasing in strength. Alternating current at the available frequency is employed for demagnetizing to a greater extent than any other current, usually in connection with an open solenoid in which the part to be demagnetized is placed and then slowly withdrawn while the current is flowing.

Methods of Electrical Analysis

10.26. Sperry Detector for Flaws in Rails. The Sperry apparatus for the detection of flaws in railway rails is installed in a self-propelled car, which is run over the track to be tested at a speed of about 6 mph. Several thousand amperes of low-voltage direct current from a special generator are fed into the track by the main brushes, which are mounted between the wheels of the rear truck. Auxiliary brushes mounted on the forward truck are used to preenergize the track so that the indications obtained will be more reliable [1002].

A flaw of sufficient magnitude to cause a deviation in the direction of current in the rail will produce a corresponding deviation in the direction of the magnetic field around the rail, which is determined by detector coils mounted on the rear truck between the main brushes and located just above the rail, one ahead of the other. The currents induced in these coils are opposed to one another and are normally in balance. When unbalanced by the influence of any defect passing under one or the other of the two coils, the current generated is amplified sufficiently to operate a recording device. A continuous record is made on a moving tape of all deviations in the magnetic field and the location of the rail causing such deviations. The detector coils also control a paint gun which makes a spot of paint on the rail at the locations of any appreciable magnetic deviations.

When the record tape shows that a flaw has been located, the car is stopped and the magnitude of the flaw is determined by a drop-of-potential method using portable equipment. Transverse fissures, compound fissures, horizontal split heads, and vertical split heads are the defects most commonly located.

10.27. Sperry Detector for Flaws in Tubes. The electrical detection of flaws in cylindrical metal tubes is accomplished by a different method from that used for rails. It consists in causing a controlled electric current to flow through the tube in a direction transverse to the direction of a possible flaw. Any flaw in the path of the current will increase the resistance of the circuit and produce a difference in the value of the current which can be measured by suitable instruments [1041].

Most flaws in tubing extend in a longitudinal direction, so the induced current is made to flow circumferentially in the tube. This is done by passing an alternating current through a set of energizing coils surrounding the tube, inducing currents which flow circumferentially in the tube.

A simplified diagram showing a typical application of the method is presented in Fig. 10.15. The two tube lengths shown, one sufficiently perfect to be taken as a standard or reference tube and the other the tube length under test, are surrounded by identical energizing coils carrying the same alternating current. Two identical test coils also surround the tubes and are in electrical balance when no defect is present. A flaw in the tube under test upsets the balance of the test circuit as shown by the indicator.

In commercial installations the tubes under test are passed through the equipment at a rate of approximately 50 fpm, and the machine is arranged so that the presence of a defect will stop the automatic-feed rolls. The defective tube is then marked.

This method is applicable to tubes of any material that can conduct an electrical current, irrespective of whether it is magnetic or nonmag-

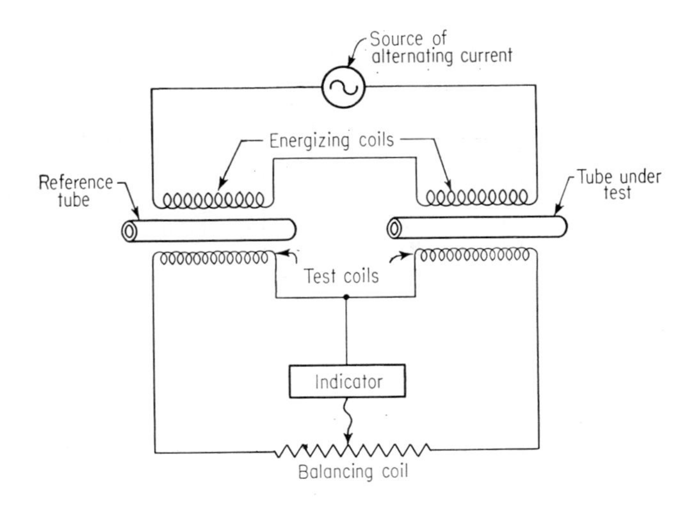

***Fig. 10.15.* Simplified diagram of Sperry electric detector for flaws in tubes.** (*From Knerr, Ref.* 1041.)

netic. Thus it overcomes one of the principal limitations of the magnetic methods.

Machines of this general type, with certain refinements in the electrical test circuits, are said to be capable of detecting flaws $\frac{1}{8}$ in. long, extending halfway through the wall of a tube of ordinary low-carbon steel, having a diameter as small as $\frac{5}{8}$ in. and a wall thickness of 0.04 in.

Metal parts and stocks appearing alike but differing in their composition, heat-treatment, etc., can be separated quickly and easily with a somewhat similar electrical instrument called a metals comparator which is effective for sorting either magnetic or nonmagnetic materials.

10.28. Ultrasonic Detector for Defects. Ultrasonic vibrations are now commonly used for locating minute internal defects in both ferrous and nonferrous metallic objects and in plastics, ceramics, etc. The frequency of the vibrations used is in the range of 100,000 to 20,000,000 cycles per sec, whereas the audible or sonic range is only 16 to 20,000 cycles per sec. Both sonic and ultrasonic vibrations or waves are transmitted through solid materials much more readily than through air; in fact, the waves initiated at one face of solid objects are reflected back when reaching any air gap in the material or when reaching the opposite face, as shown in Fig. 10.16. The test method for locating defects makes use of this phenomenon by electronically determining the relative times for the ultrasonic waves to be reflected back from the defect and from the opposite face.

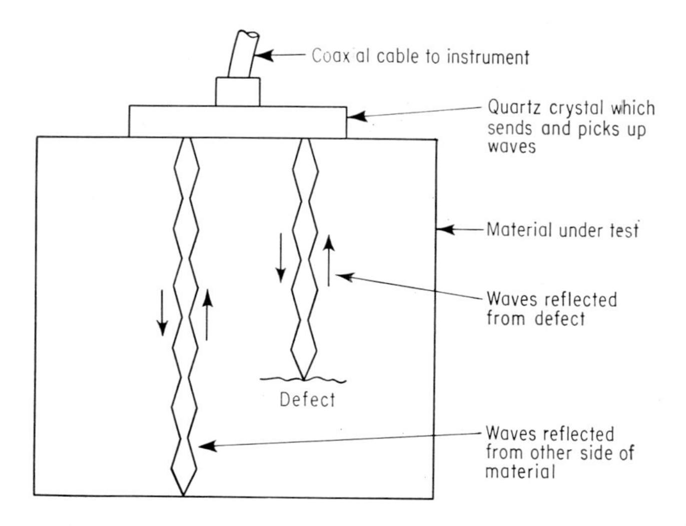

Fig. 10.16. **Detection of defects by supersonic waves. Position of defect is located on oscilloscope pattern which depends on time interval for waves to be reflected.**

Ultrasonic waves are usually produced by the piezoelectric effect, which is the production of mechanical deformation in certain crystals, such as quartz, when placed in electrical fields. An alternating voltage produces mechanical oscillations of the crystal in the same frequency. The probe containing the crystal is placed against the test piece, which is then subjected to these oscillations. The waves, like light, tend to travel in beams with a divergence angle determined by the ratio of the wavelength to the diameter of the source. In steel a sound at 5,000,000-cycle-per-sec frequency has a wavelength of only 0.046 in., so that for a crystal less than 1 in. in size the divergence of the beams is relatively small.

Usually one crystal probe is used for both sending and receiving. It is placed against the test piece using an oil film for better transmission of the sound waves. One cycle of the continuous operation of the equipment is about as follows: if a 5,000,000-cycle frequency is used, electrical oscillations of this frequency would be applied to the crystal for, say, 1 millionth of a second so that five waves of the 5,000,000-cycle frequency would be sent into the surface of the object. The crystal then stops sending and is ready to receive any reflected waves after they have traveled through the steel at about 245,000 in. per sec. The reflected waves vibrate the crystal, producing electrical impulses which are fed into a cathode-ray oscilloscope. This cycle is repeated 60 times per sec. The oscilloscope is timed so that the beam sweeps to the right 60 times per sec in coordination with the sending cycle. The original vibration striking the sending surface makes a sharp peak (or *pip*) at the left side of the oscilloscope screen, and the reflected waves are indicated by a pip toward the right a distance dependent upon the thickness traveled and the time required. While the crystal is receiving for 1/60 of a second there is time for a wave to travel 2000 in. and return before the next cycle begins. If the piece of steel has a defect, most of the beam striking this defect will be reflected to the crystal before the beam which reaches the opposite surface is reflected back to the crystal. The oscilloscope would then show pips not only for the entering waves and for those reflected back from the opposite surface, but also for one in between for the waves reflected from the defect, as shown in Fig. 10.17. A time (or distance) scale in the form of a square wave is constantly shown on the oscilloscope; this indicates the distance from the surface where the crystal is applied to the reflecting surface, whether the latter is the defect or the opposite side of the object. The distance scale may be varied so that 1 cycle of the wave may indicate, say, 1 in. or 1 ft.

Since the beam from the crystal is very narrow, it is necessary that every bit of the surface be covered by a progressive movement of the crystal. Furthermore, since some cracks may be parallel to the waves and therefore reflect very little of the beam, it is necessary to conduct two series of tests in which the ultrasonic beams are normal to each other. For checking the performance of ultrasonic testing equipment it is customary to use standard reference blocks having various sizes of holes drilled in them at various distances from the test surface, as specified in ASTM E 127. Other ASTM specifications cover the use of ultrasonic equipment for locating defects.

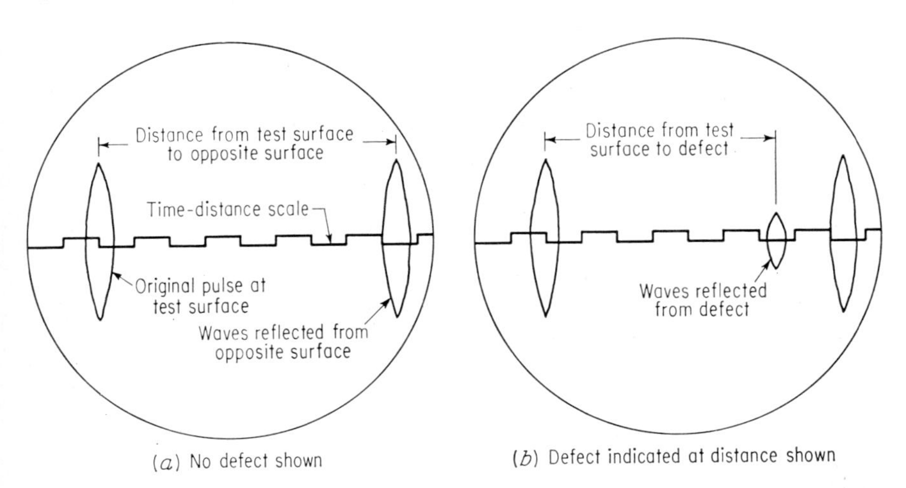

(*a*) No defect shown (*b*) Defect indicated at distance shown

Fig. 10.17. **Oscilloscope screen of ultrasonic tester showing distance from test surface to defect.**

Large forgings are inspected for internal soundness before expensive machining operations are carried out, moving strip or plate is inspected for laminations and thickness, routine inspections of locomotive axles and wheel pins for fatigue cracks are performed in place, and rails are inspected for bolt-hole breaks and other failures without the necessity of dismantling rail-end assemblies. Many industrial applications of ultrasonic inspection are of large products that could not be examined satisfactorily by any other known method.

For the use of ultrasonics in determining certain properties of materials, see Arts. 10.29 and 10.30.

OTHER METHODS FOR DETERMINING PHYSICAL CHARACTERISTICS

10.29. Thickness Tests. Magnetic methods are used for measuring the thickness of enamel, paint, and nickel coatings on metallic bases. One application is the portable electric enamel-thickness gage developed by the General Electric Company, which is used to measure the thickness of enamel or paint on a flat steel surface [1050]. The reluctance of the magnetic circuit of the sensitive gage head when placed on a coated steel surface varies with the thickness of the coating. The indicator unit connected to the gage head is calibrated to read thickness directly in thousandths of an inch.

Another type of instrument for measuring the thickness of coatings on metal has been developed at the National Bureau of Standards [1052]. The instrument employs a portable spring balance for measuring the force required to detach a permanent magnet from the surface under test. The thickness of nickel coatings on nonmagnetic base metals is determined by one form of the instrument while a second, employing a smaller magnet and a stiffer spring, is used to measure the thickness of nonmagnetic coatings on ferrous metals [1051, 1052]. The greater the thickness of a nickel coating, the greater the force required to detach the magnet from the coating, whereas for nonmagnetic coatings on iron or steel the force decreases with the thickness. Both instruments must be calibrated.

The thickness of sheet material, either ferrous or nonferrous, can be determined from one side only by (1) electric-current conduction tests, (2) the X-ray and beta-ray thickness gages which are applicable to stock moving as fast as 1800 fpm, and (3) ultrasonic methods. The latter is somewhat similar to the ultrasonic reflection method for locating defects in material. The "Sonizon" equipment shown in Fig. 10.18 uses the resonance principle and is suitable for thicknesses of any solid material up to 4 in. In this equipment a continuously varying frequency is fed to a quartz crystal held against the test piece, the mechanical resonant

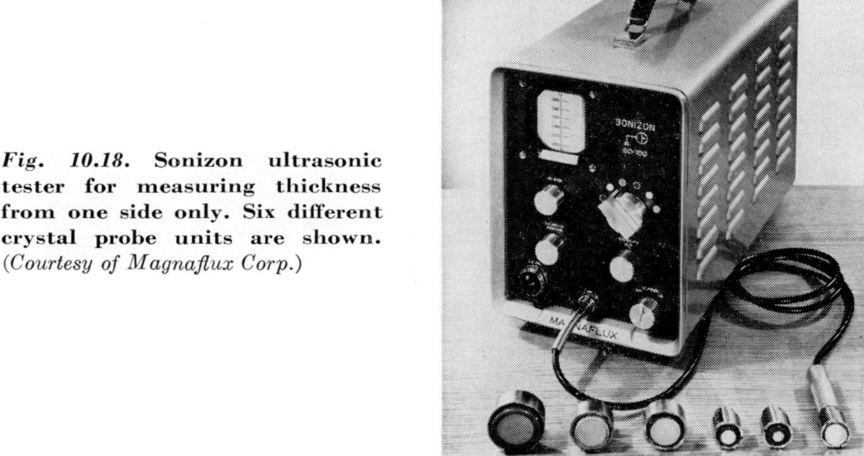

Fig. 10.18. **Sonizon ultrasonic tester for measuring thickness from one side only. Six different crystal probe units are shown.** *(Courtesy of Magnaflux Corp.)*

frequency of which varies with its thickness. When the varying frequency of the crystal hits this resonance, there is an electrical change within the crystal causing a pip on the calibrated thickness scale of the oscilloscope screen.

10.30. Sonic Materials Analyzer. The simple but valuable nondestructive qualitative technique for detecting cracks in an object by judging the emitted sound caused by vibrations due to a sharp blow from a hard object, as mentioned in Art. 10.5, may be extended to give a quantitative measure of some mechanical property.

If the induced or forced vibrations caused by a blow are transverse to the longitudinal axis of the object, the modulus of elasticity of the material can be computed by use of an appropriate flexure formula if the distance between nodes and the frequency of vibration are known. Within the audible range, comparison with sounds produced by vibrations of known frequencies is one way to make quite accurate estimates of frequency, provided the operator has good sound perception. Within or beyond the audible range, some tuned radio-frequency circuit operating an oscillograph, together with electrically produced vibrations, may be used to determine accurately the natural frequency of an object being tested. A so-called "sonic" or "dynamic modulus of elasticity" is determined by this method, which has been applied to such materials as concrete; dynamic values of the modulus correlate with those obtained by the slower static-loading method. However, the latter method is often inapplicable, since values of the modulus of some materials are influenced by inherent plastic flow.

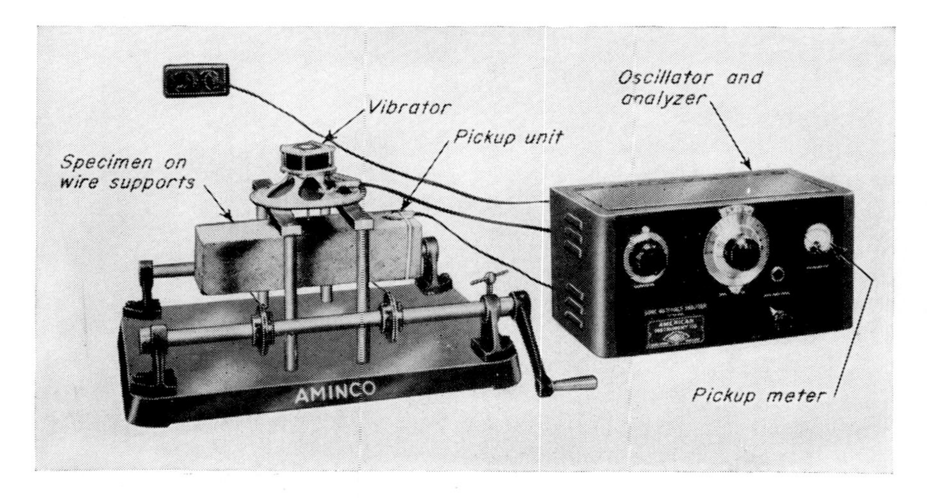

Fig. 10.19. **Sonic materials analyzer with concrete prism under test for modulus of elasticity.** (*Courtesy of American Instrument Co.*)

One form of sonic materials analyzer is shown in Fig. 10.19. A prismatic specimen is mounted on knife-edges or wire supports located at $0.224L$ from each end, where L is the total length of the specimen. By trial, the frequency of the vibrator is adjusted to resonance with the vibration of the specimen. This condition is recognized by the amplitude of vibration being greater than for any other frequency of vibration. Then [see 1062, 1066, 1067]:

$$E = \rho \left(\frac{2\pi L^2 N}{m^2 k} \right)^2$$

where E = modulus of elasticity, psi
L = length of specimen, in.
N = natural frequency, cycles per sec
$\rho = w/g$
w = weight per unit volume, pcf
g = acceleration of gravity
k = radius of gyration of section, in.
m = dimensional factor depending on order of mode and on conditions of restraint (for the fundamental flexural mode of a slender bar freely supported, $m = 4.73$ approximately)

Changes in the modulus of elasticity can be used as a criterion for estimating the effects of certain test conditions imposed on a specimen of a material like concrete, such as continued exposure to the deteriorating influence of alternating cycles of freezing and thawing [1066]. A progressive measure of deterioration can then be obtained by periodic determinations of the modulus of elasticity, since the test is nondestructive.

The sonic method also has been proposed for use for determining the modulus of elasticity of structural members or parts under load. An ultrasonic pulse applied at one face of the concrete is received by a sensitive unit at the opposite face. The time taken by the pulse to pass through the concrete, as measured by an electronic timing circuit, is used to compute the modulus of elasticity which is taken as a measure of the compressive strength [1064, 1065].

For the use of electronic waves in detecting defects in materials, see Art. 10.28.

10.31. Concrete Test Hammer. A nondestructive impact test for determining the hardness and the probable compressive strength of concrete in a structure is made using the Schmidt concrete test hammer [1068]. The tubular unit is pressed against the surface of the concrete to be tested, as shown in Fig. 10.20, causing a spring-loaded hammer inside the tube automatically to strike the concrete. The rebound of the hammer

Fig. 10.20. **Calibrating a Schmidt concrete test hammer. The test cylinder must be supported solidly to prevent any movement while using the hammer; its compressive strength is then determined for correlation with the hammer reading.** *(Courtesy of Pacific Coast Aggregates.)*

after impact is indicated on a scale The rebound number thus obtained, in combination with a calibration curve, can be used to obtain a fairly good value of the compressive strength of the concrete. It serves as a convenient substitute for test cylinders cured at the site for evaluating the compressive strength of the structure at early ages, or for cores cut from the hardened concrete.

10.32. Hardness Tests. Hardness tests have been discussed in Chap. 7. Such tests are usually classified with destructive tests, but in many cases they are, in effect, nondestructive. Owing to the general relationship that exists between hardness and the physical properties of materials, hardness tests are commonly made on parts to be used in service to control heat-treating operations and to assure satisfactory physical characteristics in materials. The test may be made on surfaces located so that the indentation and all accompanying strain effects will be removed by later machining. If this is not possible, care should be exercised to see that the size, location, and nature of the indentation will not be objectionable or result in later damage.

10.33. Surface Roughness. In various industrial operations it is necessary to produce smooth metallic surfaces for good appearance, low friction in bearings, etc. There are several crude methods of determining roughness. *Touch inspection* involves moving one's fingernail along the surface at a speed of about 1 in. per sec. Irregularities as small as 0.0005 in. can be felt by this method. *Visual inspection* with the aid of illuminated magnifiers is limited to rougher surfaces, and the results vary with the observer. *Microscopic inspection* involves a comparison of the test surface with a standard finish. Since only a small portion of the surface is visible at one time, the determination of a reliable average requires many observations.

Direct numerical measurement of surface roughness is now possible with electronic instruments which measure the average height of the surface irregularities. Such instruments are the Profilometer (Physics Research Co.) shown in Fig. 10.21 and the Surface Analyzer (Brush Instruments). The tracer unit which is moved over the surface contains a diamond-tipped stylus. In the Profilometer this stylus is spring-loaded and moves up and down with the irregularities as the tracer head is moved. These mechanical fluctuations are converted into corresponding electrical fluctuations as a tube connected to the stylus moves up and down within the coil of an electrical system. The electrical fluctuations are amplified to operate the instrument which shows the average height of surface roughness of the part.

10.34. Proof Tests. Many types of so-called proof tests are used. Perhaps the most common examples are the proof tests of chain, eyebolts,

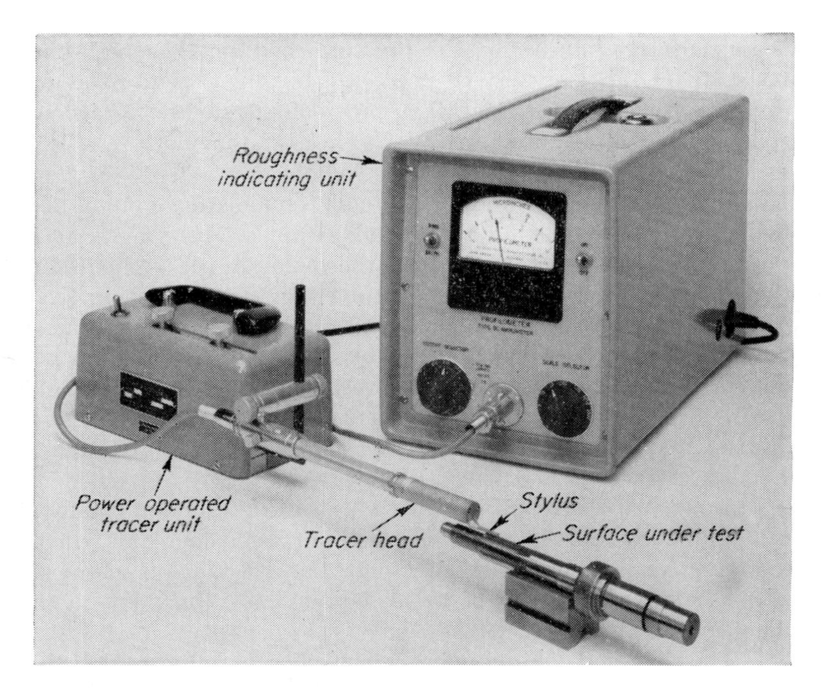

Fig. 10.21. **Profilometer for measuring surface roughness.** (*Courtesy of Physics Research Co.*)

wire ropes, crane hooks, and other parts used for lifting. The applied proof load is usually well above the allowable service load. The proof load applied to welded wrought-iron chain is commonly high enough to deform the links. For eyebars that are to be proof-loaded, the eyes are usually made undersize so that after loading the effect of any distortion is cut away when the eyes are enlarged to correct size.

The specifications of Lloyd's Register and the American Bureau of Shipping require that steel castings for ship parts and anchors shall be subjected to a drop test onto a hard surface and then be suspended clear of the ground and well hammered all over with a heavy sledge hammer to test the soundness of the material. The value of such a test is questionable, since it probably does not subject the casting even to normal loads. A proof tension test for anchors is often specified and appears to have considerably more merit than the dropping and hammering test.

Pressure tests of various types of tanks constitute a type of proof test of the material or the structure itself. Such tests will not only reveal defects but will also show regions of low strength, for they will undergo excessive deformations. The pressure applied is commonly some multiple of the design pressure, such as $1\frac{1}{2}$ times the service pressure. Water is much safer than air as the pressure medium. For high-temperature tests, steam or hot oil may be employed, but their use is rather dangerous. While under load the container or pipe being tested may be struck with a specified weight of hammer.

10.35. Moisture in Wood. Although the moisture content of wood can be determined by cutting out a small sample and drying in an oven, frequently it is desired to determine the moisture content without cutting the wood. This can be done within a few seconds using an electrical moisture meter. Such equipment is suitable for sorting lumber on the basis of moisture content and can test finished woodwork in place without serious damage to the wood. Two types of meters are available [1070], one determining the moisture content by measuring the electrical resistance of the wood while the other measures the electrical capacity of the wood.

Resistance Meters. Below the fiber-saturation point (at about 25 percent moisture) the electrical resistance of wood increases greatly as the moisture content decreases. Although the resistance is affected by temperature and the species of wood, corrections supplied with the instrument are easily made.

The electrical contact points are needles mounted so that they can be easily driven into the wood and then withdrawn after testing. For wood which is not of a uniform moisture condition, the meter indicates the moisture content at or near the points of the needles, since the wood becomes a better conductor at greater depths from the surface where the moisture content is higher. Thus it is possible to determine the moisture content at any distance from the surface by driving the electrodes to that depth. In general, the moisture content at a depth of one-fifth the thickness of the material is at about the average value for the wood. For thin material use is made of surface-plate electrodes (instead of needle electrodes), which make contact with opposite surfaces of a board. For both types of electrodes the actual measurement is made by balancing the resistance between the electrodes with known resistances.

Capacity Meters. The electrical capacity of wood varies directly with the moisture content. Temperature effects are negligible as is also the species of the wood. Although the capacity method is excellent for determining the weight of water in wood, it is not possible to convert this to a percentage without knowing the weight or specific gravity of the wood. In practice, this value is usually taken as the average for the species.

10.36. Stress Determination by Photoelasticity. One problem which arises in the design of solid structural members and machine elements is the magnitude and distribution of stresses throughout the part. Of particular interest are the zones of stress concentration and

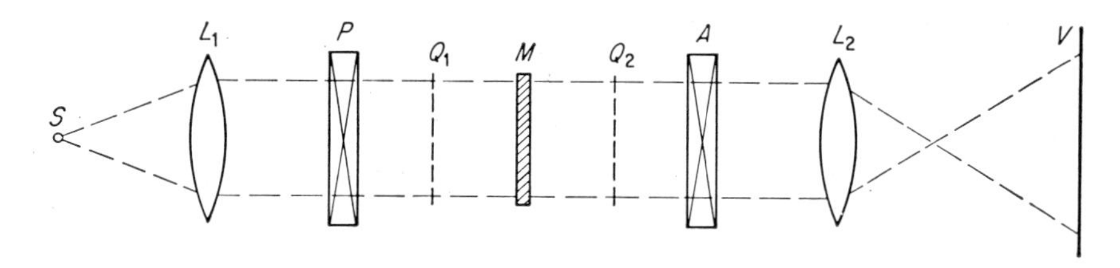

Fig. 10.22. **Schematic arrangement of photoelastic apparatus.**

magnitudes of the maximum principal and shear stresses. Stress concentrations may occur due to the way in which the external forces are applied or due to the shape of the part.

For conditions of plane stress in an isotropic material, a means of experimentally determining the stress distribution or stress concentration within the range of elastic behavior is by photoelastic analysis.* In this technique a model of the part is made of a suitable transparent material, such as transparent annealed Bakelite or a plastic designated CR-39. By passing polarized light through the loaded model it becomes birefringent (i.e., doubly refractive), and a pattern of colored or light and dark bands or *fringes* can be observed, which pattern is a function of the stress distribution in the part. These observations, together with other supplemental measurements, enable a determination of the state of stress at points of interest or throughout the part.

A simplified arrangement of apparatus for making a photoelastic analysis is shown schematically in Fig. 10.22. A variety of arrangements of optical systems and of component parts are in use. The model in which the stress distribution is to be studied, shown at M, is in a beam of parallel polarized light. The light from source S has been polarized by the polarizer P. The polarizing device may be a Nicol or similar prism, but a sheet of material called *Polaroid* is now often used. The rays through the polarizer and the model are made parallel by a lens system L_1. After leaving the model, the light passes through another polarizing unit at A, called the analyzer; the plane of polarization of the analyzer is set at right angles to the plane of polarization of the polarizer. Another lens system at L_2 serves to project the emerging light pattern on a viewing screen V. With the polarizer and analyzer "crossed" at right angles, if there is no model in the polariscope, and if the polarizing units are fully efficient, the beam of light is extinguished and no light reaches the screen.

The method depends, essentially, on two experimentally demonstrated facts. When polarized light of some given wavelength is passed through a strained transparent plastic body: (1) the incident ray of

* Some procedures for three-dimensional photoelastic analysis have been developed. See Ref. 463, pp. 928–976.

polarized light at each point is broken up into two component rays of light vibrating in mutually perpendicular planes which are parallel to the directions of the principal stresses and, (2) each component ray is transmitted through the transparent solid material with a velocity that is a linear function of the principal stress with which it is associated.

The process of composition and resolution of the rays of light passing through the system is illustrated diagrammatically in Figs. 10.23*a* and

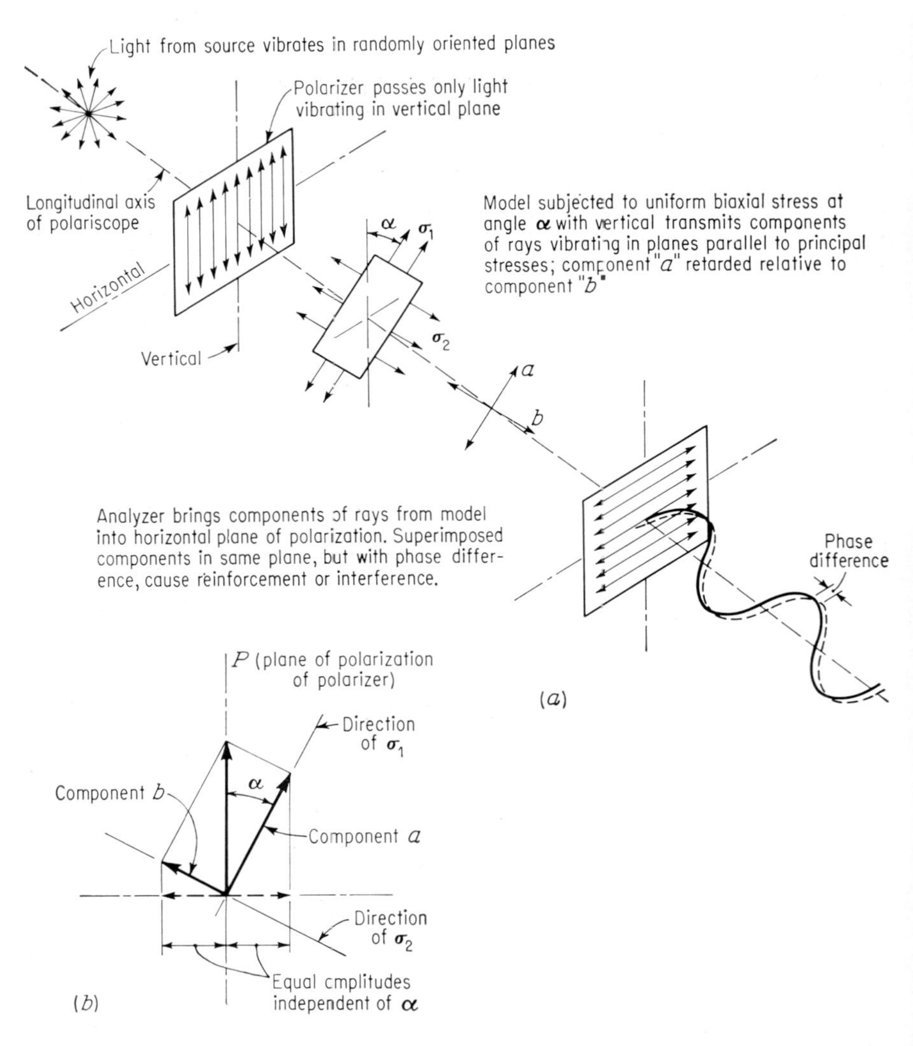

Fig. 10.23. (*a*) **Course of light rays through photoelastic polariscope.** (*b*) **Relationships between components of light ray passing through system (diagrammatic end view).**

The two component rays emerge from the stressed material out of phase and vibrate in planes at right angles to each other.

When these component rays are now passed through the second polarizing device (the analyzer), whose plane of polarization is at right angles to the initial incident polarized ray, the component vibrating light waves are again made to vibrate in the same plane (the plane of polarization of the analyzer); however, while the emerging components vibrate in the same plane, and while the waves now have the same amplitude, they are still out of phase by an amount proportional to the difference in the principal stresses at the point through which the ray passed in the stressed solid.

When the components are out of phase by one-half wavelength ($\frac{1}{2}\lambda$) or its odd integer multiple ($1\frac{1}{2}\lambda$, $2\frac{1}{2}\lambda$, etc.) the ray will be extinguished. For other phase differences the resultant ray will be partially or completely transmitted.

If the entire beam of light (all rays) passing through the stressed solid is now projected on some viewing screen, a pattern of light and dark bands may be observed; each dark band, called a *fringe,* is the locus of points having the same difference in principal stress (or constant maximum shear stress). If white light is used, each band is made up of a sequence of colors of the spectrum; lines of the same color (representing the locus of points having the same principal stress difference) are called *isochromatic* lines.

If the stressed solid is subjected to a uniform state of stress throughout, the image on the screen would be all dark, all light, or all the same color, depending on the magnitude of the difference in the principal stresses and the kind of light used.

Successively larger differences between the two principal stresses produce correspondingly larger phase differences in the final superimposed beam. Corresponding to the odd-integer multiples of $\frac{1}{2}\lambda$ are a series of fringes (dark bands, or spectral bands) of successively *higher order*. For a given material and thickness thereof, the stress difference for one fringe order can be determined by stressing a sample of the material of which the model is made under uniform uniaxial stress.

Then, beginning at a corner where both principal stresses are zero, the number of fringes to a given point can be counted and a measure of the principal stress difference at that point determined. If there is no corner at which to begin, then first locate an *isotropic* point (i.e., where both principal stresses are equal). This point will be where the black isoclinic lines described below are the same for both white and monochromatic light. This point is also one where the difference of the principal stresses is zero. Therefore, it can be used as a starting point to count the number of fringes to any other point. In this way a mapping of the principal stress differences may be done.

One way to find the *sum* of the principal stresses is to measure the change in thickness (Δt) of the model under stress. The change is a function of the sum of the principal stresses (σ_1 and σ_2), modulus of elasticity E, and Poisson's ratio μ. Thickness measurements are usually made by very precise calipering, although light interference caused by reflection of light from the front and rear surfaces of the model has been used. A calibration factor is determined by stressing the model material in uniform tension: $\sigma_1 + \sigma_2 = \left(\frac{E}{\mu t}\right) \Delta t = C\ \Delta t$.

From the sums and the differences of the principal stresses, the magnitudes of the principal stresses may be computed for each point of interest in the model. If

$$\sigma_1 + \sigma_2 = K_1$$

and

$$\sigma_1 - \sigma_2 = K_2$$

then

$$\sigma_1 = \frac{K_1 + K_2}{2} \qquad \text{etc.}$$

When the *directions* of the principal stresses are parallel to the planes of polarization of the polarizer and analyzer, the light is extinguished. In an image of a stressed model, the locus of points having the direction of the principal stresses parallel to the planes of polarization appears as a black line or zone; such lines of constant direction or inclination of the principal stresses directions are called *isoclinic* lines. By rotating both the polarizer and analyzer, the direction of the principal stresses throughout the stressed model may be determined. When white light is used, the isoclinic lines are more readily distinguishable from the stress fringes. In photographing the image of a stressed model, using monochromatic light to avoid confusion of the isoclinic lines with the stress fringes, quarter-wave plates* are inserted in the light beam on both sides of the model (at Q_1 and Q_2 in Fig. 10.22), eliminating the isoclinics.

If the models are small, they must be carefully shaped and annealed. They should have uniform thickness and flat surfaces.

Good quantitative stress determination by photoelastic analysis requires careful technique, precise work, and good equipment. However,

* The quarter-wave plates transform the plane polarized beam into two mutual perpendicular polarized components, one being $\frac{1}{4}\lambda$ out of phase with the other, producing *circularly polarized* light.

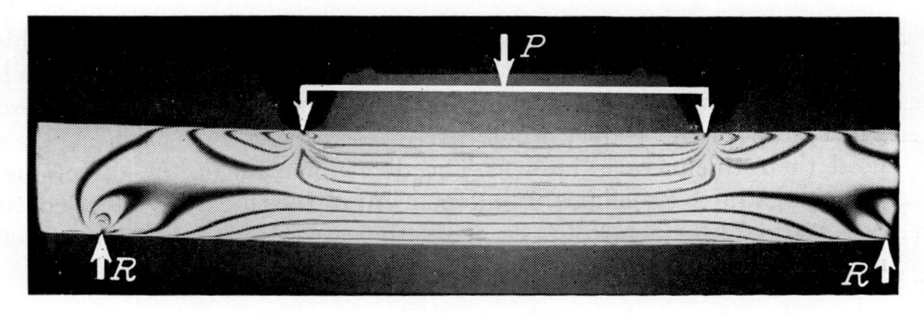

Fig. 10.24. **Fringe pattern for part of a simple-span rectangular beam subjected to two concentrated loads.**

inexpensive apparatus and fairly crude procedures can often give helpful qualitative pictures of the nature of stress concentrations.

Figure 10.24 shows a fringe photograph of a model of a simple-span beam subjected to two concentrated loads. The closeness of the fringes at a support and at the two load points indicates that high stress concentrations exist there.

10.37. PhotoStress. A special photoelastic method of determining stresses in actual structures or components of any material while under load, on flat or curved surfaces, is called *PhotoStress* [1092]. It acts as an infinite number of strain gages of practically zero gage length, uniformly distributed over the surface to be studied.

In this method, liquid PhotoStress plastic is applied and allowed to polymerize directly on the part, or flat or contoured sheets of transparent PhotoStress material are bonded to the part, which has been previously made reflective by spraying the surface with aluminum paint.

When a load is applied to the object, strains developed in the bonded plastic coating cause it to become birefringent, as noted in Art. 10.36. When illuminated with polarized light, the birefringent plastic coating displays a pattern of colored fringes which is a function of the strain distribution of the entire surface and from which peaks of strain can be determined in terms of their location, direction, sign, and magnitude.

Two sets of bands are visible in the strained plastic coating when observed through a reflection polariscope. Black bands, or *isoclinics,* give the directions of the principal strains. The colored fringes, or *isochromatics,* give the magnitude of maximum shear strain by making only one reading of fringes with light going under normal incidence through the plastic, and give separate values of principal strains by making two readings of fringes, one with light going under normal incidence through the plastic and the other under oblique incidence.

Because the birefringence in the plastic is proportional to the thickness as well as the strain, the thickness must be accurately known. When using the liquid plastic, its thickness can be measured with commercial gages or with the proper PhotoStress instrument, but when sheets are used their thickness is known. The determination of the directions of principal strains (isoclinics) is independent of the thickness of the coating. See Art. 4.23 for a novel type of photoelastic stress-gage used to determine strains in an element without use of a polariscope.

Chapter 11.

ANALYSIS AND PRESENTATION OF DATA

11.1. The Problem of Transmission of Information. It may be taken almost as axiomatic that, at least for engineering purposes, no data are of value until they are put in a form that can be readily understood and utilized. The particular form in which data should be summarized and the extent to which they should be interpreted obviously will depend upon the intended audience. However, there are a number of commonly used procedures for analyzing and reporting data which may be generally applied.

Some of the considerations that may be mentioned in connection with the problem of analysis of data are the following:

1. *Reduction of "Raw" Data.* The units in which the data are recorded are conditioned by the kind of measurements that are made, e.g., loads may be measured in pounds, and temperatures may be measured in terms of electrical resistance. Because most data have meaning only in

comparison with similar data, the quantitative measures obtained from a test are reduced to values whose units are acceptable as a basis for comparison; thus loads are reduced to stresses, say, in pounds per square inch, and deformations to strains. Further, the reliability of the data is conditioned by the errors of measurement. In reducing the data, corrections may have to be applied for systematic errors, and in order to express the final data in the appropriate number of significant figures, an estimate should be made of the accidental errors inherent in the measurements and of the effect of these errors on the accuracy of the reduced values.

2. *Summary of Data.* Although in the simple case the summary of test results may merely amount to setting down a few readily comprehensible facts and figures, in connection with large-scale operations there may be accumulated masses of data that are so numerous and variable that it is practically impossible for the mind to digest and evaluate them in unassembled form. Advantage may then be taken of well-known statistical procedures for summarizing the data.

3. *Study of Relationships between Variables.* After the data have been reduced and assembled in manageable form, the final step in the analysis is usually to seek or to develop relationships between the variables involved or between the data obtained from a particular test and previously obtained data or some theory. In many instances the skill with which this is done depends upon the capacity, the ingenuity, and the background of the analyst. Common devices employed in studying relationships are tabulations, graphs, bar charts, and correlation diagrams; the procedure is usually to hold constant (in so far as is known or is possible) all variables except two whose relationship is investigated.

Some aspects of the problem of the presentation of data are

1. The manner in which the information is summarized. Tables, figures, and the written word may be used.

2. The kind, scope, and make-up of the report. Engineering data are transmitted in the form of reports or technical papers. Their preparation involves consideration of the use of language in written form, the mechanics of preparing and reproducing the report on paper, and the interpretation of the facts and relationships to meet the needs or capacity of the person or class of persons for whom the report is intended. It should involve thought and intelligence in its preparation.

11.2. Variations in Data. Practically all data derived from tests are subject to variation. The results of a test on a single sample involve measurements that are subject to variation. The results of a test on a series of similar samples show variation between samples. After the measurements have been corrected for the effects of systematic errors, it is usually found that the variations in adjusted or corrected measurements follow a chance distribution. For large numbers of data, variations in measurements and measures of properties have been found to coincide closely with variations computed from theoretical considerations. When the data are few, the coincidence is often not so good, but for convenience the concepts developed from the theory of probability (for many numbers) are applied and afford a fairly workable means of summarizing and utilizing data. A few statistical concepts that appear to find useful application to handling data from testing work are summarized in the following paragraphs. In the condensing and presentation of homogeneous but involved results, statistical methods can be applied only to data in numerical form.

11.3. Grouping of Data. The first step in an analysis involves grouping of data. When the time of occurrence is important, a chronological sequence is sometimes used and data are presented as a time series. Frequently such data appear in complex cyclical form, and their analysis requires highly specialized techniques. In materials testing, observations involving time of occurrence such as determinations of plastic flow (creep) involve only one cycle, and determinations of progressive deterioration of materials subjected to alternate cycles of freezing and thawing are sometimes presented as time series but not in the same way as are the more widely known economic time series.

The nearest approach to the well-recognized time series is the "control chart"; however, here the emphasis is not on the uniformity of the time scale but on the sequence of the sample. The application of statistical methods and criteria to the control of quality of manufactured products by means of the control chart is of considerable interest and value in connection with the problem of inspection. The principal features of the control chart are discussed in Art. 11.8.

Some data may require geographic grouping, but the grouping of most data in materials testing is according to magnitude. The symmetrical arrangement of data according to magnitude results in what is technically known as a "frequency distribution." Sometimes the latter series is considered as being divided into two types, one consisting of n measurements of a given characteristic of n different pieces, the other of n measurements of a given characteristic of a *single* piece. An example of the first type is some quality of the material itself, such as the tensile strength of wire in a given coil; the second, some statistic of a single piece, such as its diameter. Both of these types consist of homogeneous data. In some statistical parlance they come from a common parent population, and each is not a single-valued constant but rather a frequency-distribution function.

On first thought the diameter of a tensile specimen might not appear to be a distribution function, but it should be remembered that the surfaces of test specimens cannot be given a costly ground and polished finish, so the measurements of the diameter at various points are usually made only to thousandths of an inch; hence measurements along the gage length do not show much variation. Finer measurements would tend to emphasize the variability due to a combination of indeterminate causes. This means that each item is subject to accidental or chance variations that are composed of the inherent variability of the characteristic itself in addition to errors of measurement.

In the testing of materials the accuracy of machines, strainometers, gages, and other measuring devices is usually maintained between known limits, and the variations in measurements due to parallax, lost motion, and inertia effects are kept to minimum values by proper design and use. Furthermore, these limits (errors) are small compared with the usual variations in property measures from sample to sample. When the errors become larger and systematic as may be caused by temperature changes, either computed corrections are directly applied or a procedure is used to eliminate them; for example, strain-gage readings with portable instruments such as the Whittemore type are automatically corrected by referring to check readings taken on a reference or unstressed "standard bar." Data on materials testing may be considered as being predominately affected by variability of the characteristic observed.

The first operation performed on the raw data is to arrange the items according to magnitude, usually in ascending order. This is sometimes referred to as an array or an ungrouped frequency distribution. Merely by inspection, the minimum and the maximum values can be selected, and by a simple computation, the median or middle value and the range can be determined. Although data in this form are still unwieldy, it is possible to make a detailed study of the array by dividing it into equal parts, such as quartiles (4 parts), deciles (10 parts), or percentiles (100 parts), just as the median divides it into two parts.

Instead of a grouping based on the number of parts as just described, the subdivisions may be based on the variable being measured. These groups are sometimes called cells and class or step intervals. After the length of the interval has been decided upon, the number of items in each interval, usually called the frequency, is then determined. Often the relative frequency, which is the number in each interval divided by the total number of items, is used. This fraction of the total number is an important characteristic, especially when applied to the percentage below, outside, or beyond a specified limit, thus giving the fraction defective. The interval boundaries should be established so that each item will lie within some interval and not indefinitely on a boundary line. A convenient score or tally sheet is indicated in Table 11.1. When there are a large number of items, 13 to 20 class intervals are recommended [1140]. Too many intervals may show an irregular distribution. By using larger intervals (few divisions) the appearance of the distribution can often be

***Table* 11.1. Score sheet for weights of coating on 100 sheets of galvanized iron**

Weight of coating, oz.	Tally	Frequency
1.275–1.324	II	2
1.325–1.374	卌 I	6
1.375–1.424	卌 II	7
1.425–1.474	卌 卌 IIII	14
1.475–1.524	卌 卌 IIII	14
1.525–1.574	卌 卌 卌 卌 II	22
1.575–1.624	卌 卌 卌 II	17
1.625–1.674	卌 卌	10
1.675–1.724	III	3
1.725–1.774	卌	5
Total number of items.		100

improved. When the total number of items is less than 25, such a presentation is of little value.

In the usual graphical presentation the frequencies, actual or relative, are plotted as ordinates to an arithmetical scale on the center line of each interval. When successive plotted points are connected by straight lines, the chart is called a frequency polygon. A different design would result by drawing a wide line or bar along the center line of each interval from the base to the plotted points. The diagram is then known as a frequency bar chart. If instead of filling in the bars they are left open as in Fig. 11.1, the diagram is designated as a frequency histogram. All these different forms are widely used.

A variant presentation, either tabular or graphical, in which a running total of the items in each successive interval or cell is computed is called a cumulative frequency distribution. The table or diagram could be labeled to read number or percentage of items "greater than" or "less than" marked values. In the graphical presentation the variable under consideration is usually plotted on the abscissa, and when both abscissa and ordinate are arithmetic, the cumulative distribution takes a peculiar form commonly called an ogee curve. This curve becomes a straight line for a normal distribution when the ordinates are plotted to a so-called probability scale.

Since the frequency diagram is merely a plot of the grouped frequency tabulation, it cannot yield more information than the tabulation itself, although it is sometimes advantageous to show the relationships in graphical form.

The distribution commonly encountered in materials testing is a bell-shaped curve resembling the theoretical normal frequency curve which

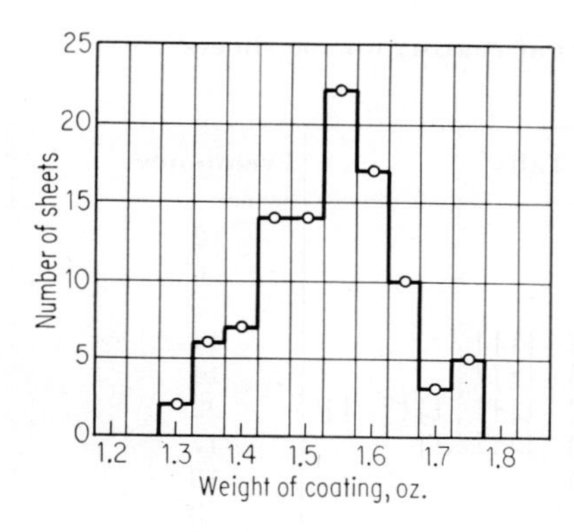

Fig. 11.1. **Frequency histogram.**

it seems was first used by Demoivre in 1733, but is known by other names as a probability curve, curve of error, Gaussian error curve, or Laplace-Gaussian curve. If the distribution curve has only one peak, it is said to be unimodal. Many times it lacks bilateral symmetry and is skewed to the left or right (asymmetrical). Nonhomogeneous data, say from two different parent populations or universe, might produce a bimodal distribution, but homogeneous data take other characteristic shapes such as J- or L-shaped curves which are extremely asymmetrical, or possibly a U-shaped curve which could be symmetrical in that it has a point of minimum frequency between higher frequencies at both ends of the range.

This discussion is restricted to unimodal frequency distributions which are so characteristic of data on the physical and mechanical properties of materials. In fact the general recognition that the true magnitude of a property or quality of a material is a distribution function, thus involving the concept of probability (which in itself is considered one of the most important advances in the field of statistics) is a decidedly modern viewpoint.

11.4. Central Tendency. A measure of central tendency, or tendency to be grouped about a central value, is called an *average* which purports to summarize the data by locating this typical value. The most significant averages are the arithmetic mean, the median, the mode, the geometric mean, and the harmonic mean.

The arithmetic mean is the most widely used of all criteria of central tendency. It is the average of everyday speech. The usual symbol is $\bar{X}$ (bar X), and the formula is frequently written

$$\bar{X} = \frac{1}{n} \sum_{i=1}^{n} X_i$$

which is identical with

$$\bar{X} = \frac{X_1 + X_2 + \cdots + X_n}{n}$$

For the following five items 2, 3, 5, 8, and 9,

$$\bar{X} = \frac{2 + 3 + 5 + 8 + 9}{5} = \frac{27}{5} = 5.40$$

The arithmetic mean is a calculated average (contrasted with the median) and is affected by every item but is greatly distorted by unusually large values at the extremes.

The median is the value of the middle item in an array. In a distribution where the central values are closely grouped, it is typical of the data since it is unaffected by unusual terminal values. It is not so well known, however, as the arithmetic mean. For the previously given array (2, 3, 5, 8, and 9), the median is 5.

The mode is the value that occurs most frequently. In an ideal (smooth) frequency distribution the modal value is located by the maximum or highest ordinate of frequency. When a large number of items are used, it is the most typical average. It is an average *position* and is unaffected by large items near the ends of the range. Since the mode is not the mid-point of the modal interval (class or cell) in grouped frequency distribution, it must be computed by some method such as "moments of force," moving averages or by smoothing the frequency distribution when a limited amount of data is available. In the array 2, 3, 5, 8, and 9 no items are repeated; therefore there is no mode.

The geometric and harmonic means are rarely used in connection with materials-testing data. For their computation and meaning Refs. 1101 to 1104 and 1140 may be consulted.

11.5. Dispersion. An inspection of the frequency diagram will furnish a qualitative indication of an important characteristic—the dispersion, scatter, or variation about the average value. A crude measure of this deviation from the average is the *range* or *spread.* Although it is easily determined, since it is merely the actual difference between the maximum and minimum items in the distribution, it is dependent upon only two values that are of unusual occurrence (low frequency), particularly when n is large. So the range may give misleading indications of distribution.

A criterion that considers the location of every item, rather than only the two extremes, would be more meaningful. An obvious measure would be the average distance of the items from some measure of central tendency. If the arithmetic mean and plus and minus distances were used, such an average would be zero; however, this difficulty can be overcome by disregarding signs.

The differences between an average and the various items are called deviations, and the

$$\text{Average deviation} = \frac{\sum_{i=1}^{n} (X_i - \bar{X})}{n}$$

The quantity $(X_i - \bar{X})$ is here taken to be the absolute value of a deviation disregarding signs. If the median were used the average deviation would always be smaller than when using the arithmetic mean $(\bar{X})$.

Perhaps the most generally used measure of dispersion is the *standard deviation,* the square root of the average of the squares of the deviations of the numbers from their average $\bar{X}$. It is a special arrangement of the average deviation and is usually designated by the symbol σ (small Greek letter sigma). The following equation gives its most general form:

$$\sigma = \sqrt{\frac{\sum_{i=1}^{n} (X_i - \bar{X})^2}{n}} = \sqrt{\frac{\sum_{i=1}^{n} X_i^2}{n} - \bar{X}^2}$$

A glance at the general form makes the reason obvious for the often used designation, root-mean-square deviation. Studies of statistics since 1900 have shown that when the number of items is small (say less than 25) it is more nearly correct to divide by the "number of degrees of freedom," usually $(n - 1)$, instead of the total number of items n.

For both grouped and ungrouped data the computation of sigma should be carried out in tabular form. For large masses of data certain short cuts, such as adding or subtracting and/or dividing all items by suitably chosen constants to reduce the magnitude of all items, are desirable. Also, careful planning of the work based on some appropriate form of the general expression (as already presented) is worthwhile. Tables of squares and square roots of numbers aid in the computation.

The standard deviation, since it uses second powers of all items, places more emphasis on widely dispersed items than does the average deviation. For a normal distribution the average deviation is 0.7979σ. This relationship can be applied with fair approximation to slightly skewed distributions.

An important characteristic of the standard deviation is the number of items or the area included between ordinates erected on each side of the center of the distribution curve at a distance of one, two, and three standard deviations. These numbers or areas for the normal (theoretical) probability curve are 68.26, 98.46, and 99.73 percent, respectively. Even in the actual analysis of materials-testing data it is unlikely that less than 96 percent of the values for a frequency distribution of more than 100 items will be within the range of three standard deviations on each side of the arithmetic mean.

Since significant comparisons of dispersions cannot be made by using absolute measurements, a variation expressed as a ratio or percentage must be determined. The most commonly used is the standard deviation on a percentage basis and is usually called v, the *coefficient of variation.*

$$v = \frac{\sigma}{\bar{X}} 100$$

It should be noted that when $\bar{X}$ is very small or even zero, as it could be in the case of temperature measurements, the resulting v would be quite misleading.

11.6. Skewness. When a frequency distribution departs from a theoretical normal shape, it is said to be skewed or asymmetrical. Absolute and/or relative measures of this lack of symmetry can be based on the values or locations of the several measures of central tendencies, all of which are identical for symmetrical distributions. Symmetry is naturally measured about the high point of the frequency polygon—the modal value. The relative location of the arithmetic mean, since it is affected by the magnitude of extreme items, offers an excellent criterion of the *skewness factor* k in either the simple form

$$k = \frac{\text{mean} - \text{mode}}{\text{standard deviation}}$$

or an alternate form which gives a more exact value

$$k = \frac{\sum_{i=1}^{n} (X_i - \bar{X})^3}{n\sigma^3}$$

For symmetrical curves k is obviously zero. Since the scale of the variable is usually arranged so that the values increase toward the right, then for curves skewed to the right, i.e., with the excess tail to the right, the larger values on that side will cause the arithmetic mean to be larger than the mode and to be located to its right with a resulting plus sign for k. Hence a negative coefficient k means the curve is skewed to the left.

11.7. Statistical Summaries. The meaning and significance contained in a mass of measurements are made clear by their interpretation, which in turn is facilitated by a well-arranged presentation. However, the accuracy of the data itself is in no way altered by the method of presentation or subsequent study and interpretation.

Since the investigator has no way of knowing what subsequent use may be made of the data, supporting evidence regarding the test specimens themselves should first be presented in brief, clear, unequivocal statements. These should include even more information than at the time may seem relevant, since later developments may prove the value of all collateral evidence presented regarding the history of the specimens, including their selection and preparation, and the grade and character of material. In the use of control charts such extraneous data as the day of the week and the weather should not be ignored.

The next information should deal with the test itself and the conditions under which it was made. This includes the measurements together with adequate description of the methods used to eliminate constant errors and reduce the size of chance or accidental errors as well as observations on the general and specific test conditions and procedures, giving especial attention to the matter of special difficulties and their treatment so as to present evidence of all precautions taken to establish controlled conditions and secure reliable, trustworthy data.

In the presentation of materials-testing data itself, the essential information is generally contained in four statistics, namely, the number of items n, the arithmetic mean $\bar{X}$, the standard deviation σ, and the coefficient of variation v. In general, significant measures of skewness cannot be obtained when $n < 250$, and in many instances a graphical frequency polygon will give sufficient information on this property. Sometimes the function p, which gives the percentage outside a certain limit such as the defective fraction, is of importance. However, it is obvious that any single statistic cannot present a complete picture of the manner in which the original data were distributed; hence, the ones to use depend upon the purpose of the investigation and the use to be made of the test results.

11.8. Control Charts. The pioneer writer on the subject of scientific control of engineering materials and products was W. A. Shewhart [1140]. Little was written previous to the appearance of his publications, which have formed the basis for widely accepted techniques of measuring and predicting the control of the quality of manufactured products. The value of these methods is in the detection of assignable causes for variations in quality of the final product. Only a brief statement is feasible for this text.

It is practically impossible to attain a given value of a quality in each successive manufactured article because the quality is a variable and the change in its magnitude is a frequency distribution.

The variation in the magnitude of some statistic of a measurable property such as tensile strength can be used as a criterion of quality. Most frequently used criteria are the arithmetic mean, standard deviation, fraction defective, and range.

Values of a given function of quality, such as the arithmetic mean of the tensile strength of successive samples, each containing an equal number of items, say five, are plotted as ordinates against a scale of

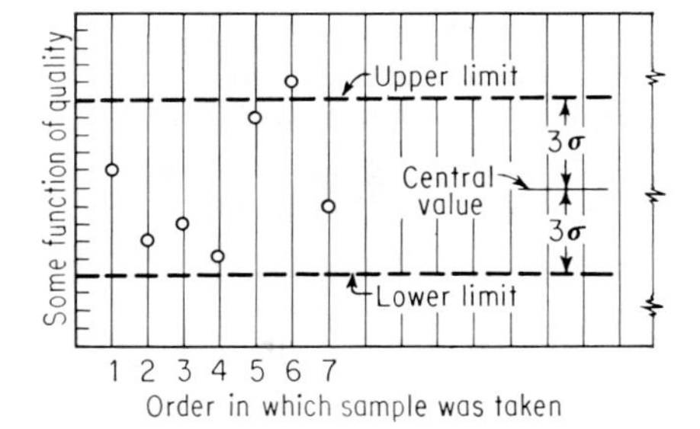

Fig. 11.2. **Control chart.**

abscissa giving the numerical sequence of samples increasing in the customary way from left to right, as shown in Fig. 11.2.

The control chart presents the data so that their consistency and regularity can be seen at a glance. The important feature of the chart, however, is the controls that appear as lines parallel to the abscissa on the chart and show the limits of the variability in the quality being measured which should be left to chance. These limits are commonly set at three standard deviations on both sides of the central value. This is not a theoretical determination but one that has been generally accepted as indicating lack of control, since 99.17 percent of the area of the normal frequency polygon is included between ordinates erected at these limits. Stating this in terms of probability, there is only 1 chance in about 370 that a single sample will fall outside these limits. Furthermore, these limits have been found to be satisfactory in industrial applications. The limits of three standard deviations vary with the number of items in each sample, being larger—i.e., farther from the central value—for small numbers. Tables are available which give the limits for the different statistics such as average, standard deviation, and range for each sample size from 2 to 25 items. Formulas and coefficients for the computation of control limits, when samples contain more items, are also available in many texts on statistics.

When the control chart is used in connection with a standard, the limits are established with respect to the specified value, but if no standards are given, the limits are determined on the basis of the data itself as they are accumulated. In both of these cases the location of the plotted points outside the established limits is taken as an indication of the existence of a causative factor.

11.9. Correlation. The statistical technique known as correlation has relatively little use in materials testing despite its wide adoption in other fields and its prominence in publications relating to statistics. However, the elements of the method are used by engineers, many of whom do not concern themselves with the commonly used statistical terms applied to the devices employed to express simple relationships. In order to study a relationship of a group of paired measurements, the obvious procedure would be the construction of a chart with arithmetic rectangular axes. This is known as a *scatter diagram,* and the line representing the best fit is the *regression* line; if the line were straight, its general form would be $Y = mx + b$, where m would be the *regression coefficient.* If all points were on the regression line, i.e., if there were no spread or variation normal to the line, the correlation would be perfect and the *coefficient of correlation* would be unity, i.e., 1, the sign depending on the slope of the line. For a straight regression line a wide scatter would decrease the coefficient of correlation. The letter r is used to designate the coefficient of simple linear correlation and is sometimes referred to as the Pearsonian coefficient. There are appropriate procedures for computing the location of the regression line, such as the method of least squares, which makes the squares of the deviations of the points from the line a minimum. The general procedure is to sketch it in freehand and let the diagram tell the story without the computation of aforementioned statistics.

The well-known scatter diagrams in materials testing are the strength-moisture and strength-density diagrams for wood and the strength-hardness relationship for steel. The latter is probably most widely used in inspection and control work and wherever destructive tests are not feasible. Strength is usually plotted on the Y axis and hardness, the independent variable, on the X axis as in Fig. 11.3. The heavy dashed lines equally spaced on both sides of the regression line can be placed so as to indicate any desired probability limits. For the example given, hardness H indicates that the chances are even (1 to 1) that the tensile strength will be between s_1 and s_2 because the limits are placed $\pm 0.667\sigma$ (standard deviations) each side of the central value s. This value, $\pm 0.667\sigma$, is known as the probable error (see Art. 11.10) and is represented by ss_1 and ss_2. In the frequency distribution shown to the right, the open area is

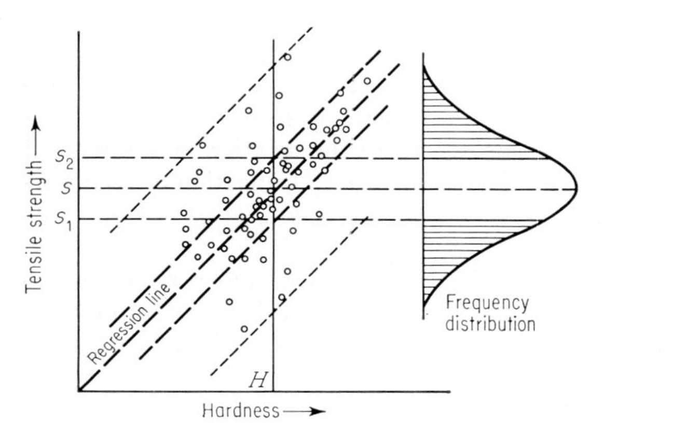

***Fig. 11.3.* Scatter diagram.**

equal to that shown cross-hatched, each being one-half the total. The frequency polygon shows, however, that the most likely deviation (error) is zero, or, saying it another way, the most likely or probable strength, indicated by H, is the central (modal) value s, but it must be remembered that the hardness of a given specimen is in itself subject to errors of measurement.

11.10. Errors. It is assumed that the student will have had courses such as surveying or physical measurements before he undertakes the materials-testing course and thus has some familiarity with the theory of errors. It is desirable to point out here two things: the relation of certain terms often used in connection with the discussion of errors to terms used in the general field of statistics, and the method of determining the error of a computed quantity.

The criterion of central tendency of a series of measurements of like quantities, which differ because of accidental errors only, is taken as the arithmetic mean, which is the most probable value of the measurements. In the theory of errors, the deviations from the mean are usually referred to as the "residuals."

Commonly used criteria of dispersion (which serve as measures of precision) are the *average error* (same as average deviation) and the *probable error*. The probable error of a single measurement in a series of measurements (of equal weight) is computed from the expression $0.6745 \sqrt{\Sigma(v^2)/(n-1)}$, where the v's are the residuals and n is the number of observations. The quantity under the square-root sign is seen to be the standard deviation for a limited number of observations. The probable error is *not* the error most likely to occur, but simply marks the limit within which the chances are 50-50 that any error taken at random will fall. The probable error of the mean of a series of measurements is found by dividing the probable error for a single observation by $\sqrt{n}$. The standard deviation provides just as useful and significant a measure of dispersion as the probable error and might just as well be used except that, in this country at least, custom has made workers in the field of physical measurements familiar with the latter factor. For comparative purposes, the *relative error* or *precision ratio* is commonly taken as the ratio of the probable error to the value of the quantity measured, expressed as a percent; this corresponds to the coefficient of variation.

The probable error of a quantity calculated from the combination of several independently measured quantities is derived from the theory of least squares. If $u = f(x, y, z, \ldots)$, and if R_u is the probable error of a quantity u, R_x of quantity x, etc., then

$$R_u^2 = \left(\frac{\partial u}{\partial x} R_x\right)^2 + \left(\frac{\partial u}{\partial y} R_y\right)^2 + \left(\frac{\partial u}{\partial z} R_z\right)^2 + \cdots$$

This may be applied to several specific cases as follows:

1. Product u of x and a constant C: $R_u = CR_x$
2. Area of a circle, u, from measured diameter x:

$$R_u = \sqrt{\left[\frac{\partial(\pi x^2)}{\partial x} R_x\right]^2} = \frac{\pi D}{2} R_x$$

3. Sum u of like quantities $x, y, z, \ldots$:

$$R_u = \sqrt{R_x^2 + R_y^2 + R_z^2 + \cdots}$$

If the probable error of each quantity is the same,

$$R_u = R_x \sqrt{n}$$

where n is the number of quantities

4. Modulus of elasticity E from measurements of P, the applied load; L, the gage length; A, the cross-sectional area; and e, the deformation:

$$R_E = \sqrt{\left(R_P \frac{L}{Ae}\right)^2 + \left(R_L \frac{P}{Ae}\right)^2 + \left(R_A \frac{-PL}{A^2e}\right)^2 + \left(R_e \frac{-PL}{Ae^2}\right)^2}$$

The relative error may be used instead of the probable error in the above equations.

11.11. Reports. Although discussion of this large subject is beyond the scope of this text, a few suggestions may be given. A report as a whole should be planned to meet the needs of the individual for whom it is intended. It should be clear in meaning and readable in form. Attention should be given to (1) format or the mechanics of make-up, i.e., the method of reproduction, kind of type, and size of page; (2) style and composition; and (3) the way in which the data are presented. References such as 1150 and 1151 may be consulted. Directions for students' reports of laboratory tests are given in Part 2.

11.12. Tables. The presentation of factual material in tabular form economizes space and facilitates the comprehension of the scope and range as well as the significance of the data. Often more varied information can be condensed into a table than can be shown in a chart or diagram, and whether used in the text or in an appendix, tables should have adequate and clearly stated titles, preferably so brief that they can be understood on the first reading. The column headings should be complete and contain units for the values in each bank or column; these units should not be attached to items in the body of the table. In tables of some length, ease of reading is improved by grouping the lines, such as by the use of horizontal dividing lines or increasing the spaces between every third or fifth line.

11.13. Figures. Illustrations used in connection with an accompanying text are referred to as figures (abbreviation, Fig.). In reports on materials tests these figures are usually actual photographs of such items as testing machines, instruments, apparatus, and specimens, or original line drawings, or some direct reproduction such as a blueprint or a blackline print or photostat. In the latter case the size of the reproduction may be greater or less than the original and the size of lettering and weight of line on the original must be proportioned in accordance with any contemplated change in size of the reproduction so as to make it clear and easily read.

The line drawings most generally used in reports on materials testing are usually referred to as graphs, charts, or curve sheets, although the designation *diagram* is frequently employed as in this text, e.g., the "stress-strain diagrams."

Unless special sizes are required, 8½ by 11-in. cross-section, graph, or coordinate sheets are used. These are available in many varieties of rulings, printed from accurately made plates, and are not to be confused with ordinary quadrille sheets. A widely used ruling is 10 by 10, ten lines to the inch with every fifth line heavy; another is millimeter paper similarly ruled.

The choice of scales depends upon the range in the data and the purpose of the graph. Sometimes the "over-all" picture is given by a small-scale drawing, often called a thumbnail sketch or key drawing, and the significant or important part of the curve is drawn to a larger scale magnifying or emphasizing the important part of the relationship. The approximate scale is the range in data to be shown, divided by the number of principal divisions, which should preferably be subdivided into five or ten lightly marked scale divisions. The exact scale should be selected so that the smallest division will be some simple part of the scale and not an awkward fractional part. Well-chosen scales make the lines on the cross-section paper of assistance in reading values of random points on a curve. Beginners often overlook the rather obvious necessity of placing the principal scale divisions on the heavy lines of the graph paper. The principal values should be lettered in the margins, if the latter are adequate.

Stresses are usually scaled on the Y axis as ordinates and strains and deflections on the X axis as abscissas. The plotted points should be fine pencil dots enclosed in some appropriate symbol such as a circle, triangle, or square. Only the symbols, however, are inked (using drawing instruments), and the inked curve or graph is extended to and not through them. If the diagram is to be used for computing (reading values off the chart instead of from tables), the experimental points are not shown, and the curve is drawn as a continuous line of lighter weight than that used when merely showing a general relationship; in the latter case it should be heavy so as to stand out clearly from the grid because it is the important part of the illustration. When experimental points are shown, the use of a smooth averaging curve is recommended rather than a broken line that would result from joining points by straight lines. Some mechanical aid, either a flexible ruler or a template, commonly called an irregular curve, should be used in inking the line.

If the entire drawing is specially prepared, the ink grid lines are not drawn through any symbols or lettering, and usually only the principal lines, corresponding to the heavy lines on prepared graph paper, are shown.

Avoid possible distortion and unintentional overemphasis which arises by using only a partial scale, i.e., one that does not run continuously from zero. If it is inexpedient to show the entire scale, insert a conspicuous note or break the scale to indicate that it is not complete.

Since the purpose of illustrations is to aid in the presentation of information, the matter of appearance and clearness should be given special attention in the preparation of charts and graphs.

The lines on cross-section paper usually have equidistant spacing, i.e., they are on an arithmetic scale. Many types of graph paper are available not only in different arithmetic rulings but in semilog, which is logarithmic on one axis and arithmetic on the other, and logarithmic on both axes, in either one or more cycles, i.e., units of 10. Log rulings are also known as ratio rulings. Other special rulings, such as probability and reciprocal, may sometimes be advantageously employed. If special rulings are used, this fact should be stated; for example, "Age, days (log scale)."

Emphasis can sometimes be more effectively accomplished by plotting deviations from an assumed or theoretical relationship, especially if the latter is linear, instead of increasing the scale of one of the variables. Although the plotting of deviations has doubtless been in use for a long time, it seems to have been introduced into materials testing by L. B. Tuckerman in connection with determinations of proportional limit and modulus of elasticity [211].

The following instructions refer to the construction of graphs for use in students' reports required in Part 2.

Materials such as soft steel undergo very little strain up to the proportional limit but develop large strains beyond the proportional limit. For materials of this type plot two stress-strain diagrams using the same data, one diagram including the entire data and a second diagram covering only the elastic range. Plot this second diagram so that the line makes an angle of approximately 60° with the strain axis. This second diagram permits the more accurate determination of certain properties.

Some materials, such as steel and wood, follow Hooke's law of proportionality of stress and strain over a considerable portion of their strength, whereas the stress-strain diagrams for such materials as cast iron and concrete may depart from a straight line even at very low stresses. Therefore, for ductile metals and wood, make the lower part of the diagrams straight lines even though these lines appear to depart somewhat from the plotted points. To determine quickly the approximate location for a straight line through average values, use an ink line on a separate piece of white paper placed underneath the transparent cross-section paper.

The first two or three observed strains may be erratic, and the straight line of the lower portion of the diagram may therefore not pass through the origin. The origin may be shifted so that the curve will pass through it, but this is not usually done. However, in determining the modulus of elasticity of the material, it is convenient to draw a line through the origin and parallel to the straight portion.

Make an appropriate title to describe the graphs. Show the material tested, problem number, party number, student's name, and the date. If more than one kind of symbol or type of line is used, mark them clearly along each curve. If this cannot be done, show a legend or key to all symbols.

Chapter 12.

PRINCIPLES OF INSPECTION

12.1. Some Aspects of the Inspection Problem. Material produced by a given manufacturing process is always, to some degree, of variable quality. A material that is satisfactory for a given type of service usually has requirements as to minimum level of quality, sometimes as to range of quality. In order to assure the presence of desired quality, a product is examined with the object of passing material that conforms to stated requirements and rejecting material that fails to conform. This is the essence of the inspection procedure.

In some instances, the acceptance-rejection operation is extremely simple and may be largely automatic; for example, in one method of sorting steel balls for ball bearings, the balls are dropped on a hard steel block in such a way that acceptable balls rebound into one bin while the culls, which do not rebound to sufficient height, fall into another bin. In general, however, inspection involves much more than the simple process of separating acceptable from substandard material on the basis of some physical or numerical requirement. The problems that arise have many ramifications: the quality requirements may be incapable of exact statement, so that there enters the element of judgment and arbitrary decision on the part of the inspector; it may be feasible to examine only selected samples of a lot, so that there are involved problems of sample selection and the relation of sample to entire lot; the inspector's acts involve human relationships, legal procedures, financial problems, and sometimes human safety—phases of inspection that have not always received appropriate attention.

Those concerned with the problem of inspection may be typified as the *purchaser*, the *producer*, and the *inspector;* for convenience in the following discussion, reference will be made to the three in these terms, even though in specific cases somewhat different relationships may be involved. The purchaser may be the ultimate owner or consumer, may be an intermediate agent in a long chain of technologic processes of business transactions, or may represent the management group of a manufacturing concern. The producer may be the actual manufacturer or constructor, the agent of a manufacturing concern, or a general contractor. Except as noted in Art. 12.2, the inspector is an agent of the purchaser; it is the inspector's function to determine compliance with the purchaser's requirements (generally stated in the form of specifications) and to initiate the steps in the enforcement of the requirements.

In many instances the ultimate purchaser is a nontechnical individual or is an organization such as a business corporation or a governmental department; here the representative of the purchaser may be an engineer, architect, chief of inspectors, or some technically trained executive. Many specifications designate the individual who has the responsibility for final decision on matters relating to the purchaser's policy and interests as the *contracting officer*—a term that will be used here to typify this functionary.

Because the inspector is in a sense the effective arbiter with respect to compliance with most requirements, it has often been assumed that the inspection was at fault when materials, machines, or structures later rendered poor performance in service. It has often been assumed by the purchaser that all he has to do is to select some kind of specification, hire an inspector, and forget about the whole thing until the final product is in use. Barring dishonest practices on the part of producers, and barring dishonest or incompetent inspection, perhaps nothing has contributed more to unsatisfactory relationships and results in the control of materials than these views. A sufficiently broad concept or perspective of the problem of materials control is needed, if the function of inspection is to be properly fulfilled.

The following requirements are considered pertinent to the general problem of obtaining, efficiently and satisfactorily, materials that are adequate for a given type of service:

1. *Intelligent Design.* This involves not only the proportioning of parts on the basis of correct mechanical principles but also the selection of appropriate materials (see Art. 1.3).
2. *Adequate Specifications.* This involves appropriate definitions of quality, unambiguous requirements, and enforceable provisions (see Art. 1.9).
3. *Reliable Production.* This involves the selection of a manufacturer or constructor who is honest and whose facilities for production are such that control of quality is possible.
4. *Competent Inspection.* This involves the setting up of an organization suitable to the kind of work involved, the selection of capable personnel, and the delegation of sufficient authority.

These requirements are admittedly idealized, but they should serve to point out that inspection is but one step in the production scheme. On the other hand, a good design may lose its value, good specifications their effectiveness, or production its efficiency if the inspection is inadequate.

In making provision for inspection, the following general factors should receive consideration:

1. The extent to which inspection is necessary, as governed by the variation in quality of materials that can be tolerated without too great loss or danger to life and property.
2. The cost of inspection, as related to the value of the materials obtained.
3. The types of inspection that are necessary, as related to effectiveness of control of quality.
4. The kind of inspection organization, as concerns the authority, number, and qualifications of the individual inspectors.

Another aspect of inspection, which is not often enough recognized but has valuable possibilities, is the use of inspection as a source of information on the basis of which better contracts and specifications can be written. The inspector is in a position to observe the shortcomings and defects in the specification, such as too close tolerances, requirements that add materially to cost but do not contribute appreciably to quality, and ambiguous phrases that lead to needless disputes. It is suggested that inspectors' reports include comments on such topics and that use be made of them.

Although the general principles involved in the inspection of all kinds of work are largely the same, the organization and procedures differ with the nature of the scheme of manufacture. Certain general patterns of practice have developed, for instance, in the construction industry (which might be thought of as manufacturing in the field); the inspection of concrete construction is a typical example [1206]. Other general patterns of practice have developed in connection with plant and factory production of finished materials; instructions for the inspection of naval and ordinance materials typify practice in this field [1203, 1204]. The many suggestions and rules for good inspection contained in Refs. 1201, 1203, 1204, 1206, and 1208 have been freely drawn upon in the preparation of the following articles.

12.2. Types of Inspection. For convenience, types of inspection are here classified on two bases: (1) as regards the place or point in the production scheme where inspections can be performed and (2) as regards the kind of duties or operations performed. Over the range of the materials-producing industries, the type of inspection necessarily varies widely; it depends upon the kind of material produced, the use to which it is to be put, and the detail with which it is specified.

As regards point of inspection, raw or crude materials may be inspected at the source of supply, semifinished or finished materials may be inspected during the course of processing or manufacture, or finished products may be examined for the quality of the component materials after receipt by the purchaser. Sometimes two or all three of these points of inspection are coincident. In general, it is considered preferable to conduct inspection work before shipment rather than after receipt of materials that are purchased in large quantities.

Two of the principal kinds of inspection operation performed in connection with materials are checking the composition or physical properties of material (which implies the conduct of chemical or physical tests) and checking form, finish, and dimensions (which involves visual examination, measurements, and gaging). In conjunction with these operations, the materials inspector may be required to check quantities, costs, and payments.

It is sometimes stated that the purpose of an inspection is to ensure "good workmanship"; this usually refers to finish and dimensions, but in a larger sense may often be taken to include assurance of correct composition or special treatment, because in many processes these are under control of the workman. Workmanship-type requirements usually necessitate inspection during manufacture. In some industries, the materials or products are graded by the producer; here the function of the inspector may be simply to observe the operations and to see that specified grades are shipped to the purchaser he represents.

In addition to the type of inspection with which this discussion is principally concerned, i.e., acceptance of specified material, an inspection procedure may be maintained by a manufacturer for the purpose of controlling his process. The application of statistical methods to the analysis of inspection data and thus to the control of manufacturing operations has made considerable progress in mass-production industries in recent years.

12.3. Inspection Organization. There is a variety of kinds of organization for inspection. In the simplest case the inspection may be performed by one man, sometimes the purchaser himself. However, for a project of any magnitude or for an organization that uses a variety of materials, a department is usually set up specifically to handle the inspection work. Sometimes an inspection department is headed by a chief of inspectors, sometimes by an engineer of materials and tests (or the equivalent) who may perform other functions besides directing inspection. On a construction project the inspection group may sometimes be responsible to the design (administrative) office, sometimes to the construction (field or resident) engineers. On large work there may be an inspector or inspectors assigned to cover one material or one phase of a process only; highly specialized work, such as certain types of testing and radiographic examinations, requires specially trained and qualified men; in addition there may be general inspectors who act in a supervisory capacity or cover some of the less specific phases of the work.

In the organization of an inspection department the selection of the right man to head it is of considerable importance. He should be given sufficient time and authority so that he can enlist a capable staff and develop a plan of procedure. The funds available for the inspection work should be commensurate with the kind of job to be done, if expected results are to be forthcoming. The authority of the inspection group should be clearly delineated.

12.4. Qualifications of the Inspector. Many arguments have taken place on the relative merits of the "practical" and the "technical" type of inspector, the practical type being one who comes up from the ranks of the workmen in a given industry (and by implication is unhampered by academic training), and the technical type being one who has been trained in a technical school. Probably the ideal inspector would be a combination of the two, and this ideal is approached by many men who have been willing to take the pains to acquire some technical knowledge or to make the effort to obtain practical experience, either through actual apprenticeship or keen observation during a training period under more experienced inspectors.

The general professional qualifications for an inspector are usually summed up by saying that he should have both practical experience and some knowledge of the engineering principles involved in the job to which he is assigned, or, as it is sometimes expressed, he should know "the how and the why." This is true for the man who is responsible for the general phases of the inspection work; however, where specialized knowledge or training is the primary requirement, a general engineering education is often entirely unnecessary. It is desirable for any inspector to be acquainted with common practice in the field in which he works. He should be able to understand the type of specification used in his field; he may be required to interpret contracts and drawings.

The kind of information that an inspector should possess for work in a few fields may be mentioned. On the general inspection of metals he must know that test coupons are at best but indicators of the general characteristics of the metal, owing in most cases to their favorable location with regard to defects. The inspector should be familiar with the physical characteristics indicated by a fractured test bar or coupon, i.e., the nature and appearance of the break as affected by degree of homogeneity, grain structure, and the more commonly known defects. Familiarity with the several types of testing equipment and testing procedure may also be essential; this includes knowledge of how such equipment should be calibrated, the effects of improper test procedures on the results, the effects of size and shape of test piece, and other variables of test upon the indicated properties. For inspection of machined metal parts the inspector must be familiar with gages and their use (see Arts. 4.30 to 4.34).

Certain traits of character are highly desirable, if not necessary, in an inspector. He should be observant and have a proper sense of proportion that will enable him to give the greater attention to the more important matters. He should have a personality that commands the respect of the workmen and the officers with whom he deals. He should be able to cooperate without having to assume the "good-fellow" attitude. He should be straightforward and prompt. He should be possessed of good judgment, tact, a sense of fairness, and the ability to act firmly and impartially. He should understand his responsibility and be faithful to it. It is doubtful whether a man possessing all these characteristics actually exists; the degree to which an inspector does possess them, however, may determine his success.

12.5. Inspector Training. Inspectors should preferably be given a period of training and apprenticeship under the guidance of an experienced man. The extent of training will naturally vary with the background of experience the new man brings to his job.

Technical information should be secured and filed accessibly and used by inspectors in an organization in increasing their professional knowledge. In establishing this file, information may be segregated under the materials and subjects to which it pertains. In large inspection organizations, such as those maintained by departments of state and Federal government, there may be a number of important cases involving special processes, unusual methods, faulty workmanship, office decisions, or other phases that, if filed with the contracts, are soon forgotten. Copies of such records belong in the information file. In addition, samples of test specimens that have failed should be kept for comparison with satisfactory specimens for purposes of instruction and for use in standardizing methods of inspection and test.

Inspectors should be urged to improve their technical education by pursuing study or reading courses and attending lectures and meetings of technical societies. When practicable to do so, they should avail themselves of the educational facilities in their district. Technical schools have data, personnel, and facilities that can be utilized to great advantage. Inspectors should also have access to and avail themselves of specific information pertaining to their particular problems by calling upon those government agencies best qualified to render aid along the line desired.

Inspectors should be warned that their prestige is lessened when they conduct their inspections by reading from the specifications. The specific details must be studied beforehand, and only a brief reference to a notebook, which may contain pertinent extracts, can be made on the job without "losing face" with the workmen.

12.6. Relations with Superior Officers. The inspector should explicitly carry out his instructions both as furnished by the specifications and as stated verbally. He should avoid any discussions with a superior in the presence of others.

Once assigned to active duty the inspector should have the loyal support of his superior officer. If it is felt that the inspector has exceeded the limits of his authority or failed to exercise good judgment, the superior officer should never argue with or reprimand him before the producer, i.e., the contractor, his foremen, or any of the latter's workmen. Every effort should be made to maintain the spirit, self-confidence, and morale of the inspector.

12.7. Responsibility. The inspector's primary responsibility is to see that work assigned to his charge is executed in accordance with the plans and specifications, except as variations are permitted in writing by the contracting officer. He is responsible for thorough knowledge of the specifications, and for the exercise of good judgment. He may be responsible for the use of specified methods as well as materials, because it is often practically impossible from the visual examination of completed work to ascertain whether or not such work has been done in a satisfactory manner and with satisfactory materials; after a lot of material or a product has been accepted, it may be too late to hold the producer responsible. The inspector may be responsible for a number of assigned duties such as the preparation of reports. He should keep a detailed diary of his observations throughout the work, noting particularly all warnings and instructions given to the producer. He also has certain ethical responsibilities in the way of safeguarding the purchaser's general interests.

12.8. Authority. The inspector should be authorized to

1. Prohibit production until the preliminary conditions (such as the inspection of raw materials and preparation of forms in concrete work) can be fulfilled and until inspection can be provided for production operations
2. Forbid the use of materials, equipment, or workmanship that does not conform to specifications
3. Stop any work that is not being done in accordance with the contract drawings and specifications
4. Require the removal or repair of faulty construction or of construction performed without inspection and not capable of being inspected after completion

In the first three cases above, the inspector is usually authorized to take direct action, reporting immediately thereafter to the contracting officer. However, he should stop work only as a last resort and when it is evident that unsatisfactory work will result from continuing operations. In the last case the inspector should obtain approval of the contracting officer before acting; one reason for this is that the correction of a defect by the removal of the defective part may result in weakness at the joint between the old and the new work. This is especially true for concrete work and for the repair of defective castings.

On minor points not covered by instructions, it will usually be necessary for the inspector to exercise personal judgment and to make decisions, thus settling as many problems as possible at the job. Matters of general policy or major points not specifically covered by instructions should be brought to the attention of the contracting officer. An inspector should attempt to anticipate problems likely to arise and secure decisions in advance.

When the producer contests an inspector's decision or interpretation, the inspector will recommend the suspension of work that would be affected by the decision, pending a settlement by the producer and the contracting officer.

12.9. Relations with the Producer. So long as the requirements of the specifications are fulfilled, the producer is entitled to complete the work at the lowest possible cost. By cooperating with the producer in any way not inconsistent with the purchaser's interests, the inspector will aid in reducing the cost of construction. Any change in the specifications that would make possible a saving for both purchaser and producer must be made by mutual agreement; the producer, however, customarily initiates the action by applying to the purchaser through the inspector. If departures from the letter but not the spirit of the specifications are proposed by the producer and appear to be necessary, the inspector may arrange to accept the work tentatively and refer the matter to the contracting officer.

Inspections should be made promptly when requested. Conditions that will obviously lead to unsatisfactory work should be anticipated whenever possible and in any event should be pointed out as the earliest opportunity in order to avoid waste of materials and labor. The work of inspection should be arranged to cause as little delay and extra work as possible. No demands should be made that are not in accordance with the specifications.

The inspector should familiarize himself with the producer's methods. Because methods affect results, the inspector should be able to recognize those which will not produce acceptable work and to suggest correct ones in their place. If the specifications permit a choice of methods, the inspector may advise but should not arbitrarily demand that a given method be employed. An appreciation of the producer's position by the inspector will go far toward making for satisfactory relations between them.

Based on experience, inspectors *may* offer information, suggestions, and advice in regard to the details of production which might be of aid in complying with the provisions of the contract to the best advantage of all concerned. Such advice should be given with the understanding that it does not relieve the producer of his responsibility in meeting the terms of the contract. In giving information, it is unethical to divulge confidential processes, methods, and practices of one producer to a competitor. The inspector should not attempt to "run the job." Arguments should be avoided and disputed questions referred promptly to the contracting officer for decision.

The inspector should maintain an impersonal, agreeable, and helpful attitude toward the producer and his employees, avoiding friction if possible. However, he should avoid familiarity and should accept no personal favors from or obligate himself to the producer. By dealing fairly and by recognizing and commending good work, he may usually secure the friendly cooperation of the workmen. He should take the attitude that changes he suggests are for the benefit of the work and are not made to show his authority. He should not indiscriminately or unjustly criticize the producer's organization.

Regular and definitely understood methods of communication should be established between the inspection force and responsible representatives of the producer. Although proper record of all definite action is essential, communication by telephone and by personal conference should be resorted to whenever desirable in the interest of promptness. Oral instructions should be confirmed by memoranda wherever possible, both as a means of establishing records and of commanding attention. Correspondence and personal contact should be courteous at all times. Instructions should be given only to the authorized representatives of the producer, usually the superintendent or foreman, except in minor and routine matters and to an extent agreeable to the producer's organization. The inspector usually deals directly with subcontractors unless his instructions are disregarded; he should then immediately refer the matter requiring correction to the general contractor who is legally responsible.

Warnings are given in the form of a caution that the faulty work will not be acceptable under the specifications. A good start is important; an incorrect method is more easily corrected the first time it is practiced than after it has been in use. In cases where rejections are made, the producers should be informed courteously as to why the material is unacceptable, with any suggestions the inspector may care to offer in order to facilitate the production of acceptable material. Although he should approach the inspection work without prejudice, the inspector should be watchful of any attempts intentionally made to substitute inferior materials, to hide defects, or to resort to questionable methods, and such attempts should be promptly reported.

Because specifications can often be given various different interpretations, it is always good policy for the inspector and producer to hold a conference preceding the actual work and to come to an understanding on questionable requirements of the specifications.

Acceptable samples or standards of surface finish, color, or similar characteristics to be determined by visual inspection should be agreed on before starting the work.

Effort should be made to establish an arrangement with the producer's inspection department, if such exists, whereby the latter will reject manifestly unsatisfactory materials and not present them for inspection. This will save much time and extra work. Where practicable, a mutual exchange of confidence and information between the inspector and the producer's inspection department is highly desirable. There is often considerable difference in the results obtained from producers where cooperative effort exists and where such is lacking.

There are, unfortunately, all classes of producers, from those who are conscientious and honest to those who are unscrupulous. Until known to the contrary, however, it may be assumed that the producer will take pride in his work and will endeavor to give satisfaction. Natural dishonesty cannot well be controlled except by experience and refusal to let a contract to a producer who has already proved dishonest. Even if the producer is honest, it is exceedingly difficult to secure good work if he is operating at a loss. A contract should be let only when prospects of a reasonable profit are assured. Even then, a foreman, through a desire to cut expenses, may sometimes resort to evasion of specification requirements, the importance of which he does not appreciate.

The inspector should not form habits of procedure which might be anticipated by the workmen. Inspection of the various details and operations should be at irregular intervals, a precaution that is frequently overlooked. The inspector should be on the job to forestall hasty and unsatisfactory work which is most likely to occur at the beginning or end of a working period. He should always be present when critical work is being done, and, in the absence of specific instructions, he must decide whether an operation requires constant or intermittent witnessing.

In any event, test specimens should not be taken except under the direct supervision of the inspector, who will mark them properly for later identification.

Inspectors need not waive their rights for claims against the producer on account of injuries sustained while performing their duties, provided they have exercised due caution and have strictly observed all plant or shop safety regulations.

12.10. Contract Is Inspector's Guide. On all work the contract with attached general and detailed specifications, including all pertinent drawings, prescribes the conditions that must be met by the producer. The drawings generally prescribe the details of layout, dimensions, tolerances, and certain material requirements for various items; in numerous cases detailed requirements are stated on the drawings. In many cases, standards of the SAE, ASME, and/or ASTM may be included in the specifications by reference. In any case, the inspector is bound by the provisions of the contract, which includes all accompanying documents. Attention is especially directed to the provisions that appear in many standard specifications for materials, such as those of the ASTM, which specifically cover such items as method of sampling, number of samples or tests, access to work to be inspected, rejections, and retests.

Before being used for inspection purposes, copies of contracts should be examined carefully to ensure that they contain the information essential to a clear understanding of the inspection and test requirements, and for omissions and errors; particular attention should be given to discover features contained in the contract which are in addition to or at variance with the requirements of drawings and specifications. Specifications furnished with the contract should also be checked carefully to ensure that the issue intended for use therewith has been provided. If approved drawings, sketches, cuts, or samples are essential and not furnished, they should be requested. Should any conflict arise between the requirements of the contract proper, the drawings, the detail specifications, and the general specifications, then the requirements of the contract, the drawings, the detail specifications, and the general specifications generally prevail in the order named. A different order of precedence may be specifically indicated in the contract.

Where no specifications exist for some material and authorization for the use thereof does not appear to lie within the jurisdiction of the inspector, specific approval must be obtained from the contracting officer.

When permissible tolerances or limiting values are not specified, the inspector should so reasonably and fairly interpret variations in measurements that his judgment will be sustained by higher authority if his decisions are questioned.

12.11. Credentials. Before going on a new job the inspector should always be provided the credentials with which to establish his identity and authority. Frequently a letter of introduction to the producer's representative with whom inspection problems are to be discussed will be sufficient. If the request is not made in the letter of introduction, the inspector should request the facilities that he may require and that are to be furnished by the producer for the conduct of the inspection work; there should also be an agreement on what members of the producer's organization may be considered to be responsible agents to whom minor instructions may be given.

In case he is reporting for duty under an inspector already at a job, the letter of introduction should be addressed to the inspector in charge.

Part 2.

INSTRUCTIONS AND PROBLEMS IN TESTING AND INSPECTION

GENERAL INSTRUCTIONS FOR LABORATORY WORK

LABORATORY DUTIES

Party Organization. The students will be grouped in parties of about four men each. On successive periods during the course, each man in a party will perform the duties of recorder, operator, observer, and computer. The schedule of problem assignments will be made at the beginning of the term, and each student is to report to the laboratory fully prepared to carry out each assignment.

Instructions to Recorder. The recorder is the leader of the party, since he is in a position to supervise the general progress of the work. Before the work begins, he should understand all that is to be done and should so plan the work that the problem will be completed within the allotted time. Although each member of a party will be held responsible for adequate preparation before the work begins and for the performance of his duties in the laboratory, the responsibility for the satisfactory execution of the problem rests upon the recorder. He will give directions to other members of the party and will see that all the required data are obtained. Specific instructions to the recorder follow:

Prepare a sheet with appropriate headings for the entry of all data if data sheets are not provided by the instructor. Make a carbon copy for the instructor and for each member of the party. On the data sheet show such general information as the party number, party personnel, problem number, special equipment used, and the date. Before the test is begun, secure the instructor's approval of (1) the layout of the data sheet with all preliminary data recorded and (2) the test setup. Record all data as soon as observed. Do not record any data on other sheets with the intent of transferring them later to the regular data sheet. Record the speed of the movable head of the testing machine or the rate of load application. Draw a sketch of any special apparatus and describe its operation. Draw a sketch of any ruptured specimens and describe the character of the fracture. When the data sheet is completed, submit it to the instructor for approval, while accompanied by the entire party.

Instructions to Operator. Check over all equipment, except the measuring devices used by the observer. Under the supervision of the instructor, become familiar with the equipment so that no delays or difficulties will arise during the test. If a mechanical machine having a scale beam and movable poise is being used, set it to read zero before beginning the test as follows: adjust the recoil nuts so that they are just loose, set the poise at zero, and then balance the scale beam with the counterpoise. For a dial-type machine, select the proper load dial, and set at zero (see Chap. 4).

For tension tests see that the proper grips are ready for use and make sure that they bear for their full length against the specimen and also against the head of the testing machine (see Fig. 5.9). For compression tests see that a compression plate and spherical bearing block are in position (see Fig. 5.27). During the test, do not apply load so rapidly that the load and strain observations are of doubtful accuracy.

In using a hydraulic machine move the piston slightly out of the cylinder before positioning the movable crosshead. Then, if the crosshead should accidentally stall against the specimen, it can be released by lowering the piston.

If the grips remain stuck at the end of a test, use a hammer to strike the sides of the specimen. Do not strike the grips or grip handles. After the test, clean up the machine and its surroundings. Check all tools and equipment, and leave them in proper order. Report any lost or damaged equipment to the instructor immediately. Before leaving the machine, make certain that the motor has stopped. A mechanical machine should be left out of gear, and a hydraulic machine should be left with the unloading valve open and the loading valve closed.

Instructions to Observer. Check all measuring apparatus and see that it is in proper order. Study the action of any strainometer to be used and become familiar with its operation. Note its range. Determine how to convert strainometer readings to unit strains. Attach the strainometer to the specimen and center the latter in the testing machine. Set the strainometer to read zero at zero load, or some predetermined small load, and check to see that most of its range is available for use. Report all measurements and observations to the recorder. Keep the specimen constantly under observation in order to determine critical points in the test.

Instructions to Computer. Make any preliminary computations that may be necessary. Determine the probable maximum critical values to be observed and select suitable increments for the development of adequate stress-strain curves. Secure approval of these by the instructor. Assist the observer to take readings during the test.

REPORTS

Arrangement of Reports. Use the following general arrangement of subject matter in the report of a test, unless otherwise informed by the instructor:

Statement of the problem
Materials, apparatus, and/or methods of testing
Summary of test results—including tables, diagrams, and some discussion of detailed results
Discussion of the general features of the test and the findings
Appendixes, to include complete or sample computations, data sheet, and other special supporting evidence if available

The detailed arrangement and the headings used may vary somewhat with the particular type of problem on which a report is made. The report should be as brief as possible, consistent with the inclusion of all important facts, observations, and discussions.

Under the statement of the problem give a clear statement of the objectives of the test, the general method of attack, and the type of material (or apparatus) tested.

Under the section on materials, apparatus, and methods of testing, identify or describe the material tested, include a listing or brief description of important pieces of apparatus, and briefly state the principal features of the test procedure. Use sketches where possible. Standardized and well-known methods and apparatus may be incorporated by reference to ASTM Specifications.

Under summary of test results, show results wherever possible in tabular or graphical form. Introduce tables and diagrams by brief appropriate statements in the text and point out the nature of the particular data obtained. Give thought to clearness and readability in arrangement of tables (see Chap. 11, under Tables, for general rules for construction of tables). Draw clear, accurate, forceful charts and diagrams (see Chap. 11, under Figures for general rules for construction of diagrams). See also Chap. 2, Art. 2.5. Assign figure numbers and table numbers and refer to figures and tables by their numbers. In general each table and figure should be sufficiently complete to show clearly the relationships intended without the necessity of searching through the text for explanations.

Under discussion of findings state in general terms the results of the particular test, and draw attention to salient facts discovered. Make reference to figures and tables if necessary. Compare the results of the test with data given in this text or in the technical literature; give references as footnotes or as a bibliography in an appendix. Describe the behavior of the material tested. Criticize the results or the test procedure. Draw definite conclusions concerning the quality of the material, and state whether or not the material is acceptable under stated specification requirements. Discuss various aspects of the problem by answering questions listed at the end of the instructions for each problem.

In an appendix show computations that support and illustrate the steps in the reduction of the data. Where the computations merely involve repetitions of the same operations, show a sample computation. The slide rule may be used unless otherwise instructed. In stating formulas, state where they were obtained and what the symbols mean. Also in an appendix, bind a carbon copy of the laboratory data sheet.

Submission of Reports. Each student will submit a separate report on each problem or group of related problems assigned for a given period. Reports for a given assignment will be bound together in a suitable folder.

Consult the instructor or the bulletin board for due dates on reports.

PROBLEMS

PROBLEM 1. INSPECTION OF THE TESTING LABORATORY AND STUDY OF THE SCOPE OF MATERIALS TESTING

Object. To acquaint the student with the place of materials testing in the field of engineering; to introduce the problems encountered in the testing, inspection, and preparation of adequate specifications for engineering materials; and to familiarize the student with the types of equipment used in testing.

Preparatory Reading. Chapters 1, 2, and 4.

Assignment. During the period allotted to this problem, an introductory lecture will be given on materials testing, and the student will, under the guidance of the instructor, make a survey of the facilities and equipment of the laboratory.

Report. For this problem prepare an informal report of the essay type, discuss the scope of materials testing, and classify, according to some logical scheme, the equipment and facilities inspected in the laboratory.

Discussion. 1. What are the principal objectives in the testing of engineering materials?

2. Discuss the relative merits of various types of testing machines. What is the most common type of machine used in the testing of engineering materials? Why is it used so commonly?

3. What item other than load is often progressively observed for a specimen under test? What equipment is used to aid these observations?

4. Why is it essential that standard methods of testing be used?

5. What considerations are involved in the selection of materials for machines and structures?

6. Discuss the function of inspection.

7. What is meant by the "significance" of a test?

8. What factors are involved in the design of a test?

9. Discuss the problem of specification of materials.

10. What are the functions and objectives of the ASTM?

PROBLEM 2. STUDY AND CALIBRATION OF A TESTING MACHINE

Object. To study the construction and operation of a testing machine; to determine the errors in the loads indicated by the weighing system and the sensitivity for each load applied.

Preparatory Reading. Articles 4.3 to 4.8, incl.

Special Apparatus. Standard weights, proving ring, or other calibration devices.

Procedure. 1. Study the construction of the testing machine and note how it is used to apply tensile or compressive loads. Draw a sketch to show how the load is applied and measured.

2. To the limit of space available on the platen of the testing machine, or to the limit of weights available, apply standard weights in increments, as assigned by the instructor, and record the corresponding loads indicated by the machine, provided the machine responds to this method of loading.

3. Arrange proving levers in the machine, and to the limit of space available on the baskets or to the limit of weights available, apply load by increments, as assigned, and record the indicated loads. Keep equal amounts of load on each basket.

4. Before placing the proving ring in the testing machine, turn the dial of the ring at least six revolutions away from contact with the reed mechanism. This is done so that when applying the preload the reed will not contact the anvil as it might be bent and cause a large change in the zero reading. Place the proving ring in a vertical position as shown in Fig. 4.12, with the lower boss resting on a hardened steel plate and a soft steel plate bearing on the upper boss. After adjusting the machine to read zero load, preload the ring to the limit specified by the instructor and then unload.

Obtain the initial reading of the ring by turning the dial until the anvil almost touches the reed hammer. Vibrate the reed by deflecting the hammer about ½ in. to one side and releasing; then slowly turn the dial until the anvil touches the hammer, which condition will be indicated by a buzzing sound. Read the dial to 0.1 division and observe the temperature. Apply load in 10 approximately equal increments, but before each increment withdraw the anvil from the reed. For each increment record the reading of the calibration device and the load indicated by the machine. If the machine has a lever weighing system, always bring the beam to balance by advancing the poise as for increasing loads.

The difference between the initial and subsequent dial readings, after correcting for the temperature deviation from 70°F (see Art. 4.8) and multiplying by the ring factor, gives the true load.

5. Observe the sensitivity of the machine for each load applied by all the methods given above. The measure of sensitivity is the increase in the applied load required to produce an observable movement of the load-indicating device. Small auxiliary weights may be placed on the platen of a mechanical machine and some types of hydraulic machine for this determination.

Report. 1. Compute the error in each indicated load as a percentage of the true load. For indicated loads that are too high, the error is positive. Plot a curve of errors as ordinates against indicated loads as abscissas. On this graph show the maximum error permitted by the ASTM. Also plot a curve showing the sensitivity at each load.

2. Indicate in which portions of the loading range covered by the test the machine meets the ASTM requirements for accuracy. Draw conclusions regarding the suitability of the machine for ordinary testing in so far as accuracy and sensitivity are concerned. If the machine was calibrated for only a portion of its capacity load, discuss the probable inaccuracies for loads above the calibration limit.

Discussion. 1. What effect might the position of the load on the table have on the indicated load (*a*) for a lever machine and (*b*) for a hydraulic machine?

2. Name in order of relative accuracy four methods of calibrating a testing machine. State the limitations of each method.

3. What may cause a change in calibration of (*a*) a hydraulic machine and (*b*) a lever machine?

4. Is any error introduced by the use of heavy base plates inserted below the specimen in a compression test? Explain.

5. How does the Emery hydraulic machine differ from the other types of hydraulic machines? What are its advantages and disadvantages?

6. Discuss the relative accuracy and sensitivity of the following types of testing machine: (*a*) simple hydraulic, (*b*) Emery hydraulic, and (*c*) lever.

7. Does the ASTM permit a blanket acceptance or rejection of a testing machine as a result of its calibration? Explain.

8. Distinguish clearly between accuracy and sensitivity. Is a sensitive machine always accurate? Can an accurate machine be sluggish?

9. What is the most simple method of correction of the error in (*a*) a lever machine and (*b*) a hydraulic machine, so that the indicated loads will meet the ASTM requirements for accuracy?

10. Discuss the effect of temperature changes of say 30°F upon the accuracy of testing machines.

PROBLEM 3. CALIBRATION OF A STRAINOMETER

Object. To determine the multiplication ratio of a strainometer, which is defined as the ratio of the movement shown by the indicating device to the corresponding displacement between the points of attachment to a specimen. Two determinations will be made: (1) by measurement of the critical dimensions with a steel scale and (2) by use of a calibrator.

Preparatory Reading. Articles 4.9 to 4.11, 4.29.

Special Apparatus. Extensometer or compressometer, and calibrator.

Procedure. 1. Study the general construction of the strainometer and note how it may be used to measure strains. Draw a sketch to show how the indicator is actuated.

2. Determine the multiplication ratio from the careful measurement of critical distances with a steel scale.

3. Measure the gage length between contact points in the two yokes.

4. Attach the strainometer to the calibrator, securing one yoke to the fixed end and the other yoke to the movable end; then remove the spacer bars. Adjust the strainometer and the calibrator to accommodate the full range of the strainometer.

5. Record corresponding readings from the extensometer and from the calibrator at the beginning and at about 10 points over the range of the strainometer.

Report. 1. Compute the value of the multiplication ratio from (*a*) the measurements with the scale and (*b*) the average of the dial readings on each instrument.

2. Using each of these multiplication ratios, compute the constant by which one unit of the strainometer dial readings must be divided to obtain the corresponding unit strain.

3. If a calibration record is available for the calibrator itself, compute the values by which the strainometer readings must be divided to obtain the corresponding unit strain at each increment of strainometer reading. Plotting readings as abscissas, prepare a graph showing the variation in these values over the range of the strainometer.

Discussion. 1. What precautions were taken in the construction of the dial indicator and strainometer to prevent (*a*) backlash effects and (*b*) damage due to excessive strain?

2. What possible errors may exist in the strainometer conversion factors that would not be detected by checking at only 10 points of the range?

3. Discuss the relative merits of axial strainometers having one, two, or three dials.

4. A tensile specimen is tested using an extensometer that is inaccurate. Explain which of the following properties will be in error:

a. Proportional limit
b. Yield strength
c. Yield point
d. Ultimate strength
e. Modulus of elasticity
f. Modulus of resilience

PROBLEM 4. TENSION TEST OF STEEL

Object. To determine the strength and several elastic and nonelastic properties of a ductile steel, to observe the behavior of the material under load, and to study the fracture. The specific items to be determined are

1. Elastic strength in tension
   a. Proportional limit
   b. Upper yield point
   c. Lower yield point
   d. Yield strength for an offset of 0.1 percent
2. Tensile strength
3. Ductility
   a. Final elongation in each inch of length
   b. Percentage elongation in 2 in.
   c. Percentage elongation in 8 in.
   d. Percentage reduction of area
4. Modulus of elasticity
5. Modulus of resilience
6. Modulus of toughness
7. Type and character of fracture

Preparatory Reading. Chapter 2; Arts. 5.1 to 5.7; Appendix A.
Specimen. Low-carbon steel or as specified.
Special Apparatus. Extensometer with 8-in. gage length.
Procedure. 1. Determine the average cross-sectional dimensions of the specimen with a micrometer caliper. Scribe a line along the bar and with a center punch lightly mark an 8-in. gage length symmetrical with the length of the bar. Also mark each inch of length within the gage length and make other marks 1 in. outside the gage length at each end.

2. Firmly grip the upper end of the specimen in the fixed head of the testing machine. Place the specimen so that the punch marks face the front of the machine.

3. Measure the gage length and determine the multiplication ratio of the extensometer. Determine the value of the divisions on the dial indicator. Firmly attach the extensometer to the specimen so that its axis coincides with that of the specimen, and remove the spacer bar. Adjust the testing machine and the extensometer to read zero, setting the latter so that most of its range will be available. Grip the lower end of the specimen, taking care not to jar the extensometer.

4. Select suitable increments of strain to secure at least 15 readings below the probable proportional limit. Apply load at a slow speed, and take simultaneous observations of load and strain without stopping the machine. Continue loading until the yield point is passed; in a lever-poise machine keep the beam balanced at all times so as to determine both upper and lower yield points. Stop the machine (but hold the load) to remove the extensometer.

5. Again apply the load continuously. When the gage length has increased 0.1 in. in the 8-in. gage length, as measured with dividers and scale, observe the load. Thereafter, for each 0.2 in. increase in gage length, observe the load. For this part of the test the speed of the machine may be increased to about 0.2 in. per min. Record the maximum and breaking loads.

6. Remove the broken specimen from the machine. If the specimen is jammed in the grips, use a hammer to strike the sides of the specimen. Do not strike the grips or grip handles.

7. Observe the location and character of the fracture, and measure the dimensions of the smallest section. Fit the broken parts together and measure the gage length and the intervals between intermediate punch marks.

Report. 1. Plot a stress-strain diagram for the test in accordance with general instructions. On a second sheet plot a graph showing the elongation in each inch of gage length as ordinates and the number of the inch divisions as abscissas. Indicate the location of the fracture on this graph.

2. Compute all properties called for. For the percentage elongation in 2 in., use the 2-in. length most nearly centered on the fracture.

Discussion. 1. Did the test specimen used conform with ASTM Standards in shape and dimensions?

2. Was the test carried out in accordance with ASTM Standards?

3. State whether or not you would accept the material as satisfactory and in conformity with ASTM Specification A 7 (see Appendix A). Indicate wherein it may have failed to meet requirements.

4. What is the relation between nominal and actual stress for loads beyond the yield point?

5. Why is it necessary to state the gage length when reporting the percentage of elongation?

6. Discuss the variation in the percentage elongation with size and shape of bar.

7. How can the work required to rupture the specimen be determined?

8. How are the properties of steel, such as yield point, tensile strength, ductility, modulus of elasticity, and work to rupture, affected by changes in carbon content?

9. What errors may be introduced if the axis of the extensometer and that of the specimen do not coincide?

10. Are wedge grips suitable for tests of brittle material? Explain.

11. Distinguish clearly between proportional limit and elastic limit.

12. Distinguish between yield point and yield strength.

13. What are the advantages of a stress-strain diagram over a load-elongation diagram for showing the results of the test?

PROBLEM 5. TENSION TEST OF WIRE AND WIRE ROPE

Object. To determine the strength and efficiency of a wire rope and the strength and ductility of the wires of which it is made. The specific items to be determined are

1. Details of construction of wire rope including its diameter, number of strands, number and diameter of wires in strand, type of center (hemp or steel), lay (regular or lang, right or left), and pitch of wires and strands
2. Tensile strength and percentage elongation of each wire in one strand
3. Tensile strength and efficiency of wire rope

Preparatory Reading. Chapter 2; Arts. 5.1 to 5.7; Appendix C.
Specimens. One socketed specimen of ½-in. wire rope (or as specified) and one strand from the same coil.
Special Apparatus. Wire extensometer with 8-in. gage length.
Procedure. *Rope Test.* 1. Draw a sketch of the wire rope showing both cross section and side view. Measure the pitch of the wires and the strands and observe the lay. Take complete notes so that a brief description of the construction can be included in the report. Measure the diameter of the rope, which is defined as the diameter of the circle that will just enclose it.

2. Secure the sockets in the heads of the testing machine and apply load at a moderate rate of speed. Record the load when the first wire breaks, the maximum load, and the character and location of the fracture.

3. Pull the specimen apart, then measure the nominal size of the center and note the kind of material from which it is made.

Test of Wire. 1. Unravel the extra strand of rope, noting its construction.

2. Grip one of the wires in one head of the wire-testing machine. Pull the other end of the wire taut with a pair of pliers and tighten the grips in the second head, taking care that the clear length of wire between the grips is at least 12 in. (*Note.* A low-capacity universal testing machine with flat grips may also be used for this test.)

3. Apply a load that will stress the wire to not more than one-fourth of its ultimate strength and then measure and record the average diameter of the wire.

4. Increase the load so as to bring the stress to 75,000 psi. Attach the strainometer and set the dial at zero.

5. Apply the load continuously at a speed of the movable head of not more than 1 in. per min. Record the maximum load and the elongation observed at the instant of rupture. The break must occur within the gage length, but not closer than ¼ in. to the jaws, or a retest must be made.

6. Test each wire in the strand, following the procedure as outlined.

Report. 1. To the percentage elongation obtained from the extensometer add 0.25 to allow for the elongation that occurred before application of the extensometer.

2. Tabulate the wire test results, showing for each wire the diameter, maximum load, tensile strength, and percentage elongation. Compute the average tensile strength and percentage elongation of the wires tested.

3. Compute the efficiency of the rope and include all properties called for.

Discussion. 1. Are the strength and elongation characteristics of the wires reasonably uniform?

2. In what respects, if any, did the rope and wire tests differ from API requirements?

3. Why is the efficiency of a wire rope always less than 100 percent?

4. What is the purpose of the hemp center in a wire rope?

5. Name the principal factors that tend to cause a short life of a wire rope in service.

6. What are the possible causes of single-strand breaks in wire-rope tensile-strength tests?

7. What are some of the practical difficulties involved in the determination of the modulus of elasticity of wire rope?

PROBLEM 6. COMPRESSION TEST OF WOOD PARALLEL TO THE GRAIN

Object. To study the action of wood under compressive loading parallel to the grain and to determine some of its physical and mechanical properties as follows:

1. Moisture content
2. Approximate specific gravity, both as received and when dry
3. Annual growth rings per inch
4. Percentage of summerwood
5. Percentage of sapwood
6. Elastic strength
   a. Proportional limit
   b. Yield strength at 0.05 percent offset
7. Ultimate strength
8. Modulus of elasticity
9. Modulus of resilience

Preparatory Reading. Chapter 2; Arts. 5.8 to 5.14; Appendix D.
Specimen. Clear wood, 2 by 2 by 8 in.
Special Apparatus. Compressometer having 6-in. gage length.
Procedure. 1. If the ends are not plane and at right angles to the axis of the specimen, reject and obtain a second specimen; note any defects in the specimen. Measure the cross section and length of the specimen to the nearest 0.01 in. and weigh to the nearest gram. Determine the average number of annual growth rings per inch, the percentage of summerwood, and the percentage of sapwood.

2. Determine the gage length and multiplication ratio of the compressometer. Determine the strain corresponding to the least reading of the dial. Attach the compressometer to the specimen and remove the spacer bars. Center the specimen on the table of the testing machine, using a machined bearing block at the lower end and a spherical bearing block at the upper end. Adjust the compressometer dial to read zero and make certain that most of its range is available.

3. Apply the load continuously at a speed of about 0.024 in. per min until failure. Read the compressometer dial and the load at intervals of about 2000 lb. Draw a sketch, in perspective, indicating the grain of the wood and the manner of failure.

4. Cut a moisture sample about 1 in. in length from the specimen near the failure. Remove all splinters and weigh to the nearest 0.1 g. Place in a drying oven controlled at 103 ± 2°C. After the moisture specimen has dried to constant weight, which may take 2 to 4 days, again weigh it.
Report. 1. Construct a stress-strain diagram and mark the proportional limit and the yield strength on the curve.

2. Calculate the moisture content as a percentage of the oven-dry weight. Calculate all other values called for.
Discussion. 1. What is the effect of moisture on the strength of wood in compression?

2. What is the relation between specific gravity and strength?

3. Compare the proportional limit strength, compressive strength,

and the modulus of elasticity of the specimen tested with the values for the same species presented in Appendix D. Explain how the moisture content and specific gravity of the sample may account for any differences noted in the mechanical properties.

4. What effect does the time element have in the loading of wood?
5. What is the effect of the percentage of summerwood on strength?
6. Would a large test piece having the same slenderness ratio be expected to exhibit the same unit strength as the specimen here tested? Explain.
7. Explain how any nonhomogeneity, such as a knot near one face of a wood specimen, may produce the equivalent of eccentric loading.

PROBLEM 7. COMPRESSION TEST OF WOOD PERPENDICULAR TO THE GRAIN

Object. To study the action of wood under compressive loading perpendicular to the grain and to determine some of its physical and mechanical properties as follows:

1. Moisture content
2. Approximate specific gravity, both as received and when dry
3. Annual growth rings per inch
4. Percentage of summerwood
5. Percentage of sapwood
6. Elastic strength
   a. Proportional limit
   b. Yield strength at 0.05 percent offset
7. Modulus of elasticity

Preparatory Reading. Chapter 2; Arts. 5.8 to 5.14; Appendix D.
Specimen. Clear wood, 2 by 2 by 6 in.
Special Apparatus. Two-dial strainometer or equivalent strainometer equipment.
Procedure. 1. Measure the cross section and length of the specimen to the nearest 0.01 in. and weigh to the nearest gram. Determine the average number of annual growth rings per inch, the percentage of summerwood, and the percentage of sapwood. Note any defects in the specimen.

2. Place the specimen flatwise on the base plate of the special two-dial strainometer, with a radial surface uppermost. Note the width of this surface. Place a 2-in.-wide loading plate across the middle third of the specimen, with the long axis of the plate normal to the long axis of the specimen. Place a spherical bearing block on top of the loading plate.

3. Study the action of the strainometer and determine the deflection corresponding to an estimated tenth of the smallest graduated division.

If a strainometer is not available, arrange two dial micrometers to measure the movement of the loading plate.

4. Check the position of the two dials to see that one is on each side of the specimen and equidistant therefrom, with the contact points bearing against the underside of the loading plate.

5. Apply the load continuously at a speed of about 0.012 in. per min. Observe both dials at load increments of 200 lb until a total average deflection of 0.1 in. is observed. Draw a sketch, in perspective, indicating the grain of the specimen and the manner of failure.

6. Cut a moisture specimen about 1 in. in length from the part adjacent to the loaded middle third. Remove all splinters and weigh to the nearest 0.1 g. Place in a drying oven controlled at 103 ± 2°C. After the moisture specimen has dried to constant weight, which may take 2 to 4 days, again weigh it.

Report. 1. Plot a graph with stresses on the loaded area as ordinates and mean strains as abscissas. Mark the proportional limit and the yield strength on the curve.

2. Calculate the moisture content as a percentage of the oven-dry weight. Calculate all other values called for.

Discussion. 1. Compare your results with the values presented in Appendix D, noting any differences in moisture content and specific gravity.

2. Of what significance is the compressive strength of wood perpendicular to the grain?

3. Why is the ultimate compressive strength perpendicular to the grain never determined?

PROBLEM 8. COMPRESSION TEST OF CONCRETE

Object. To study the behavior of concrete under compressive loading, and to determine the following physical and mechanical properties:

1. Proportional limit
2. Yield strength at 0.01 percent offset
3. Compressive strength
4. Initial tangent modulus of elasticity
5. Secant moduli of elasticity at stresses of 500, 1000, 1500, and 2000 psi
6. Weight per cubic foot

Preparatory Reading. Chapter 2; Arts. 5.8 to 5.14; Appendix E.
Specimen. 6 by 12-in. concrete cylinder.
Special Apparatus. Compressometer.
Procedure. 1. From the instructor obtain data regarding the kinds and proportions of the constituent materials, the water-cement ratio and

the consistency of the mix, the curing and storage conditions, and the age of the specimen. From these data predict the ultimate strength. However, if scheduled by the instructor, the students may cast concrete cylinders for testing at various ages to develop the relationship of age, compressive strength, and modulus of elasticity.

2. Determine the mean diameter of the cylinder at its mid-section, and its average length, making measurements to 0.01 in. Weigh the specimen to 0.01 lb.

3. Cap each end of the specimen with the material provided and a machined metal capping plate. If hydrostone is used, mix sufficient material with water to produce a mixture of fairly stiff consistency. Place a trowelful of the paste in the center of the capping plate or cylinder and work it out to the edge by rotating the cylinder against the plate, or vice versa. Allow the cap to set ½ hr before testing the specimen with plates attached.

4. Study the action of the compressometer, note its gage length and multiplication ratio, and determine the strain corresponding to the least reading of the dial. Attach the compressometer to the central portion of the specimen, and remove the spacer bars.

5. After the caps have hardened, center the specimen in the testing machine and center the spherical bearing block on top of the specimen. Centering operations should be carried out by actual measurements.

6. Adjust the compressometer dial to read zero and make sure that most of its range is available.

7. Apply load continuously at a speed of about 0.02 in. per min, or about 10 psi per sec, and read the compressometer after each load increment of about one-twentieth of the estimated ultimate load. Consult the instructor regarding this value. If the strain approaches the range of the dial, hold the load and set the hand back a full revolution by means of the adjusting screw. At a load of three quarters of the estimated ultimate load remove the compressometer. Thereafter apply the load continuously. Record the maximum load.

8. Draw a sketch to show the type of failure.

Report. 1. Plot a stress-strain diagram. Draw a smooth curve through the plotted points. Note that the curve may not pass through the origin of the graph. Mark the proportional limit and yield strength (0.01 percent offset) on the diagram. Determine the several moduli of elasticity specified.

2. Compute the compressive strength and the unit weight of the concrete.

3. Tabulate the test results in a suitable form.

Discussion. 1. Compare your results with the range of values indicated in Appendix E.

2. Discuss briefly the important facts disclosed by the test.

3. What factors affect the development of strength of concrete?

4. How is the compressive strength affected by moisture content at time of test?

5. How is the modulus of elasticity affected by age and by moisture content at time of test?

6. Why is the compression test the one most frequently made for concrete?

7. What was the purpose of using the spherical bearing block in this test?

8. List the various precautions which should be taken in positioning the spherical bearing block.

9. Are the strength correction factors given in Table 5.5 for concrete specimens having height-diameter ratios below 2 rational or empirical values?

PROBLEM 9. COMPRESSION TESTS OF SMALL WOODEN COLUMNS

Object. To determine the relationship of slenderness ratio to the ultimate unit load resisted by small pin-ended wooden columns.

Preparatory Reading. Articles 5.8 to 5.14; chapter on columns in textbook on mechanics of materials.

Specimen. Small, clear, straight-grained wooden column of uniform cross section, preferably about 0.75 by 1.5 by 50 in.

Special Apparatus. Two pin-ended bearing plates.

Procedure. 1. Mount the pin-ended bearing plates on the ends of the column with the axis of the knife-edges or pins parallel to the longer dimension. Measure the length of the column from center to center of pins and measure the cross-sectional dimensions of the column. Compute the probable maximum load by Euler's equation.

2. Center the end fixtures on the column and center the column in the testing machine. Apply load slowly and note the maximum load when a slight lateral deflection occurs. Release the load, shift both end fixtures slightly toward the convex side of the column as developed during the previous loading, and reload. If the buckling load is not higher than before, take the first value as the maximum column load. If the second load is higher than the first, the column was not properly centered during the first loading. Repeat these operations until by trial the maximum buckling load is obtained.

3. Cut an 8-in. length from the column and test the remaining 42-in. length as in 2 above. Cut off an additional 22-in. length and test the remaining 20-in. length. Also, test a 30-in. length of similar stock.

4. Test the 8-in. length, which was cut from the full-length column, as a short compression block with square ends.

Report. 1. Compute Young's modulus by substituting the maximum load carried by the longest column in Euler's equation for pin-ended columns. For each column compute the theoretical value of P/A by Euler's equation and compute the actual P/A. Specify the method of failure (buckling or crushing) for each column.

2. Plot a graph of the theoretical values of P/A against the slenderness ratio for each column. On the same sheet and using the same origin plot a graph of the actual values of P/A against the slenderness ratio.

3. Determine the equation of a straight line which is tangent to Euler's curve and which passes through the point corresponding to the ultimate strength at a slenderness ratio of zero. Determine the corresponding straight-line equation for a factor of safety of 10.

Discussion. 1. From your tests, what is the minimum slenderness ratio for which Euler's equation appears to be applicable? Is the straight-line equation satisfactory for lower values?

2. What effects do the end conditions of Euler columns have upon the buckling load?

3. Why is the Euler equation not used to any great extent for the design of actual structural columns?

4. Discuss the principal differences between the column specimens tested and those used in ordinary construction.

5. What property of a material determines the load that (*a*) a short compressive member and (*b*) a slender compressive member may carry?

6. In all columns, the actual stress is a combination of direct stress plus a bending stress. For what limiting conditions may (*a*) the direct stress and (*b*) the bending stress be neglected without appreciable error?

PROBLEM 10. FLEXURE TEST OF CAST IRON

Object. To determine the strength and stiffness of a cast-iron flexure bar. The specific items to be determined are

1. Yield load for an offset deflection of 0.02 in.
2. Fiber stress at yield load
3. Maximum load
4. Modulus of rupture
5. Maximum deflection
6. Flexural secant moduli of elasticity at fiber stresses of 10,000 and 20,000 psi
7. Type and character of fracture

Preparatory Reading. Chapter 2; Arts. 6.8 to 6.15; Appendix A.
Specimen. Gray cast-iron flexure bar.
Special Apparatus. Deflectometer.

Procedure. 1. With calipers and scale, measure several diameters of the bar at the mid-section, and record the maximum, minimum, and mean values.

2. If the load-measurement mechanism is connected to the end reaction rather than to the part that applies the center load, determine whether it indicates the reaction or the center load by suspending a known weight from the part that serves as one reaction.

3. Fix the supports on the machine at 12, 18, or 24 in. apart as assigned. Place the bar in position with its minimum diameter vertical. Mark the top of the bar for identification after rupture. Place the deflectometer in position.

4. Apply center loads at a uniform rate. Record loads and deflections at 50-lb increments up to a stress of 20,000 psi and at 100-lb increments thereafter. Note the maximum load, maximum deflection, and the manner of failure. Measure the vertical and horizontal diameters at the fracture.

Report. 1. Plot a graph showing loads as ordinates and center deflections as abscissas. Determine the yield load for an offset deflection of 0.02 in.

2. For bars that are not of standard size (0.875, 1.20, or 2.00 in. in diameter) compute corrected yield loads and corrected maximum loads and deflections for standard round bars by use of the factors listed in the accompanying table.

In case of slightly elliptical bars obtain the load correction factor as a ratio of the section modulus of the actual bar to that of the standard size bar as follows: Square the depth of the bar at the fracture, multiply by the width, and divide the product by the cube of the diameter of the standard or nominal size bar. Obtain the deflection correction factor from the table, using the vertical diameter of the test bar.

3. Compute yield stress, modulus of rupture, and secant modulus of elasticity using actual loads and actual cross section. Note that for an elliptical section $I = (\pi/64)bd^3$ in which b = breadth and d = vertical diameter at the fracture.

4. Compute the approximate probable tensile strength of the material.

5. Tabulate the values derived from the test in an appropriate form. Compare the results with values presented in Appendix A.

Discussion. 1. What is the effect of variations in span length on the modulus of rupture and the modulus of elasticity of the flexure bar?

2. Is the modulus of rupture a true ultimate fiber stress? Explain.

3. Discuss the shape of the load-deflection diagram for cast iron. What effect does it have upon the computed modulus of elasticity?

4. Does the strength of the flexure-test specimen represent the actual strength of iron in the corresponding castings? Explain.

5. List the advantages of the flexure test over tension or compression tests for cast iron.

Correction factors for round flexure test bars*

In order to correct to the standard diameter, the breaking load and deflection obtained in testing the bar shall be divided by the respective correction factors.

Test bar *A*, 0.875 in. in diameter			Test bar *B*, 1.20 in. in diameter			Test bar *C*, 2.00 in. in diameter		
Diameter of test bar, in.	Correction factor		Diameter of test bar, in.	Correction factor		Diameter of test bar, in.	Correction factor	
	Load	Deflection		Load	Deflection		Load	Deflection
0.825	0.838	1.061	1.10	0.770	1.091	1.90	0.857	1.053
0.830	0.853	1.054	1.11	0.791	1.081	1.91	0.871	1.047
0.835	0.869	1.048	1.12	0.813	1.071	1.92	0.885	1.042
0.840	0.885	1.042	1.13	0.835	1.062	1.93	0.899	1.037
0.845	0.900	1.036	1.14	0.857	1.053	1.94	0.913	1.032
0.850	0.916	1.029	1.15	0.880	1.043	1.95	0.927	1.026
0.855	0.933	1.023	1.16	0.903	1.034	1.96	0.941	1.021
0.860	0.949	1.017	1.17	0.927	1.026	1.97	0.955	1.015
0.865	0.966	1.012	1.18	0.951	1.017	1.98	0.970	1.010
0.870	0.983	1.006	1.19	0.975	1.009	1.99	0.985	1.005
0.875	1.000	1.000	1.20	1.000	1.000	2.00	1.000	1.000
0.880	1.017	0.994	1.21	1.025	0.992	2.01	1.015	0.995
0.885	1.034	0.989	1.22	1.051	0.984	2.02	1.030	0.990
0.890	1.051	0.983	1.23	1.077	0.976	2.03	1.046	0.985
0.895	1.069	0.978	1.24	1.103	0.968	2.04	1.061	0.980
0.900	1.087	0.972	1.25	1.130	0.960	2.05	1.076	0.976
0.905	1.106	0.967	1.26	1.158	0.952	2.06	1.092	0.972
0.910	1.125	0.962	1.27	1.185	0.945	2.07	1.109	0.967
0.915	1.143	0.956	1.28	1.214	0.938	2.08	1.125	0.962
0.920	1.162	0.951	1.29	1.242	0.930	2.09	1.141	0.957
0.925	1.181	0.946	1.30	1.271	0.923	2.10	1.158	0.952

* Based on ASTM A 48.

6. How does white cast iron differ from gray cast iron? What are its chief uses?

7. What alloying materials are commonly used to improve cast iron? What benefits are derived from their use?

PROBLEM 11. FLEXURE TEST OF WOOD

Object. To determine the mechanical properties of wood subjected to bending, to observe the behavior of the material under load, and to study the failure. The specific items to be determined are

1. Proportional limit stress in outer fiber
2. Modulus of rupture
3. Modulus of elasticity
4. Average work to proportional limit, inch-pounds per cubic inch
5. Maximum shearing stress
6. Average total work to ultimate load, inch-pounds per cubic inch
7. Type of failure

Preparatory Reading. Chapter 2; Arts. 6.8 to 6.15; Appendix D.
Specimen. Clear wood, 2 by 2 by 30 in.
Special Apparatus. Deflectometer, beam supports, and loading block.
Procedure. 1. Measure and weigh each specimen, count the number of annual growth rings per inch, and estimate the percentage of sapwood and of summerwood. Make a sketch of each specimen in perspective, showing any defects and direction of rings on end sections.

2. Mark the center and end points for a 28-in. span. Draw right-section lines through these points, using a square. Drive small brads in one side of the beam at mid-depth, over each end support. Set the beam supports for a 28-in. span so that they can move away from the center of span as the lower fibers elongate due to their being in tension. Place the specimen in position so that the tangential surface nearest the pith will be face up. Place the deflectometer on the brads at the supports and attach it to the center of span. See that the deflectometer fits close to the beam, and bend up the two end brads slightly so that vibrations will not displace it.

3. Adjust the deflectometer and the testing machine to read zero.

4. Apply the load continuously through a standardized wooden loading block at mid-span at the rate of 0.10 in. per min or 400 lb per min, taking simultaneous load and deflectometer readings for increments of load that will give at least 20 readings below the ultimate. Secure reading at the ultimate load if possible.

5. Sketch the appearance of the failure.

6. Cut a moisture sample about 1 in. in length from the specimen near the ruptured section. Remove all splinters and weigh to the nearest 0.1 g. Place in a drying oven controlled at 103 ± 2°C. After the moisture specimen has dried to constant weight, which may take 2 to 4 days, again weigh it.

Report. 1. Plot a diagram showing the relation between applied loads (ordinates) and center deflections (abscissas).

2. Determine the moisture content, approximate specific gravity both as received and when dry, and all other properties listed. Summarize results in tabular form and compare with similar values shown in Appen-

dix D. Discuss the several factors which may account for the difference between values.

Discussion. 1. Are the values obtained from tests on small clear beams applicable to beams of large size? Explain.

2. Is the modulus of elasticity in bending the same as in compression or tension parallel to the grain? Explain. Would you expect it to be higher or lower?

3. What other factors besides fiber stress in bending are taken into account in the design of wood beams?

4. Discuss the relationship between the modulus of rupture of a small wood beam and the ultimate compressive strength of a short block loaded parallel to the grain.

5. How does the speed of load application affect the observed strength?

6. What method may be employed for determining the moisture content of wood, other than the one used here?

7. Would you expect the neutral axis to remain at the mid-depth of a rectangular wood beam as the ultimate load is approached? If not, which way would it shift, and why?

8. Considering elastic resilience, how would you rate wood in comparison with other structural materials?

9. Discuss the relative importance of shearing and fiber stresses as the span length increases.

10. What are the advantages of measuring center deflections as in this test instead of measuring the change in distance between the beam and the table of the testing machine?

11. If each of the following factors is doubled in turn, while all others remain constant, show the effect upon both the elastic strength and the elastic stiffness of a beam: (*a*) width, (*b*) depth, (*c*) span, (*d*) modulus of elasticity.

PROBLEM 12. COMPRESSION, FLEXURE, AND ABSORPTION TESTS OF BRICK

Object. To determine the compressive strength and modulus of rupture of brick, to inspect typical fractures, and to determine the percentage of absorption.

Preparatory Reading. Chapter 2; Arts. 5.8 to 5.14, 6.8 to 6.15; Appendix F.

Specimens. Two each of two types of brick which have been dried to constant weight at 230 to 239°F.

Special Apparatus. Special supports for transverse testing, chisel, hammer, capping plates, trowel and pan.

Procedure for Flexure Test. 1. Note the brand and type of each brick. Measure each specimen to 0.01 in. and weigh to 0.01 lb. With a square and pencil, mark the lines of support and center of 7-in. span.

2. Place the brick flatwise in position on its supports with a 1/4 by 1 1/2-in. steel plate between the upper knife-edge and the specimen. Apply the load at a speed of not over 0.05 in. per min or 2000 lb per min. Record the maximum load and note the character of fracture.

Procedure for Compression Test. 1. Using a mason's chisel, square up one half of each brick ruptured in the flexure test. Measure the dimensions of the resulting specimen.

2. Coat the two opposite broad surfaces with shellac; when they are dry, bed them in a thin layer of capping material on metal capping plates. Allow the caps to set at least 1/2 hr before testing the brick. (For commercial tests the ASTM requires a minimum setting period of 16 hr.)

3. Center a specimen in the machine with a spherical bearing block centered on top, and apply the load at any convenient speed up to one-half the expected maximum load. Apply the remaining load in not less than 1 nor more than 2 min. Record the load at which the first crack appears and the maximum load. Sketch the form of break.

Procedure for Absorption Test. 1. Mark and weigh the half brick left from the flexure test. Submerge the specimens in water at a temperature of between 60 and 86°F. After 24 hr, take the specimens from the water, remove the surface moisture with a damp cloth, and weigh a second time.

Report. Calculate the specified values, summarize them in tabular form, and compare with values in Appendix F.

Discussion. 1. What is the significance of each of the three tests?

2. In what grade do the bricks belong?

3. For flexure tests of brick why is it necessary to provide special supports that may rotate transversely to fit the bottom of the brick irrespective of how it may be warped?

4. Is the compressive strength of the individual brick a criterion of the compressive strength of a wall made with these bricks? Explain.

5. What is the purpose of the ASTM requirement that the bearing surfaces of bricks for compression tests be coated with shellac before they are capped?

6. What is the purpose of the requirement in some specifications for the absorption test that the brick be immersed in boiling water? Would you expect your values to be lower, the same, or higher if you had followed such specifications?

7. What does a high absorption indicate? What does a low absorption indicate?

8. Do test results indicate any relation between absorption and strength?

9. Can water absorption of a brick be used as a measure of its resistance to freezing and thawing?

10. What effect has repressing of a brick before burning upon its strength and absorption?

11. What effect has the degree of burning upon the quality of a brick?

12. List the requirements of good building brick.

PROBLEM 13. DETERMINATION OF STRESS IN AN I BEAM BY USE OF MECHANICAL AND ELECTRICAL STRAIN GAGES

Object. To familiarize the student with the use of both mechanical and electrical strain gages, to determine stresses in a beam by means of such gages, and to compare the results with those indicated by the common theory of flexure. The following items will be determined:

1. Strains on several gage lines
2. Experimental location of the neutral surface
3. Stresses computed from the observed strains
4. Corresponding stresses computed from the theory of flexure

Preparatory Reading. Articles 2 1 to 2.6, 4.9 to 4.19, 6.8 to 6.15.
Specimen. I beam with holes for mechanical gage, and with SR-4 resistance wire gages bonded in position and wired to selector switch.
Special Apparatus. A mechanical strain gage with reference bar of the same material as the beam, SR-4 strain indicator, dummy SR-4 gage, and deflectometer.
Procedure. 1. Arrange the beam so that loads will be applied at two points symmetrical with the center of the span. Measure the span length, depth, flange width, and web thickness of the beam. Determine the distance of each of the mechanical and electrical gage lines from the top of the beam. Draw sketches showing locations of loads, reactions, gage lines, etc. Arrange the deflectometer for use and set it to read zero.

2. Measure the gage length of the standard bar to within 0.01 in. Determine the initial reading of the strain gage on the reference bar after cleaning the holes with a pointed piece of soft wood. Take five observations, pressing the points of the gage gently but firmly into the holes and removing them after each observation. Make sure that the gage is held normal to the surface of the bar, and endeavor to apply the same pressure each time an observation is made. Exert no lateral or longitudinal pressure on the gage. Note that the heat of the hand may change its length and thus affect the reading; avoid errors due to this by frequent checks on the standard bar. The reference bar should be near or on the I beam in order that temperature conditions for one will be the same as for the other. Consider the initial reading as the mean of

five observations whose maximum range does not exceed 0.00005 in. Record average readings, using the following form.

Column headings for mechanical-strain-gage data and computations

Beam load, lb	Gage line No.	Distance from top of beam, in.	Reference-bar reading	Gage-line reading	First difference, reference bar and gage line	Second difference, zero to full load	Unit strain	Experimental stress, psi	
								Using experimental *E*	Using assigned *E*

3. Take strain-gage readings on each of the gage lines with no load on the beam, following the same procedure as in 2 above. Observe and record readings on the reference bar after observing each two gage lines. Repeat any gage-line observations if the reference-bar readings before and after the gage-line readings differ by more than 0.00003 in.

4. Examine the SR-4 strain indicator and prepare to take readings using the following sequence of operations:

a. Connect the two common leads from the selector switch on the SR-4 gage circuit to the two binding posts on the strain indicator marked "measuring gage."

b. Connect the two leads from the compensating or "dummy" gage to the two posts marked "compensating gage."

c. Consult your instructor concerning the gage factor of the gages used, and set the gage-factor pointer at this value.

d. Turn the contact switch on the strain indicator from "off" to "on" about 10 sec before taking any readings, to allow the tubes to warm up. Turn off when not taking readings to conserve the batteries.

e. Turn the gage selector switch to a particular SR-4 gage. Since the indicator is probably not in balance for this SR-4 gage, the galvanometer needle will immediately whip from zero to one extreme or the other of its range. Bring the galvanometer needle back to zero by adjusting the knob on the strain-measuring scale, or by adjusting the reference switch, or more likely by adjusting both simultaneously.

f. Determine the strain reading at zero load for each SR-4 gage by moving the gage selector switch from position to position, and at each position proceed as outlined above.

5. Observe deflection dial reading.

6. Calculate the load required to produce an extreme fiber stress as assigned and check with the instructor. Apply half of this load; observe

the resulting beam deflection and the strain indicator reading for each SR-4 gage while maintaining the load.

7. Apply the full load. Again observe the beam deflection and the strain indicator reading for each SR-4 gage. In addition make mechanical strain-gage observations on each mechanical gage line. Finally, read all SR-4 gages at zero load.

8. Use the accompanying form for recording SR-4 gage readings and computing results.

SR-4 Strain measurements and computations for I beam

SR-4 gage No.	Distance from top of beam, in.	Zero load						Half load, ______ lb			Full load, ______ lb		
		Before loading		After unloading		Average		Ref.	Scale	Unit strain $(f - e)$	Ref.	Scale	Unit strain $(h - e)$
		Ref.	Scale	Ref.	Scale	Ref.	Scale						
a	*b*	*c*		*d*		*e*		*f*		*g*	*h*		*i*

Report. 1. Compute the modulus of elasticity of the beam based upon the observed deflection and the formula for deflection of a beam loaded as in this problem.

2. Use the following procedure in calculating strains along each mechanical gage line: (*a*) Subtract the reference-bar reading from the corresponding observation on each gage line on the beam at no load to obtain the "first differences." These give the datum for each gage line. Note whether these differences are plus or minus. (*b*) Compute similar first differences for full load. (*c*) Subtract "first differences" for zero load from first differences for full load to obtain "second differences." These give the total strain along each gage line corrected for temperature and strain-gage variations by means of the reference-bar readings. The reference bar, being of the same material as the I beam, should undergo the same changes in length due to variations in temperature, and hence an actual record of temperatures will be unnecessary. (*d*) Compute strains and the corresponding stresses based upon the computed modulus of elasticity. Call these the experimental stresses. Indicate whether stresses are tensile or compressive. (*e*) Also compute stresses based upon an assumed modulus of elasticity as assigned by the instructor. (*f*) Repeat *d* and *e* based upon SR-4 strains.

3. Calculate the theoretical stresses existing along the gage lines, using the known load on the beam and the formula for flexure in beams.

4. For corresponding locations on the I beam, tabulate theoretical stresses and those computed from both types of strain-gage observations. Comment on variation or similarity of the stresses for a particular location.

5. Determine the experimental location of the neutral axis as follows: Prepare a graph by plotting distances of the several mechanical gage lines from top of beam as ordinates and unit strains at the various depths as abscissas. Note that some strains are plus and others are minus. Draw a straight line that averages these points and note at what depth it indicates the neutral axis. Compare this with the theoretical location. Repeat for the SR-4 gage lines. Plot the two curves on one sheet of graph paper.

Discussion. 1. Does the plotted graph of strains vs. distances from the top of the beam substantiate the usual assumption that a plane section before bending remains a plane section after bending?

2. Explain discrepancies between theoretical and experimental results and discuss the possible sources of error in the test.

3. What advantages were derived from the symmetrical loading, considering shearing effects and the variations of stress along the gage lines?

4. For the following assumed conditions outline procedures for correcting for a temperature rise of 15°F after the zero-load strain-gage readings but before the full-load observations are made: (*a*) For all mechanical strain-gage observations on a structural-steel member, corresponding observations were made on a reference bar of invar steel having a zero coefficient of expansion and having the same temperature as the member. (*b*) Conditions as in *a*, but the reference bar is made of structural steel.

5. List the advantages and disadvantages of mechanical strain gages and SR-4 gages in comparison with an ordinary strainometer fixed in position for the following tests: (*a*) determination of axial strains in a 1-in.-diameter tensile bar; (*b*) determination of strains at many points in a structure for one load condition; (*c*) determination of strains at many points for several load conditions and extending over a considerable period of time.

PROBLEM 14. FLEXURE TEST OF STEEL CANTILEVER BEAM

Object. To determine the extreme fiber stresses and the deflections at various points of a cantilever beam under load.

Preparatory Reading. Articles 4.13 to 4.19, 6.8 to 6.15; chapters on stresses and deflections of beams in textbook on mechanics of materials.

Specimen. Steel cantilever beam, about $\frac{7}{8}$-in. square by 46-in. span, with SR-4 gages bonded to top and bottom surfaces close to the support.

Beam must be welded to a substantial plate and anchored so that full fixity can be developed at support.

Special Apparatus. Hanger for applying loads, ten 5-lb weights, two dial gage deflectometers, and SR-4 strain indicator.

Procedure. 1. Measure the width and depth of the beam. Compute the moment arm for a maximum load of 50 lb and a maximum fiber stress of 20,000 psi. Attach the hanger for the loads at this distance (full span) from the support.

2. Mount the dial gage deflectometers to measure deflections at the full span and half span of the beam. Observe deflection readings for zero load.

3. Measure the distance from the face of the support to the two SR-4 gages. Connect these gages to the strain indicator—the top gage to the active gage and the lower gage to the compensating-gage terminals. This hookup will give strains double the actual value (see Art. 4.18). Set the proper gage factor on the strain indicator and observe the strain for zero load.

4. Apply 10 loads in 5-lb increments. For each load observe the deflections at the full-span and half-span locations, and observe the strains indicated by the SR-4 gages.

Report. 1. Taking Young's modulus as 29,000,000 psi, compute the fiber stresses corresponding to the SR-4 strain readings. For the 25- and 50-lb loads compute (*a*) theoretical values of stress $\sigma = Mc/I$ at location of SR-4 gages, (*b*) deflection at end of span using $y = PL^3/3EI$, and (*c*) deflection at mid-span using $y = 5PL^3/48EI$.

2. Plot two graphs, using the same origin, showing the relationship of loads as ordinates to

a. Stresses computed from observed strains

b. Theoretical stresses computed from flexure formula

3. Plot four graphs on another sheet (using loads as ordinates) for

a. Observed deflections at end of span

b. Computed deflections at end of span

c. Observed deflections at mid-span

d. Computed deflections at mid-span

4. For the 25- and the 50-lb loads compute the percentage of error in theoretical values relative to the observed values for (*a*) fiber stress, (*b*) deflection at end of span, and (*c*) deflection at mid-span.

Discussion. 1. Discuss the sources of experimental errors in this problem.

2. Using the moment-area method, derive the equation for the deflection at the end of span.

3. State the assumptions and limitations of the theory used for determining stresses and deflections in this beam.

4. How do beams in actual construction differ from those considered in developing the theory of flexure for beams?

5. Both bending moment and shear cause deflections of a beam. Considering these two factors, is the deflection due to shear more important for a short span or for a long span? Explain.

6. Compute the end-span deflection caused by shear for the maximum load on the beam tested. What percentage is this of the maximum deflection as observed?

PROBLEM 15. FLEXURE TEST OF REINFORCED - CONCRETE BEAM

Object. To determine the mechanical properties of reinforced concrete when subjected to transverse loads and to compare the results of application of the straight-line and ultimate-strength methods for evaluation of the load capacity of a beam. The following items will be determined:

1. k calculated by straight-line formula $= k_1$
2. Experimental $k = k_2$ at $f_c = 0.45f'_c$, using trial computation to determine proper load to make $f_c = 0.45f'_c$
3. Probable ultimate moment based on safe moment at $f_c = .45f'_c$, using $k = k_2$ and factor of safety $= 2$
4. Probable ultimate moment based on ultimate strength method
5. Maximum shearing stress in concrete when $f_c = 0.45f'_c$ and at maximum load using j from straight line formula
6. Maximum bond stress on main steel when $f_c = 0.45f'_c$ and at maximum load using j from straight-line formula

Preparatory Reading. Chapter 2; Arts. 6.8 to 6.15; Appendix E; flexure, shear, and bond in text on reinforced concrete.
Specimen. Reinforced-concrete beam as assigned.
Special Apparatus. Flexure strainometer and deflectometer.
Procedure. 1. From the instructor obtain data regarding the kinds and proportions of constituent materials, water-cement ratio, concrete consistency, curing and storage conditions, the age of the specimen, and f'_c and E_c at time of test. If the latter two values are not available, determine them from tests of a corresponding concrete cylinder. Weigh the beam. Measure its length, midspan width, overall depth, and effective depth. Note the amount and type of steel reinforcement bars. Using a square and pencil, mark the center of the span, points of support, third points of span, and strainometer contact points on concrete and steel on opposite sides of and equidistant from the center of the span. Place the beam on its supports.

2. Measure the distance between the upper and lower dial-gage actuating rods of the strainometer, the distance between upper and lower points in contact with the beam; and the distance from each set of contact

points to the nearest actuating rods. Attach the strainometer collars to the beam so that (*a*) the tension dial is below and the compression dial is above the beam, (*b*) the lower contact points bear firmly against the reinforcement bars, all four points at the same height above the bottom of the beam, and (*c*) the upper points bear firmly on the concrete. See that the collars are symmetrical with the beam cross section. Note the distance from the top of the beam to the upper actuating rod. Balance the testing machine at zero after the test beam (without loading equipment) is in position.

3. Place the deflectometer beneath the beam at the center of the span and set at zero. Remove the strainometer spacer bar and set the strainometer dials to read zero. Then place the third-point loading equipment in position after having determined its weight.

4. Apply the load continuously at the rate of about 0.05 in. per min or 500 lb per min. Take strainometer readings to 0.0001 in. and deflection readings to 0.001 in. for load increments of 200 lb, or as assigned. Make a sketch of the beam, and on it sketch the cracks as they appear. Note beside each crack the load at which it was first observed, and in a similar manner show the progress of the crack as further loads are applied. Determine the maximum load. Make a sketch of the ruptured beam and note the type of failure.

5. Carry the ruptured beam to the waste pile and remove the steel bars, taking care not to bend them. Straighten them if bent and return them to the laboratory.

Report. 1. Graphically determine the position of the neutral surface at each load, assuming that a plane section before bending is a plane section after bending. Plot a diagram, using loads as ordinates and values of k as abscissas.

2. Plot a second diagram with loads as ordinates and deflections as abscissas. The beam probably has no true proportional limit; nevertheless draw the lower portion of the load-deflection curve as a straight line through the mean of the plotted points until they deviate from it. Draw a smooth curve through the remaining points. Correct the straight portion of the graph to pass through the origin.

3. Plot a third diagram using loads as ordinates and strains in the steel as abscissas. On the same sheet but with the origin shifted along the X axis plot a fourth diagram, using loads as ordinates and strains in the extreme fiber of the concrete as abscissas. Note that strainometer readings do not give these strains directly.

4. Submit a summary of results for all required values. In computing actual moments consider that due to the weight of the beam and the loading equipment.

Discussion. 1. Explain the shifting of the position of the neutral surface and the causes of initial and final failure.

2. Describe and draw a sketch to show the most probable type of failure of a reinforced-concrete beam for the following conditions: (*a*) ample tensile steel, well anchored, but no web reinforcement; (*b*) ample tensile steel, well anchored, and adequate web reinforcement; and (*c*) low steel ratio but adequate web reinforcement.

3. Compare the probable ultimate moments based on the straight-line stress analysis and the ultimate-strength method with the observed ultimate moment.

4. If the horizontal length of the neutral axis of a beam tends to shorten because of curvature, explain why the beam supports must be arranged to rock away from, rather than toward, the center of the beam.

5. Why is it unnecessary to attach the flexure strainometer at the extreme compression fibers to obtain a measure of the strains of such fibers?

PROBLEM 16. DIRECT SHEAR TEST OF STEEL

Object. To make a shear test of steel approximating the conditions of shear existing in rivets, pins, and bolts, and to determine the average strength in single and double shear.

Preparatory Reading. Chapter 2; Arts. 6.1 to 6.7; Appendix A.

Specimens. One or more grades of steel in the form of round rods.

Special Apparatus. Shear tool for steel.

Procedure. 1. With a micrometer caliper determine the mean diameter of the specimen.

2. Fix the specimen in the shear tool so that it will be in single shear, and apply load at a slow speed until rupture takes place. During the progress of the test note whether or not there is a yield point. Determine the maximum load and the character of the fracture.

3. In the same manner test the same specimen in double shear after placing the specimen so that it does not extend more than 1 in. beyond edge of movable shear block.

Report. 1. Calculate and compare the unit shearing strengths found in single and double shear. Compare with values given in Appendix A.

Discussion. 1. Suggest an explanation for any tendency for either unit single-shear or unit double-shear strength to be the greater.

2. Were the test specimens subjected to any stress other than that of direct shear? If so, was it of equal importance in the single- and double-shear tests?

3. Although ASTM specifications for rivets require tension and cold-bend tests, they do not require any shear test. Can this be justified?

4. Would this test be satisfactory for brittle materials? Explain.

5. In what other manner may the shearing strength of a steel rod be determined?

6. In a flexure test of a homogeneous rectangular beam, what is the value of the maximum shearing stress in terms of the average stress, and where does it occur?

PROBLEM 17. TORSIONAL SHEAR TEST OF STEEL

Object. To determine the behavior of a ductile steel when subjected to torsion, and to obtain the following torsional properties:

1. Shearing proportional limit using the torsion formula
2. Probable true shearing proportional limit
3. Yield strength at an offset of 0.001 rad per inch of gage length (using the torsion formula), or as assigned
4. Shearing modulus of rupture
5. Probable true shearing strength
6. Modulus of rigidity
7. Average energy absorbed per unit volume at true proportional limit
8. Approximate percentage elongation in outer fiber at failure
9. Probable tensile proportional limit
10. Probable tensile strength

Preparatory Reading. Chapter 2; Arts. 6.1 to 6.7; Appendix A.
Specimen. Steel rod.
Special Apparatus. Torsion testing machine and troptometer.
Procedure. 1. With a micrometer caliper determine the mean diameter of the specimen near its mid-length. Assuming the shearing proportional limit as 0.6 the tensile proportional limit and the shearing modulus of rupture as equal to the tensile strength, compute loading increments that will give at least 10 observations below the proportional limit, several close together near the proportional limit, and at least 10 beyond the proportional limit.

2. Note the gage length and least reading of the troptometer. Securely clamp the instrument to the specimen, making certain that the axes of the troptometer and the test piece coincide and that the troptometer is in proper position for ease of reading.

3. Adjust the torsion machine to read zero and then insert the specimen into the two heads. See that each end is centered inside each head. Gradually bring the grips in the heads to a firm equal bearing, taking care not to displace the specimen. If in tightening the grips they produce some torque, operate the machine in forward or reverse so that it will be reduced to zero.

4. Remove the troptometer spacer bar and set the instrument to read zero.

5. Apply load at a slow speed. Take readings of torque and twist simultaneously without stopping the machine. After the specimen shows definite signs of yielding, apply the load at higher speed until failure occurs. Note the character of the fracture.

Report. 1. Plot two diagrams from the same origin, showing the relation between torque in inch-pounds as ordinates against twist in radians per inch of gage length as abscissas. One diagram will extend to the yield strength with a slope of not more than 60° with the strain axis. The second diagram will show the curve for the entire test.

2. Determine the torques at (*a*) proportional limit and (*b*) yield strength, and mark them on the first diagram.

3. Compute the quantities required. Tabulate them and compare with such values as are given in Appendix A.

Discussion. 1. Discuss the feasibility of determining the shearing strength of brittle materials by the torsional method.

2. Are there any indications that the specimen was subjected to other than shearing stresses during the test? If so, what were they, what was their cause, and what was their probable effect on the results?

3. List the relative advantages and disadvantages of tubular and solid cylindrical torsion specimens for determinations of shearing strength.

4. Which of the properties determined in this test is of most significance in the selection of steel for coil springs? Why?

5. Explain why ductile materials under torsional stress shear on a right section, whereas brittle materials fracture on a helicoidal section.

6. Why is the torsion formula not applicable to noncircular cross sections?

7. If a given round steel bar having proportional limits in shear and tension of 36,000 and 60,000 psi, respectively, is to be used to absorb energy without undergoing plastic deformation, would it absorb more energy if used as a torsion member or as an axially loaded tension member? Explain.

PROBLEM 18. TEST OF HELICAL SPRING

Object. To observe the load-deflection relationship of a helical spring and to determine some of its physical characteristics.

Preparatory Reading. Articles 6.1 to 6.5; chapter on helical springs in textbook on mechanics of materials.

Specimen. Helical spring of either tension or compression type.

Procedure. 1. With a micrometer caliper measure the diameter of the spring stock at several locations. With a spring caliper measure the outside or inside diameter of the helical spring. Measure the gage length corresponding to a whole number of turns on each of two opposite sides of the spring.

2. Based on the proportional limit in shear of the material, as designated by the instructor, compute the load capacity of the spring. Report the value to the instructor before proceeding further.

3. Apply loads in increments of about one-tenth the load capacity of the spring and observe the deflection within the gage length on each of two opposite sides at each load. Observe for both increasing and decreasing loads and note any permanent set which may remain at zero load.

Report. 1. Compute the average deflections per inch of gage length and plot a load-deflection diagram.

2. Compute the following values:

a. The spring modulus per turn which is the load required to deflect one turn of the spring elastically 1 in.

b. The modulus of rigidity of the spring stock using the equation for deflection of a helical spring due to torsional stresses only

c. The modulus of rigidity based on the equation for deflections due both to torsional and transverse shearing stresses

d. The torsional shearing stress at maximum load and the average transverse shearing stress at this load

e. The elastic energy stored in the spring per turn at the maximum load based on the load-deflection data; repeat, based on the torsional stresses in the spring.

Discussion. 1. If the deflection due to flexural effects had been allowed for in the calculation of the modulus of rigidity, would the result have been higher or lower than the values determined above? Explain.

2. Compare the computed values of the modulus of rigidity for the spring with the value reported in Appendix A.

3. State where in a transverse section of the spring do the maximum torsional and transverse shearing stresses occur.

4. To determine the maximum shearing stress in the spring, should the torsional and transverse shearing stresses be added directly? Explain.

5. Why do the two values for elastic energy which were computed fail to agree?

6. With average values of the modulus of elasticity of steel in tension and torsion taken as 30,000,000 and 12,000,000 psi, respectively, it can be shown that for a given allowable stress the elastic energy which can be stored in a tension rod is greater than for a rod in torsion. However, a rod in tension is not a desirable device for absorbing energy loads. Explain.

7. Explain how it would be possible to (*a*) use a calibrated helical spring to apply a desired sustained compressive load to a concrete specimen for creep studies over a long period of time, (*b*) check on any loss of compressive load due to creep of the specimen, and (*c*) finally determine any loss in load due to permanent set of the spring.

PROBLEM 19. SHEAR TEST OF WOOD PARALLEL TO THE GRAIN

Object. To determine the shearing strength of wood parallel to the grain.

Preparatory Reading. Chapter 2; Arts. 6.1 to 6.7; Appendix D.

Specimen. Standard 2 by 2-in. wood shear specimens.

Special Apparatus. Shear tool.

Procedure. 1. Measure each specimen to 0.01 in. and weigh to 0.1 g. Count the number of annual growth rings per inch. Determine the percentage of sapwood and of summerwood. Make sketches in perspective, showing the position of annual rings. Note whether the shear is radial or tangential or some intermediate condition. Note any defects.

2. Place the specimen in position in the shearing tool. Adjust the crossbar at the rear of the tool so that (*a*) the specimen will not twist when the load is applied, (*b*) its axis is vertical, and (*c*) the lower end rests evenly on its support.

3. Center the shearing tool in the testing machine and adjust the machine to read zero load. Bring the movable head into contact with the shearing tool at slow speed; otherwise failure may occur before intended. Apply the load at a speed of 0.024 in. per min. Record the maximum load. Make a sketch of the ruptured specimen.

4. Determine the moisture content by drying to constant weight the piece sheared from the specimen.

Report. 1. Tabulate results in suitable form.

2. Compare your results with those given in Appendix D.

Discussion. 1. How is the shearing strength of wood parallel to the grain affected by the direction of the shear plane with respect to the direction of the annual growth rings?

2. Why is wood never tested in shear perpendicular to the grain?

3. A factor of safety of three is often considered satisfactory for steel; would the same value be satisfactory for wood? Explain.

PROBLEM 20. HARDNESS TESTS OF STEEL

Object. To study the Brinell, Rockwell, and Shore-scleroscope hardness testers and to determine the hardness numbers of assigned specimens of steel.

Preparatory Reading. Chapter 7; Appendix A.

Specimens. Steel specimens as assigned. Both top and bottom surfaces of specimens should be clean and smooth.

Special Apparatus. Brinell tester with measuring microscope and possibly with depth gage, Rockwell machine and Shore scleroscope.

Procedure for Brinell Test. 1. Examine the depth gage on the Brinell machine. Note the value of the smallest dial division in terms of depth of ball impression, and report to instructor for check.

2. Place the specimen upon the anvil so that its surface will be normal to the direction of the applied load. Raise the anvil with the handwheel until the specimen just makes contact with the ball. See that the ball is at least 1/4 in. from the edge of the specimen.

3. Read the depth gage or set the dial to read zero.

4. Apply load by means of the hand pump until the yoke and attached weights rise and float. Be sure that the yoke does not rise more than 1/2 in. Use a slow, steady stroke in lifting the yoke. Maintain the full load for 20 sec.

5. Release the load but make certain that ball still contacts specimen. Record the depth-gage reading. Remove the specimen from the machine.

6. Check the scale in the field of view of the measuring microscope with the standard scale provided for that purpose. Measure the diameter of impression left by the ball, to the nearest 0.01 mm.

7. Make five independent hardness determinations on each specimen.

8. After the test has been completed, place a dummy specimen on the anvil to protect it.

Procedure for Rockwell Test. 1. Examine the Rockwell hardness tester and get it ready for the test. Use a 1/16-in.-diameter ball indenter. Move the operating handle as far forward as it will go and see that the weight marked "100 kg" is in position.

2. Place the specimen upon the anvil of the machine. Raise the anvil and the test specimen by means of the elevating screw, until the specimen comes in contact with the ball. Continue to raise the work slowly until the initial load is applied and the pointer is within plus or minus five scale divisions of its upper vertical position. Turn the bezel of the gage until the mark B 30, which is also designated by a red arrow and by the word "Set," is directly behind the pointer.

3. Release the operating handle so as to apply the major load. Allow the operating handle to move without interference until the major load is fully applied. This is accomplished when the weight arm is completely free from the control of the dashpot. Immediately after the major load has been fully applied, gently bring back the operating handle to its latched position.

4. Read the position of the pointer on the red or B scale of the dial, which gives the Rockwell hardness number.

5. Make five tests on each specimen.

Procedure for Scleroscope Test. 1. Examine the scleroscope, and to facilitate the reading of the rebound of the hammer, place it on the bench so that light will be reflected from the scale. Adjust the leveling screws so that the plumb bar will hang freely in the center of the ring at its lower end.

2. Using the central portion of the specimen so that it will be practically balanced on the anvil, clamp it in position and maintain a slight pressure on the clamp handle during the test.

3. Release the hammer and note the scale reading at the top of the hammer on the first rebound. Lift the hammer back into position. Make five determinations on each specimen. Do not let the hammer fall twice on the same spot.

Report. 1. Compute the average of the five values of diameter of impression of each specimen using the Brinell method and report the hardness number as taken from Table 7.7. Also compute the hardness from the depth of Brinell impression.

2. Determine the probable tensile strengths of the materials from which the specimens were made.

3. Determine the average Rockwell hardness number of each specimen and the probable corresponding Brinell number.

4. Determine the average scleroscope hardness number of each specimen and the probable corresponding Brinell number.

5. In a single table present the results of tests by all three methods.

Discussion. 1. Is the depth of impression or the diameter of impression measured in the ASTM Standard Brinell hardness test?

2. Why do the Brinell hardness numbers obtained from the depth of indentation usually differ from those computed from the diameter of indentation?

3. Do the hardness numbers of unlike materials give a satisfactory basis for comparing their hardness?

4. Would you expect the Brinell hardness numbers for loads of 500, 1500, and 3000 kg to vary or remain constant for any given specimen? Explain.

5. How long should the load be maintained in the Brinell test for hardness of steel according to ASTM Specifications? Why is a definite loading period essential?

6. Why is a minor load applied before setting the Rockwell depth-measuring dial?

7. How would the lack of verticality of the glass tube affect the scleroscope hardness numbers?

8. Discuss the probable effect of (*a*) surface roughness and (*b*) an oily surface of the specimen on the hardness numbers by each method.

9. How would compressible material between the specimen and the anvil affect the Rockwell and Brinell hardness numbers? Explain.

10. If a Brinell or Rockwell impression was made close to the edge causing a side bulge, would the hardness number be greater or less than its true value?

11. If a Brinell or Rockwell impression was made on too thin a specimen, how would the observed hardness be affected if the anvil is (*a*) harder than the specimen and (*b*) softer than the specimen?

12. In what units are the Brinell, Rockwell, and scleroscope numbers expressed?

PROBLEM 21. IMPACT FLEXURE AND TENSION TESTS OF STEEL

Object. To study an impact-testing machine of the Charpy type and to determine the relative impact resistance of various steels, in the form of notched-bar flexure specimens and in the form of plain and grooved tension specimens.

Preparatory Reading. Chapter 8; Appendix A.

Specimens. Standard notched flexure specimens and plain and grooved tension specimens of steels as assigned.

Special Apparatus. Charpy impact machine.

Procedure for Notched-bar Flexure Test. 1. Measure the lateral dimensions of the specimen at a full section and at the notch to thousandths of inches. If necessary, by means of the templet, adjust the anvils of the machine so that they are centered with respect to the pendulum, and so that the clear distance between them is 1.575 in., or 40 mm.

2. Note the weight of the pendulum and the radius of its center of gravity (they may be stamped on the pendulum, although for some machines energy values in foot-pounds are shown directly on the graduated scale). Find out from the instructor whether or not these values are for the pendulum with tension specimen holder attached; if so, the latter should be attached to the pendulum. With the pendulum freely suspended, adjust the lever actuating the friction pointer so that the pointer reads zero.

3. Place the specimen accurately in position on the anvils. The notch should be on the side of the specimen farthest from the striking edge of the pendulum and directly in line with it. Raise the pendulum to its upper position, and then let it fall to rupture the specimen.

4. Record the angle of rise of the pendulum, estimating to tenths of degrees (or observe the energy to rupture). With the pendulum freely suspended, see whether or not the pointer again reads zero.

5. Note the shape of the fractured surface, its inclination with respect to the axis of the piece, its texture and its relation with respect to the notch.

6. Without a specimen in the machine obtain data for determining energy losses, as outlined in Art. 8.9.

Procedure for Tension Test. 1. Measure and record the dimensions of the specimens.

2. Adjust the pointer to read zero with the pendulum hanging vertically.

3. Attach the specimen to the back of the pendulum and then attach the hammer block to the specimen. Engage all threads of the adapters at these connections. Carefully adjust the specimen so that the hammer block strikes both anvils squarely and simultaneously, since otherwise large bending and vibratory stresses may be produced in the specimen.

4. Break the specimen and determine the angle of rise of the pendulum (or observe the energy to rupture).

Report. 1. Calculate the effective angle of fall of the pendulum, and compute the corrected angle of rise for each test (if the scale does not give energy values directly). Corrections will be made for friction and windage but not for the energy used in imparting a velocity to the specimen.

2. Compute the energy of rupture of each specimen. Note that in the tension test the specimen and hammer block are attached to the pendulum, and consequently the energy of the falling pendulum and attached specimen and block is greater than that of the pendulum alone. From a study of the kinetic energy of the machine, show whether or not this additional weight will need to be taken into consideration in determining the energy absorbed by the specimen.

3. Prepare a tabulation of results.

Discussion. 1. Discuss the significance and advantages of impact tests in comparison with static tests.

2. Why are impact flexure specimens notched?

3. Discuss the effect of the following factors upon the results of impact tests:

a. Characteristics of the notch
b. Velocity of the hammer
c. Type of testing machine
d. Temperature of specimen

4. Why is it necessary that impact tests be standardized?

5. Discuss the relative impact tensile resistance of plain and notched specimens of the same grade of steel and having the same breaking area. Is the relationship affected by the characteristics of the steel?

6. Can the absolute impact resistance of a specimen be determined by the test procedure used? Explain.

7. Discuss the relative advantages of long vs. short bolts of a given size subject to impact loading.

8. What physical property of a material is determined by means of an impact test?

PROBLEM 22. CHARACTERISTICS OF FRESH CONCRETE

Object. To observe characteristic properties of fresh concretes of a given water-cement ratio and given consistency but of variable aggregate gradation. The following properties will be considered:

a. Cohesiveness
b. Proportion of sand
c. Troweling workability
d. Slump and flow
e. Unit weight
f. Water gain
g. Cement factor, sacks of cement per cubic yard of concrete
h. Cost of concrete materials per cubic yard of concrete

Preparatory Reading. Appendix E.
Materials. Cement and aggregate as assigned.
Procedure. 1. Two parties working together will make a total of six mixes, using a constant water-cement ratio of 0.55 by weight, 3 to 4-in. slump, and ¼ to ¾-in. gravel for coarse aggregate, or as assigned. The kind of sand and percentage by weight of total aggregate will be as follows:

Mix No.	Percentage of sand in total aggregate	
	Coarse sand passing No. 4 sieve	Fine sand passing No. 16 sieve
1	30	
2	35	
3	40	
4	45	
5	50	
6	30	10

2. Blend each stock pile separately. The instructor will advise concerning its moisture condition and its gradation. For each batch use the equivalent of 30 lb of mixed saturated surface-dry aggregate, corrected for the effective absorption or free water of the aggregate. Place the aggregates in the mixing pan, add 5 lb of cement, and mix thoroughly while dry. Add $5 \times 0.55 = 2.75$ lb (1246 ml) of water corrected for the absorption or free water of the aggregate, and mix wet. Check the slump (see 3*d*). The batch will probably be too dry, so add water and cement in the proper ratio of 0.55, mixing thoroughly after each addition, until a 3 to 4-in. slump is obtained. Keep a record of all quantities of materials added to the mix.

3. All members of both parties will examine each mix for the six following characteristics, will agree upon the ratings assigned, and will record the results for all mixes in neat tabular form:

a. Cohesiveness. Note whether the concrete tends to hang together well or whether it tends to crumble readily. Rate as high, normal, or low.

b. Proportion of Sand. If the pieces of coarse aggregate cannot be embedded in the mortar without excessive tamping, the mix is undersanded; if they sink into the mortar without tamping, the mix is oversanded. Rate as undersanded, normal, or oversanded.

c. Troweling Workability. Work the concrete with a trowel. If it works smoothly and with little effort, the troweling workability may be called good. Rate as good, fair, or poor.

d. Slump. Dampen the slump cone and place on a smooth, moist, rigid base. Place the newly mixed concrete in the mold in three layers, each approximately one-third the volume of the mold. Rod each layer with 25 strokes of the ⅝-in. tamping rod. Distribute the strokes over the cross section of the mold, each stroke just penetrating into the underlying layer.

After rodding the top layer, strike off the surface of the concrete with a trowel, leaving the mold exactly filled. Clean the surface of the base outside the mold of any excess concrete. Immediately remove the mold from the concrete by raising it in a vertical direction. If the pile topples sideways, the test should be remade.

Measure the slump by determining the difference between the height of the mold and the height of the vertical axis (not the maximum height) of the specimen. Clean the mold thoroughly.

e. Flow. Fill the flow cone in two layers, tamping each layer 25 times. Strike off the top, remove the mold, then raise and drop the flow table ½ in., 15 times in about 15 sec.

$$\text{The flow in percent} = \frac{\text{spread diameter} - 10 \text{ in.}}{10} \times 100$$

f. Unit Weight. Weigh an empty 6- by 12-in. cylinder and base plate combined, measure the diameter and height of cylinder to 0.01 in., fill the cylinder in three equal layers, tamp each layer 25 times to exclude air voids, strike off the top evenly, and weigh again. The unit weight will be computed from the volume and weight of concrete in the cylinder.

g. Water Gain. Make a conical depression in the full top area of the concrete contained in the cylinder to collect any water rising to the surface. Place a sheet-metal cover over the top of the mold to prevent evaporation. About ½ hr after filling the mold, note the accumulation of water on the surface. Record as excessive, moderate, or low. As a rough guide, water gained in amounts less than about 3 ml will be considered as low; water gained in amounts greater than perhaps 10 ml will be considered excessive.

Report. Each party will compute the values for mixes that they made. Each student will report on all six mixes.

1. Compute the weight of the fresh concrete for each mix in pounds per cubic foot and in pounds per cubic yard.

2. Calculate the percentage by weight of water, cement, fine aggregate, and coarse aggregate in each mix. From these percentages and the unit weight of concrete, calculate the yield in cubic feet of concrete per

sack of cement, the cement factor in sacks per cubic yard of concrete, and the amount, in tons, of each of the aggregates (on a saturated, surface-dry basis) required for 1 cu yd of concrete in place.

3. Compute the cost of materials per cubic yard of each mix, assuming the following unit prices: cement, $1.60 per sack; sand, $6.80 per ton; gravel, $6.50 per ton. Neglect the cost of the mixing water.

4. Prepare a table in which are summarized the results of the observations and calculations for all six mixes.

5. Plot a diagram showing percentage of sand as abscissas vs. cement factor as ordinates for mixes 1 to 5.

Discussion. 1. Draw conclusions from your test results regarding the optimum percentage of sand for the several mixes of a given water-cement ratio and consistency.

2. Comparing the results for mixes 3 and 6, discuss the effect on workability of using the fine sand.

3. Approximately what differences in strength would you expect between the six mixes? Explain.

4. In selecting the best of several mixes of a given water-cement ratio and given consistency, why is it logical to select the one giving the greatest yield?

5. Discuss the general workability (considering flow, cohesiveness, and troweling workability) of the series of six mixes, remembering that all are of about the same consistency.

6. For concretes of a given water-cement ratio and given consistency, what would be the effect upon yield of overwashing a fine aggregate so that it has few fine particles? Explain.

7. What would be the effect upon water gain under similar circumstances? Explain.

PROBLEM 23. TRIAL MIX PROPORTIONING OF CONCRETE

Object. To give the student some experience with the application of this method, so that he may appreciate its advantages and limitations.

Preparatory Reading. Appendix E.

Materials. Cement and saturated surface-dry aggregate as assigned.

Procedure. Each of the several parties working on this problem will make a trial batch using the net water-cement ratio and slump assigned by the instructor. The following values are suggested:

W/C, by wt	*Slump, in.*
0.50	1–2
0.50	4–5
0.60	1–2
0.60	4–5

If more than four parties are available, a given set of conditions should be assigned to more than one party so that the parties working on a given assignment may check one another. Each trial batch will be made using 6 lb of cement and the proper amounts of fine and coarse aggregates for best combination of workability and yield. For this work weigh out 20 lb of each aggregate in a separate covered container, then deposit about half of each weighed lot into the mixing pan. Follow the procedure of Problem 22 for mixing a batch. Based upon your studies and previous experience with concrete mixes, use a suitable trial ratio of fine to coarse aggregate, and add in sufficient amounts to produce the desired slump. Check the weights of all materials used.

Determine (1) cohesiveness, (2) proportion of sand, (3) troweling workability, (4) unit weight, and (5) water gain as in Problem 22. Clean all equipment. Exchange the following values for your mix for corresponding items obtained by all other parties from their mixes:*

1. Water-cement ratio, by weight
2. Actual slump, in inches
3. Mix, parts by weight
4. Percent sand in aggregate
5. Sacks of cement per cubic yard of concrete
6. Gallons of water per cubic yard of concrete
7. Remarks on bleeding, cohesiveness, etc.

Report. Each student will report on all mixes.

1. Make comparisons and draw conclusions from these tests regarding the influence of the several variables upon the characteristics of the mixes.

2. Would large batches of concrete made in the field be expected to have the same workability as the small batches having the same proportions prepared in the laboratory?

3. Why is it desirable to use saturated surface-dry aggregates in trial mix designs? Would it be possible to use aggregates which are not saturated surface-dry? Explain.

PROBLEM 24. EFFECT OF WATER-CEMENT RATIO AND CEMENT CONTENT UPON QUALITY AND ECONOMY OF CONCRETE MIXTURES

Object. To familiarize the student with the general characteristics of concrete and concrete materials and with laboratory methods of manufacture and test of concrete specimens. In particular, to determine the effect of various water-cement ratios and cement-aggregate ratios upon

* It is suggested that, if possible, these items be entered in a tabulation on a bulletin board or blackboard for ease of exchange and review.

the consistency and yield of the fresh concrete and upon the strength and cost of the hardened material.

Preparatory Reading. Appendix E.

Materials. Cement and aggregate as assigned.

Procedure. 1. Mold four 3 by 6-in. (or one 6 by 12-in.) concrete cylinders of each mix shown in the following schedule. The work may be assigned to more than one party, but all students will report on all four mixes. The coarse aggregate will be limited to ¾ in. maximum size so that 3 by 6-in. cylinders may be used.

Mix No.	Proportions by weight		Percentage of sand in total aggregate, by weight	Slump, in.	Type of mix illustrated
	Cement	Aggregate			
1	1	3	45	3–4	Rich, normal slump
2	1	6	50	3–4	Normal; workable
3	1	6	50	¼–½	Dry; unworkable
4	1	6	50	7–8	Wet; high W/C

2. Blend each stock pile separately. The instructor will advise regarding its moisture condition. Measure out the required quantities for a batch (about 33 lb of saturated surface-dry materials). Place them in the mixing pan and turn the dry mixture several times with the trowel to obtain a uniform color. Take care not to spill any of the ingredients. Add water gradually and mix thoroughly. Determine the amount of water used by weighing the water container and its contents before and after mixing.

3. Determine the consistency of the mix by the slump test. If the concrete is found to be too dry, add more water to produce the desired slump, recording the water used as before. Finally, determine the flow.

4. Each student will observe the general characteristics of each mix, making note of its cohesiveness and troweling workability. Cohesiveness determines whether the concrete hangs together well or whether it crumbles readily. Rate as high, normal, or low. To determine troweling workability, work the concrete with a trowel. If it works smoothly and with little effort, the troweling workability may be called good. Rate as good, fair, or poor.

5. Determine the mean diameter and mean height of the cylindrical forms to the nearest 0.01 in. Weigh the empty molds together with their base plates. Fill the molds completely in three layers, rodding each layer 25 times. Smooth off the tops evenly and weigh the filled cylinders together with their molds and base plates.

6. Repeat the above procedure for the remaining batches. Leave an identification slip with each group of cylinders.

7. At the assigned test age, measure and weigh each of the cylinders previously molded. Cap the rough ends and test the cylinders in compression. Record the maximum load. Draw a sketch of the characteristic fracture.

Report. 1. Calculate the water-cement ratio for each mix on the basis of saturated surface-dry aggregates. Express the water-cement ratio in terms of weight and gallons per sack.

2. From the net weight of concrete in the mold at the time of casting, calculate for each mix the unit weight of the fresh concrete in pounds per cubic foot and in pounds per cubic yard.

3. Calculate the percent (by weight) of water, cement, fine aggregate, and coarse aggregate in each freshly placed mix. From these percentages, calculate for each mix the yield in cubic feet of concrete per sack of cement, the cement factor in sacks per cubic yard of concrete, and the amounts, in tons, of each of the various aggregates required for 1 cu yd of concrete.

4. Compute the weight of hardened concrete at time of test in pounds per cubic foot, and the compressive strength for each concrete cylinder.

5. Compute (*a*) the total cost of materials per cubic yard of each mix and (*b*) the total cost of materials per cubic yard per kips per square inch of compressive strength. Assume the following prices of materials: cement, $1.60 per sack; sand, $6.80 per ton; gravel, $6.50 per ton. Neglect the cost of mixing water.

6. Prepare a table in which are summarized the results of your calculations made above and of your observations on the fresh batches.

7. Prepare a graph to show the relation between strength and water-cement ratio. Note that the test results for the cylinders having the ¼ to ½-in. slump may have no relation whatsoever to the curve.

Discussion. 1. State the "water-cement–ratio" law or principle. Does it apply to all mixes? Explain.

2. What factors, in addition to the water-cement ratio, affect the strength of concrete?

3. What considerations besides strength affect the selection of a water-cement ratio?

4. Why is it that a rich mix may be weaker than a leaner mix, both mixes being of the same lots of cement and aggregate?

5. What is the effect of age of the concrete upon the water-cement ratio–strength curve?

6. If you were in charge of construction and found that certain portions of the structure require a wetter consistency for proper placement, how would you modify the mix to maintain a uniform quality of hardened concrete?

7. Discuss the cost of concrete of a given required strength as affected by the consistency.

8. For concretes of equal consistency, what effect does the variation in richness of mix and in water-cement ratio have on (*a*) cohesiveness and (*b*) troweling workability?

9. How may the consistency of a concrete be varied without changing the water content?

10. Discuss the characteristic fracture of your test specimens.

PROBLEM 25. DEMONSTRATION OF ENTRAINED AIR IN CONCRETE

Object. To acquaint the student with (1) the characteristics of air-entrained concrete in comparison with plain concrete, and (2) the pressure method for the measurement of air content.

Preparatory Reading. Appendix E.

Procedure. *Preparation of Concrete Mixes.* In this demonstration the instructor will make two concrete mixes using the same nominal cement content, one without entrained air and the other with entrained air. Also, a preliminary half-batch mix of concrete without air should be made to coat the inside of the drum mixer, as otherwise some of the first batch would be lost. For the air-entrained concrete a slightly lower sand content can be used. An effort will be made to have the slump of the two mixes as nearly equal as possible.

The instructor will give the students full information on each mix, including kind and amounts of materials used and the moisture condition of the aggregates.

Preliminary determinations on the mixes will include their slump, Kelly ball penetration, and weight of $\frac{2}{10}$ cu ft of concrete.

Air Content by Pressure Method. This method depends on an application of Boyle's law. Pressure is applied to a known volume of concrete, and the reduction in volume measured. Since the air entrained in the concrete is the only significantly compressible ingredient, the observed reduction in volume is due to compression of the air. The amount of air is shown on the calibrated scale of the apparatus.

The procedure for the test, as conducted by the instructor, will be as follows: Fill the calibrated container (lower part of the assembly) in three equal layers, rodding each layer with 25 strokes of the steel tamping rod. Follow the rodding of each layer by tapping the sides of the bowl smartly 15 times with the mallet. Slightly overfill the bowl, and strike off any excess concrete by sliding the strike-off bar across the top flange with a sawing motion until the bowl is just level full.

Thoroughly clean the flanges of the bowl and the conical cover. Assemble the apparatus, making a watertight connection. Introduce water into the head and graduated tube assembly to about the halfway mark on the tube. Avoid turbulence and the inclusion of air in this operation.

Incline the assembly about 30° from vertical, and, using the bottom of the bowl as a pivot, describe several complete circles with the upper end, simultaneously tapping the conical cover lightly to remove any entrapped air bubbles above the concrete. Fill the water column to the zero mark. Close the tube at the top and apply the required pressure. Read the air content. Release the pressure and read the scale again. The actual air content is the difference between the two readings on the scale. Repeat these observations for a check.

Compression Tests. After measuring the air content, four 3 by 6-in. cylinders will be cast for each mix and tested at the age of 14 days.

Report. *Reduction of Data.* The report on this test will include all observed data and the following items for each mix:

1. Unit weight
2. Percent sand, by weight of total aggregate and by solid volume of total concrete mix; use specific gravity of 2.60 for sand and 2.70 for gravel, or as assigned
3. Net W/C by weight
4. Net W/C by volume, in gallons per sack
5. Water content, pounds per cubic yard
6. Yield, cubic feet per sack
7. Cement factor, sacks per cubic yard
8. Compressive strength

Discussion. Write a one-page discussion on air entrainment based on the demonstration and reading assignment.

APPENDICES

A. PROPERTIES OF FERROUS METALS

A.1. General Composition. Ferrous metals are principally iron-carbon alloys containing small amounts of sulfur, phosphorus, silicon, and manganese. Some are alloyed with nickel, chromium, molybdenum, vanadium, or other elements to alter their physical and mechanical properties.

The three common forms of ferrous metal are steel, cast iron, and wrought iron. Steel is essentially a solid solution of carbon in iron. Since iron at room temperature will not hold in solution more than about 1.7 percent carbon by weight, this becomes the upper theoretical limit of carbon in steel; however, commercial steels rarely contain more than 1.2 percent carbon.

Wrought iron is practically a low-carbon steel except that it contains a small amount of slag, usually less than 3 percent. The carbon content is generally less than 0.10 percent, and the small particles of slag that are thoroughly distributed throughout the metal appear as long fibrous elements owing to the rolling operations employed in its manufacture; these slag fibers serve to distinguish it from steel having the same carbon content.

Commercial cast irons contain 2.2 to 4.5 percent carbon. Slow cooling of cast iron results in the separation of as much as 0.9 of the carbon in the form of graphite, the remainder being chemically combined with iron to form iron carbide or cementite. Ordinary cast iron is called "gray" cast iron because of the color of its fracture. Rapid cooling of molten cast iron does not permit the separation of the graphite, and most of the carbon remains combined with the iron. Such material is called "white" cast iron, since a fractured surface has a characteristic silvery white metallic color.

The approximate carbon content for the principal classes of ferrous metals, designated by commonly used names, is presented in Table A.1.

***Table* A.1. Carbon content of principal ferrous metals**

Material	Carbon, percent by weight	Material	Carbon, percent by weight
Wrought iron........	Trace to 0.09	Tool steel.........	0.61 to 1.20
Boiler steel..........	0.10 to 0.15	Special cast iron....	1.21 to 2.20
Structural steel......	0.16 to 0.30	Cast iron..........	2.21 to 4.50
Machine steel........	0.31 to 0.60		

A.2. Manufacture of Iron and Steel. Iron ores, consisting chiefly of iron oxide, are deoxidized and partially purified in a blast furnace to produce pig iron. In this operation coke is used both as a reducing agent and a fuel, and limestone is used as a fluxing agent. Some of the pig iron from the blast furnace is remelted and used for making iron castings, but most of it is made into steel.

The refinement of pig iron to produce steel is usually accomplished in a Bessemer converter, an open-hearth furnace, or an electric furnace. The open-hearth process is used more than any other, but the electric furnace is being used in many recent installations, particularly for the production of high-quality steel. These two latter methods have an advantage over the Bessemer method, since they can be used with any grade of pig iron or mixture of pig iron and scrap steel, whereas the Bessemer method is suitable only with special grades of pig iron. Some high-grade steels are made from steel scrap or high-grade iron by the crucible process. In all methods of making steel the principal object is the decrease in the amount of the impurities to within specified limits. Excess carbon is removed as carbon dioxide gas; the oxides of other impurities form a slag on top of the molten steel. The Bessemer converter uses no additional heat as do the open-hearth and electric furnaces; instead, the air blown through the steel oxidizes the impurities and thus generates enough heat to keep the charge liquid.

A.3. Forming Operations. When molten steel is cast directly into its final form, the resulting product is coarsely crystalline and relatively brittle and weak. However, this condition can be improved by suitable heat-treatment of the product. Casting the steel into ingots that are subsequently hot-worked to the desired shape greatly improves the properties of the finished steel.

Most steel is hot-rolled to its final shape, but many shapes that are not capable of being rolled are formed by a forging process. This operation produces a denser mass by closing minute cavities and reducing the grain size. Industrial forming operations on cold metal are known as "cold-working." In general, cold work might be said to be any plastic deformation carried on below the lowest critical temperature. Such cold-forming operations as rolling, cupping, and drawing produce strain hardening, which is accompanied by an appreciable increase in tensile strength and a marked decrease in ductility of the metal, as shown in Fig. A.1.

A.4. Heat-treatment. The heat-treatment of steel is a heating and subsequent cooling procedure for the purpose of improving certain desirable properties of the

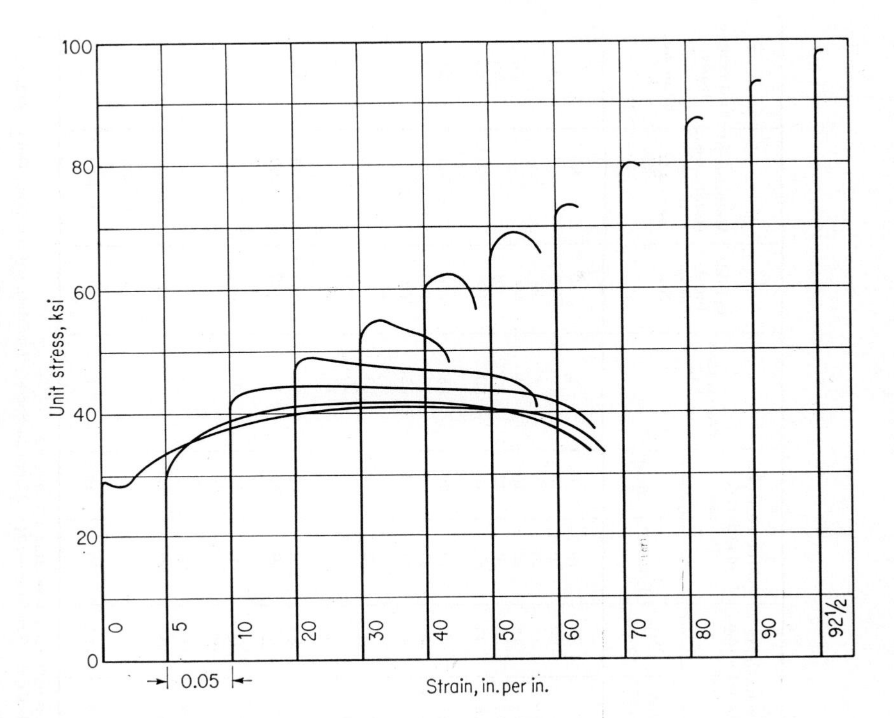

Fig. A.1. **Effect of cold rolling on stress-strain diagrams for ingot iron. Numbers on diagrams indicate percent reduction by cold rolling.** (*From Metals Handbook—1948, American Society for Metals, Cleveland, Ohio, 1948.*)

metal. The three principal operations classified as heat-treatment are annealing, hardening, and tempering. Annealing involves heating the steel to about 1400 or 1600°F and maintaining this temperature long enough for sufficient refinement of grain size to occur to produce softness, to relieve any internal strains caused by cold work or rapid cooling, or to eliminate any coarse crystalline condition due to cooling from excessive temperatures. The heating is followed by slow cooling, which leaves the steel in a strain-free condition and with the fine grain size attained during the heating period. The annealed steel is weaker but more ductile and tougher than the unannealed steel. When steel is heated as in annealing but cooled rather rapidly, usually in still air, it is said to be "normalized." Such steel also has a refined structure, free of the principal internal strains.

Hardening of steel is accomplished by use of the same heating process as when annealing and then cooling by sudden immersion in some quenching medium such as oil, water, or brine. In the quenched condition, the hardened steel is usually so brittle that its practical usefulness is greatly restricted. To correct this condition the steel is usually tempered by reheating to some temperature between 400 and 1000°F, which is well below that of the preliminary heating, followed by either rapid or slow cooling. This operation is sometimes called drawing. It produces a steel that is somewhat softer but much tougher than the steel quenched at the higher temperature only. Tempered steel is generally superior to hardened steel, since ordinarily it can be made hard enough for many uses yet tough enough so that it will not shatter in service.

Table A.2. **Mechanical properties of iron and steel***

Material	Strength in tension, ksi		Compressive yield strength,† ksi	Strength in torsional shear, ksi		Modulus of elasticity, 10^6 psi		Elongation in 2 in. percent	Brinell hardness No.	Modulus of toughness, in.-lb per cu in.	Endurance limit, reversed bending, ksi
	Yield strength†	Ultimate		Yield strength†	Ultimate	Tension	Shear				
Gray cast iron	...	20	35	...	37	15	6	1	130	80	11
White cast iron	...	60	100	...	60	20	8	...	400		
Nickel cast iron, 1.5 % nickel	...	45	60	...	...	20	8	1	200		
Malleable iron	33	50	33	19	48	25	10	14	120	......	26
Ingot iron, annealed, 0.02 % carbon	24	42	21	15	30	30	12	45	70	......	26
Wrought iron, 0.10 % carbon	30	50	30	18	35	27	10	30	100	14,000	25
Steel, 0.20 % carbon:											
Hot-rolled	40	60	40	24	45	30	12	35	120	16,500	31
Cold-rolled	60	80	60	36	60	30	12	15	160	12,000	40
Annealed castings	35	60	35	21	45	30	12	25	130		
Steel, 0.40 % carbon:											
Hot-rolled	42	70	42	25	55	30	12	25	135		
Heat-treated for fine grain	60	90	60	36	75	30	12	25	190		
Annealed castings	35	65	35	21	55	30	12	15	130		
Steel, 0.60 % carbon:											
Hot-rolled	63	100	63	37	80	30	12	15	200	12,300	50
Heat-treated for fine grain	78	120	78	47	100	30	12	15	235	15,000	55
Steel, 0.80 % carbon:											
Hot-rolled	73	120	73	44	105	30	12	10	240		
Oil-quenched, not drawn	125	180	125	75	150	30	12	2	360		
Steel, 1.00 % carbon:											
Hot-rolled	83	135	83	50	115	30	12	10	260	11,000	60
Oil-quenched, not drawn	140	220	140	84	185	30	12	1	430	2,000	100
Nickel steel, 3.5 % nickel, 0.40 % carbon, max hardness for machinability	150	170	150	90	140	30	12	12	350	14,000	76
Silicomanganese steel, 1.95 % Si, 0.70 % Mn, spring tempered	130	174	130	78	115	30	12	1	380	21,000	

Note. Most steels depend on heat treatment as well as composition to develop particular mechanical properties.
* Based on F. B. Seely, *Resistance of Materials*, Wiley, New York, 1947; and *Metals Handbook—1948*, American Society for Metals, Cleveland, Ohio, 1948.
† At 0.2 % set.

A.5. Mechanical Properties. Some of the mechanical properties of several types and grades of ferrous metals are presented in Table A.2.

One typical specification for carbon steel is ASTM A 7. It does not specify a carbon content but states, "This specification covers carbon steel shapes, plates and bars of structural quality for use in the construction of bridges and buildings and for general structural purposes." To be acceptable, the material should conform to the following tensile requirements:

Tensile strength, psi	60,000–72,000
Yield point, min, psi	33,000
Elongation in 8 in., min, percent	21
Elongation in 2 in., min, percent	24

Another widely used specification is that of the Society of Automotive Engineers [5] in which various carbon and alloy steels are classified and the permissible ranges in the principal constituents are stated for each class. In this system the first digit indicates the type to which the steel belongs; thus 1 indicates a carbon steel; 2, a nickel steel; and 3, a nickel-chromium steel. In the case of the simple alloy steels the second digit generally indicates the approximate percentage of the predominant alloying element. Usually the last two digits indicate the carbon content in "points" or hundredths of a percent. For example, an SAE 1020 steel is a carbon steel containing 0.20 percent of carbon and corresponds to a structural grade of steel.

Direct-shear tests on undriven steel rivets having 0.13 percent carbon using the device shown in Fig. 6.5*d* averaged 46,280 psi in a single shear and 42,140 psi in double shear [613]. This steel had a tensile yield point, by drop of beam, of 37,900 psi and an ultimate tensile strength of 57,600 psi with an elongation of 36.1 percent.

A number of tests on 1-in.-diameter bars of structural steel using a shear tool similar to that shown in Fig. 6.5*a* gave strengths in single shear of about 45,600 psi and in double shear of about 44,200 psi.

EFFECT OF VARIOUS FACTORS ON PROPERTIES OF STEEL

The quality and mechanical properties of steel are affected by (1) the method of manufacture, (2) the composition, (3) mechanical work, and (4) the heat-treatment.

A.6. Effect of Compositions of Carbon Steels. *Carbon* is employed as the controlling constituent in regulating the properties of both common steels (so-called plain carbon steels) and alloy steels. Its most important influence is on the strength, hardness, and ductility of the metal. Beginning with pure iron, which is soft and ductile, additions of carbon to normally cooled steel increase the hardness and strength and decrease the ductility. For each 0.1 percent of carbon added until "eutectoid" composition (0.84 percent carbon) is reached, the proportional limit is raised nearly 6000 psi, the tensile strength is increased about 10,000 psi, and the percentage of elongation is reduced about 5 percent, as shown in Fig. A.2. Above this eutectoid point further additions of carbon result in increasing hardness and strength but more rapidly increasing brittleness.

The yield point, as shown in Fig. A.3, and yield strength, as measured by the set method, increase with an increase in carbon content. However, only steels with carbon contents below about 0.40 percent have well-defined yield points.

Carbon has no appreciable effect on the stiffness of steel. The modulus of elas-

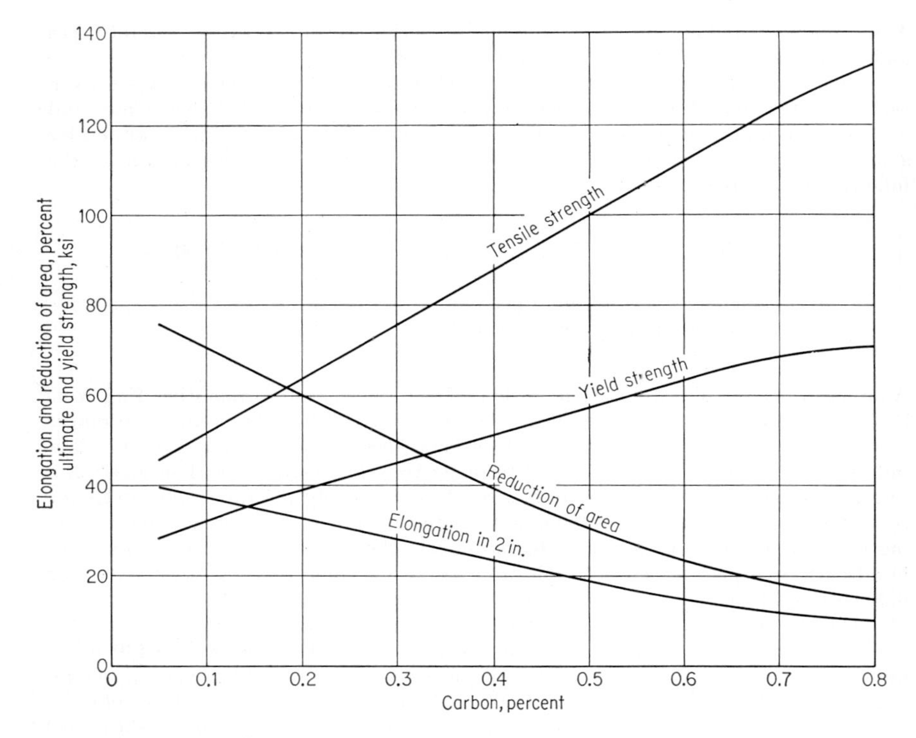

Fig. A.2. **Effect of carbon on tensile properties of hot-worked carbon steels.** (*From Sisco, Ref.* 1302.)

ticity is practically the same (30,000,000 psi) for all grades of steel, as shown in Fig. A.3, but is slightly lower for wrought iron (27,000,000 psi).

Manganese in steel has a strong affinity for oxygen and sulfur; it tends to eliminate these harmful components by withdrawing them into the slag. That which is left in the steel after the removal of oxygen and sulfur forms the carbide, Mn_3C, which causes it to roll and forge better and also slightly increases the tensile strength and hardness.

Silicon is a deoxidizing agent when added to steel and tends to diminish blowholes in ingots and castings. It is desirable because it increases both ultimate strength and elastic limit without decreasing ductility. In common steels each 0.1 percent of silicon increases the Brinell hardness number about 6; this shows it to be about one-third as effective as carbon in increasing hardness. Since silicon has a marked tendency to prevent the solution of carbon in iron, care must be exercised in treating steels having high carbon and silicon contents to avoid prolonged heating at high temperatures, since otherwise graphite may be formed.

Oxygen causes steel to be both "hot-short" (brittle at red heat), and "cold-short" (brittle at ordinary temperatures). Overblown bessemer steel without deoxidation contains only 0.15 percent oxygen; yet this renders the metal unfit for use. Oxygen combines with iron, forming iron oxide, which appears as microscopic inclusions; these inclusions cause the metal to be known as "dirty" steel, which is weak, especially under fatigue and impact stresses.

Sulfur in the form of iron sulfide causes steel to be hot-short, but it can be partially neutralized by manganese so that it is comparatively harmless. However, most specifications commonly limit the sulfur content to about 0.05 percent. Since a high sulfur content produces a steel that is more easily machined, rods for making screws commonly have a high sulfur content.

Phosphorus causes steel to be cold-short, although some evidence has been given to prove that up to 0.1 percent it does not produce harmful brittleness. Most steel specifications limit phosphorus, as they do sulfur, to about 0.05 percent.

Aluminum is a powerful deoxidizer, for it facilitates the escape of gases from molten steel. It is commonly added to plain carbon steel as well as to various alloy steels.

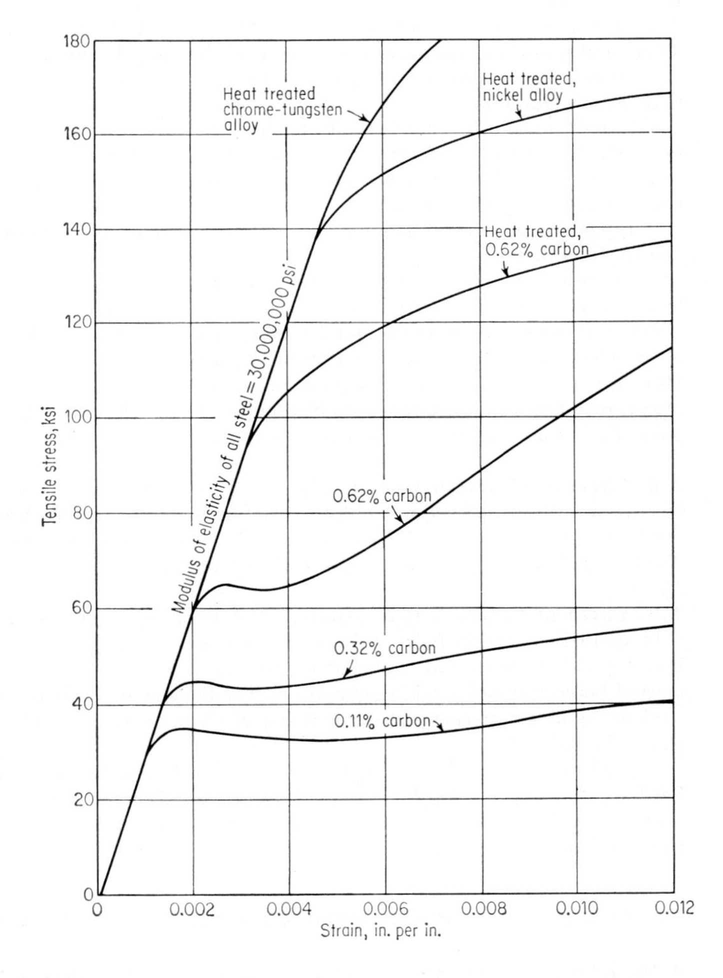

Fig. A.3. **Stress-strain diagrams for various steels.** (*From Muhlenbruch, Ref.* 112.)

A.7. Effect of Alloying Elements. *Chromium* in steel forms a stable carbide. It produces stainless and heat-resisting iron and steel and makes steel hard and strong. Chromium is principally used with other alloying elements such as nickel and vanadium.

Manganese gives steel a fine-grained structure and when present in 0.8 to 1.5 percent carbon steels, in amounts from 10 to 14 percent, it produces extreme hardness and resistance to abrasion.

Molybdenum is, next to carbon, the most effective hardening agent used in the manufacture of steel. It facilitates effective heat-treatment and causes the alloy to retain its hardness and cutting edge at high temperatures.

Nickel produces a fine-grained crystalline structure when added to steel, makes quenching effective to greater depths, and adds strength with little loss of ductility. It is useful in many types of stainless and heat-resisting steels. Low-nickel steels (2 to 4 percent) are used without heat-treatment for structural purposes and machine parts. Their best qualities, however, are made available only by proper heat-treatment.

Silicon, in amounts up to 1.75 percent, increases the yield strength of steel with little loss of ductility.

Tungsten produces increased strength of steel, but this is accompanied by brittleness, which unfits it for structural uses. It makes steel suitable for permanent magnets, self-hardening tool steels, and high-speed steels, which retain their hardness and cutting edge at high temperatures.

Vanadium gives steel high elastic and tensile strengths. It produces a fine-grained, "clean" metal.

An indication of the typical effects of alloys on the mechanical properties of steel is given in Table A.2. Perhaps the highest tensile strength obtainable with present commercial alloy steel rods is about 290,000 psi, using heat-treated silicon-molybdenum steel, which in its normalized condition has a tensile strength of about 124,000 psi. However, the tensile strength of plain carbon steel having 0.80 percent carbon, when drawn into fine wires, approaches 500,000 psi. This is doubtless the maximum tensile strength of commercial steel available at the present time.

A.8. Effect of Heat-treatment. In general, the superior qualities of a high-carbon steel or an alloy steel are not obtained by simple additions of these elements to the steel. It is only by proper heat-treatment that the desired qualities are brought out. Table A.2 shows the beneficial effects of certain heat-treatments.

A.9. Effect of Forming Operations. Hot-working processes leave the metal with a comparatively strain-free condition and fine-grained structure producing a steel of higher yield strength and tensile strength and of considerably greater ductility than existed before the hot-working operation. The beneficial effects of hot-working will be evident upon comparing some hot-rolled steels with castings of the same steel, as shown in Table A.2.

Cold-working produces a higher yield strength, tensile strength, and indentation hardness (not scratch hardness) than hot-working, but the plasticity and formability decrease owing to the lowered ductility. This is shown in Table A.2 by the results for the steel having 0.2 percent carbon.

EFFECT OF VARIOUS FACTORS ON PROPERTIES OF CAST IRON

A.10. Effect of Carbon. Carbon has a very important effect upon cast iron, since not only the amount but also the form in which the carbon exists affect the strength, hardness, brittleness, and stiffness of the metal. In gray cast iron, some of it is in the graphitic form as free carbon and the metal is weak and soft, whereas in white cast iron the carbon is in the form of cementite, which makes the iron strong and so hard that it can be finished only by grinding. The differences between several properties of these two types of cast iron are shown in Table A.2. The strength and hardness of white cast iron may be three to four times that for gray iron, but the modulus of elasticity in tension is usually less than twice as great. The tensile strength and flexural breaking load for several classes of gray-iron castings are presented in Table A.3.

Table* A.3. Requirements for gray-iron castings

Class no.	Tensile strength, min, psi	Flexural breaking load at center, min, lb		
		0.875-in. diam, 12-in. span	1.2-in. diam, 18-in. span	2.0-in. diam, 24-in. span
20	20,000	900	1800	6,000
25	25,000	1025	2000	6,800
30	30,000	1150	2200	7,600
35	35,000	1275	2400	8,300
40	40,000	1400	2600	9,100
50	50,000	1675	3000	10,300
60	60,000	1925	3400	12,500

* Based on ASTM A 48.

A.11. Effect of Annealing. The annealing of white cast iron changes the combined carbon to a finely divided, free, amorphous form composed of small rounded particles called temper carbon. The resulting metal is malleable cast iron. It is weaker, softer, and much more ductile than white cast iron.

A.12. Effect of Alloying Elements. *Nickel* is often added to improve the properties of cast iron. Nickel cast iron may be divided into three classes. In the first, called nickel cast iron, the nickel, which is in amounts of ½ to 3 percent, decreases the presence of hard spots in the metal and gives uniform machinability. In the second class, commonly known as chilled cast iron, the nickel content ranges from 3 to 5 percent. Usually some chromium and manganese are used together with the nickel, and with or without heat-treatment they produce a metal having a surface with a high resistance to abrasion. Its Brinell hardness number ranges from 600 to 750. When the nickel content is high, say 12 to 20 percent, the metal is called corrosion-resistant cast iron because of its ability to give good service under conditions causing corrosion.

Chromium when added to cast iron is essentially a carbide former. It increases the "chill" (the hard outer shell) hardness and tensile strength at elevated temperatures up to certain limits. So-called chrome-nickel alloys of cast iron are quite common. *Molybdenum* has somewhat the same effect on cast iron as chromium, although it is less drastic in its action.

B. PROPERTIES OF NONFERROUS METALS AND ALLOYS

B.1. General Characteristics. Nonferrous metals and their alloys are an important group of engineering materials. Some have a high strength-weight ratio, whereas others have good antifriction qualities and resistance to corrosion, and still others are suitable for die-casting and extrusion. Heat-treatment does not generally improve their properties to the same extent as it does those of steel and its alloys. Cold-working, however, quite effectively increases the yield strength of most nonferrous metals. The mechanical properties of the principal nonferrous metals are listed in Table B.1.

B.2. Copper. Annealed copper is relatively weak, but many of its properties such as strength and fatigue resistance may be improved by the addition of alloying elements and subsequent heat-treatment. It is one of the best electrical conductors besides being highly resistant to atmospheric corrosion.

B.3. Nickel. The principal use of nickel is as an alloy for the purpose of improving the properties of copper, iron, and steel. Its mechanical properties approximate those of medium-carbon steel, and it is much stronger, tougher, and stiffer than any other commercial nonferrous metal. It is highly resistant to corrosion, retains its strength at elevated temperatures, and has the ability to develop these properties in metals with which it is alloyed. As pure nickel, however, it is too costly for general use.

Table **B.1. Mechanical properties of nonferrous metals***

Metal	Tensile yield strength, psi†	Tensile strength, psi	Tensile modulus of elasticity, 10^6 psi	Elongation in 2 in., percent	Brinell hardness No.	Weight, lb per cu in.
Copper, 0.25 in. thick:						
Annealed, 0.05 mm grain	10,000	32,000	16	45	47	0.320
Hard	45,000	50,000	16	12	105	0.320
Nickel:						
Hot-rolled	25,000	75,000	30	45	110	0.319
Hard drawn	120,000	140,000	30	2	...	0.319
Zinc:						
Cast	....	6,000	11	1	...	0.260
Hard-rolled sheet	5,000	24,000	12	35	...	0.260
Aluminum:						
Sand cast, 1100-F	6,000	11,000	9	22	...	0.097
Annealed sheet, 1100-O	5,000	13,000	10	35	23	0.097
Hard sheet, 1100-H18	21,000	24,000	10	5	44	0.097
Magnesium:						
Cast	600	13,000	6	6	30	0.063
Extruded	1,200	28,000	6	8	35	0.063
Rolled	3,000	25,000	6	4	40	0.063

* Based on *Metals Handbook—1961* [143].
† Yield strength at approximately 0.2 percent set.

B.4. Zinc. The ordinary mechanical properties of zinc do not have much significance because it tends to creep at low stresses and temperatures. It has good resistance to corrosion, and when used as a protective coating on iron and steel plate, the product is known as galvanized iron.

B.5. Aluminum. Aluminum is one of the relatively strong, lightweight metals of construction, and only gold is more malleable. Its value as an electrical conductor, even though its conductivity is twice as good as copper of the same length and weight, is greatly reduced by its low resistance to fatigue stresses. Pure aluminum has good resistance to atmospheric corrosion but is readily attacked by strong alkalies and certain acids. Since its mechanical properties can be considerably improved by alloying it with small amounts of other metals, most aluminum is used in the alloyed state.

B.6. Magnesium. Magnesium is the lightest metal available in sufficient quantities for use in engineering construction. It is not used unalloyed. Humid air produces a self-healing film of magnesium hydroxide on its surface and thus protects the underlying metal from further oxidation.

B.7. Beryllium. Beryllium is about the same in weight as magnesium, but it has a much higher strength-weight ratio, which together with a tensile modulus of elasticity of 27×10^6 psi (some references show it as high as 42.7×10^6 psi [143]) gives it great possibilities in lightweight construction. In fact it has one of the highest known stiffness-weight ratios, but high cost and poor workability limit its present use. Its principal value lies in its use as an alloy, especially with copper.

B.8. Copper-base Alloys; Brasses and Bronzes. Copper is used as the principal metal in a variety of nonferrous alloys. Brass, made of 60 to 90 percent copper

Table **B.2. Mechanical properties of heavy nonferrous alloys***

Alloy	Approximate composition, percent	Tensile yield strength,† psi	Tensile strength, psi	Tensile modulus of elasticity, 10^6 psi	Elongation in 2 in., percent	Shear strength, psi	Rockwell hardness No.	Weight, lb per cu in.
Free-cutting brass:								
Annealed	Copper 61.5; zinc 35.5; lead 3	18,000	49,000	12	53	30,000	F68	0.30
Quarter hard, 15% reduction		45,000	56,000	12	20	33,000	B62	0.30
Half hard, 25% reduction		52,000	68,000	14	18	38,000	B80	0.30
High-leaded brass (0.04 in. thick):								
Annealed, 0.050-mm grain	Copper 65; zinc 33; lead 2	15,000	47,000	12	55	33,000	F66	0.30
Extra hard		62,000	85,000	15	5	45,000	B87	0.30
Red brass (0.04 in. thick):								
Annealed, 0.070-mm grain	Copper 85, zinc 15	10,000	39,000	12	48	31,000	F66	0.31
Extra hard		61,000	78,000	15	4	44,000	B83	0.31
Aluminum bronze:								
Sand cast	Copper 89; aluminum 8; iron 3	28,000	75,000	...	40	......	.....	0.30
Extruded		37,500	82,000	18	25	......	.....	0.30
Beryllium copper:								
A (solution annealed)	Copper 97.9; beryllium 1.9; nickel 0.2	....	70,000 ±	18	35	......	B60 ±	0.32
HT (hardened)		150,000	200,000	18	2	......	C42	0.32
Manganese bronze (A):								
Soft annealed	Copper 58.5; zinc 39; iron 1.4; tin 1; manganese 0.1	30,000	65,000	13	35	42,000	B65	0.30
Hard, 15% reduction		60,000	82,000	15	25	47,000	B90	0.30
Phosphor bronze, 5% (A):								
Annealed, 0.035-mm grain	Copper 95, tin 5	22,000	49,000	13	57	......	B33	0.32
Extra hard, 0.015-mm grain		92,000	94,000	17	5	......	B94	0.32
Cupro-nickel, 30%:								
Annealed at 1400°F	Copper 70; nickel 30	20,000	55,000	22	45	......	B37	0.32
Cold drawn, 50% reduction		78,000	85,000	22	15	......	B81	0.32

* Based on *Metals Handbook—1961* [143].

† Yield strength at 0.2% set or 0.5% elongation.

Table B.3. Mechanical properties of light nonferrous alloys*

Alloy	Approximate composition, percent	Tensile yield strength,† psi	Tensile strength, psi	Tensile modulus of elasticity, 10^6 psi	Elongation in 2 in., percent	Shear strength, psi	Rockwell hardness No.	Fatigue limit for reversed bending, psi	Weight, lb per cu in.
Aluminum alloy 2024:	Aluminum 93; copper 4.5; magnesium 1.5; manganese 0.6								
Temper 0		11,000	27,000	10.6	20	18,000	H90	13,000	0.100
Temper T36		57,000	72,000	10.6	13	42,000	B80	18,000	0.100
Aluminum alloy 2014:	Aluminum 93; copper 4.4; silican 0.8; manganese 0.8; magnesium 0.4								
Temper 0		14,000	27,000	10.6	18	18,000	H92	13,000	0.102
Temper T6		60,000	70,000	10.6	13	42,000	B83	18,000	0.102
Aluminum alloy 5052:	Aluminum 97; magnesium 2.5; chromium 0.25								
Temper 0		13,000	28,000	10.0	30	18,000	H82	16,000	0.096
Temper H38		37,000	42,000	10.0	8	24,000	E85	20,000	0.096
Aluminum alloy 5456:	Aluminum 94; magnesium 5.0; manganese 0.7; copper 0.15; chromium 0.15								
Temper 0		23,000	45,000	....	24	28,000	...	......	0.092
Temper H321		37,000	51,000	....	16	30,000	...	......	0.092
Aluminum alloy 7075:	Aluminum 90; zinc 5.5; copper 1.5; magnesium 2.5; chromium 0.3								
Temper 0		15,000	33,000	....	17	22,000	E65		
Temper T6		73,000	83,000	....	11	48,000	B90	23,000	
Magnesium alloy AM100A:	Magnesium 90; aluminum 10; manganese 0.1								
Cast, condition F		12,000	22,000	6.5	2	18,000	E64	10,000	0.066
Cast, condition T61		22,000	40,000	6.5	1	21,000	E80	10,000	0.066
Magnesium alloy AZ63A:	Magnesium 91; aluminum 6; zinc 3; manganese 0.2								
Cast, condition F		14,000	29,000	6.5	6	18,000	E59	11,000	0.066
Cast, condition T6		19,000	40,000	6.5	5	20,000	E83	11,000	0.066

* Based on *Metals Handbook—1961* [143].

† Yield strength at 0.2 % set.

and 10 to 40 percent zinc, is the one most used. For resistance to corrosion by water, the red brasses having a high copper content are preferred. For highest strengths a combination of approximately 2 parts copper to 1 zinc, termed "standard" or "high brass," is best. Brasses suitable for hot-working, by forging, rolling, or extruding, usually contain only about 60 percent copper. Typical mechanical properties of several brasses are shown in Table B.2.

Ordinary bronze consists of copper with less than 20 percent tin. The tin hardens the copper but does not appreciably increase its tensile strength. With more than 25 percent tin, the alloy is weak and either brittle or soft. Very small percentages of phosphorus added to bronze produce a phosphor bronze that has good resistance to corrosion, excellent strength, and fair ductility.

A number of so-called bronzes contain very little or no tin; they are not true bronzes. Manganese bronze is essentially brass with a little tin and manganese. It has good mechanical properties and high resistance to corrosion. Aluminum bronze is copper with 5 to 10 percent aluminum. It also has good mechanical properties and high resistance to corrosion and is a good antifriction metal.

The most effective alloy of copper contains about 2.5 percent beryllium; besides retaining many of the desirable properties of copper, the alloy attains great strength. In the annealed condition it is as strong as mild steel, and proper heat-treatment will increase its tensile strength to nearly 200,000 psi (see Table B.2). Its high endurance limit, 35,000 to 44,000 psi, is maintained even under corrosive conditions.

B.9. Monel Metal. Monel metal is an alloy of about 66 percent nickel, 28 percent copper, and small amounts of iron, manganese, and sometimes aluminum. It is highly resistant to corrosive liquids. It is very ductile, tough, and strong and retains its strength at high temperatures better than most other commercial ferrous and nonferrous alloys.

B.10. Aluminum Alloys. The addition of small amounts of copper to commercial aluminum serves to harden the latter and to increase its strength considerably. One such alloy is known as duralumin. By suitable heat treatment and aging its mechanical properties can be made to approach those of mild steel (see Table B.3). The addition of small percentages of magnesium to aluminum also increases the strength. Such alloys may not be so strong as the aluminum-copper ones, especially when hard-rolled, but they are slightly lighter in weight.

B.11. Magnesium Alloys. When magnesium is alloyed with 5 to 10 percent aluminum and small amounts of manganese, it produces an exceptionally light-weight metal that possesses good mechanical properties (see Table B.3). The use of magnesium alloys in the construction of certain airplane parts, where high strength is not essential, is increasing.

B.12. Die-casting Alloys. Alloys of zinc are most commonly used for die-casting, but alloys of aluminum, magnesium, and copper are also suitable. The zinc alloys have the highest strength and ductility.

C. PROPERTIES OF WIRE ROPE

C.1. Types and Characteristics of Wire Ropes. A wire rope is a cord made of strands of wire of any ductile metal such as iron, steel, or bronze, twisted together. Ordinary rope is composed of a number of strands, each containing a group of wires, usually laid around a hemp center. It is made in two stages, the first of which is the forming of the strands. Briefly, the individual wires, which have been previously wound on bobbins, are twisted together in a predetermined pattern in a machine to form a strand. The wires pass to the front end of the stranding machine where they pass through holes in a twister-head plate from which they are brought together and pulled through a round hole in a hardened steel die. The effect of revolving the cage carrying the bobbins and twister-head plate in the stranding machine and pulling the strand out at a uniform rate is to lay one or more layers of wires around a center or core in helical paths. The strands are wound on haul-off drums. In the second stage, strands instead of wires are wound on larger bobbins that are mounted in a rope closing machine, and the finished rope is taken off as was the strand. Thus it is seen that the wires and the strands are merely sprung into position and that the finished rope is under initial stress.

Another process, known as preforming, is accomplished by properly bending the wires and the strands in the closing operation. That the usual torsional forces exerted in the ordinary wire rope are removed can be easily demonstrated by cutting a finished preformed rope, which maintains its shape without the use of the usual seizings. The preforming is accomplished by passing the wires over rollers (placed between the twister-head plate and the closing die in the stranding machine) of such diameter and position, relative to one another, that the wires assume the helical shape they will occupy in the finished strand. In a similar operation the strand is made to assume the exact helical shape it will occupy in the finished rope. Preformed wire ropes are said to have much longer life owing to the reduction of internal friction, and also the worn ropes are more easily handled since the broken wires do not stick out as they do in the ordinary worn rope. In fact it is very difficult to detect broken wires in a preformed rope, and hence it is likely that such a rope might be kept in service long after it should have been retired since the number of broken wires is the principal criterion for the determination of the time of replacement.

Most rope is made right lay, in which the strands are laid around the center, like right-hand screw threads. However, it is supplied left lay without extra cost.

The terms *regular lay* and *lang lay* refer to the type of rope construction in which for regular lay (either left or right lay) the wires in the strands and the strands in the rope are laid in opposite directions, thus making the wires on the outside of the rope lie parallel with the axis of the rope. This is standard construction. In lang-lay construction both wires and strands are laid in the same direction, with the result that the wires, which now lie at an angle with the axis of the rope, have a greater exposed length and hence the wearing qualities of the rope are increased. The lang-lay rope, although it has a high resistance to abrasion if not damaged, is not generally used because it is more liable to curl and kink and is difficult to handle, often being called a cranky line. Therefore, regular lay is more widely used.

When the rope is to resist long-continued static loading in such applications as guying for derricks, ship rigging, stacks, and small bridge cable, the only bending stresses will be those incidental to the anchorages and splices. For such applications flexibility is not required, and hence ropes having few wires and a long pitch have the least cost for the required strength. The standard construction for this kind of service is 6 by 7 (six strands of seven wires each) and listed variously as coarse lay, haulage, transmission, and standing rope.

When wire ropes are subjected to bending as when they are passed over sheaves, flexibility is an important consideration. It is obtained in ropes made of numerous small wires. In addition to bending stresses, wire ropes often carry dynamic loads such as in mine hoists where starting, acceleration, and vibration produce additional stresses so that strength with moderate flexibility, since the sheaves are usually large, is important. When sheaves are of small diameter and the loads relatively light, flexibility becomes more important, and the classification in most catalogs is hoisting rope. The standard hoisting rope is 6 by 19. Extra flexible types are 6 by 37, and 8 by 19 ropes.

The pattern or arrangement and number of wires in the strand, the fundamental rope unit, is influenced by the use to be made of the rope, which in turn determines the properties the rope must possess. The seven-wire strand pattern is a single central wire surrounded by a layer of six other wires, all of the same diameter. The 19-wire strand uses three different diameters of wire in the strand.

The hemp center is designed to support and lubricate the wire strands. Obviously it is impossible to put enough lubricant in a rope during construction to last for the life of the rope. Internal friction will be reduced and resistance to corrosion increased if the rope is periodically treated with a lubricant that will adhere to the metal and penetrate to the hemp core. Such treatment is especially necessary if the rope is subjected to acid fumes or acid or salty water.

Several types of wire are used in the manufacture of wire rope, as shown in the accompanying tabulation, the carbon contents and the tensile strengths of the several types increasing in the order named. Some special ropes that must resist corrosion are made of brass, bronze, or stainless steel.

Grade of wire	*Carbon, percent*
Iron	0.05–0.15
Traction steel	0.20–0.50
Mild plow steel	0.40–0.70
Plow steel	0.65–0.80
Improved plow steel	0.70–0.85

Long life for a wire rope depends upon quality of wire, suitability of construction of rope for the given use, avoidance of high bending stresses by use of large diameter sheaves, avoidance of wear and distortion of rope by use of proper size of grooves in sheaves, avoidance of overloading or kinking, avoidance of lateral crushing on drums or sheaves, adequate use of correct lubricants, and prevention of corrosion and abrasion [143].

The diameter of sheave used in wire rope installations has an important influence on the life of the rope. Therefore it is customary to use minimum sheave diameters of 18 times the rope diameter for 6 by 37 ropes and 42 times the rope diameter for 6 by 7 ropes. However, whenever possible, it is desirable to use sheaves having diameters at least 50 percent larger than these minimum sizes.

In determining the required strength of a wire rope the factors to be considered include the nominal load, the speed at which the load is to be applied, the rate of acceleration and deceleration, the vibration, the additional load due to friction, and the decrease in strength due to bending around sheaves or drums. A factor of safety is finally applied; this varies from 4 to 10, depending upon the relative necessity for safety.

C.2. Wire-rope Specifications. Commonly used specifications are those of the American Petroleum Institute, which were developed to cover the application and use of wire rope in the oil industry. There are no ASTM test procedures or specifications for wire rope.

General notes and tolerances taken from the API specifications [6] are

1. The pitch or lay of a 6 by 7 rope shall not exceed eight times its nominal diameter.

2. The pitch or lay of a 6 by 19, 6 by 37, or 8 by 19 wire rope shall not be more than 7¼ times its nominal diameter.

3. In like positioned wires total variations of wire diameter shall not exceed the following values:

Wire diameter, in.	*Total variation, in.*
0.018–0.029	0.0015
0.030–0.059	0.0020
0.060–0.099	0.0025
0.100–0.130	0.0030

4. The total number of wires selected and tested shall be equal to the number of wires in any one strand. They shall be selected from all strands of the rope so as to use, as nearly as possible, an equal number from each strand. Thus the specimens

selected would constitute a complete composite strand exactly similar to a regular strand in the rope.

5. Specimens for tension tests of wires shall be at least 18 in. long, and the distance between the grips of the testing machine shall be at least 12 in. The speed of the testing machine shall not exceed 1 in. per min. Any specimen breaking within ¼ in. from the jaws shall be disregarded and a retest made.

6. In making torsion tests the distance between the jaws of the testing machine shall be 8 in. when the wire is straight, and the wire shall be twisted at a uniform speed not to exceed sixty 360° twists per minute. Tests in which breakage occurs within ⅛ in. of the jaw shall be discarded.

7. The finished rope specimen for tension tests shall be at least 3 ft between sockets for ropes up to 1 in. in diameter, at least 5 ft between sockets for larger ropes, and the sockets shall be for a ¼ in. larger diameter rope than the one under test.

The API specifications list the minimum acceptable tensile loads and the minimum number of twists for a wide range of wire sizes, for both plow steel and improved plow steel. They also list the required breaking strengths for many sizes in each of several types of rope constructions. Table C.1 shows these values for 6 by 7 bright wire ropes having a fiber core. For galvanized ropes the breaking strengths are somewhat lower.

It should be noted that these specifications are not severe, especially in the requirement for tensile strength of the finished rope, but when the individual wires must meet the tensile strength and twisting requirements, undesirable material is easily eliminated. Furthermore, these specifications do not cover fatigue testing, which is now being used by progressive manufacturers in an attempt to improve the quality of their product. The endurance of wire rope when subjected to bending stresses is greatly influenced by the design, with respect to arrangement and wire sizes, as well as by materials and fabricating processes. It is a very important test, but it will be a difficult one for the consumer to adopt.

***Table* C.1. API Strength requirements for 6 by 7 bright wire ropes**

Nominal diameter, in.	Breaking strength, kips			
	Plow steel		Improved plow steel	
	Minimum	Nominal	Minimum	Nominal
½	17.5	17.9	20.1	20.6
⅝	27.1	27.8	31.0	31.8
¾	38.6	39.6	44.3	45.4
⅞	52.1	53.4	59.9	61.4
1	67.3	69.0	77.4	79.4

C.3. Properties. Typical properties, including the endurance limit of six types of wires, are presented in Table C.2.

The efficiency of a wire rope is computed by dividing its actual tensile strength by the sum of the strengths of the individual wires and multiplying this ratio by

Table C.2. **Typical properties of wire*†**

Type	Carbon, percent	Tensile strength, psi	Twists in length of 100 diam	90° reversed bends	Endurance limit, psi	Endurance ratio
Iron	0.06	85,000	94	86	54,000	63
Toughened steel	0.36	158,000	34	86	71,000	45
Cast steel	0.41	201,000	44	88	91,000	45
Mild plow steel	0.59	215,000	33	52	73,000	34
Plow steel	0.72	275,000	38	64	105,000	38
Improved plow steel	0.71	284,000	34	84	89,000	31

* A. V. de Forest, and L. W. Hopkins, "The Testing of Rope Wire and Wire Rope," *Proc. ASTM*, vol. 32, pt. II, 1932.

† Diameter of wire, 0.043 in.

100 to convert to a percentage. It is always less than 100 and is usually about 80 to 85. Three factors influence this value:

1. The longer the pitch of the wires and strands and the smaller the rope, the more nearly parallel are the wires to the axis of the rope and the greater the efficiency.
2. The more uniform the quality of the steel wires, the higher the efficiency because then they tend to fail simultaneously rather than progressively.
3. A uniform distribution of load to the several wires has the same effect as uniform quality in developing a high efficiency. To attain this condition all wires should be of the same length, be under the same initial stress, and have the same general position with respect to the axis of the rope. This latter condition is not realized in ropes with steel centers. In short test specimens the wires may not be of the same length, thus tending to cause one strand to fail before the other strands are carrying an equal load; the computed efficiency in such a case would be low.

The modulus of elasticity of wire rope is difficult to determine because the rope tends to twist as it is loaded. This difficulty can be overcome by the use of a special extensometer arranged to compensate for the torsional displacement of the two attachment collars. Care must be taken to keep the collars securely clamped to the rope as the diameter of the latter becomes considerably smaller as load is applied. The modulus of a new rope may vary from 4,000,000 to 8,000,000 psi. It becomes successively greater with each cycle of loading until the strands settle themselves in the hemp center and with respect to one another. The modulus of elasticity may ultimately reach a value of 14,000,000 psi. This tendency of increasing modulus is less pronounced in a rope having a steel center. The modulus of a rope having a steel center is somewhat greater than for one of the same diameter having a hemp center.

Resistance to abrasion is proportional to the tensile strength and size of outer wires in the strand. Radial crushing strength is inversely proportional to the number of strands and the number of wires in the strand. It also depends on the type of center.

D. PROPERTIES OF WOOD

Common native woods are obtained from (1) *conifers or cone-bearing* trees which produce so-called "soft" woods, although some pines are very hard, and (2) *broadleaf or deciduous* trees which produce so-called "hard" woods, although basswood is quite soft.

D.1. Structure and Growth. Wood is an organic substance composed principally of cellulose (about 60 percent) and lignin (about 28 percent).

The unit of wood structure is called the *cell,* a general term for diverse wood elements such as fibers, pores or vessels, and rays. The characteristic size, shape, wall thickness, structure, and arrangement of cells in various species account for variations in weight and strength. Fibers (longitudinal cells) are composed principally of cellulose cemented together by lignin. They are of various sizes from $\frac{1}{24}$ to $\frac{1}{8}$ in. in length, with diameters about $\frac{1}{100}$ of their length; they have pointed ends and are usually placed with their long dimensions lengthwise with the tree. Those radially oriented are called *rays.* Cells are the principal constituent of wood.

Trees grow in diameter and height by adding new layers of wood cells. Growth in both conifers and broad-leaved trees occurs in the same way, i.e., by the formation of new cells in the *cambium layer,* just beneath the bark. During each growing season new wood is formed in a continuous layer, directly on the previous growth, covering the entire tree. No growth occurs in the wood already formed.

Annual growth rings or growth layers, usually concentric about the innermost ring or pith, can be observed on any cross section of a tree. They are caused by variations in rate of growth as a result of seasonal differences in moisture and temperature. An annual growth ring usually consists of two bands, one darker colored than the other, the difference being quite distinct in most native woods. *Springwood* (early wood) is the inner light-colored layer in the ring. It contains cells of larger openings and thinner walls. *Summerwood* (late wood) is the outer dark-colored layer in the ring. In most native woods with well-defined differences in seasonal growths, summerwood is heavier, harder, and stronger but shrinks more than springwood. The difference in appearance between the beginning and end of a growing season is usually well-defined in *ring-porous* and *nonporous* woods. In ring-porous woods, such as white oak, large pores develop during the rapid growth of the springwood. In nonporous woods, such as pine, the apparent difference in color of the layers is principally due to differences in specific gravity. *Diffuse-porous* woods and woods from trees that grow continuously throughout the year, as in the tropics, have no well-defined rings.

Rate of growth is measured by the number of annual growth rings per inch counted along a line perpendicular to the rings across a right section of the tree. It is quite variable, even in the same species, as shown by old slowly growing original-growth redwoods, called virgin trees, which may take 60 years to add an inch of wood, and second-growth redwood which may show only four rings per inch.

Heartwood is the physiologically inactive inner portion of a tree. Usually it is darker colored than *sapwood* which in an old tree is the relatively narrow band of wood (about 1½ to 3 in. wide in Douglas fir) between the heartwood and the bark, and which contains the active cells that participate in the life processes of the tree. In living trees heartwood is more susceptible to decay than sapwood, but in cut lumber it is less susceptible. There is no consistent difference in strength or weight between dry sapwood and heartwood.

Grain is a term usually employed to indicate the direction of fibers with respect to the main axis or the surface of a piece of wood, as in cross, diagonal, edge, and flat or slash grain. It also refers to the arrangement of the fibers themselves as in straight, curly, or spiral grain.

Defects are any irregularities in wood that decrease its strength, durability, or utility, such as knots, checks (seasoning cracks), shakes (separation of annual rings in the living tree), cross grain, decay, wane (bark or lack of wood on corner or edge of piece), warp, and pitch pockets.

Besides these specific defects, to which might be added certain inherent weaknesses such as relatively low strength perpendicular to the grain, wood has other "over-all" shortcomings such as hygroscopicity, combustibility, and susceptibility to boring organisms such as teredos and termites and attack by fungi and other destructive parasitic growths.

PHYSICAL CHARACTERISTICS

D.2. Density, Specific Gravity, and Weight. Wood substance (the cell-wall material) has a specific gravity of about 1.5 irrespective of species. Either specific gravity or unit weight of a specimen affords an index of the amount of wood substance in the piece. Specific gravity of wood varies from 0.3 to 0.9; the weight of structural woods varies between 20 and 40 pcf.

D.3. Moisture. "Imbibed" water is in the cell walls; "free" water is in the cell cavities. The cell walls must be saturated before free water can exist. The moisture content at the critical stage when the cell walls are saturated but no free water exists is called the *fiber-saturation point,* which for most woods varies from about 25 to 30 percent. *Moisture content* is expressed as the percentage of water, by weight, on the basis of wood dried at a temperature of 103 ± 2°C. The moisture content for four characteristic conditions is green, 30 to 250 percent; air dry, 12 to 15 percent; kiln dry, 6 to 7 percent; and oven dry, 0 percent. Wood is hygroscopic; so its (imbibed) moisture content tends to come to equilibrium with that of the surrounding atmosphere.

Moisture is usually determined by *oven drying* or by an *electrical resistance* method. *Oven drying* is slow but reliable; it is not applicable if the wood cannot be cut for a sample. It is widely used in testing laboratory practice for which the moisture sample should be a clear, solid piece (no knots or splinters), about 1 in. long in the direction of the grain and taken from the test specimen near the break. The sample should be left in a well-ventilated oven until dried to constant weight. Portable *electrical-resistance* instruments have small terminals that can be embedded in the wood without seriously injuring it; the terminals are connected to a battery circuit and a resistance meter. Factors for converting observed resistances to moisture content have been determined for each species of wood. This method is most satisfactory for moisture contents from about 7 to 24 percent and for lumber less than 2 in. thick. It is rapid and often used as a basis for sorting lumber.

D.4. Shrinkage. Shrinkage of wood is due to drying of the cell walls. No shrinkage occurs as the moisture content is reduced to the fiber saturation point, but it attains a maximum value when the wood is oven dry; the shrinkages of air-dry and kiln-dry wood are about one-half and three-fourths the oven-dry shrinkage, respectively. Shrinkage is a reversible phenomenon, as the imbibing of water by the cell walls causes wood to swell. Tangential shrinkage (parallel to the growth rings) is approximately double the radial shrinkage (normal to the growth rings); longitudinal shrinkage is practically zero. The predominance of radial seasoning checks is due to the high ratio of tangential to radial shrinkage. Shrinkage is a direct function of the specific gravity, the approximate percentages of radial, tangential, and volumetric shrinkages being respectively 9, 17, and 28 times the specific gravity. For Douglas fir the percentages are 4, 7.5, and 12, respectively. Warping of lumber is due to unequal shrinkage of various portions. This may result from unequal loss of moisture from various parts of the piece or to variations in structure throughout the piece.

MECHANICAL PROPERTIES

D.5. Summary of Mechanical Properties. A summary of the principal mechanical properties for a few representative native woods is given in Table D.1. Standard types of specimens and methods of testing were used. Since the values of modulus of elasticity shown were derived from beam tests and include some effect of shear deflection, they should be increased by about 10 percent for application to problems involving axial loads.

Wood is strongest in tension parallel to the grain, but values for this property are not given in Table D.1 since it is difficult to design end connections to develop it. On the other hand, wood is weakest in tension across the grain. Values of these two tensile strengths for air-dry Douglas fir are about 14,000 and 300 psi, respectively.

D.6. Effect of Various Factors on Strength and Stiffness. The greatest influence on the growth of trees is exerted by soil, climate (moisture and temperature), and growing space. Hence, clear wood of any species will exhibit variations in characteristic mechanical properties, such as strength and stiffness, because of changes in the uncontrollable but easily determined factors of specific gravity and closely related percentage summerwood and rate of growth. Other important factors, under more

Table **D.1. Strength and stiffness of air-dry wood**[a,b]

Commercial name	Specific gravity	Weight, pcf	Static bending[c]			Impact bending,[c] height of drop causing failure, 50-lb tup, in.	Compression parallel to grain[d]		Compression perpendicular to grain, stress at proportional limit, psi[e]	Shear parallel to grain, max strength, psi[f]
			Fiber stress at proportional limit, psi	Modulus of			Stress at proportional limit, psi	Maximum strength, psi		
				Rupture, psi	Elasticity, 1000 psi					
Ash, Oregon	0.55	34	7,000	12,700	1,360	33	4,100	6,040	1,540	1,790
Cedar, western red	0.33	21	5,300	7,700	1,120	17	4,360	5,020	610	860
Douglas fir (coast)	0.48	30	8,100	11,700	1,920	30	6,450	7,420	910	1,140
Hemlock, western	0.42	26	6,800	10,100	1,490	26	5,340	6,210	680	1,170
Hickory, true	0.73	46	10,900	19,700	2,180	75	.....	8,970	2,310	2,140
Locust, black	0.69	43	12,800	19,400	2,050	57	6,800	10,180	2,260	2,480
Maple, red	0.54	34	8,700	13,400	1,640	32	4,650	6,540	1,240	1,850
Oak, white	0.67	42	7,900	13,900	1,620	39	4,350	7,040	1,410	1,890
Pine, ponderosa	0.40	25	6,300	9,200	1,260	17	4,060	5,270	740	1,160
Pine, longleaf	0.58	36	9,300	14,700	1,990	34	6,150	8,440	1,190	1,500
Redwood (virgin)	0.40	25	6,900	10,000	1,340	19	4,560	6,150	860	940
Spruce, Sitka	0.40	25	6,700	10,200	1,570	25	4,780	5,610	710	1,150

[a] *Wood Handbook*, Forest Products Laboratory, U.S. Department of Agriculture, 1955.
[b] All specimens of clear straight-grained wood having 12 percent moisture content.
[c] 2 by 2 by 30-in. specimen on 28-in. span, center load.
[d] 2 by 2 by 8-in. specimen, 6-in. gage length.
[e] 2 by 2 by 6-in. specimen, 4 sq in. under load.
[f] 4 sq in. under load. Shear strength across the grain about five times that parallel to the grain.

or less control, are size and shape of specimen, test procedure such as rate and duration of loading, and the amount of moisture in the wood. When comparing test results with other reported values, these various factors should be taken into account.

The specific gravity–strength relationship within any species is very definite; the correlation is even high for all wood, irrespective of species. Approximate values are given in Table D.2.

***Table* D.2. Specific gravity–strength relationship*†**
(Regardless of species)

Property	Moisture condition	
	Green	Air dry (12%)
Static bending:		
Fiber stress at proportional limit, psi	10,200 $G^{1.25}$	16,700 $G^{1.25}$
Modulus of rupture, psi	17,600 $G^{1.25}$	25,700 $G^{1.25}$
Modulus of elasticity, psi	2,360,000 G	2,800,000 G
Total work, in.-lb per cu in.	103 G^{2}	73 G^{2}
Impact bending:‡		
Height of drop at failure, in.	114 $G^{1.75}$	95 $G^{1.75}$
Compression parallel to grain:		
Proportional limit, psi	5,250 G	8,750 G
Ultimate strength, psi	6,730 G	12,200 G
Modulus of elasticity, psi	2,910,000 G	3,380,000 G
Compression perpendicular to grain:		
Proportional limit, psi	3,000 $G^{2.25}$	4,630 $G^{2.25}$
Shearing strength:		
Parallel to grain, psi	2,750 $G^{1.33}$	4,000 $G^{1.33}$

* *Wood Handbook,* Forest Products Laboratory, U.S. Department of Agriculture, 1955.

† G represents the specific gravity of oven-dry wood, based on the volume at the moisture condition indicated. It varies between about 0.3 and 0.9.

‡ 50-lb tup.

Percentage summerwood, when the difference between it and the adjacent springwood is well marked as in Douglas fir, is a good index of strength, the more summerwood, the greater the strength. This follows from the close relationship between percentage of summerwood and specific gravity.

Rate of growth is not so reliable a criterion of strength as is specific gravity, although rapid growth (fewer rings per inch) is generally indicative of good strength in some hardwoods such as oak. However, in softwoods, the wood grown at a normal rate is usually the strongest.

Large wood beams and columns always have some defects such as checks and knots, whereas the small standard specimens are always of clear straight-grained wood; therefore, the effects of size and defects must be considered together. The possible variations in defects which may occur in large timbers make it impossible to assign specific reduction factors for any particular defects, but the strength of large timbers is considerably lower than that for small clear test pieces.

Moisture variations above the fiber saturation point do not affect the strength properties, but decreases in moisture below that condition produce increases in strength

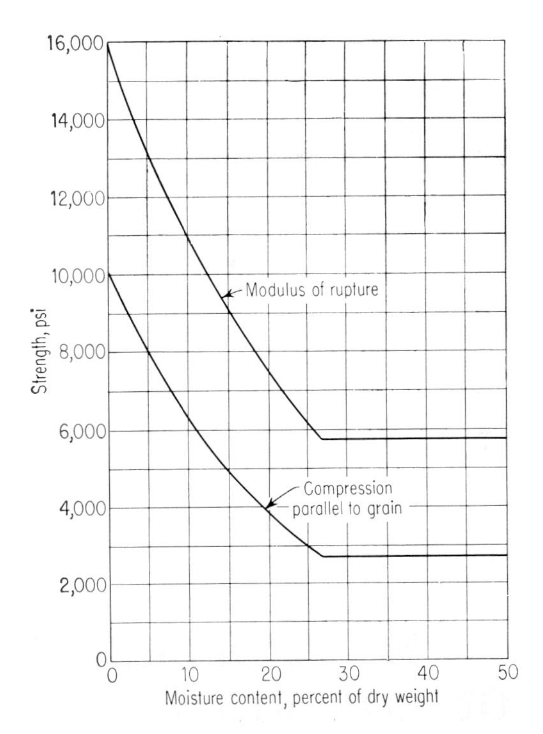

Fig. D.1. **Effect of moisture on strength of clear specimens of Sitka spruce.** (*From Markwardt and Wilson, Ref.* 597.)

and stiffness of small, clear specimens. This relation is shown in Fig. D.1 for the compressive strength parallel to the grain and the modulus of rupture for clear specimens of Sitka spruce. Stiffness and other mechanical properties show somewhat similar relationships with moisture content, although the percentage effects are not the same. These relations do not hold for large timbers, since any increase in strength due to seasoning is offset by defects that develop during the drying process.

Rate of loading and period over which the load is sustained have a marked effect upon the strength of wood. Under long-time loading, beams have been broken at stresses only slightly greater than 50 percent of the ultimate strength as given by tests in which the load was applied within a few minutes. In compression parallel to the grain, failure under long-time loading may occur at about the proportional limit stress as found in a test of short duration.

In compression tests perpendicular to the grain, the strength is influenced by the direction of the load with respect to the growth rings. The proportional limit strength is greatest when the load is applied normal to the rings (on a tangential surface), next highest when applied parallel to the rings (on a radial surface), and lowest when applied at 45° to the direction of the rings; the relative values for these three conditions are about 2, 1½, and 1 [1330].

The shearing strengths parallel to the grain for standard specimens which are given in Table D.1 are for shear planes tangent to the growth rings. Shear strength on radial planes is slightly greater.

E. PROPERTIES OF CONCRETE

E.1. Composition of Concrete Mixtures. Concrete may be thought of as a mass of inert filler material (aggregate) held together by a matrix of binder (cement paste). The properties of the concrete, both in the freshly mixed and in the hardened state, are intimately associated with the characteristics and relative proportions of these components.

The solid portion of the hardened concrete is composed of the aggregate and a new product which is the result of a chemical combination of cement with water. The remaining portion of the space occupied by a given volume of concrete is composed of free water and air voids. The latter are comparatively small, usually not over 1 or 2 percent in ordinary freshly made concrete. After any considerable period, the amount of free water depends upon the extent of combination which has taken place with the cement and the loss due to evaporation.

The binder material, the cement-water paste, is the active component and has three main functions: (1) to provide lubrication of the fresh plastic mass, (2) to fill the voids between the particles of the inert aggregates and thus to produce watertightness in the hardened product, and (3) to give strength to the concrete in its hardened state. The properties of the paste depend upon the characteristics of the cement, the relative proportions of cement and water, and the completeness of chemical combination between the cement and water, which chemical reaction is termed *hydration*. The extent of hydration of the cement requires time, favorable temperatures, and the continued presence of moisture. The period during which concrete is definitely subjected to these conditions is known as the "curing" period. On construction work, this period may vary from 4 to 10 days; in the laboratory, a common curing period is 28 days. Good curing is essential for the production of quality concrete.

The aggregate has three principal functions: (1) to provide a cheap filler for the cementing material, (2) to provide a mass of particles that are suitable for resisting the action of applied loads and abrasion, and (3) to reduce the volume changes resulting from the setting and hardening process and the drying of the cement-water paste. The properties of concretes resulting from the use of particular aggregates depend upon (1) the mineral character of the aggregate particles, particularly as related to strength, elasticity, moisture volume changes, and durability; (2) the surface characteristics of the particles, particularly as related to workability of the fresh concrete and bond within the hardened mass; (3) the grading of the aggregates, particularly as related to the workability, density, and economy of the mix; and (4) the amount of aggregate in unit volume of concrete, particularly as related to strength and elasticity, and to volume changes due to drying and to cost.

MATERIALS FOR CONCRETE MAKING

E.2. Cement. Portland cement is produced by burning in a rotary kiln, almost to the point of fusion, a properly proportioned mixture, the essential components of which are lime and clay. The kiln product, known as "clinker," is subsequently ground, with a small amount of gypsum, to a desired degree of fineness. The function of the gypsum is to retard the time of setting, as without gypsum the cement would develop a flash set. For an average cement an oxide analysis shows that the CaO content is about 64 percent, the SiO_2 content is about 22 percent, and the Al_2O_3 content is about 5 percent. During the burning of the raw materials they react to form four complex compounds. These compounds with their common abbreviated designations and their approximate percentages in a normal (Type I) cement are tricalcium silicate (C_3S), 45 percent; dicalcium silicate (C_2S), 27 percent; tricalcium aluminate (C_3A), 11 percent; tetracalcium alumino-ferrite (C_4AF), 8 percent.

Altogether there are five types of portland cement; they differ primarily in their compound composition but also in their fineness. These types and their uses are as follows:

Type I. For use in general concrete construction when the special properties specified for the four other types are not required

Type II. For use in general concrete construction exposed to moderate sulfate action, or where moderate heat of hydration is required

Type III. For use when high early strength is required

Type IV. For use when a low heat of hydration is required

Type V. For use when high sulfate resistance is required

Four physical tests are usually prescribed for the acceptance of a cement. These are (1) fineness, (2) soundness, (3) time of set, and (4) strength of a mortar made with the cement. Pertinent ASTM Specifications are C 109, C 115, C 151, C 190, C 191, and C 126.

The fineness is expressed in terms of the "specific surface," i.e., the surface area of the particles, in square centimeters, contained in 1 gram of cement. This measure of fineness is usually determined by measuring the air permeability of a specially compacted bed of cement. Other methods involve (1) a sedimentation procedure employing Stoke's law regarding the settlement of fine particles, or (2) a determination of the turbidity (using a photoelectric cell) of a specified cement-kerosene mixture. An average commercial present-day cement has a specific surface of about 2800 to 3200 sq cm per g as measured by the air permeability method. Since greater fineness induces more rapid hydration, many of the commercial high-early-strength cements are simply produced by fine grinding.

The soundness or volume constancy of a cement is determined by measuring the expansion of a cement bar which occurs while it is cured for 5 hr in an autoclave at a maximum steam pressure of 295 psi. To be acceptable this expansion must not exceed 0.50 percent.

To ensure sufficient time to place concrete while it remains plastic, a minimum limit is imposed upon the time of "initial" set, which may be taken as the condition of the mass when it begins to stiffen appreciably. ASTM Specifications require that it should not take place within 1 hr. Depending on the test used to determine it, initial set usually takes place within 2 to 4 hr. To ensure a cement that will harden for use, a maximum limit is imposed upon the time of "final" set. ASTM Specifications require that final set occur within 10 hr. With many commercial cements final set occurs within 5 to 8 hr. The condition of initial and final set is determined by the penetration of standard needles or rods into a "neat" (straight cement) paste of specified consistency.

Strength tests are made on tension briquets (sometimes on 2-in. compression cubes) made with Ottawa standard sand. The mortar is of 1 part cement to 3 parts standard sand by weight, and the tests are made at specified ages. For a Type I cement to be acceptable under the ASTM Specifications, briquets made and stored under "standard" curing conditions which prescribe moist air for 1 day followed by water storage thereafter, all at a temperature of 70°F, must possess tensile strengths exceeding 275 psi at 7 days and 350 psi at 28 days. For compression cubes made of a 1:2.75 graded standard sand mortar the corresponding required compressive strengths are 2100 and 3500 psi.

E.3. Admixtures. Small amounts of various admixtures are added to some concrete mixes to modify some of their characteristics in both the fresh and hardened states. Some admixtures act as wetting agents to increase the fluidity of the mix or to permit a reduction in the water content for a given fluidity. Other admixtures tend to entrain minute globules of air which serve (1) to lubricate the mass in the same way as wetting agents and (2) to increase markedly the resistance of the concrete to freezing and thawing.

E.4. Aggregates. Aggregates may be classified as to source, as to mode of preparation, and as to mineralogical composition.

With reference to source, aggregates may be either natural or artificial. Natural aggregates may be the result of weathering and the action of running water, producing sands and gravels, or may have been prepared by crushing, giving "stone" sands and crushed stone. Natural aggregates result from the attrition or disintegration of any or all of the rock types: igneous, sedimentary, or metamorphic. Artificial aggregates are usually lightweight materials produced by burning special clays or shales in a rotary kiln.

Material passing a No. 4 screen is classified as sand or fine aggregate, although some fine aggregates may be considerably finer than this size. There are usually available for concrete work two or three gradations of fine aggregate and several sizes (size groups) of coarse aggregates; e.g., ¼ to ½ in., ¼ to ¾ in., ¾ to 1½ in., 1½ to 2½ in., etc.

The principal qualifications of aggregates for concrete are that they be clean, hard, tough, strong, durable, of the proper gradation, and nonreactive with cement.

Fortunately, in this country, there is usually little difficulty in securing materials that meet these requirements. The question of obtaining satisfactory materials is largely a matter of cost. Pertinent ASTM Specifications are C 33 and C 130.

Properties of an aggregate which it is desirable to know in order to proportion a concrete mix and calculate batch quantities are specific gravity, moisture content, gradation, and weight per cubic foot.

The *specific gravity* is determined by displacement methods and is based upon the gross volume of the aggregate particle, including voids. The specific gravity on this basis is called the "bulk" specific gravity. The specific gravities of many gravels in current use average about 2.65. The specific gravity multiplied by 62.4, the unit weight of water, gives the weight of one solid cubic foot of material; this is sometimes called the "solid unit weight."

The *moisture content* of an aggregate is based upon its oven-dry weight, although it is the free or surface moisture in excess of the absorption causing the saturated, surface-dry condition which usually is the important value. The various moisture conditions are (1) oven dry, all moisture, external and internal, driven off by heating at 212°F; (2) air dry, no surface moisture on the particles, some internal moisture, but not saturated; (3) saturated surface dry, no free or surface moisture on the particles, but all voids within the particles filled with water; (4) damp, or wet, saturated and with free or surface moisture on the particles. The excess of (3) over (1) is the absorption capacity, or total possible internal moisture of the aggregate.

In a concrete mix, the free or surface moisture becomes a part of the mixing water and must be taken into account in determining the quantity of water to be added to a batch. Also, a dry aggregate absorbs some of the mixing water. Hence, in the proportioning and batching of concrete mixes, all calculations are referred to the saturated surface-dry condition. The excess or deficiency of moisture to produce this condition may be determined in either of two ways: (1) by completely drying out and applying the absorption capacity as a correction or (2) by displacement methods employing the specific gravity of the aggregate based on a saturated surface-dry condition. In the latter method either a specially constructed flask or a pycnometer is employed. Approximate absorption capacities and surface-moisture contents of common aggregates are given in Tables E.1 and E.2. The finer the aggregate, the more free water it can and usually does carry.

The film of surface moisture on the particles of a fine aggregate holds them apart so that the gross or bulk volume is increased. This phenomenon is called "bulking," and when batching quantities for a concrete mix by volumetric methods, it must be taken into account if the proper proportions of ingredients are to be obtained. For ordinary concrete sands, maximum bulking occurs when the moisture content is about 4 to 7 percent by weight, and the increase in volume may be as much as 25 to 40 percent. Larger percentages of water tend to decrease the bulking until, when the moisture is roughly 12 to 20 percent by weight, the sand is completely submerged or

***Table* E.1. Approximate absorption capacities of various aggregates**

Aggregate	*Absorption capacity, percentage by weight*
Average concrete sand	0 to 2
Granitic-type rocks	0 to ½
Average gravel and crushed limestone	½ to 1
Sandstone	2 to 7
Lightweight porous materials	Up to 25

***Table* E.2. Approximate surface moisture on ordinary aggregates**

Aggregate	Approximate surface moisture	
	Percentage by weight	Gallons per cubic foot
Moist gravel and crushed stone	½–2	1/16–¼
Moist sand	1–3	⅛ –⅓
Moderately wet sand	3–5	⅓ –½
Very wet sand	5–10	½ –1

"inundated" and occupies about the same space it did in the dry, loose condition. The bulking of coarse aggregates is negligible.

Gradation of particles is determined by a "sieve analysis." A suitable gradation of the combined aggregate in a concrete mix is necessary in order to obtain proper workability and density. Harsh or otherwise deficient gradings usually result in concretes of inferior quality or poor economy. For many types of work, if properly sized commercial materials are available, the aggregates may be combined in arbitrary proportions, based upon experience. However, on important jobs, it has been found advantageous to analyze the materials for their particle-size distribution (by means of a series of sieves) and proportion the various sizes so as to secure a satisfactory gradation of materials from fine to coarse. Fortunately, provided proper adjustments are made in the mix, a fairly wide latitude in grading may be tolerated without seriously affecting the properties of the resulting concrete. For concretes used in building construction, the approximate proportions of fine to coarse aggregate are given in Table E.3.

***Table* E.3. Percentage of sand in total aggregate for various maximum sizes of aggregate**

Maximum size of aggregate, in.	Percentage of sand in total aggregate, by weight*†	
	Range for various job materials	Recommended maximum
⅜	55–75	60
¾	40–60	55
1	40–55	50
1½	35–50	45
2	.....	40

* For aggregates having bulk specific gravities within range of about 2.5 to 2.8.
† The finer the sand or the richer the mix, the lower the percentage required.

The *weight per cubic foot* of aggregates is used for computing quantities in batching by volume and in estimating quantities of materials. It is determined by weighing the aggregate required to fill a container of known volume under specified conditions of compaction or moisture content: (1) loose or compact, and (2) dry or damp.

Reactive aggregates are those which contain certain forms of silica which combine with the alkali in the cement to cause differential expansions of the concrete. This leads to crazing and cracking and other forms of disintegration of the concrete. Petrographic examinations and long-time expansion tests can be used to evaluate a new source of aggregates. If any reactive silica is present, only a low-alkali cement should be used. However, any appreciably reactive aggregates should not be used under any circumstances.

THE MANUFACTURE OF CONCRETE

In common with the production of other materials, the problem of the manufacture of concrete is to obtain the best product for the least cost, using the materials that are available. The term "best" concrete here implies one that has the necessary and desired properties, such as workability and homogeneity in the fresh material, and strength, watertightness, durability, and volume constancy in the hardened product. The securing of the desired properties, as well as the economy of the work, involves the selection of a suitable combination of materials and the control of the process of manufacture until the concrete is the proper age. Pertinent ASTM Specifications are C 31, C 39, C 124, C 143. See also Ref. 8.

E.5. Design of Concrete Mix. The first step in the manufacture of concrete, after materials have been selected, is the design of the mix. This usually resolves itself into proportioning the ingredients (including the water) for a given strength, within general limits imposed by the cement content and workability, which in turn are dictated by general experiences with regard to design requirements and to weathering. Strength has been found to be a fair index of the most desirable properties of concrete; impermeability or watertightness is the next most important criterion since, other factors being the same, the greater the impermeability of a concrete, the greater its durability. The principal methods of proportioning in use at the present time are

1. By arbitrary proportions of cement, sand, and coarse aggregate, without any preliminary tests.

The proportions of a concrete mix are stated either by weight or by volume, e.g., a 1:2:3½ mix means 1 part cement, 2 parts sand, and 3½ parts coarse aggregate. This method is now used only on relatively small jobs or where the character and gradation of the materials vary within rather narrow limits.

2. By using a fixed cement content per cubic yard of concrete, a specified grading of combined aggregates, and a specified consistency of fresh concrete. The specifications are the result of experience with concrete for a particular class of work.

3. By trial mixes, using a specified strength (or water-cement ratio) and specified consistency.

This method is the most flexible, and maximum economy may be derived by its use, although it requires facilities for testing and some experience with concreting methods. Briefly, the trial-mix method of proportioning involves making a series of trial batches and tests from which a water-cement ratio vs. strength relation, and the relation between proportions and consistency, is established for the given materials. From these data, the quantities for the desired mix may be computed. Water-cement ratio strength diagrams for both portland-cement concrete and high-early-strength concrete are given in Fig. E.1, and recommended consistency, cement content, and aggregate size for various classes of work are given in Table E.5.

4. By the ACI calculation method of proportioning based on data tabulated from observation of a large number of trial mixes.

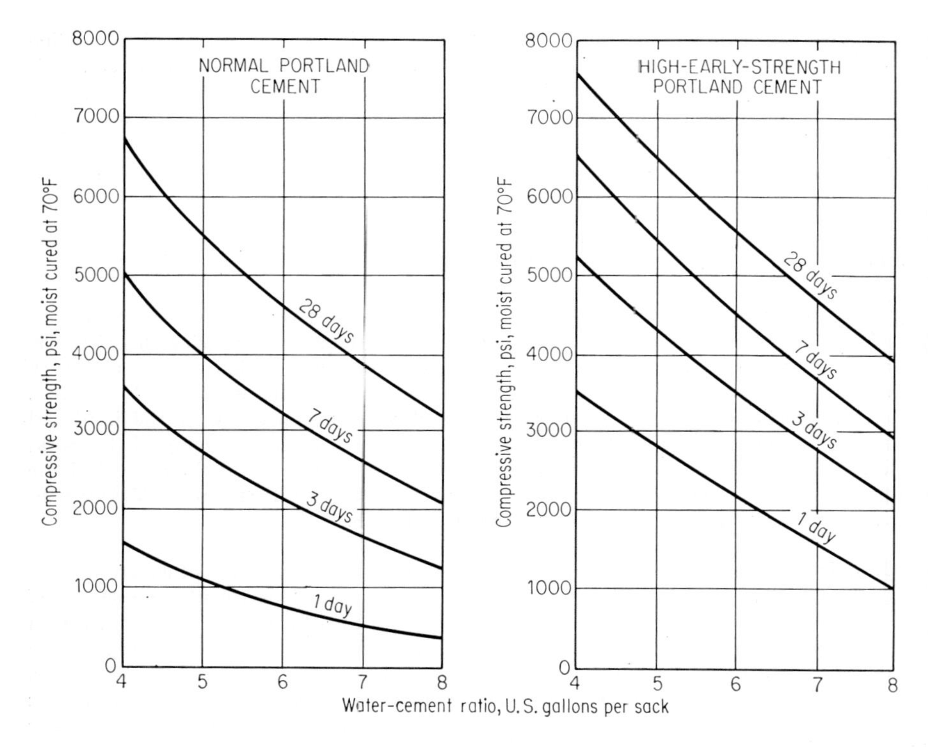

Fig. E.1. **Water-cement ratio vs. compressive strength at various ages for normal (Type I) and high-early-strength (Type III) portland cements.** (*From Portland Cement Assn., Ref.* 1342.)

The calculation of batch quantities, yield factors, and similar quantities when using methods 2, 3, or 4 is best done by a consideration of the solid volumes. The solid volume of an aggregate, for example, is the sum of the spaces occupied by all the individual particles in a given quantity. The difference between the gross or bulk volume of a given weight of aggregate and the solid volume is the void space. The solid volume may be computed by the following equation:

$$\text{Solid volume, cu ft} = \frac{\text{weight, lb}}{\text{specific gravity} \times 62.4}$$

In the following illustrative example, the concrete produced by a one-sack batch and the cement requirement per cubic yard of concrete are computed.

Example of Concrete-mix Calculations:

Given: Proportions, 1:2.5:3.5 by weight. W/C = 6 gal per sack.

To find: Weights of materials for one-sack batch, yield, and cement factor.

Solution: As 1 cu ft = 7.48 gal, 1 gal water = 8.33 lb and 1 sack cement = 94 lb, then

$$\text{W/C of 6.00 gal per sack} = \frac{6.00}{7.48} \text{ or 0.80 by volume}$$

$$= \frac{6.00 \times 8.33}{94} \text{ or 0.53 by weight}$$

	Water	Cement	Sand	Gravel
Proportions, by weight	0.53	1.00	2.50	3.50
Weights for one-sack batch, lb	0.53 × 94 = 50	94	2.50 × 94 = 235	3.50 × 94 = 329
Specific gravity (assumed)	1.00	3.10	2.65	2.65
Solid volume for one-sack batch, cu ft $\left(\text{solid volume} = \frac{\text{weight}}{62.4 \times \text{sp gr}}\right)$	0.80	0.49	1.42	1.99

Total solid volume = 4.70 cu ft
Yield, assuming 1 percent air voids = 4.70 × 1.01 = 4.75 cu ft

$$\text{Cement factor} = \frac{27.00}{4.75} = 5.68 \text{ sacks per cubic yard}$$

At the present time, on almost all work of any importance, the aggregates and cement are in bulk and are batched by weight. On small jobs the cement is supplied in 94-lb sacks, and the batches are such as to use integral multiples of one sack. The quantity of water should be, and usually is, carefully controlled. Correction is made at the mixer for the amount of free water contained in wet aggregates or absorbed by dry aggregates. On the small job, if batching is done by volume, the bulking of the aggregates, due to moisture, must be taken into account.

E.6. Mixing, Placing, and Curing. Machine mixing is always done in batch mixers, usually of the revolving-drum type. Concrete is often batched at a central plant and transit mixed while en route to the job in special truck mixers. Although mixers have now been developed for the satisfactory handling of small batches, hand mixing is often resorted to in the laboratory.

Methods of transporting concrete from the mixer to the forms vary to suit the job. Wheelbarrows, buggies, buckets, and pumps are used. The precaution to be observed is the prevention of segregation of the component materials of the mix.

The compaction of concrete in place is commonly accomplished by the use of power-driven vibrators. Superior results can be obtained with vibratory tamping over hand tamping if the mix has been designed for this purpose and proper technique in the use of the vibratory equipment is observed.

The last step, and an exceedingly important one in the manufacture of concrete, is the curing. It has been pointed out that hydration of the cement will take place only in the presence of moisture and at favorable temperatures. Depending upon the type of construction, moist conditions may be maintained by retention of forms, sprinkling, ponding, paper covers, moist coverings such as wet burlap and straw, or by use of a sealing compound such as an asphalt or wax coating. These curing methods retain in the concrete mass the free water, an excess of which, over and above that required for just complete hydration, is required for plasticity of the fresh concrete. The effect of curing conditions on the compressive strength of concrete is shown in Fig. E.2.

The optimum temperature range for the curing of ordinary concretes appears to be between 70 and 100°F. Concrete is usually seriously damaged by temperatures below freezing, the hardening process is considerably slowed by temperatures below 50°F, and temperatures much in excess of 120°F have sometimes resulted in concretes whose strength retrogressed at later ages.

PROPERTIES OF CONCRETE

E.7. Consistency. Consistency, a property of the fresh concrete, is an important consideration in the securing of a workable concrete that can be properly com-

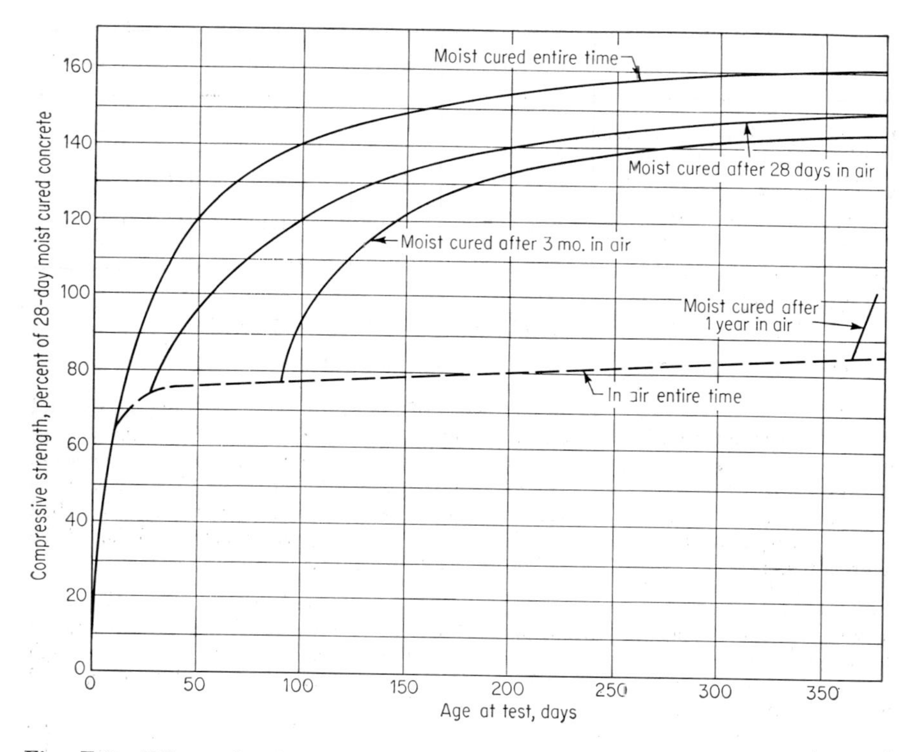

***Fig. E.2.* Effect of curing conditions on the compressive strength of concrete. Specimens were smaller than most concrete members, so resaturation was possible.** (*From Portland Cement Assn., Ref.* 1342.)

pacted in the forms. Workability is a relative term referring to the comparative ease with which concrete can be placed on a given type of work, e.g., concrete considered workable for pavement construction may be entirely unworkable for a thin wall or narrow beam. The term *consistency* relates to the state of fluidity of the mix and embraces the range of fluidity from the driest to the wettest mixtures. Three common tests to determine consistency are the "slump" test, the "ball penetration" test, and the "flow" test. These tests give only rough measures of this property, particularly the first two, but they do give satisfactory measures of consistency for most practical work with this material.

The slump test is made by measuring the subsidence, in inches, of a pile of concrete 12 in. high, formed in a mold that has the shape of the frustum of a cone (ASTM C 143).

The ball penetration test is made by measuring the settlement of a 6-in. steel ball (weighing 30 lb with its handle) into the surface of the concrete (ASTM C 360). Tests have shown that approximately two times this settlement equals the slump for the same concrete.

The flow test is made by jogging a specified pile of concrete formed in a truncated cone on a metal table which is raised and dropped ½ in. 15 times in about 15 sec and noting the spread of the pile as a percentage of the original formed 10-in. diameter (ASTM C 124).

For convenience various degrees of wetness of a mix may be roughly classified as dry, stiff, medium, wet, and sloppy. A concrete is said to be of medium or plastic consistency when it is just wet enough to flow sluggishly—not so dry that it crumbles or so wet that water or paste runs from the mass. The range in values of slump and flow corresponding to these arbitrary degrees of consistency is indicated in Table E.4. The values for each class are not to be taken as absolute. The ranges in flow corresponding to the ranges in slump are approximate only, corresponding roughly to what can be obtained with fairly well-graded sand-gravel aggregates and moderate cement contents; the relation between slump and flow varies with the mobility of the mix. Consistencies employed on various types of concrete construction are shown in Table E.5.

The principal factors affecting consistency are

1. Relative proportions of cement to aggregate. For given cement-water pastes the more aggregate that is crowded into the paste, the stiffer is the resulting concrete.
2. Water content of mix. For fixed proportions of cement and aggregate, the more water, the more fluid the resulting mix tends to be. There is a limit to the amount of water that a harsh mix will hold without serious segregation of the aggregate from the cement.
3. The gradation of the aggregate. For a fixed water-cement ratio, and a fixed aggregate-cement ratio, the finer the gradation, the stiffer the mix. In a given volume of aggregate, the more finely divided the particles are, the greater the surface area of the particles; hence the more paste required to coat them to produce a given consistency.
4. The shape and surface characteristics of the aggregate particles. Angular particles or those with rough surfaces require a greater amount of paste for the same mobility of mass than is necessary for smooth well-rounded particles.
5. The fineness and type of cement and the kind and amount of admixture. These factors may affect the fluidity of the paste and thus the consistency of the concrete.

***Table* E.4. Approximate range in slump and flow of concrete for various degrees of consistency**

Consistency	Slump,* in.	Flow, percent	Remarks
Dry	0–1	0–20	Crumbles and falls apart under ordinary handling; can be compacted into rigid mass under vigorous ramming, heavy pressure, or vibration, but unless care is used exhibits voids or honeycomb
Stiff	½–2½	15–60	Tends to stand as a pile; holds together fairly well but crumbles if chuted; with care and effort can be tamped into solid dense mass; satisfactory for vibratory compaction
Medium	2–5½	50–100	Alternate terms: plastic, mushy, quaking. Easily molded, although some care required to secure complete compaction
Wet	5–8	90–120	Pile flattens readily when dumped; can be poured into place
Sloppy	7–10	110–150	Grout or mortar tends to run out of pile, leaving coarser material behind

* Ball penetration is approximately half the slump.

***Table* E.5. Consistency, cement content, and aggregate size employed on various types of concrete construction**

Type of construction	Typical structures	Consistency	Cement factor, sacks per cu yd	Maximum size aggregate, in.
Massive	Dams, heavy piers, large open foundations	Stiff	2½–5	3–6
Semimassive	Piers, heavy walls, foundations, heavy arches, and girders	Stiff; medium	4–6	2–4
Pavement	Road surfaces, heavy slabs, moderately heavy footings	Stiff; medium	4½–6	1½–2½
Heavy building	Large structural members, small piers, medium footings. Wide to moderately wide spacing of reinforcement	Medium wet	5–7	1–2
Light	Small structural members, thin slabs, small columns, heavily reinforced sections, closely spaced reinforcement	Wet	5½–7	⅜–1

E.8. Bleeding. The tendency for water to rise to the surface of freshly placed concrete is known as "water gain," or "bleeding." It results from the inability of the constituent materials to hold all the mixing water. As a result of water gain the top portion of a lift becomes overly wet, and water films accumulate under the particles of coarse aggregate and under horizontal reinforcement bars. Concrete subject to water gain is not so strong, durable, or impervious as properly designed concrete. Water gain can be at least partially controlled by making a workable concrete mix with a minimum amount of water, higher cement content, and natural sands having an adequate percentage of fines.

E.9. Strength. The strength of concrete is taken as an important index of its general quality. Strength tests are relatively simple to make, and since strength is a first requisite for the structural designer, this property is one that is most frequently determined. Tests are commonly made in compression and flexure and occasionally in tension. The compression test of a 6 by 12-in. cylinder at age 28 days, after moist storage at a temperature of 73°F, is a "standard" test (ASTM C 31, C 39).

The compressive strength of concrete, made and tested under standard conditions, may ordinarily vary from 1500 to 4000 psi. No difficulty should be encountered, under present-day conditions, in securing strengths in excess of 2500 psi at 28 days.

The tensile strength of concrete is roughly 10 percent of the compressive strength, and the flexural strength of plain concrete, as measured by the modulus of rupture, is about 15 to 20 percent of the compressive strength.

The principal factors affecting strength are

1. Water-cement ratio. It has been shown by the results of various investigations that the water-cement ratio may be taken as the most important factor in controlling strength. The effect of the water-cement ratio upon strength of *workable* mixtures, for standard curing conditions, is shown in Fig. E.1, in which the water-cement ratio is expressed in gallons of water per sack of cement. In practice, the quality of concrete

is usually rated by its compressive strength at the age of 28 days, the relation of water-cement ratio to strength ordinarily being established by trial for each particular set of job conditions.

2. Age. The strength of moist concrete generally increases with age. For normal cements, stored under standard conditions, an empirical relation that has been used is

$$s_{28} = s_7 + 30\sqrt{s_7}$$

in which s_{28} = 28-day strength, s_7 = 7-day strength in pounds per square inch.

An illustration of the compressive strength at various ages is shown in Fig. E.1.

3. Character of the cement. Both fineness of grinding and chemical composition of the cement affect the strength of concrete, particularly at early ages. Fine cement and high proportions of tricalcium silicate (C_3S) and tricalcium aluminate (C_3A) promote high strength at early ages. The two groups of curves of Fig. E.1 show the influence of character of the cement upon strength.

4. Curing conditions—moisture and temperatures. The greater the period of moist storage (as shown in Fig. E.2), and the higher the temperature, for a temperature range of 40 to 100°F, the greater the strength at any age.

5. Moisture content of concrete at time of test. The higher the moisture content, the lower the strength.

6. Richness of mix and character of the aggregate. These factors largely affect strength through their influence upon the water-cement ratio required to produce the desired consistency.

E.10. Modulus of Elasticity. The stress-strain diagram for concrete, as determined from the ordinary compression test, is a curved line. The slower the rate of loading, the sharper the curvature due to plastic deformation which takes place with time. The modulus of elasticity may therefore be expected to exhibit considerable variability. Furthermore, concrete is not strictly speaking truly elastic, so that successive loadings may be expected to result in different values of the observed modulus.

The "secant" modulus of elasticity is determined from the slope of a straight line drawn through the origin of the stress-strain curve and some point on the curve, say, at 800 psi or at some percentage of the strength, as shown in Fig. 2-6. This secant modulus is the one most commonly used when no qualification is given. The slope of a line tangent to the stress-strain diagram at the origin is sometimes used; this is called the "initial-tangent" modulus. The secant modulus for ordinary concrete ranges from about 2,000,000 to 5,000,000 psi although values as low as 1,000,000 and as high as 6,000,000 psi are not infrequent.

The factors that influence the strength of concrete similarly influence the modulus of elasticity. The following observations indicate the general influence of certain variables upon the modulus of elasticity: (1) The stronger the concrete, the higher the modulus. (2) The modulus increases with age, sometimes to a very marked degree. (3) The drier the concrete at time of test, the lower the modulus. Wet concrete is stiffer although often weaker. (4) For the same consistency and cement factor, the larger the maximum size of aggregate, the coarser the grading and the higher the modulus.

For use in reinforced concrete design, the ACI Building Code (Ref. 8) assumes that the modulus of elasticity of concrete is given closely enough by the relation

$$E_c = w^{1.5}\, 33\sqrt{f'_c}$$

where E_c = modulus of elasticity, psi
f'_c = compressive strength of concrete, psi
w = weight of concrete, lb per cu ft, for range of 90 to 155 lb per cu ft

E.11. Durability and Watertightness. A factor that has been too little considered in the design of concrete mixtures is durability. The action of weather in the deterioration of concrete structures is due in part to the cyclic expansion and contraction under changing moisture and temperature conditions, in part to the expansive force of ice crystals as they are formed in the pores of the concrete, and in part to the leaching of soluble compounds from the mass by water. It follows, therefore, that a relatively watertight concrete is a more weather-resistant concrete.

Although the principal factors that affect strength also affect watertightness, important variables are richness of mix, gradation of aggregate (particularly in the fine particles), compaction of the concrete in place, and curing.

The tests for watertightness and durability do not yield so definite results as do strength tests, but they serve to compare different concretes on the same basis. The permeability test is made by subjecting one surface of a slab or block of concrete to a head of water and measuring the amount that passes into or through the concrete. An accelerated durability test is made by subjecting concrete test specimens to cycles of freezing, thawing, drying, and wetting and then noting the extent of deterioration by (1) testing the specimen in compression, and comparing the results with those from companion specimens that have been stored under standard conditions, (2) measuring the progressive expansion of the concrete resulting from its disintegration, or (3) noting reductions in the sonic modulus of the concrete (see Art. 10.30).

Many tests have shown the marked improvement in the resistance of concrete to freezing and thawing when it contains about 4 percent of air entrained in the form of minute globules. These air cells serve to relieve the internal pressures exerted by the ice, which tends to form in the pores of the concrete, and thereby prevent deterioration of the concrete. Because the reduction in compressive strength may become appreciable when the air content exceeds about 4 percent, success in the use of entrained air requires careful measurement of the ingredients and control of the concrete-making process to regulate the amount of entrained air.

Various methods for measurement of the volume of entrained air have been developed, and three have been standardized by the ASTM. Two of the latter methods involve a comparison of the volume of a batch of air-entrained concrete with the solid volume of the same batch as computed from the summation of the solid volumes of the several components of the mix (ASTM C 138) or as determined by displacement of the mix under water (ASTM C 173). Both these methods are subject to appreciable errors unless controlled very carefully.

A third method (ASTM C 231) has been developed in which the reduction in volume of the air voids in a sample of the fresh concrete (and hence a reduction in volume of the concrete) is effected by application of water pressure to the sample, the quantity of air being determined from a consideration of Boyle's law. In general, this pressure method gives percentages of air about one higher than the others and is the method most commonly used.

E.12. Other Properties. Other properties that have not been considered but which may be of importance under special conditions are volume changes due to variation in moisture content, creep under sustained stress, wear, density, extensibility, fatigue, and thermal properties. These have been dealt with extensively in the technical literature, although the results of tests have not been adequately correlated. Active research is constantly adding to the knowledge and understanding of the properties of concrete.

F. PROPERTIES OF BRICK

Bricks constitute an old and important class of building material. Ease and relatively low cost of manufacture, durability of the properly made product, and moderate strength, all have contributed to their wide usage. Bricks are used not only for building construction but also to a limited extent for other kinds of construction.

The use of ordinary brick in structures has been limited to bearing walls, curtain walls, low piers, and short-span arches, chiefly because of the relatively low strength of the mortar and the inability of the masonry to resist tensile forces. Improvements in the quality of masonry mortars and recent developments in reinforced-brick masonry permit more efficient designing and greatly extend the scope of use of this material.

F.1. Definition. A brick is generally a solid prismatic unit about 2½ by 4 by 8 in. The unqualified term *brick* is used to designate a ceramic product. Other materials, such as cement and lime mortars and adobe, are formed in brick sizes and used without subsequent burning. In this text the discussion will be limited to clay building brick.

PROPERTIES AND TESTS

F.2. General Requirements. Good bricks for structural purposes should conform to the following requirements under visual inspection: they should be of compact, nonlaminated structure; reasonably uniform in shape; free from cracks, chipped corners or edges, warpage, kiln marks, large pebbles, balls of clay and particles of free lime; and not soft as the result of underburning.

A rough test for hardness and freedom from cracks is made by striking a brick with a hammer while held by hand, and noting the sound emitted: a high-pitched metallic sound indicates a hard crack-free unit, whereas a characteristic dull, quickly damped sound is produced by an underburned or cracked brick.

Evidence relative to the degree of burning may also be obtained by scratching the surface of a brick with a knife; a well-burned product can be scratched only with difficulty. A rough indication of relative toughness may be obtained by breaking the brick with a hammer. Color of the surface and of the core may be an index to the relative quality of bricks from the same plant.

F.3. Compressive Strength. The compression test for clay building brick is made by testing a half brick flatwise. The brick should be thoroughly dry, and the bearing surfaces should be coated with shellac before capping with plaster of paris to prevent the absorption of moisture that would decrease the strength.

From a series of tests on building bricks from many parts of the United States, reported by McBurney and Lovewell [1351], the weighted average compressive strength of brick tested was in excess of 7000 psi; the weighted average of all hard samples was about 7400 psi and of all "salmon" (soft) brick, about 4100 psi. Other sources indicate a range in compressive strength from about 1500 to over 20,000 psi for bricks of various types and grades. Lent estimates that over 75 percent of all bricks produced in the United States have an average compressive strength in excess of 3000 psi [1350]. Recent ASTM (minimum) Specification requirements are shown in Table F.1.

The compressive strength, as well as other properties, is considerably improved by repressing the brick before burning. The repressing operation produces a denser unit in which the component parts are bonded together better.

Table* F.1. Specification requirements for clay building brick

Grade†	Min compressive strength, tested flatwise, psi (average of 5 bricks)	Max water absorption, 5-hr boiling test, percent (average of 5 bricks)	Max saturation coefficient‡ (average of 5 bricks)
S.W.	3000	17	0.78
M.W.	2500	22	0.88
N.W.	1500	(No limit)	(No limit)

* Based on ASTM C 62.

† S.W. = severe weather, M.W. = moderate weather, N.W. = normal weather, for interior masonry only.

‡ Ratio of absorption by 24-hr cold-water submersion test to absorption by 5-hr boiling test.

F.4. Flexural Strength. The flexural strength of brick is taken as the modulus of rupture of the brick tested flatwise on a simple span under central load. Although bricks that have a high flexural strength also usually have a high compressive strength, the relationship is indefinite.

The ASTM specifies that the test be made on a whole brick on knife-edge rocker supports over a 7-in. span; the central load is applied through a steel bearing strip,

$\frac{1}{4}$ by $1\frac{1}{2}$ in. in section (ASTM C 67). Various tests show that the modulus of rupture may vary from about 200 to 3000 psi. The flexure test is no longer required for building brick.

F.5. Absorption. Water absorption of brick is often taken as a measure of porosity, which in turn is considered to be indicative of (1) possible leakage through the brick and (2) tendency toward disintegration when the moist bricks are subject to alternate freezing and thawing. In general, a porous brick cannot be expected to be so resistant to the action of loads or to be so durable under the action of weather or other exposure as a more dense brick.

These observations appear to be sound if applied to one make of brick, but it is a fact that the soft- or medium-burned brick having relatively high absorption from one plant may outperform the hard-burned brick having relatively low absorption from some other plant, owing to inherent differences in the materials and methods of manufacture.

Variations in absorption characteristics are caused by differences in the clay used, the method of forming the brick, and the burning conditions. Clays used cover a wide range of compositions, varying in absorptive properties. The moisture content at time of forming and the type of molding machine affect the density and hence the absorption of the product. Repressing may be expected to lower the absorption, since a more dense, compact structure is produced. A high burning temperature tends to produce, at least at the surface, a dense, relatively impervious shell resulting in low absorptive characteristics.

Various test procedures have been used to determine the absorption of brick. One method standardized by the ASTM is as follows: five half bricks are dried to constant weight at 230 to 239°F. When cool, each brick is weighed to nearest 0.5 g and immersed in soft water at 60 to 86°F for 24 hr. The surfaces are then wiped with a damp cloth and the bricks quickly weighed. The absorption is calculated on the basis of the oven-dry weight (ASTM C 67).

By another method of test the procedure is similar to that above, except that after immersing the bricks for 24 hr the water is heated to boiling in 1 hr and then boiled continuously for 5 hr, after which it is allowed to cool to 60 to 86°F within 16 to 18 hr. The surfaces are then wiped and the bricks weighed as above (ASTM C 67). This method produces a higher absorption as the bricks are left immersed for a longer period of time and there is a tendency, because of the boiling, to eliminate air trapped in the voids, thus permitting deeper penetration of water into the brick.

For a given make of brick, relationships appear to exist between the results obtained by the different methods of determining absorption. Since the rates of absorption differ for different kinds of brick, such relations are in general only approximate. Absorption tests made for several periods of immersion at normal temperatures show that common bricks absorb about half as much in the first minute as in a 48-hr period.

The absorption of brick varies between 1 and 25 percent, although it will usually be less than 20 percent for good common building brick. The maximum allowable values of absorption of various classes of building brick are indicated in Table F.1.

The saturation coefficient is taken as the ratio of absorption after 24-hr immersion in cold water to that after an additional 5-hr boiling test. The ASTM Specification requirements for clay building brick are shown in Table F.1.

F.6. Durability. The resistance of brick to large changes in moisture and temperature conditions is evaluated by a freezing-and-thawing test during which the bricks are subjected to many cycles of freezing and thawing while saturated and to several cycles of wetting and drying. Two slightly different procedures are specified in the ASTM method (ASTM C 67). The loss in weight is used in rating their resistance.

Some engineers are of the opinion that the durability of a brick is a direct function of its absorption. Although this may be true in some cases, it should not be accepted as a generality.

BRICK MASONRY

F.7. Factors Affecting Behavior. The following factors affect the behavior of brick masonry as regards strength, stability, and durability: (1) workmanship, (2) type of construction, (3) regularity in form and size of brick, (4) physical properties of the brick, (5) properties of the mortar, and (6) relative behavior of the brick and mortar. Although quality of the brick has to do with the behavior of the masonry, it does not, in all instances, control.

Since the strength of masonry is less than that of well-made brick, the compressive strength of the individual brick is seldom a significant factor. This property serves principally as one convenient way of classifying brick of a given type.

The most important factors leading to the failure of brick masonry under compressive loads are (1) incompleteness of the mortar bedment and (2) failure and subsequent lateral flow of the mortar, both of which cause flexural stresses in the brick. The stronger the mortar, the less the tendency for the development of excessive flexural stresses in the brick; with a given mortar the stronger the brick in flexure, the greater the strength of the masonry. The transverse strength of brick may then be considered as being of considerable significance.

Tests of brick piers show that transverse failure of the individual brick commonly begins when about 40 to 60 percent of the maximum load has been applied. Piers usually fail by splitting into a number of slender columns. Piers with the bricks laid edgewise show greater strengths under test than those with bricks laid flatwise. This would indicate that thicker brick than ordinary would be desirable for heavily loaded structures.

G. PROPERTIES OF PLASTICS

G.1. Types and General Characteristics. The term *plastics* is generally used to designate synthetic resins which constitute the most important type of organic plastics. Also included in this classification of materials are cellulose derivatives, protein materials, and natural resins. Less frequently such materials as rubber and glass are called plastics.

The synthetic-resin plastics have received major attention by the chemist. Where modified by the addition of suitable fillers, these plastics are readily molded into lightweight products of sufficient strength and durability to maintain dimensional stability under moderately deteriorating conditions of heat, moisture, and sunlight. The simplicity of manufacture makes possible large-scale production of accurately sized parts having distinctive or imitative surface effects. Furthermore, since they are uniformly colored throughout, plastics do not have the shortcomings of surface-colored materials.

Synthetic-resin plastics are commonly known under familiar trade names such as Bakelite, Catalin, Beetle, Lucite, Lustron, and Vinylite. Their engineering uses have included many applications in the electrical, aeronautical, and automotive industries and in various other fields.

Celluloid is a typical cellulose derivative. These plastics are easily formed into tough thin sheets having excellent flexibility.

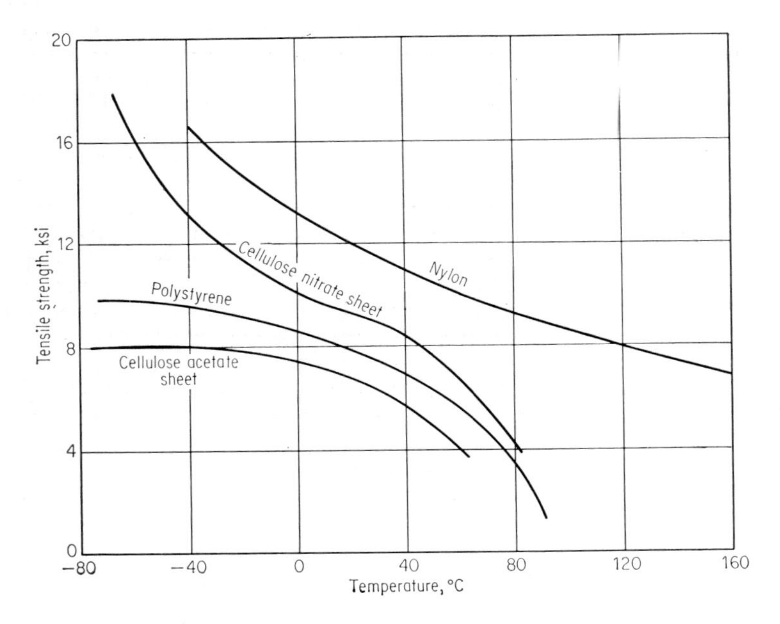

Fig. G.1. **Effect of temperature on tensile strength of thermoplastics.** (*From ASTM Symposium on Plastics, Ref.* 1364.)

Casein, from skimmed milk, and soybean plastics are typical of the protein group. Their principal defect, a low resistance to warping, is caused by their relatively high water absorption.

The natural resins, although made from shellac, rosin, asphalt, or pitch, are marketed under their common or under various proprietary names. A typical filler for these materials is asbestos, but irrespective of the filler the resulting plastics are not so strong and dimensionally stable as are the synthetic compounds.

Plastics harden in three different ways which give rise to the qualifying adjectives chemosetting, thermosetting, and thermoplastic. The first two types are plastics of permanent hardness which, once formed, are unaffected by ordinary solvents and do not soften upon heating. As the name implies, chemosetting plastics are the result of chemical reactions between the constituent compounds at room temperatures, whereas the chemical reaction in the thermosetting plastics requires heat. The thermoplastic compounds produce plastics of only temporary hardness at normal room temperatures. These plastics again become soft upon being heated. The hardening cycle can be repeated many times. Thermoplastics when in the cool, hardened state are soluble in certain solvents.

Some plastics contain fillers such as wood flour or macerated fabrics which provide properties not possessed by the binder. These fillers may also be used to lower the cost of the plastic. Plasticizers (usually an oily liquid) may also be used to improve some properties of plastics.

G.2. Manufacturing Methods. Plastic products are either cast, molded, or laminated. Molding pressures may be several tons per square inch, and temperatures of molding may be as high as 450°F. The laminated construction results from pressing together heated layers of glass fibers, cloth, or paper that have been saturated with plastic material. When hardened, this material is dense and tough with excellent mechanical and dielectric properties. It can be easily machined into such products as gears and low-pressure friction bearings.

G.3. Properties. The range of the properties of plastics is very considerable. Depending on their compositions and reinforcing materials, they can be extremely weak or very strong. Fillers and plasticizers greatly change the properties of plastics. They can modify brittle plastics, like phenolic resin, which are difficult to mold and are rather costly, so that they are more moldable and less expensive to use. Laminates and reinforced plastics made of plastics combined with sheet or fibrous materials have properties superior to those of either material alone. The laminates make use of paper, fabric, wood, and other sheet materials, whereas the reinforced plastics usually incorporate glass fibers in the form of mats or woven fabrics.

One of the first engineering uses for large amounts of plastics was for electric insulation. Their other characteristics of good resistance to acid corrosion and mechanical abrasion, together with high strength-weight ratios, greatly widened their field of use.

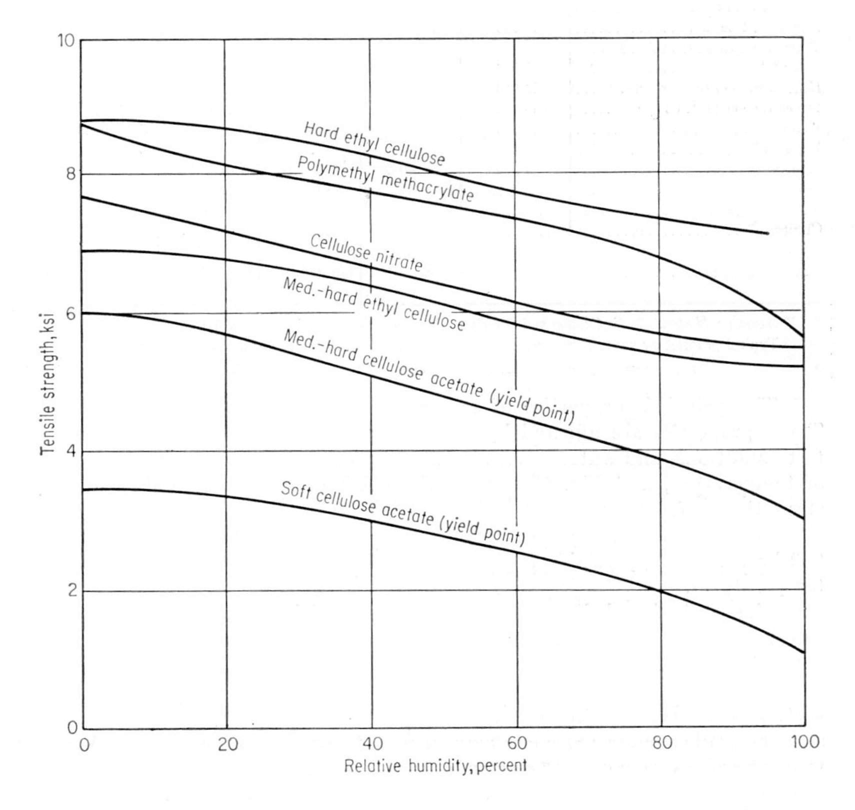

Fig. G.2. **Effect of relative humidity on tensile strength of thermoplastics.** (*From ASTM Symposium on Plastics, Ref.* 1364.)

Table **G.1. Properties**

Properties	ASTM test method	Phenol formaldehyde compounds: Molded (Bakelite, Durez, Durite)†: No filler	Wood-flour filler	Macer-ated-fabric filler	Cast, no filler (Catalin, Mar-blette)†	Urea formal-dehyde molding com-pound, alpha cellulose filler (Beetle, Plaskon)†
Specific gravity	D 792	1.25–1.30	1.32–1.55	1.36–1.43	1.30–1.32	1.47–1.52
Tensile strength, ksi	D 638, D 651	7–8	6.5–8.5	3.3–9	6–9	7–13
Elongation, %	D 638	1.0–1.5	0.4–0.8	0.37–0.57	1.5–2.0	0.5–1.0
Modulus of elasticity, 10^5 psi	D 638	7.5–10	8–12	9–13	4–5	15
Compressive strength, ksi	D 695	10–30	23–36	15–30	12–15	25–35
Flexural strength, ksi	D 790	12–15	8.5–12	8.5–15	11–17	10–16
Impact strength, ft-lb per in. of notch, Izod test	D 256	0.20–0.36	0.24–0.60	0.75–8.0	0.25–0.40	0.25–0.35
Hardness, Rockwell	D 785	M124–M128	M100–M120	M95–M120	M93–M120	M115–M120
Thermal expansion, in./in. $\times 10^{-5}$ per °C	D 696	2.5–6.0	3.0–4.5	1–4	6–8	2.7
Heat distortion temperature, °F	D 648	240–260	260–340	250–300	165–175	270–280
Dielectric constant, 60 cycles	D 150	5–6.5	5–9	6.5–15	6.5–7.5	7.0–9.5
Dissipation factor, 60 cycles	D 150	0.06–0.10	0.05–0.30	0.08–0.45	0.10–0.15	0.04
Water absorption, 24 hr, ⅛-in. thickness, %	D 570	0.1–0.2	0.3–1.0	0.04–1.75	0.3–0.4	0.4–0.8
Molding qualities	....	Fair	Excellent	Good to fair	....	Excellent
Machining qualities	....	Fair to good	Fair to good	Fair to good	Excellent	Fair
Clarity	....	Transparent, translucent	Opaque	Opaque	Transparent to opaque	Translucent to opaque
Color possibilities	....	Dark	Limited	Limited	Unlimited	Unlimited

* Based on *Modern Plastic Catalog and Directory*, Breskin Publishing Corp., New York.
† Typical proprietary names. Properties shown are for the general group, not any particular plastic.

The principal properties of several common plastics are given in Table G.1. These properties are affected by compositions, molding temperature and pressure, test conditions, and water content of absorptive plastics. Figure G.1 shows the effect of temperature, and Fig. G.2 shows the effect of relative humidity on the tensile strength of a few thermoplastics.

Certain plastics, such as phenolic resins, have been heated as high as 475°F for 600 hr without appreciable decrease in tensile or flexural strengths at normal room temperatures.

The method and direction of molding should be known when interpreting test results because of the marked effect of direction of molding pressure on the mechanical properties.

G.4. Plastic-type Adhesives. Structural adhesives capable of withstanding high stresses are becoming more common in many industrial applications, even replacing welding and riveting of light-gage metallic components. Some are suitable for bonding wood or concrete members and even for bonding rubber to metal. The various organic synthetic resins in common use have a wide range of mechanical properties depending upon the composition of the base material and also upon the additives which may

of plastics*

Melamine formaldehyde molding compound, alpha-cellulose filler (Melmac, Plaskon)†	Cellulosic molding compounds		Vinyl chloride-acetate molding compound, rigid (Vinylite)†	Polyester molding compound, glass-fiber filler (Plaskon)†	Methyl methacrylate resin molding compound (Lucite, Plexiglas)†	Polystyrene molding compound (Bakelite, Catalin)†	Epoxy resin reinforced with glass cloth
	Cellulose acetate (Fibestos, Lumarith)†	Cellulose nitrate (Celluloid, Nitron, Pyralin)†					
1.47–1.52	1.27–1.34	1.35–1.40	1.34–1.45	1.90	1.18–1.19	1.05–1.06	1.8
7–13	1.9–8.5	7–8	7.2–9	5–9	7.5–10	5–9	50
0.6–0.9	6–50	40–45	....	....	2–10	1.0–3.6	
13	0.86–4.0	1.9–22	3.5–4.1	16–20	4.5	4–6	250
25–43	13–36	22–35	9.9–11.3	20–25	12–18	11.5–16	60
10–16	2–16	9–11	12.6–14.5	14–17	13–17	8–16	70
0.24–0.35	0.4–5.2	5.0–7.0	0.4–0.7	12–24	0.4–0.5	0.25–0.50	20
M110–M125	R50–R125	R95–R115	....	....	M85–M105	M65–M90	
4.0	8–16	8–12	6.9–18.5	1.0–3.0	9	6–8	9
400	115–235	140–160	140–155	400	160–195	160–210	250
7.9–9.5	3.5–7.5	7.0–7.5	3.2–3.3	5.2–6.0	3.5–4.5	2.4–2.6	
0.03–0.08	0.01–0.06	0.09–0.12	0.01	0.02	0.04–0.06	0.000	
0.1–0.6	1.7–6.5	1.0–2.0	0.07–0.08	0.15–0.20	0.3–0.4	0.03–0.05	
Excellent	Excellent	Good	Excellent	....	Excellent	Excellent	
Fair	Good	Excellent	Good	Fair	Excellent	Fair to good	
Translucent	Transparent to opaque	Transparent to opaque	Transparent to opaque	Opaque	Transparent	Transparent	
Unlimited	Unlimited	Unlimited	Unlimited	Limited	Unlimited	Unlimited	

be combined with it to improve certain characteristics. Their high strength results from the polymerization of their molecules to form long chains. They are commonly sold under trade names such as Cycle Weld, Scotch Weld, etc., but are essentially based on epoxy, phenolic, polyester, allyl, acrylic, polyvinyl, polystyrene, or similar resins used singly or in combination. Their strength characteristics are determined in accordance with ASTM Standards D 897, D 950, and D 1002.

H. SELECTED REFERENCES AND SOURCES OF INFORMATION

SEC. 0. GENERAL SPECIFICATIONS FOR MATERIALS AND FOR METHODS OF TESTING

1. American Society for Testing and Materials (abbr. ASTM), 1916 Race St., Philadelphia, Pa. *ASTM Standards*. Special compilations of selected standards are also issued, e.g., "Standards for Students of Engineering." All ASTM specifications carry an identifying designation, e.g., C 150–52. The initial letter indicates the general group to which the specification belongs:

   A: Ferrous metals
   B: Nonferrous metals
   C: Cementitious, ceramic, concrete and masonry materials
   D: Highway materials, asphaltic materials, petroleum products, and miscellaneous materials
   E: Test methods and miscellaneous subjects

   The numerals immediately following the letter are serial numbers giving the order in which the specification was adopted in a given letter group. The numerals following the dash indicate the year of adoption or of latest revision. If the letter T follows the numerals, it indicates that the specification is still in the "tentative" stage. For many purposes, only the first part of the designation, e.g., C 150, is necessary to identify the specification.

2. American Standards Association (abbr. ASA), 70 East 45th St., New York, N.Y. Promulgates many standards and specifications including materials specifications (see Art. 1.11, text). Many of its materials specifications are taken in their entirety from ASTM standards.
3. American Association of State Highway Officials (abbr. AASHO), 917 National Press Bldg., Washington 4, D.C. *Standard Specifications for Highway Materials and Methods of Sampling and Testing.* Issued at various intervals. Many of these specifications are identical with those of ASTM.
4. American Railway Engineering Association, Chicago, Ill., *AREA Manual.* Yearly supplements are issued. Of particular interest are the specifications for open-hearth steel rails and for the drop-test machine for steel rails.
5. Society of Automotive Engineers, New York, *Standard Specifications for Steels.* See Handbook of the Society.
6. American Petroleum Institute, New York, *API Specifications,* revised periodically. Of particular interest is the Specification for Wire Rope, API Std. 9A.
8. American Concrete Institute, Detroit, Mich. *Building Code Requirements for Reinforced Concrete,* 1963.
9. *Specification for the Design Fabrication and Erection of Structural Steel for Buildings,* Amer. Inst. Steel Construction, N.Y. 1961.

SEC. 100. GENERAL AND INTRODUCTORY REFERENCES

General Texts and Articles Covering a Number of Phases of Testing

101. Batson, R. G., and J. H. Hyde: *Mechanical Testing,* Vol. I: *Testing of Materials of Construction,* Dutton, New York (Chapman & Hall, London), 1922, 413 pp.
102. Michelon, L. C.: *Industrial Inspection Methods,* Harper & Row, New York, 1950.
103. Churchill, H. D.: *Physical Testing of Metals and Interpretation of Test Results,* American Society for Metals, Cleveland, Ohio, 1936, 109 pp.
104. Clark, D. S.: *Physical Testing of Metals,* American Society for Metals, Los Angeles Chapter, 1941, 84 pp. (mimeographed).
105. Cowdrey, I. H., and R. G. Adams: *Materials Testing—Theory and Practice,* 2d ed., Wiley, New York, 1935, 156 pp.
106. Foster, P. F.: *The Mechanical Testing of Metals and Alloys,* Pitman, London, 1936, 267 pp.
107. Gilkey, H. J., G. Murphy, and E. O. Bergman: *Materials Testing,* McGraw-Hill, New York, 1941, 185 pp.
108. Marin, J.: *Mechanical Behavior of Engineering Materials,* Prentice-Hall, Englewood Cliffs, N.J., 1962, 498 pp.
110. Liddicoat, R. T., and P. O. Potts: *Laboratory Manual of Materials Testing,* Macmillan, New York, 1952, 239 pp.
112. Muhlenbruch, C. W.: *Testing of Engineering Materials,* Van Nostrand, New York, 1944, 200 pp.
115. Strong, J., *et al.: Procedures in Experimental Physics,* Prentice-Hall, New York, 1939, 642 pp.
118. *The Testing of Materials, Natl. Bur. Standards Circ.* 45 (1913), 89 pp.
120. Tuckerman, L. B.: "Aircraft: Materials and Testing," *Proc. ASTM,* vol. 35, pt. II (1935), pp. 3–46.
122. Larson, L. J.: "Weld Metal as an Engineering Material and Some Methods of Testing," *Proc. ASTM,* vol. 37, pt. II (1937), pp. 22–34.
123. Pulsifer, H. B.: *Inspection of Metals,* American Society for Metals, Cleveland, Ohio, 1941, 180 pp.

131. Marin, J.: *Engineering Materials*, Prentice-Hall, New York, 1952, 491 pp.; also *Mechanical Properties of Materials and Design*, McGraw-Hill, New York, 1942, 273 pp.
132. Mills, A. P., H. W. Hayward, and L. F. Rader: *Materials of Construction*, 5th ed., Wiley, New York, 1939, 564 pp.
133. Moore, H. F., and M. B. Moore: *Textbook of the Materials of Engineering*, 8th ed., McGraw-Hill, New York, 1953, 363 pp.
134. Murphy, G.: *Properties of Engineering Materials*, International Textbook, Scranton, Pa., 1957, 464 pp.
135. Seely, F. B., and J. O. Smith: *Resistance of Materials*, Wiley, New York, 1956, 459 pp.
137. Swain, G. F.: *Structural Engineering—Strength of Materials*, McGraw-Hill, New York, 1924, 570 pp.
138. Timoshenko, S.: *Strength of Materials*, 2 vols., 3d ed., Van Nostrand, New York, 1955. See Part II, Chap. 10, "Mechanical Properties."
140. White, A. H.: *Engineering Materials*, 2d ed., McGraw-Hill, New York, 1948, 686 pp.
141. Withey, M. O., and G. W. Washa: *Johnson's Materials of Construction*, 9th ed., Wiley, New York, 1954.
142. Upton, G. B.: *Materials of Construction*, Wiley, New York, 1916, 327 pp.
143. *Metals Handbook–1961*, American Society for Metals, Cleveland, Ohio, 1961, 1236 pp.
144. *Welding Handbook*, 3d ed., American Welding Society, New York, 1950, 1579 pp.
145. Richards, C. W.: *Engineering Materials Science*, Wadsworth, San Francisco, Calif., 1961, 500 pp.
146. Seely, F. B., and J. O. Smith: *Advanced Mechanics of Materials*, Wiley, New York, 1952.
147. Shanley, F. R.: *Strength of Materials*, McGraw-Hill, New York, 1957, 783 pp.

Significance of Tests

160. Moore, H. F.: "The Significance of More Common Physical Properties of Structural Materials," *S.A.E. J.*, vol. 17, no. 2 (August, 1925), pp. 175–180.
163. Moore, H. F.: "Correlation between Metallography and Mechanical Testing," *Trans. AIMME*, vol. 120 (1936), pp. 13–35; *Univ. Ill. Eng. Expt. Sta. Reprint* 9 (1936).
164. Rosenhain, W.: "Some Principles of Testing," *Metallurgist* (supplement to *Engineer*), vol. 153 (Apr. 29, 1932), pp. 57–58; (May 27, 1932), pp. 71–72.
165. Tuckerman, L. B.: "From Material to Structure," *J. Wash. Acad. Sci.*, vol. 23, no. 5 (May 15, 1933), pp. 225–246.
166. Emley, W. E., and L. B. Tuckerman: "The Significance of Tests," *ASTM Bull.* 99 (August, 1939), pp. 11–12.
167. Hadfield, R., and S. A. Main: "Laboratory Tests in Relation to the Serviceability of Steel and Steel Products," *Proc. IATM Cong., London*, 1937, pp. 578–581.
168. Schuster, L. W.: "Mechanical Properties vs. Service Failures," *Iron Age*, vol. 141, no. 26 (June 30, 1938), pp. 30–33.
169. "Usefulness and Limitation of Various Tests as Applied to Sheet Metals," in Report of ASTM Committee E-1, *Proc. ASTM*, vol. 37, pt. I (1937), pp. 518–522.
170. *Metals—How They Behave in Service*, American Society for Metals, Cleveland, Ohio, 1939, 45 pp.
171. "Symposium on Significance of the Tension Test of Metals in Relation to Design," *Proc. ASTM*, vol. 40 (1940), pp. 501–609.

172. ASTM Committee C-9: *Report on Significance of Tests of Concrete and Concrete Aggregates,* 3d ed., American Society for Testing and Materials, 1955, 396 pp.
173. Barringer, L. E.: "Significance of Material Tests and of Design Tests upon Electrical Insulating Materials," *Proc. ASTM,* vol. 36, pt. II (1936), pp. 592–599.
175. Ashcroft, A. G.: "The Interpretation of Laboratory Tests," *ASTM Bull.* 131 (December, 1944), pp. 27–31.

Development, Use, and Enforcement of Specifications

180. Mead, D. W.: *Contracts, Specifications, and Engineering Relations,* 3d ed., McGraw-Hill, New York, 1956.
181. Tucker, J. I.: *Contracts in Engineering,* 4th ed., McGraw-Hill, New York, 1947, 341 pp.
182. Dudley, C. B.: "The Enforcement of Specifications," *Proc. ASTM,* vol. 7 (1907), pp. 19–38.
183. Warwick, C. L.: "The Work in the Field of Standardization of the American Society for Testing and Materials," *Ann. Am. Acad. Political and Social Sci.,* vol. 137 (May, 1928).
184. "The Economic Significance of Specifications" (a symposium), *Proc. ASTM,* vol. 31, pt. II (1931), pp. 955–995.
185. Ashkinazy, S. B.: "Materials Purchase Specifications," *ASTM Bull.* 143 (December, 1946), pp. 48–51. Good bibliography.
186. "Advantages of Standards in Purchasing," *ASTM Bull.* 173 (April, 1951), p. 21.

History of Testing

190. Girvin, H. F.: *A Historical Appraisal of Mechanics,* International Textbook, Scranton, Pa., 1948, 275 pp.
191. Pugsley, A. G.: "History of Structural Testing," *Structural Engr.,* vol. 22, no. 12 (December, 1944), pp. 492–505.

SEC. 200. MECHANICAL PROPERTIES OF MATERIALS AND GENERAL FEATURES OF MECHANICAL TESTING

General References

For general books on properties of materials, including mechanical properties, see Refs. 131 to 147.

Stress, Strain, Strength

201. Moore, H. F.: "Hooke's Law of Stress and Strain," *Proc. ASTM,* vol. 28, pt. I (1928), pp. 38–44.
202. Moore, H. F.: "Stress, Strain and Structural Damage," *Proc. ASTM,* vol. 39 (1939), pp. 549–570.
203. Murphy, G.: "Stress-strain-time Characteristics of Materials," *ASTM Bull.* 101 (December, 1939), pp. 19–22.
205. Ramberg, W., and W. R. Osgood: "Description of Stress-Strain Curves by Three Parameters," *NACA, TN* 902 (July, 1943).
211. Templin, R. L.: "The Determination and Significance of the Proportional Limit in the Testing of Metals," *Proc. ASTM,* vol. 29, pt. II (1929), pp. 523–553.

For "strain-difference" method of plotting, see especially discussion by Tuckerman, pp. 538–546.

212. Winlock, J., and R. W. E. Leiter: "Some Observations on the Yield Point of Low-carbon Steel," *Trans. ASME*, vol. 61, no. 7 (October, 1939), pp. 581–587.
213. "Yield Strengths Corresponding to Small Percentages of Set," in Report of ASTM Committee E-1, *Proc. ASTM*, vol. 37, pt. I (1937), pp. 523–527.
214. Donnell, L. H.: "Suggested New Definitions for Proportional Limit and Yield Point," *Mech. Eng.*, vol. 60, no. 11 (November, 1938), pp. 837–838.
215. "Stresses in Overstrained Materials," *Engineer*, vol. 140, no. 3635 (Sept. 13, 1935), pp. 291–292.
216. Templin, R. L.: "Some Factors Affecting Strain Measurements in Tests of Metals," *Proc. ASTM*, vol. 34, pt. II (1934), p. 182.
220. Templin, R. L., and R. G. Sturm: "Some Stress-strain Studies of Metals," *J. Aeronaut. Sci.*, vol. 7, no. 5 (March, 1940), pp. 189–198.
221. MacGregor, C. W.: "Relations between Stress and Reduction in Area for Tensile Tests of Metals," *Trans. AIMME*, vol. 124 (1937), pp. 208–226.
222. Nadai, A.: "Plastic Behavior of Metals in the Strain Hardening Range—I," *J. Appl. Phys.*, vol. 8, no. 3 (March, 1937), pp. 205–213.
223. Davis, E. A.: "Plastic Behavior of Metals in the Strain-hardening Range—II," *J. Appl. Phys.*, vol. 8, no. 3 (March, 1937), pp. 213–217.
224. MacGregor, C. W.: "A Two-load Method of Determining the Average True Stress-strain Curve in Tension," *J. Appl. Mechanics*, vol. 6, no. 4 (December, 1939), pp. A156–A158.
225. MacGregor, C. W.: "The True Stress-strain Tension Test—Its Role in Modern Materials Testing," *J. Franklin Inst.*, vol. 239, no. 2 (August, 1944) and no. 3 (September, 1944).
226. "Symposium on Significance of the Tension Test of Metals," *Proc. ASTM*, vol. 40 (1940), pp. 501–609.
230. Seely, F. B.: "Ideal and Practical (Test) Relations between Elasticity and Plasticity, Tenacity, and Brittleness," *Proc. IATM First Cong.*, Zurich, 1931, vol. 2.
231. Nádai, A.: *Plasticity*, McGraw-Hill, New York, 1931, 349 pp. Also *Theory of Flow and Fracture of Solids*, McGraw-Hill, New York, 1950, 567 pp.
232. Houwink, R.: *Elasticity, Plasticity, and Structure of Matter*, Cambridge, New York, 1937, 376 pp.
233. Bridgman, P. W.: *Studies in Large Plastic Flow and Fracture*, McGraw-Hill, New York, 1952, 362 pp.
234. Hill, R.: *Mathematical Theory of Plasticity*, Oxford, New York, 1950, 356 pp.
235. Prager, W.: *Theory of Perfectly Plastic Solids*, Wiley, New York, 1951, 264 pp.
236. Westergaard, H. M.: *Theory of Elasticity and Plasticity*, Harvard University Press, Cambridge, Mass., 1952, 176 pp.
240. Gough, H. J.: "Crystalline Structure in Relation to Failure of Metals—Especially by Fatigue," *Proc. ASTM*, vol. 33, pt. II (1933), pp. 3–114.
241. Bridgman, P. W.: "Considerations on Rupture under Triaxial Stress," *Mech. Eng.*, vol. 61, no. 2 (February, 1939), pp. 107–111.
244. Marin, J.: "Failure Theories of Materials Subjected to Combined Stresses," *Trans. ASCE*, vol. 101 (1936), pp. 1162–1194.
245. Gensamer, M.: *Strength of Metals under Combined Stresses*, American Society for Metals, Cleveland, Ohio, 1941, 106 pp.
260. Norris, C. B.: "The Elastic Theory of Wood Failure," *Trans. ASME*, vol. 61, no. 3 (April, 1939), pp. 259–261.
270. Haushalter, F. L.: "The Mechanical Characteristics of Rubber," *Trans. ASME*, vol. 61, no. 2 (February, 1939), pp. 149–158.

271. Yerzley, F. L.: "Properties of Rubber Revealed by Mechanical Tests," *ASTM Bull.* 99 (August, 1939), pp. 31–34.
280. Taylor, T. S.: "Plastics—Some Applications and Methods of Testing," *Proc. ASTM*, vol. 37, pt. II (1937), pp. 5–21.

SEC. 300. THE PROBLEM OF FAILURE

301. Van Vlack, L. H.: *Elements of Materials Science*, Addison-Wesley, Reading, Mass., 1959, 528 pp.
302. Frankel, J. P.: *Principles of the Properties of Materials*, McGraw-Hill, New York, 1957, 228 pp.
303. Ludwig, P.: "Gleit- und Reissfestigkeit Forschungsarbeit," *Zeit. Ver. Deut. Ingr.* vol. 295 (1927), pp. 56–61. See also *Elementen der Technologischen Mechanik* (1909).
304. Murray, W. M. (ed.): *Fatigue and Fracture of Metals*, Wiley, New York, 1952. See especially paper by M. Gensamer, "General Survey of the Problem of Fatigue and Fracture," p. 1; paper by E. Orowan, "Fundamentals of Brittle Behavior in Metals," p. 139; paper by C. W. MacGregor, "Significance of Transition Temperature in Fatigue."
305. Parker, E. R.: "Status of Ductile Ceramic Research," *ASTM Spec. Tech. Pub.* 283 (1960), p. 52.
306. Davis, H. E., and E. R. Parker: "A Study of the Transition from Shear to Cleavage Fracture in Mild Steel," *Proc. ASTM*, vol. 47 (1947), p. 483.
307. Gensamer, M.: "Fundamentals of Fractures in Metals," in Symposium on Effect of Temperature on the Brittle Behavior of Metals, *ASTM Spec. Tech. Pub.* 158 (1953), p. 164.
308. Parker, E. R.: "Theory of Brittle Fracture and Criteria for Behavior at Low Temperature," *ASTM Spec. Tech. Pub.* 158 (1953), p. 126.
309. Hoyt, S. L.: "Brittle Fracture," *ASTM Spec. Tech. Pub.* 158 (1953), p. 141.
310. Manjoine, M. J.: "Influence of Rate of Strain and Temperature on Yield Stress of Mild Steel," *J. Appl. Mech.*, vol. 11 (1944), p. A211.
311. Lubahn, J. D.: *Plasticity and Creep of Metals*, Wiley, New York, 1961, p. 111.
312. Shanley, F. R.: "Tensile Instability (Necking) of Ductile Materials," *Aerospace Eng.* (December, 1961), p. 30.
313. Popov, E. P.: *Mechanics of Materials*, Prentice-Hall, Englewood Cliffs, N.J., 1952, Chap. 13.
314. Shanley, F. R.: *Strength of Materials*, McGraw-Hill, New York, 1957, Chaps. 24, 25.
315. Timoshenko, S.: *Theory of Elastic Stability*, McGraw-Hill, New York, 1936.
316. Gerard, George: "Compressive and Torsional Buckling of Thin-wall Cylinders in Yield Region," *NACA, TN* 3726 (August, 1956).
317. Bleich, F.: *Buckling Strength of Metal Structures*, McGraw-Hill, New York, 1952.
318. de Vries, Karl: "Strength of Beams as Determined by Lateral Buckling," *Trans. ASCE*, vol. 112 (1947), pp. 1245–1271.

SEC. 400. MEASUREMENTS AND MEASURING APPARATUS

Testing Machines

403. Manufacturers of testing machines and other testing apparatus issue a variety of general and special catalogs, bulletins, and informational publications. Some of the manufacturers are:

a. Baldwin-Lima-Hamilton, Philadelphia 42, Pa.
b. Riehle Testing Machines, East Moline, Ill.
c. Tinius Olsen Testing Machine Co., Willow Grove, Pa.
d. Young Testing Machine Co., Bryn Mawr, Pa.
e. National Forge and Ordnance Co., Irvine, Pa.
f. Steel City Testing Machines, Inc., Detroit 21, Mich.
g. Scott Testers, Inc., Providence, R.I.
h. Testing Machines, Inc., New York 23, N.Y.
i. Wilson Mechanical Instrument Division, New York, 17, N.Y.
j. Sperry Products, Inc., Danbury, Conn.
k. Central Scientific Co., Chicago 13, Ill.
l. Fisher Scientific Co., Pittsburgh 19, Pa.
m. Morehouse Machine Co., York, Pa.
n. Wiedemann Machine Co., King of Prussia, Pa.

See also general texts on materials testing in Sec. 100.

Calibration of Testing Machines

410. Burgess, G. K.: "The Bureau of Standards," *Civil Eng.*, vol. 1, no. 6 (March, 1931), pp. 491–494.
411. Gibbons, C. H.: "Load-weighing and Load-indicating Systems," *ASTM Bull.* 100 (October, 1939), pp. 7–13.
412. Tuckerman, L. B., H. L. Whittemore, and S. N. Petrenko: "A New Deadweight Testing Machine of 100,000 Pounds Capacity," *J. Research Natl. Bur. Standards*, vol. 4, no. 2 (February, 1930), pp. 261–265 (R.P. 147). See also Wilson, B. L., D. R. Tate, and G. Borkowski: "Deadweight Machines of 111,000- and 10,100-lb. Capacities," *Natl. Bur. Standards Circ.* C446 (1946).
413. Moore, H. F.: "The Calibration of Testing Machines," *Proc. IATM First Cong.*, Zurich, 1931, vol. 2, pp. 482–491.
414. Moore, H. F., J. C. Othus, and G. N. Krouse: "Full Load Calibration of a 600,000-lb. Testing Machine," *Proc. ASTM*, vol. 32, pt. II (1932), pp. 778–782.
415. Wilson, B., and C. Johnson: "Calibration of Testing Machines under Dynamic Loading," *J. Research Natl. Bur. Standards*, vol. 19, no. 1 (July, 1937), pp. 41–57 (R.P. 1009).
416. Wilson, B. L., D. R. Tate, and G. Borkowski: "Proving Rings for Calibrating Testing Machines," *Natl. Bur. Standards Circ.* C454 (1946).
417. Aber, W. C., and F. M. Howell, "The Constancy of Calibration of Elastic Calibrating Devices," *Proc. ASTM*, vol. 51 (1951), pp. 1072–1086.

Loading Control of Testing Machines

420. Stang, A. H., and L. R. Sweetman: "Speed Control for Screw-power Testing Machines Driven by Direct Current Motors," *ASTM Bull.* 87 (August, 1937), pp. 15–18.
421. Bernhard, R. K.: "Automatic Speed Control for Tension and Compression Testing Machines," *ASTM Bull.* 106 (October, 1940), pp. 31–34.
422. Bernhard, R. K.: "Influence of the Elastic Constant of Tension Testing Machines," *ASTM Bull.* 88 (October, 1937), pp. 14–15.
423. Ernst, G. C., and L. W. Empey: "Testing Speed Specifications," *ASTM Bull.* 131 (December, 1944), pp. 31–33.
424. "Symposium on Speed of Testing," *Proc. ASTM*, vol. 48 (1948), pp. 1129–1199; see especially paper by L. K. Hyde, "Methods and Equipment for Controlling Speed of Testing," pp. 1191–1199.

Measurement of Strain

434. Pollard, A. F. C.: "The Mechanical Amplification of Small Displacements," *J. Sci. Instr.*, vol. 15, no. 2 (February, 1938), pp. 37–55.
435. Stang, A. H., and L. R. Sweetman: "An Extensometer Comparator," *J. Research Natl. Bur. Standards,* vol. 15, no. 3 (September, 1935), pp. 199–203 (R.P. 822).
440. Vose, R. W.: "Characteristics of the Huggenburger Tensometer," *Proc. ASTM*, vol. 34, pt. II (1934), pp. 862–876.
441. Cuykendall, T. R., and G. Winter: "On the Characteristics of the Huggenburger Strain Gage," *Civil Eng.*, vol. 10, no. 7 (July, 1940), pp. 448–450.
447. Mathews, B. H. C.: "Optical Levers," *J. Sci. Instr.*, vol. 16, no. 4 (April, 1939), pp. 124–125.
450. Clarke, T. W. K.: "A New Surface Extensometer," *J. Sci. Instr.*, vol. 12, no. 3 (March, 1935), pp. 84–91; light-lever principle.
451. Tuckerman, L. B.: "Optical Strain Gages and Extensometers," *Proc. ASTM*, vol. 23, pt. II (1923), pp. 602–610. See also Wilson, B. L., "Characteristics of the Tuckerman Strain Gage," *Proc. ASTM*, vol. 44 (1944), pp. 1017–1026.
452. Meisse, L. A.: "Improvement in the Adaptability of the Tuckerman Strain Gage," *Proc. ASTM*, vol. 37, pt. II (1937), pp. 650–654.
455. Merritt, G. E.: "The Interference Method of Measuring Thermal Expansion," *J. Research Natl. Bur. Standards,* vol. 10, no. 1 (January, 1933), pp. 59–76 (R.P. 515).
456. Saunders, J. B.: "Improved Interferometric Procedure with Application to Expansion Measurements," *J. Research Natl. Bur. Standards,* vol. 23, no. 1 (July, 1939), pp. 179–195 (R.P. 1227).
457. Vose, R. W.: "An Application of the Interferometer Strain Gage in Photoelasticity," *J. Appl. Mech.*, vol. 2, no. 3 (September, 1935), pp. A99–A102; describes a lateral strainometer.
460. Brown, R. L.: "A Level-bubble Strain Gage," *ASTM Bull.* 93 (August, 1938), pp. 16–17.
461. Davidenkoff, N.: "The Vibrating Wire Method of Measuring Deformations," *Proc. ASTM*, vol. 34, pt. II (1934), pp. 847–861.
462. Lee, G. H.: *An Introduction to Experimental Stress Analysis*, Wiley, New York, 1950, 319 pp.
463. Hetenyi, M.: *Handbook for Experimental Stress Analysis*, Wiley, New York, 1950, 1077 pp.
464. Dobie, W. B., and P. C. G. Isaac: *Electric Resistance Strain Gauges,* English Universities Press, Ltd., London, 1948, 110 pp.
465. "Symposium on Characteristics and Applications of Resistance Strain Gages," *Natl. Bur. Standards Circ.* 528 (1954).
466. Perry, C. C., and H. R. Lissner: *The Strain Gage Primer*, McGraw-Hill, New York, 1955.
467. "Symposium on Strain Gages at Elevated Temperature," *ASTM Spec. Tech. Pub.* 230 (1958).
468. Meyer, H. O.: "A Method of Waterproofing Electrical Strain Gages," *Proc. Soc. Exptl. Stress Anal.*, vol. 10, no. 1 (1952), pp. 243–245. Also see *J. Am. Concrete Inst.* (October, 1953), pp. 121–135.
469. *Strainline Gage Instruction Sheet,* Baldwin-Lima-Hamilton Corp., Philadelphia, Pa.
470. Davis, R. E., and R. W. Carlson: "The Electric Strainometer and Its Use in Measuring Internal Strains," *Proc. ASTM*, vol. 32, pt. II (1932), pp. 793–802.
471. Carlson, R. W., and D. Pirtz: "Development of a Device for the Direct Measurement of Compressive Stress," *Am. Concrete Inst.*, November, 1952, pp. 201–215.

472. Bleakney, W. M.: "Compensation of Strain Gages for Vibration and Impact," *J. Research Natl. Bur. Standards*, vol. 18, no. 6 (June, 1937), pp. 723–729 (R.P. 1005).
473. Carson, R. W.: "Measuring Elastic Drift," *Proc. ASTM*, vol. 37, pt. II (1937), pp. 661–674. Bibliography.
474. Dean, Mills, III: *Conductor and Conventional Strain Gages*, Academic Press, New York, 1962, 381 pp.
475. Norton, J. T., and D. Rosenthal: "Recent Contributions to the X-ray Method in the Field of Stress Analysis," *Proc. Soc. Exptl. Stress Anal.*, vol. 5, no. 1 (1947), pp. 71–77. See also Barrett, C. S., *Structure of Metals*, Chap. 14, "Stress Measurement by X-Rays," McGraw-Hill, New York, 1943.
476. Mathar, J.: "Determination of Initial Stresses by Measuring the Deformations around Drilled Holes," *Trans. ASME*, vol. 55 (1933), pp. 249–254.
477. de Forest, A. V., and G. Ellis: "Brittle Lacquers as an Aid to Stress Analysis," *J. Aeronaut. Sci.*, vol. 7, no. 5 (March, 1940), pp. 205–208. See also de Forest, A. V., G. Ellis, and F. B. Stern, Jr.: "Brittle Coatings for Quantitative Strain Measurements," *J. Appl. Mech.*, vol. 64 (December, 1942), pp. A184–A188.
478. Hetenyi, M., and W. E. Young: "How Brittle Lacquer Strain Analysis Aids Design," *Machine Design*, vol. 16 (1944).
479. Ellis, G.: "Practical Strain Analysis by Use of Brittle Coatings," *Proc. Soc. Exptl. Stress Anal.*, vol. 1, no. 1 (1943), pp. 46–60.

Gages and Gage Blocks

482. Boston, O. W.: *Metal Processing*, Wiley, New York, 1941, 630 pp. See chapter on measuring instruments and gages.
483. Kent's *Mechanical Engineers' Handbook*, 12th ed., vol. I, Sec. 24 on dimensional control, Wiley, New York, 1950. See also Ref. 1372, Chap. 13.
484. Judge, A. W.: *Engineering Precision Measurements*, Chapman & Hall, London, 1957.
485. Grohe, W.: *Precision Measurement and Gaging Techniques*, Tudor Pub. Co., New York, 1960.
486. Jergens, J. G.: *Gage Design*, Cleveland (1954).

Miscellaneous Instruments and Measurements

491. Franklin, C. H. H.: "Geometrical Measurement," *Eng. Inspection*, vol. 2, no. 2 (Summer, 1946), pp. 9–23.
494. Stang, A. H., R. L. Sweetman, and C. Gough: "The Areas and Tensile Properties of Deformed Concrete-reinforcement Bars," *J. Research Natl. Bur. Standards*, vol. 9, no. 4 (October, 1932), pp. 512–514 (R.P. 486).
496. Eastman, F. S.: "Flexure Pivots to Replace Knife Edges and Ball Bearings," *Univ. Wash. Eng. Expt. Sta. Bull.* 86 (1935). Also "Design of Flexure Pivots," *J. Aeronaut. Sci.*, vol. 5, no. 1 (November, 1937), pp. 16–21.
497. Young, W. E.: "An Investigation of the Cross-spring Pivot," *J. Appl. Mech.*, vol. 11, no. 2 (June, 1944), pp. A113–A120.

SEC. 500. STATIC TENSION AND COMPRESSION TESTS

Tension Tests

501. Lessells, J. M.: "Significance of the Tension Test of Metals in Relation to Design," *Proc. ASTM*, vol. 40 (1940), pp. 501–507. See also *Mech. Eng.*, vol. 62, no. 4 (April, 1940), p. 311.

502. MacGregor, C. W.: "The Tension Test," *Proc. ASTM*, vol. 40 (1940), pp. 508–534. Bibliography.
503. Seely, F. B.: "The Strength Features of the Tension Test," *Proc. ASTM*, vol. 40 (1940), pp. 535–550. Bibliography.
504. Gillett, H. W.: "The Limited Significance of the Ductility Features of the Tension Test," *Proc. ASTM*, vol. 40 (1940), pp. 551–578. Bibliography.
505. Parker, E. R., H. E. Davis, and A. E. Flanigan: "A Study of the Tension Test," *Proc. ASTM*, vol. 46 (1946), pp. 1159–1174.
506. Mitchell, N. B., Jr.: "The Indirect Tension Test for Concrete," *Materials Research and Standards*, vol. 1, no. 10 (October, 1961), pp. 780–788. Also vol. 2, no. 3 (March, 1962), p. 179.
507. Rudnick, A., A. R. Hunter, and F. C. Holden: "An Analysis of the Diametral-Compression Test," *ASTM Materials Research and Standards*, vol. 3, no. 4 (April, 1963), pp. 283–289.
510. Barba, J.: "Résistance à la traction et allongements des métaux après rupture," *Mem. soc. ing. civils France*, 1880, 1re partie, p. 682.
511. Moore, H. F.: "Tension Tests of Steel with Test Specimens of Various Size and Form," *Proc. ASTM*, vol. 18, pt. I (1918), pp. 403–421. Bibliography.
514. Lyse, I., and C. C. Keyser: "Effect of Size and Shape of Test Specimen upon the Observed Physical Properties of Structural Steel," *Proc. ASTM*, vol. 34, pt. II (1934), pp. 202–215.
517. Campbell, H. L.: "Relation of Properties of Cast Iron to Thickness of Castings," *Proc. ASTM*, vol. 37, pt. II (1937), pp. 66–67.
518. "Effect of Size of Casting on Strength of Test Bar," in Report of ASTM Committee A-3, *Proc. ASTM*, vol. 34, pt. I (1934), pp. 148–153.
519. Nadai, A., and C. W. MacGregor: "Concerning the Effect of Notches and Laws of Similitude in Materials Testing," *Proc. ASTM*, vol. 34, pt. II (1934), pp. 216–228.
520. Anderegg, F. O., R. Weller, and B. Fried: "Tension Specimen Shape and Apparent Strength," *Proc. ASTM*, vol. 39 (1939), pp. 1261–1269. See also *Ohio State Univ. Eng. Expt. Stat. Bull.* 106 (1940), pp. 4–9.
521. Frocht, M. M.: "The Behavior of a Brittle Material at Failure," *J. Appl. Mech.*, vol. 3, no. 3 (September, 1936), pp. A99–A103.
522. Frocht, M. M.: "Factors of Stress Concentration Photoelastically Determined," *J. Appl. Mech.*, vol. 2, no. 2 (June, 1935), pp. A67–A68.
526. McAdam, D. J., Jr., G. W. Geil, and R. W. Mebs: "Effect of Combined Stresses on the Mechanical Properties of Steels between Room Temperature and −188°C.," *Proc. ASTM*, vol. 45 (1945), pp. 448–481.
530. Jones, P. G., and H. F. Moore: "An Investigation of the Effect of Rate of Strain on the Results of Tension Tests of Metals," *Proc. ASTM*, vol. 40 (1940), pp. 610–624.
531. Fry, L. H.: "Speed in Tension Testing and Its Influence on Yield Point Values," *Proc. ASTM*, vol. 40 (1940), pp. 625–636. See also *Metal Progr.*, vol. 37, no. 1 (January, 1940), p. 46.
532. Davis, E. A.: "The Effect of the Speed of Stretching and the Rate of Loading on the Yielding of Mild Steel," *J. Appl. Mech.*, vol. 5, no. 4 (December, 1938), pp. A137–A140.
533. "Effect of Speed of Testing," in Reports of ASTM Committee E-1, *Proc. ASTM*, vol. 31, pt. I (1931), p. 599; vol. 32, pt. I (1932), p. 501.
534. "Symposium on Speed of Testing of Non-metallic Materials," *ASTM Spec. Tech. Pub.* 185 (1956).
535. *Symposium on High Speed Testing*, Interscience Publishers, New York, 1958, p. 12.

536. "Symposium of Speed of Testing," *Proc. ASTM*, vol. 48 (1948), pp. 1129–1200.
537. Tatnall, F. G.: "A Summary on Speed of Testing," *ASTM Bull.* 161 (October, 1949), pp. 23–28.
538. Morrison, J. L. M.: "The Influence of Rate of Strain in Tension Tests," *Engineer*, vol. 158, no. 4102 (Aug. 24, 1934), pp. 183–185. Tests at very slow rates.
539. Clark, D. S., and P. E. Duwez: "The Influence of Strain Rate on Some Tensile Properties of Steel," *Proc. ASTM*, vol. 50, p. 560 (1950).
540. Templin, R. L.: "Some Factors Affecting Strain Measurements in Tests of Metals," *Proc. ASTM*, vol. 34, pt. II (1934), pp. 182–201.
544. Smith, C. S.: "Proportional Limit Tests on Copper Alloys," *Proc. ASTM*, vol. 40 (1940), pp. 864–884.
547. "Use of the Tension Test for Judging the Suitability of Sheet Metals for Various Purposes," in Report of ASTM Committee E-1, *Proc. ASTM*, vol. 34, pt. I (1934), pp. 563–570. Bibliography.

Compression Tests

560. Lyse, I., and R. R. Litehiser: "Compressive Strength of Concrete," in *Report on Significance of Tests of Concrete and Concrete Aggregates*, 2d ed., pp. 3–8, American Society for Testing Materials, 1943.
563. Gyengo, T.: "Effect of Type of Test Specimen . . . on Compressive Strength of Concrete," *Proc. Am. Concrete Inst.*, vol. 34 (1938), pp. 269–284.
564. Blanks, R. F., and C. C. McNamara: "Mass Concrete Tests in Large Cylinders," *Proc. Am. Concrete Inst.*, vol. 31 (1935), pp. 280–302, and discussion vol. 32 (1936), pp. 234–262.
565. Gonnerman, H. F.: "Effect of End Condition of Cylinder on Compressive Strength of Concrete," *Proc. ASTM*, vol. 24, pt. II (1924), pp. 1036–1065.
566. Troxell, G. E.: "The Effect of Capping Methods and End Conditions before Capping upon the Compressive Strength of Concrete Test Cylinders," *Proc. ASTM*, vol. 41 (1941), pp. 1038–1052.
567. Liska, J. A.: "Effect of Specimen Shape on the Compressive-strength Properties of Laterally Supported Plywood Specimens," *ASTM Bull.* 133 (March, 1945), pp. 33–36.
570. Kimmich, E. G.: "Rubber in Compression," *ASTM Bull.* 106 (October, 1940), pp. 9–14.
575. Schuyler, M.: "Spherical Bearings," *Proc. ASTM*, vol. 13 (1913), pp. 1004–1018.
581. "Effect of Load Rate on Compressive Strength of Mortar Cubes," in Report of ASTM Committee C-1, *Proc. ASTM*, vol. 34, pt. II (1934), pp. 336–338.
582. Jones, P. G., and F. E. Richart: "The Effect of Testing Speed on Strength and Elastic Properties of Concrete," *Proc. ASTM*, vol. 36, pt. II (1936), pp. 380–392.
591. Endersby, V. A.: "The Mechanics of Granular and Granular Plastic Materials . . . ," *Proc. ASTM*, vol. 40 (1940), pp. 1154–1173.
592. Balmer, G.: "A General Analytic Solution for Mohr's Envelope," *Proc. ASTM*, vol. 52 (1952), pp. 1260–1271.
593. Griffith, J. H.: "Strengths of Brick and Other Structural Silicates," *J. Am. Ceramic Soc.*, vol. 14, no. 5 (May, 1931), pp. 325–355.
594. Popov, E. P.: *Mechanics of Materials*, Prentice-Hall, New York, 1952, 435 pp. See Chap. 8 for explanation of Mohr's stress circle.
596. Wilson, T. R. C.: "Strength-Moisture Relations for Wood," *U.S. Dept. Agr. Tech. Bull.* 282 (March, 1932), 88 pp.
597. Markwardt, L. J., and T. R. C. Wilson: "Strength and Related Properties of

Woods Grown in the United States," *U.S. Dept. Agr. Tech. Bull.* 479 (September, 1935), 99 pp.

SEC. 600. STATIC SHEAR AND BENDING TESTS

Shear Tests

601. "Proposed Method for Torsion Tests to Determine the Mechanical Properties of Metallic Materials under Shearing Stress," in Report of ASTM Committee E-1, *Proc. ASTM*, vol. 25, pt. I (1925), pp. 430–436.
602. Templin, R. L., and R. L. Moore: "Specimens for Torsion Tests of Metals," *Proc. ASTM*, vol. 30, pt. II (1930), pp. 534–543.
603. Sauveur, A.: "The Torsion Test," *Proc. ASTM*, vol. 38, pt. II (1938), pp. 3–20.
604. Seely, F. B., and W. J. Putnam: "Relation between the Elastic Strengths of Steel in Tension, Compression and Shear," *Univ. Ill. Eng. Expt. Sta. Bull.* 115 (1919), 42 pp.
605. Lyse, I., and H. J. Godfrey: "Shearing Properties and Poisson's Ratio of Structural and Alloy Steels," *Proc. ASTM*, vol. 33, pt. II (1933), pp. 274–292. Bibliography.
606. Borgeson, S. E.: "Flexure and Torsion Testing of Copper Wire," *Proc. ASTM*, vol. 36, pt. II (1936), pp. 249–262.
607. Draffin, J. O., and W. L. Collins: "Effect of Size and Type of Specimen on the Torsional Properties of Cast Iron," *Proc. ASTM*, vol. 38, pt. II (1938), pp. 235–248.
608. Draffin, J. O., W. L. Collins, and C. H. Casberg: "Mechanical Properties of Gray Cast Iron in Torsion," *Proc. ASTM*, vol. 40 (1940), pp. 840–848.
609. "Symposium on Shear and Torsion Testing," *ASTM Spec. Tech. Pub.* 289 (1960).
610. Ludewig, J. W.: "Torsional Moduli Variations of Spring Materials with Temperature," *Trans. Am. Soc. Metals*, vol. 22 (1934), p. 833.
611. Emmons, J. V.: "Some Physical Properties of Hardened Tool Steel," *Proc. ASTM*, vol. 31, pt. II (1931), pp. 47–76.
612. Green, O. V., and R. D. Stout: "A Study of the Influence of Speed on the Torsion Impact Test," *Proc. ASTM*, vol. 39 (1939), pp. 1292–1298.
613. Wilson, W. M., and F. P. Thomas: "Fatigue Tests of Riveted Joints," *Univ. Ill. Eng. Expt. Sta. Bull.* 302 (1938), pp. 36–41.
620. Richart, F. E.: "Shearing and Torsional Strengths of Concrete," in *Report on Significance of Tests of Concrete*, pp. 15–17, American Society for Testing Materials, Philadelphia, 1943.
621. Anderson, Paul: "Experiments with Concrete in Torsion," *Trans. ASCE*, vol. 100 (1935), pp. 949–983. See especially discussion by Gilkey.
625. Delmonte, J.: "Shear Strength of Molded Plastic Materials," *ASTM Bull.* 114 (January, 1942), pp. 25–28.
630. "Symposium on Shear Testing of Soils," *Proc. ASTM*, vol. 39 (1939), pp. 999–1122.
631. "Symposium on Direct Shear Testing of Soils," *ASTM Spec. Tech. Pub.* 131, 1952, 96 pp.
632. "Triaxial Testing of Soils and Bituminous Materials," *ASTM Spec. Tech. Pub.* 106 (1950), 303 pp.
633. Taylor, D. W.: *Fundamentals of Soil Mechanics*, Wiley, New York, 1948. See Chaps. 13–15 for analysis of shear strength of soil.
634. *Proceedings of Research Conference on Shear Strength of Cohesive Soils*, Amer. Soc. Civ. Engrs., New York, June, 1960.
See also Refs. 101, 103, 135, 141, 169, 231.

Bending Tests

657. MacKenzie, J. T., and C. K. Donoho: "A Study of the Effect of Span on the Transverse Test Results for Cast Iron," *Proc. ASTM*, vol. 37, pt. II (1937), pp. 71–87.
659. MacKenzie, J. T.: "Tests of Cast Iron Specimens of Various Diameters," *Proc. ASTM*, vol. 31, pt. I (1931), pp. 160–166.
668. Kellerman, W. F.: "Effect of Size of Specimen, Size of Aggregate, and Method of Loading upon the Uniformity of Flexural Strength Tests," *Public Roads*, vol. 13, no. 11 (January, 1933), pp. 177–184.
669. Goldbeck, A. T.: "Tensile and Flexural Strengths of Concrete," in *Significance of Tests of Concrete* . . . , 2d ed., pp. 9–14, American Society for Testing Materials, Philadelphia, 1943.
670. Wright, P. J. F.: "Effect of the Method of Test on the Flexural Strength of Concrete," *Mag. Concrete Research (London)*, no. 11 (October, 1952), pp. 67–76.
681. Beyer, A. H., trans. and abs. of paper by T. Wyss: "Cracking of Reinforcing Bars at Cold Bends," *Civil Eng.*, vol. 1, no. 14 (November, 1931), pp. 1266–1268.
685. Templin, R. L.: "Ductility Testing of Aluminum and Aluminum Alloy Sheet," *Proc. ASTM*, vol. 36, pt. II (1936), pp. 239–248.
686. MacBride, H. L.: "The Stiffness or Flexure Test," *Proc. ASTM*, vol. 37, pt. II (1937), pp. 146–159.
687. LaTour, H., and R. S. Sutton: "Improved Scale System for Stiffness Testing Machine," *ASTM Bull.* 196 (February, 1954), pp. 40–42.
690. Rader, L. F.: "Investigations of the Physical Properties of Asphaltic Mixtures at Low Temperatures," *Proc. ASTM*, vol. 35, pt. II (1935), pp. 559–580. Flexure test of chilled beams used to study cracking tendency in pavements.

See also Refs. 101, 103, 141.

SEC. 700. HARDNESS TESTS

General

701. O'Neill, H.: *The Hardness of Metals and Its Measurement*, Sherwood Press, Cleveland, Ohio, 1935.
702. Williams, S. R.: *Hardness and Hardness Measurements*, American Society for Metals, Cleveland, Ohio, 1942.
703. Lysaght, V. E.: *Indentation Hardness Testing*, Reinhold, New York, 1949.
704. "Symposium on the Significance of the Hardness Test of Metals in Relation to Design," *Proc. ASTM*, vol. 43 (1943), pp. 803–856.
706. Small, L.: *Hardness, Theory and Practice*, Service Diamond Tool Co., Ferndale, Mich., 1960.
713. Templin, R. L.: "The Hardness Testing of Light Metals and Alloys," *Proc. ASTM*, vol. 35, pt. II (1935), pp. 283–304.
715. Oberle, T. L.: "Hardness, Elastic Modulus, Wear of Metals," *Trans. SAE*, July, 1952, pp. 511–515.

See also Refs. 101, 103–106, 141, 143.

Static Indentation Tests

721. Petrenko, S. N., W. Ramberg, and B. Wilson: "Determination of the Brinell Number of Metals," *J. Research Natl. Bur. Standards*, vol. 17, no. 1 (July, 1936), pp. 59–96 (R.P. 903).
723. Heyer, R. H.: "Analysis of the Brinell Hardness Test," *Proc. ASTM*, vol. 37, pt. II (1937), pp. 119–145.

724. Heyer, R. H.: "Hardness Conversion Relationships," *Proc. ASTM,* vol. 42 (1942), pp. 708–726. See also vol. 44 (1944), pp. 1027–1050.
725. Peek, R. L., Jr., and W. E. Ingerson: "Analysis of Rockwell Hardness Data," *Proc. ASTM,* vol. 39 (1939), pp. 1270–1280.
726. Ingerson, W. E.: "Rockwell Hardness of Cylindrical Specimens," *Proc. ASTM,* vol. 39 (1939), pp. 1281–1291.
727. Kenyon, R. L.: "Effect of Thickness on the Accuracy of Rockwell Hardness Tests on Thin Sheets," *Proc. ASTM,* vol 34, pt. II (1934), pp. 229–243.
731. Heyer, R. H., and Y. E. Lysaght: "Survey of Investigations of Specimen Thickness on Rockwell Tests," *ASTM Bull.* 193 (October, 1953), pp. 32–39.
732. Sutton, R. S., and R. H. Heyer: "Correlation of Published Data for Correction of Rockwell Diamond Penetrator Hardness Tests on Cylindrical Specimens," *ASTM Bull.* 193 (October, 1953), pp. 40–41.
740. Knoop, F., C. G. Peters, and W. B. Emerson: "A Sensitive Pyramidal-diamond Tool for Indentation Tests," *Natl. Bur. Standards Research Paper* 1220 (July, 1939). See also *Natl. Bureau Standards Letter Circ.* 819.
741. Taylor, E. W.: "Microhardness Testing of Metals," *J. Inst. Metals,* vol. 74, pt. 10 (1948).

Scratch, Wear, Rebound, Dynamic Hardness Tests

750. "Proposed Method of Test for File Scratch Hardness of Metallic Materials," *ASTM Bull.* 88 (October, 1937), p. 26.
752. Herbert, E. G.: "Work-hardening Properties of Metals," *Trans. ASME,* vol. 48 (1926), pp. 705–745.
753. Shank, J. R.: "A Wear Test for Flooring Materials," *Proc. ASTM,* vol. 35, pt. II (1935), pp. 533–545.

SEC. 800. IMPACT TESTS

General

802. "Symposium on Impact Testing," *ASTM Spec. Tech. Pub.* 176 (1956).
803. "Symposium on Impact Testing," *Proc. ASTM,* vol. 38, pt. II (1938), pp. 21–156.
See also Refs. 101, 103, 106, 135, 612.

Failure Mechanism

810. McAdam, D. J., Jr., and R. W. Clyne: "The Theory of Impact Testing: Influence of Temperature, Velocity of Deformation, and Form and Size of Specimen on Work of Deformation," *Proc. ASTM,* vol. 38, pt. II (1938), pp. 112–132.
811. Jones, P. G.: "On the Transition from a Ductile to a Brittle Type of Fracture in Several Low-alloy Steels," *Proc. ASTM,* vol. 43 (1943), pp. 547–555.
812. Davis, H. E., E. R. Parker, and A. Boodberg: "A Study of the Transition from Shear to Cleavage Fracture in Mild Steel," *Proc. ASTM,* vol. 47 (1947), pp. 483–499.
813. Jones, P. G., and W. J. Worley: "An Experimental Study of the Influence of Various Factors on the Mode of Fracture of Metals," *Proc. ASTM,* vol. 48 (1948), pp. 648–662.
814. Dolan, T. J., and C. S. Yen: "Some Aspects of the Effect of Metallurgical Structure on Fatigue Strength and Notch Sensitivity of Steel," *Proc. ASTM,* vol. 48 (1948), pp. 664–695. See especially pp. 685–688. See also vol. 50 (1950), pp. 587–618.

815. McAdam, D. J., Jr., and R. W. Mebs: "The Technical Cohesive Strength and Other Mechanical Properties of Metals at Low Temperatures," *Proc. ASTM*, vol. 43 (1943), pp. 661–706.
816. Muhlenbruch, C. W.: "Elastic and Fracture Toughness Studies of a Stainless Steel," *Proc. ASTM*, vol. 49 (1949), pp. 738–756.

Velocity of Testing Effects

820. Mann, H. C.: "The Relation between the Tension Static and Dynamic Tests," *Proc. ASTM*, vol. 35, pt. II (1935), pp. 323–340.
821. Mann, H. C.: "High-velocity Tension-impact Tests," *Proc. ASTM*, vol. 36, pt. II (1936), pp. 85–109.
822. MacGregor, C. W., and J. C. Fisher: "Relations between the Notched Beam Impact Test and the Static Tension Test," *J. Appl. Mechanics*, vol. 11 (March, 1944), p. A-28.
823. Duwez, P. E., and D. S. Clark: "An Experimental Study of the Propagation of Plastic Deformation under Conditions of Longitudinal Impact," *Proc. ASTM*, vol. 47 (1947), pp. 502–522. See also discussion by L. H. Donnell and LeVan Griffis, pp. 523–532; Von Kármán, T., "On the Propagation of Plastic Strains in Solids," *Sixth International Congress for Applied Mechanics*, 1946, Paris.
824. Clark, D. S., and D. S. Wood: "The Time Delay for the Initiation of Plastic Deformation at Rapidly Applied Constant Stress," *Proc. ASTM*, vol. 49 (1949), pp. 717–737.
825. Clark, D. S., and P. E. Duwez: "The Influence of Strain Rate on Some Tensile Properties of Steel," *Proc. ASTM*, vol. 50 (1950), pp. 560–575.
826. Hoppmann, W. H., II: "The Velocity Aspect of Tension-impact Testing," *Proc. ASTM*, vol. 47 (1947), pp. 533–544. Description of the guillotine impact testing machine.
827. Johnson, J. E., D. S. Wood, and D. S. Clark: "Delayed Yielding in Annealed Low-carbon Steel under Compression Impact," *Proc. ASTM*, vol. 53 (1953), pp. 755–764.

Temperature Effects

830. "General Summary and Comparison of . . . Results . . . on Study of Low-temperature Impact Testing . . . ," in Report of ASTM Joint Research Committee on Effect of Temperature on Properties of Metals, *Proc. ASTM*, vol. 36, pt. I (1936), pp. 132–142.
831. Moore, H. F., H. B. Wishart, and S. W. Lyon: "Slow-bend and Impact Tests of Notched Bars at Low Temperatures," *Proc. ASTM*, vol. 36, pt. II (1936), pp. 110–117.
832. Armstrong, T. N., and A. P. Gagnebin: "Impact Properties of Some Low Alloy Nickel Steels at Temperature Down to −220 Degrees Fahr.," *Trans. Am. Soc. Metals*, March, 1940.
833. Crafts, W., and J. J. Egan: "Factors Affecting Notched-bar Impact Tests on Steel at Low Temperatures," *Proc. ASTM*, vol. 39 (1939), pp. 659–673.
834. Manjoine, M., and A. Nadai: "High-speed Tension Tests at Elevated Temperatures," *Proc. ASTM*, vol. 40 (1940), pp. 822–839.
835. Gillett, H. W.: "Impact Resistance and Tensile Properties of Metals at Subatmospheric Temperatures," *ASTM Spec. Tech. Pub.* 47 (1941).
836. "Symposium on Effect of Temperature on the Brittle Behavior of Metals, with Particular Reference to Low Temperatures," *ASTM Spec. Tech. Pub.* 158 (1953).

Geometry Effects in Notched-bar Testing

837. Foley, F. B.: "A Comparison of Charpy Keyhole and V-Notch Impact Test Values," *Materials Research and Standards*, ASTM (November, 1962), pp. 917–920.
840. Hoyt, S. L.: "Notch Bar Testing," *Metals & Alloys*, vol. 7 (1936), pp. 5–7, 39–43, 102–106, 140–142.
842. Curll, C. H.: "Subsize Charpy Correlation with Standard Charpy," *Materials Research and Standards*, ASTM (February, 1961), pp. 91–94.
843. Habart, H., and W. J. Herge: "Sub-size Charpy Relationships at Sub-zero Temperatures," *Proc. ASTM*, vol. 39, pt. II (1939), pp. 649–658.
844. Clark, D. S., and D. S. Wood: "The Influence of Specimen Dimension and Shape on the Results of Tension Impact Testing," *Proc. ASTM*, vol. 50 (1950), pp. 577–585.
845. Kahn, N. A., E. A. Imbembo, and F. Ginsberg: "Effect of Variations in Notch Acuity on the Behavior of Steel in the Charpy Notched-bar Test," *Proc. ASTM*, vol. 50 (1950), pp. 619–648. See also *ASTM Bull.* 146 (May, 1947), p. 66.
846. Mann, H. C.: "A Fundamental Study of the Design of Impact Test Specimens," *Proc. ASTM*, vol. 37, pt. II (1937), pp. 102–118.
847. Moser, M.: "Notched-bar Tests," *J. Appl. Mech.*, vol. 1, no. 3 (September, 1932), pp. 105–110.
848. Kahn, N. A., and E. A. Imbembo: "A Study of the Geometry of the Tension Impact Specimen," *Proc. ASTM*, vol. 46 (1946), pp. 1179–1197.
849. Raring, R.: "Load-deflection Relationships in Slow-bend Tests of Charpy V-notch Specimens," *Proc. ASTM*, vol. 52 (1952), pp. 1034–1049.

Torsion Impact Tests

850. Luerssen, G. V., and O. V. Greene: "The Torsion Impact Test," *Proc. ASTM*, vol. 33, pt. II (1933), pp. 315–333.
851. Luerssen, G. V., and O. V. Greene: "The Torsion Impact Properties of Tool Steel," *Trans. Am. Soc. Metals*, vol. 22 (1934), p. 311.
852. Luerssen, G. V., and O. V. Greene: "Interpretation of Torsion Impact Properties of Carbon Tool Steel," *Trans. Am. Soc. Metals*, vol. 23 (1935), p. 861.

Testing Techniques

860. Driscoll, D. E.: "The Charpy Impact Machine and Procedure for Inspection and Testing Charpy V-notch Impact Specimens," *ASTM Bull.* 191 (July, 1953), pp. 60–64.
861. Siemen, S. E.: "Method of Notching Impact Test Specimens," *ASTM Bull.* 139 (March, 1946), p. 45.
862. Speer, J. R.: "Improved Guide for Positioning Impact Specimens," *ASTM Bull.* 139 (March, 1946), p. 46.
863. Vanderbeck, R. W., R. W. Lindsay, H. D. Wilde, W. T. Lankford, and S. C. Snyder: "Effect of Specimen Preparation on Notch Toughness Behavior of Keyhole Charpy Specimens in the Transition Temperature Zone," *ASTM Spec. Tech. Pub.* 158 (1953).
864. Frazier, R. H., J. W. Spretnak, and F. W. Boulger: "Reproducibility of Keyhole Charpy and Tear-test Data on Laboratory Heats of Semi-killed Steel," *ASTM Spec. Tech. Pub.* 158 (1953).
865. "Methods and Definitions for Mechanical Testing of Steel Products," *ASTM Tentative Standard* A370-61T.

Cast Iron

870. ASTM Committee A-3 on Cast Iron, Subcommittee VI, "Report of Subcommittee on Impact Testing of Cast Iron," *Proc. ASTM*, vol. 33, pt. I (1933), pp. 87–129. Also vol. 49 (1949), pp. 100–109.
871. "Methods of Impact Testing of Cast Iron," *ASTM Standard* A327-54.

Plastics

890. Telfair, D., and H. K. Nason: "Impact Testing of Plastics," *Proc. ASTM*, vol. 43 (1943), pp. 1211–1225; vol. 44 (1944), pp. 993–1005. Bibliography.
891. Findley, W. N., and O. E. Hintz, Jr.: "The Relation between Results of Repeated Blow Impact Tests and of Fatigue Tests," *Proc. ASTM*, vol. 43 (1943), pp. 1226–1239.
892. Burns, R.: "A New Impact Machine for Plastics and Insulating Materials," *ASTM Bull.* 195 (January, 1954), pp. 61–62.

SEC. 900. FATIGUE, CREEP, AND LOW - HIGH TEMPERATURE TESTS

Fatigue Tests

901. Moore, H. F., and J. B. Kommers: *Fatigue of Metals*, McGraw-Hill, New York, 1927, 326 pp.
902. *Symposium on Fatigue and Fracture of Metals*, Wiley, New York, 1952.
903. Batelle Memorial Institute: *Prevention of the Failure of Metals under Repeated Stress*, Wiley, New York, 1941, 264 pp.
904. Moore, H. F., T. M. Jasper, and J. B. Kommers: "An Investigation of the Fatigue of Metals," *Univ. Ill. Eng. Expt. Sta. Bulls.* 124 (1921); 136 (1923); 142 (1924).
905. Dolan, T. J., B. J. Lazan, and O. J. Horger: *Fatigue*, Amer. Soc. Metals, 1953.
909. Kommers, J. B.: "The Effect of Overstressing and Understressing on Fatigue," *Proc. ASTM*, vol. 38, pt. II (1938), pp. 249–268.
910. Templin, R. L.: "Fatigue Machines for Testing Structural Units," *Proc. ASTM*, vol. 39 (1939), pp. 711–722.
911. Kenyon, J. N.: "Rotating-wire Arc Fatigue Machine for Testing Small Diameter Wire," *Proc. ASTM*, vol. 35, pt. II (1935), pp. 156–166.
912. Soderberg, C. R.: "Working Stresses," *Trans. ASME*, vol. 55 (1933). APM 55-16.
913. Krouse, G. N.: "High-speed Fatigue Testing Machine and Some Tests of Speed Effect on Endurance Limit," *Proc. ASTM*, vol. 34, pt. II (1934), pp. 156–164.
914. Breunich, T. R.: "Fatigue Testing: Its Machines and Methods," *Product Eng.*, vol. 24, no. 2 (February, 1953), pp. 128–134; and vol. 24, no. 3 (March, 1953), pp. 148–154.
915. Cazaud, R.: *Fatigue of Metals*, Chapman & Hall, London, 1953.
916. McAdam, D. J., and G. W. Geil: "Pitting and Its Effect on the Fatigue Limit of Steels Corroded under Various Conditions," *Proc. ASTM*, vol. 41 (1941), pp. 696–731.
917. McAdam, D. J., and R. W. Clyne: "Influence of Chemically and Mechanically Formed Notches on Fatigue of Metals," *J. Research*, Natl. Bur. Standards, vol. 13, no. 4 (October, 1934), (R.P. 725).
918. Murray, W. M. (ed.): *Fatigue and Fracture of Metals*, Wiley, New York, 1952.
919. *Symposium on Basic Mechanisms of Fatigue*, ASTM, 1958.
920. Horger, O. J., and H. R. Neifert: "Fatigue Strength of Machine Forgings 6 to 7 in. in Diameter," *Proc. ASTM*, vol. 39 (1939), pp. 723–740.

921. Wilson, W. M., and F. P. Thomas: "Fatigue Tests of Riveted Joints," *Civil Eng.*, vol. 8, no. 8 (August, 1938), pp. 513–516.
922. Wilson, W. M., *et al.*: "Fatigue Tests of Welded Joints in Structural Steel Plates," *Univ. Ill. Eng. Expt. Sta. Bull.* 327 (1941).
923. Baron, F., and E. W. Larson, Jr.: *Static and Fatigue Tests of Riveted and Bolted Joints Having Different Lengths of Grip,* Department of Civil Engineering, Northwestern University, with cooperation of Research Council on Riveted and Bolted Structural Joints of Engineering Foundation, Project V, January, 1952, 90 pp.
924. Baron, F., and E. W. Larson, Jr.: *Static and Fatigue Properties of High-strength Low-alloy Steel Plates . . . ,* Department of Civil Engineering, Northwestern University, with cooperation of Research Council on Riveted and Bolted Structural Joints of Engineering Foundation, Project VI, January, 1952, 78 pp.
925. Sines, G. H.: *Metal Fatigue,* McGraw-Hill, New York, 1959.
926. Oberg, T. T., and J. B. Johnson: "Fatigue Properties of Metals Used in Aircraft Construction at 3450 and 10,600 Cycles," *Proc. ASTM,* vol. 37, pt. II (1937), pp. 195–205.
927. Boone, W. D., and H. B. Wishart: "High-speed Fatigue Tests of Several Ferrous and Non-ferrous Metals at Low Temperatures, *Proc. ASTM,* vol. 35, pt. II (1935), pp. 147–155.
928. Templin, R. L.: "The Fatigue Properties of Light Metals and Alloys," *Proc. ASTM,* vol. 33, pt. II (1933), pp. 364–386.
929. Pope, J. A.: *Metal Fatigue,* Chapman & Hall, London, 1959.
930. Lessells, J. M., and W. M. Murray: "The Effect of Shot Blasting and Its Bearing on Fatigue," *Proc. ASTM,* vol. 41 (1941), pp. 659–673.
932. Grover, H. J., S. A. Gordon, and L. R. Jackson: *Fatigue of Metals and Structures,* Bureau of Naval Weapons, Dept. of the Navy, Washington, 1960.
933. "Symposium on Fatigue of Aircraft Structures," *ASTM Spec. Tech. Pub.* 203 (1957). Also *ASTM Spec. Tech. Pub.* 274 (1960).
934. Frankel, H. E., J. A. Bennett, and W. A. Pennington: "Fatigue Properties of High-strength Steels," Natl. Bur. Standards *Tech. News Bull.,* vol. 43, no. 4 (1959), p. 74.
935. Templin, R. L., F. M. Howell, and E. C. Hartmann: "Effect of Grain Direction on the Fatigue Properties of Aluminum Alloys," *Product Engineering,* vol. 21 (1950), p. 126.
936. Gohn, Geo. R., H. F. Hardrath, and R. E. Peterson: "Fatigue of Metals," *Materials Research and Standards,* ASTM, vol. 3, no. 2 (February, 1963), pp. 105–139.
937. Forest, P. G.: *Fatigue of Metals,* Addison-Wesley, Reading, Mass., 1963.
940. Manual on Fatigue Testing, *ASTM Spec. Tech. Pub.* 91, 1949, 86 pp.

Tests at Low Temperatures

941. Seigle, L., and R. M. Brick: "Mechanical Properties of Metals at Low Temperatures," *Trans. Am. Soc. Metals,* vol. 40 (1948), p. 813.
942. "Symposium on Effect of Low Temperatures on Materials," *ASTM Spec. Tech. Pub.* 78 (1948).
943. Teed, P. L.: *The Properties of Metallic Materials at Low Temperatures,* Wiley, New York, 1950.
944. Eldin, A. S., and S. C. Collins: "Fracture and Yield Stress of 1020 Steel at Low Temperature," *J. Applied Physics* (October, 1951).
945. Allen, N. P.: "Recent European Work on the Mechanical Properties of Metals at Low Temperatures," *Natl. Bur. Standards Circ.* 520 (1952).
946. *Symposium on Behavior of Metals at Low Temperatures,* American Society for Metals, Cleveland, Ohio, 1952.

947. "Symposium on Low-temperature Properties of High-strength Aircraft and Missile Materials," *ASTM Spec. Tech. Pub.* 287 (1960).
948. "Physical Properties of Metals and Alloys from Cryogenic to Elevated Temperatures," *ASTM Spec. Tech. Pub.* 296 (1961).
See also Refs. 830–836 inclusive.

Tests at High Temperatures

950. McVetty, P. G.: "Creep of Metals at Elevated Temperatures—The Hyperbolic Sine Relation between Stress and Strain Rate," *Trans. ASME,* vol. 66 (1943), p. 761.
951. Nadai, A.: *Theory of Flow and Fracture of Metals,* McGraw-Hill, New York, 1950.
952. Norton, F. H.: *The Creep of Steel at High Temperatures,* McGraw-Hill, New York, 1929.
953. Clark, C. L., and A. E. White: "Creep Characteristics of Metals," *Trans. Am. Soc. Metals* (December, 1936), pp. 831–868.
954. Roberts, I.: "Prediction of Relaxation of Metals from Creep Data," *Proc. ASTM,* vol. 51 (1951), p. 811.
955. Bridgman, P. W.: *Studies in Large Plastic Flow and Fracture,* McGraw-Hill, New York, 1952.
956. McVetty, P. G.: "The Interpretation of Creep Tests," *Proc. ASTM,* vol. 34, pt. II (1934), pp. 105–122.
957. Kanter, J. J.: "Interpretation and Use of Creep Results," *Trans. Am. Soc. Metals,* December, 1937, pp. 870–910.
958. Soderberg, C. R.: "The Interpretation of Creep Tests for Machine Design," *Trans. ASME,* vol. 58 (1936), pp. 733.
959. Marin, J.: "A Comparison of the Methods Used for Interpreting Creep Test Data," *Proc. ASTM,* vol. 37, pt. II (1937), pp. 258–264.
960. "Symposium on Creep and Fracture of Metals," *Proc. Natl. Physics Laboratory,* Teddington, England, 1954.
961. McVetty, P. G.: "Working Stresses for High-temperature Service," *Mech. Eng.,* vol. 56, no. 4 (March, 1934), pp. 149–154. Also *Proc. ASTM,* vol. 43 (1943), pp. 707–727.
962. *Seminar on Creep and Recovery of Metals,* American Society for Metals, Cleveland, Ohio, 1956.
963. Laks, H., C. D. Wiseman, O. D. Sherby, and J. E. Dorn: "Effect of Stress on Creep at High Temperature," *J. Appl. Mech.,* vol. 24 (1957), p. 207.
964. "Compilation of Available High-temperature Creep Characteristics of Metals and Alloys," *ASME-ASTM,* 1938.
965. "Symposium on Plasticity and Creep of Metals," *ASTM Spec. Tech. Pub.* 107 (1950), 74 pp.
966. Weaver, S. H.: "Actual Grain Size Related to Creep Strength of Steels at Elevated Temperatures," *Proc. ASTM,* vol. 38, pt. II (1938), pp. 176–181.
967. White, A. E., C. L. Clark, and R. L. Wilson: "Influence of Time on Creep of Steels," *Proc. ASTM,* vol. 35, pt. II (1935), pp. 167–186.
968. Clark, P. H., and E. I. Robinson: "An Automatic Creep Test Furnace Guide," *Metals & Alloys,* vol. 6, no. 2 (February, 1935), pp. 46–49.
969. *Symposium on Materials for Gas Turbines,* ASTM, 1946, 199 pp.
970. Cross, H. C., and J. G. Lowther: "Study of Effects of Variables on the Creep Resistance of Steels," *Proc. ASTM,* vol. 40 (1940), pp. 125–153. Also vol. 44 (1944), pp. 161–185.
971. Simmons, W. F., and H. C. Cross: "Elevated-temperature Properties of Stainless Steels," *ASTM Spec. Tech. Pub.* 124 (1952), 116 pp.
972. Finnie, I., and W. R. Heller: *Creep of Engineering Materials,* McGraw-Hill, New York, 1959.

973. Smith, G. V.: *Properties of Materials at Elevated Temperatures,* McGraw-Hill, New York, 1950.
974. Nadai, A., and M. J. Manjoine: "High Speed Tension Tests at Elevated Temperatures," *Trans. ASME,* vol. 63 (1941).
975. *Symposium on High-temperature Properties of Metals,* American Society for Metals, Cleveland, Ohio, 1951.
976. Simmons, W. F., and H. C. Cross: "The Elevated-temperature Properties of Stainless Steels," *ASTM-ASME Joint Committee Pub.* 124 (January, 1952).
977. Shahinian, P., and J. R. Lane: "The Effect of Specimen Dimensions on High Temperature Mechanical Properties," *Proc. ASTM,* vol. 55 (1955), p. 724.
978. *Symposium on Short-time High-temperature Testing,* American Society for Metals, Cleveland, Ohio, 1957.

Creep of Nonferrous Materials

990. Kennedy, R. R.: "Creep Characteristics of Aluminum Alloys," *Proc. ASTM,* vol. 35, pt. II (1935), pp. 218–232.
991. Boyd, J.: "The Relaxation of Copper at Normal and at Elevated Temperatures," *Proc. ASTM,* vol. 37, pt. II (1937), pp. 218–234.
992. Moore, H. F., B. B. Betty, and C. W. Dollins: "The Creep and Fracture of Lead and Lead Alloys," *Univ. Ill. Eng. Expt. Sta. Bull.* 272 (1935).
993. Davis, R. E., H. E. Davis, and E. H. Brown: "Plastic Flow and Volume Changes of Concrete," *Proc. ASTM,* vol. 37, pt. II (1937), pp. 317–330.
994. Lorman, W. R.: "The Theory of Concrete Creep," *Proc. ASTM,* vol. 40 (1940), pp. 1082–1102.
995. Maney, G. A.: "Concrete under Sustained Working Loads; Evidence That Shrinkage Dominates Time Yield," *Proc. ASTM,* vol. 41 (1941), pp. 1021–1034.

SEC. 1000. NONDESTRUCTIVE TESTS

1001. "Symposium on the Role of Non-destructive Testing in the Economics of Production," *ASTM Spec. Tech. Pub.* 112 (1951), 164 pp.
1002. McMaster, R. C.: *Non-destructive Testing Handbook,* vols. I and II, The Ronald Press, New York, 1959.
1003. Stanford, E. G., and J. H. Fearon: *Progress in Nondestructive Testing,* vol. I and II, Macmillan, New York, 1960.
1004. McGonnagle, W. J.: *Nondestructive Testing,* McGraw-Hill, New York, 1961, 455 pp.
1005. St. John, A., and H. R. Isenburger: *Industrial Radiography,* Wiley, New York, 1934, 223 pp.
1006. Clark, G. L.: *Applied X-rays,* McGraw-Hill, New York, 1940, 674 pp.
1007. *Symposium on Radiography,* ASTM, 1943, 256 pp. Also *ASTM Spec. Tech. Pub.* 96 (1950), 102 pp.
1008. "Symposium on Radioisotopes in Metal Analysis and Testing," *ASTM Spec. Tech. Pub.* 261 (1960).
1009. "Symposium on Applied Radiation and Radioisotopes Test Methods," *ASTM Spec. Tech. Pub.* 268 (1960).
1011. Zucker, M.: "The Technique of Inspecting Wood Poles by X-Ray," *ASTM Bull.* 106 (October, 1940), pp. 19–26.
1012. Lester, H. H.: "Some Aspects of Radiographic Sensitivity in Testing with X-Rays," *ASTM Bull.* 100 (October, 1939), pp. 33–40.
1013. Lester, H. H.: "Radiography in Industry," *ASTM Bull.* 94 (October, 1938), pp. 5–13.
1015. *ASME Boiler and Pressure Vessel Construction Code.*

1016. *General Specifications for Inspection of Material,* Appendix II: "Metals," Part F: "Radiography," U.S. Navy, 1938.

1017. "Review of the Literature of 1939 on the Testing of Materials by Radiographic Methods," *ASTM Bull.* 111 (August, 1941), pp. 31–32.

1018. *Manual of Radiography,* Eastman Kodak Company, Rochester, N.Y., 1939.

1019. "Non-destructive Tests of Welds," *American Welding Society Welding Handbook,* 1950, pp. 952–980.

1026. Zuschlag, T.: "Magnetic Analysis Applied to the Inspection of Bar Stock and Pipe," *ASTM Bull.* 99 (August, 1939), pp. 35–40.

1028. "Symposium on Magnetic Testing," *ASTM Spec. Tech. Pub.* 85 (1949), 220 pp.

1030. Doane, F. B., and C. E. Betz: *Principles of Magnaflux,* Magnaflux Corp., Chicago, 1941, 388 pp.

1032. McCune, C. A.: "Magnaflux Inspection of Railroad Steel Parts," *Proc. Southern and Southwestern Railway Club,* July, 1940.

1033. *Symposium on Magnetic Particle Testing,* ASTM, 1945, 122 pp.

1040. *Symposium on Electronic Methods of Inspection of Metals,* American Society for Metals, Cleveland, Ohio, 1946.

1041. Knerr, H. C.: "Electrical Detection of Flaws in Metal," *Metals & Alloys,* vol. 11, no. 10 (October, 1940).

1042. Keinath, G.: "The Measurement of Thickness," *Natl. Bur. Standards Circ.* 585 (1958).

1050. "Electric Thickness Gage," *Instruments,* vol. 8 (1935), p. 341.

1051. Breuner, A.: "Magnetic Method for Measuring the Thickness of Nickel Coatings on Nonmagnetic Base Metals," *J. Research,* Natl. Bur. Standards, vol. 18 (1937), p. 565 (R.P. 994).

1052. Breuner, A.: "Magnetic Method for Measuring the Thickness of Nonmagnetic Coatings on Iron and Steel," *J. Research,* Natl. Bur. Standards, vol. 20 (1938), p. 357 (R.P. 1081).

1059. *Symposium on Electronic Methods for Inspection of Metals,* American Society for Metals, Cleveland, Ohio, 1947, 189 pp.

1060. "Symposium on Ultrasonic Testing," *ASTM Spec. Tech. Pub.* 101 (1951), 140 pp.

1061. Kesler, C. E., and T. S. Chang: "A Review of Sonic Methods for the Determination of Mechanical Properties of Solid Materials," *ASTM Bull.* (October, 1957), p. 40.

1062. Mason, W. P.: "Motion of a Bar Vibrating in Flexure, Including the Effect of Rotatory and Lateral Inertia," *J. Acoust. Soc. Am.,* vol. 6 (1935), p. 246.

1063. Powers, T. C.: "Measuring Young's Modulus of Elasticity by Means of Sonic Vibrations," *Proc. ASTM,* vol. 38, pt. II (1938), pp. 460–469.

1064. Obert, L.: "Sonic Method of Determining the Modulus of Elasticity of Building Materials under Pressure," *Proc. ASTM,* vol. 39 (1939), pp. 987–998.

1065. Jones, R.: *Non-destructive Testing of Concrete,* Cambridge University Press, 1962, 101 pp.

1066. Hornibrook, F. B.: "Application of Sonic Method to Freezing and Thawing Studies of Concrete," *ASTM Bull.* 101 (December, 1939), pp. 5–8.

1067. Thomsen, W. T.: "Measuring Changes in Physical Properties of Concrete by the Dynamic Method," *Proc. ASTM,* vol. 40 (1940), pp. 1113–1129. Bibliography.

1069. Greene, G. W.: "Test Hammer Provides New Method of Evaluating Hardened Concrete," *Proc. ACI,* vol. 51 (1955), pp. 249–256.

1070. Dunlap, M. E.: *Electrical Moisture Meters for Wood,* Forest Products Laboratory, U.S. Department of Agriculture, February, 1939, 5 pp.

1090. Frocht, M. M.: *Photoelasticity,* vols. I and II, Wiley, New York, 1948.
1091. Coker, E. G., and L. N. G. Filon: *A Treatise on Photoelasticity,* Cambridge University Press, 1931.
1092. Zandman, Felix: "Photostress, Principles and Applications," *Handbook,* Society of Non-Destructive Testing, 1959.

SEC. 1100. ANALYSIS AND PRESENTATION OF DATA

1101. Yule, G. U., and M. G. Kendall: *An Introduction to the Theory of Statistics,* 14th ed., Griffin, London, 1950, 701 pp.
1102. Waugh, A. E.: *Elements of Statistical Method,* 3d ed., McGraw-Hill, New York, 1952, 531 pp.
1103. Arkin, H., and R. R. Colton: *An Outline of Statistical Methods,* 4th ed., Barnes & Noble, New York, 1950, 224 pp.
1105. Simon, L. E.: *An Engineers' Manual of Statistical Methods,* Wiley, New York, 1941, 231 pp.
1110. Worthing, A. G., and J. Geffner: *Treatment of Experimental Data,* Wiley, New York, 1943, 342 pp.
1120. "Symposium on Application of Statistics," *ASTM Spec. Tech. Pub.* 103 (1950), 42 pp.
1130. *Symposium on Usefulness and Limitations of Samples,* ASTM, 1948, 45 pp.
1131. "Symposium on Bulk Sampling," *ASTM Spec. Tech. Pub.* 114 (1952), 72 pp.
1132. Tanner, L., and W. E. Deming: "Some Problems in the Sampling of Bulk Materials," *Proc. ASTM,* vol. 49 (1949), pp. 1181–1186.
1139. "Manual on Quality Control of Materials," *ASTM Spec. Tech. Pub.* 15-c (1957).
1140. Shewhart, W. A.: *Economic Control of Quality of Manufactured Product,* Van Nostrand, New York, 1931, 501 pp.
1142. "Symposium on Statistical Quality Control," *ASTM Spec. Tech. Pub.* 66 (1946), 15 pp.
1150. Sypherd, W. O., and S. Brown: *The Engineer's Manual of English,* Scott, Foresman, Chicago, 1933, 515 pp.
1151. Nelson, J. R.: *Writing the Technical Report,* 3d ed., McGraw-Hill, New York, 1952, 356 pp.
1160. *American Standards Drawings and Drafting Room Practice,* ASA Standard Z14.1-1946, American Standards Association, 1946, 53 pp.

SEC. 1200. INSPECTION

1201. DuPuy, DeG.: "Can the Inspector Be an Umpire," *Eng. News-Record,* vol. 126, no. 15 (Apr. 10, 1941), p. 552.
1202. King, A.: *Engineering Inspection Practice,* Chemical Pub. Co., 1944.
1203. *General Inspection Manual,* Ordnance Department, U.S. Army, March, 1938, 63 pp.
1204. *Instructors to Inspectors of Naval Material,* Naval Inspection Service, U.S. Navy Department.
1205. Freeman, P. J.: *Control of Material and Construction of Steel and Concrete Bridges,* American Institute of Steel Construction, New York, 1934.
1206. Kelly, J. W.: *ACI Manual of Concrete Inspection,* American Concrete Institute, Detroit, Mich., 1957, 240 pp.
1207. *Concrete Pavement Inspector's Manual,* Portland Cement Association, Chicago, 1959, 51 pp.

1208. Thompson, J. E.: *Inspection Organization and Methods,* McGraw-Hill, New York, 1950.
1210. Vogdes, J.: "Control of Materials on a Large Building Operation," *ASTM Bull.* 98 (May, 1939), pp. 31–34.
1211. Murphy, G.: "Sampling of Materials," in *Selected ASTM Standards for Students in Engineering,* pp. 288–293, American Society for Testing Materials, Philadelphia, 1950.
See also Ref. 1340; and references in Section 0.

SEC. 1300. PROPERTIES OF MATERIALS

Metals

1301. Bullens, D. K.: *Steel and Its Heat Treatment,* vol. I: *Principles,* vol. II: *Tools, Processes, Control,* vol. III: *Engineering and Special-purpose Steels,* 5th ed., Wiley, New York, vol. I, 1948, 489 pp; vol. II, 1948, 289 pp; vol. III, 1949, 607 pp.
1302. Sisco, F. T.: *Modern Metallurgy for Engineers,* Pitman, New York, 1948, 499 pp.
1303. Stoughton, B., A. Butts, and A. M. Bounds: *Engineering Metallurgy,* 4th ed., McGraw-Hill, New York, 1953, 479 pp. Chapter on metallurgical inspection and testing.
1304. *Steel Castings Handbook,* Steel Founders' Society of America, Cleveland, Ohio, 1960, 680 pp.
1305. Keyser, C. A.: *Basic Engineering Metallurgy,* Prentice-Hall, Englewood Cliffs, N.J., 1952, 384 pp.
1306. Mondolfo, L. F.: *Engineering Metallurgy,* McGraw-Hill, New York, 1955, 397 pp.
1307. Burton, M. S.: *Applied Metallurgy for Engineers,* McGraw-Hill, New York, 1956.

Wire Rope

1322. de Forest, A. V., and L. W. Hopkins: "The Testing of Rope Wire and Wire Rope," *Proc. ASTM,* vol. 32, pt. II (1932), pp. 398–412.
1323. *Wire Rope and Fittings,* catalogue, John A. Roebling's Sons Co.; see also Ref. 6.

Wood

1330. *Wood Handbook,* Forest Products Laboratory, U.S. Department of Agriculture, 1955, 528 pp.
1332. Desch, H. E.: *Timber, Its Structure and Properties,* Macmillan, New York and London, 1938, 169 pp.
See also Ref. 596.

Concrete

1340. *Concrete Manual,* 7th ed., U.S. Bureau of Reclamation, Denver, 1963, 642 pp.
1341. Troxell, G. E., and H. E. Davis: *Composition and Properties of Concrete,* McGraw-Hill, New York, 1956, 434 pp.
1342. *Design and Control of Concrete Mixtures,* 10th ed., Portland Cement Assn., Chicago, Ill., 1952, 68 pp.
1343. "Symposium on Effect of Water-reducing Admixtures and Set-retarding Admixtures on Properties of Concrete," *ASTM Spec. Tech. Pub.* 266 (1960).
1344. Ferguson, P. M.: *Reinforced Concrete Fundamentals,* Wiley, New York, 1958.
See also Refs. 8, 172, and 1206.

Brick

1350. Lent, L. B.: *Brick Engineering*, vol. I: *Physical Properties of Brick and Brickwork*, Common Brick Mfrs. Assn., Cleveland, Ohio, 1929, 92 pp.
1351. McBurney, J. W., and C. E. Lovewell: "Strength, Water Absorption and Weather Resistance of Building Bricks Produced in the United States," *Proc. ASTM*, vol. 33, pt. II (1933), pp. 636–650.
1352. McBurney, J. W.: "Water Absorption of Building Brick," *Proc. ASTM*, vol. 36, pt. I (1936), pp. 260–271.

Plastics

1360. "Symposium on Plastic Testing and Standardization," *ASTM Spec. Tech. Pub.* 247 (1959).
1361. Delmonte, J.: *Plastics in Engineering*, 3d ed., Penton Publishing Co., Cleveland, Ohio, 1949, 646 pp.
1362. *Modern Plastics Catalog and Directory*, Breskin Publishing Corp., New York. See also *Modern Plastics*, especially vol. 17, no. 2 (October, 1939).
1363. Rahm, L. F.: *Plastic Molding*, McGraw-Hill, New York, 1933, 246 pp.
1364. "Symposium on Plastics," *ASTM Spec. Tech. Pub.* 59 (1944), 200 pp.; see also *ASTM Bull.* 91 (March, 1938), p. 3.
1365. "Symposium on Plastics in Buildings," *Building Research Institute Pub.* 337, National Research Council (1955).
1366. "Symposium on Reinforced Plastics for Rockets and Aircraft," *ASTM Spec. Tech. Pub.* 279 (1960).
1367. "Symposium on Testing Adhesives for Durability and Permanence," *ASTM Spec. Tech. Pub.* 138 (1953).
1368. *Symposium on Adhesion and Adhesives, Fundamentals and Practice*, Wiley, New York, 1954.
1369. "Symposium on Adhesion and Adhesives," *ASTM Spec. Tech. Pub.* 271 (1961).
See also Ref. 280.

General

1370. Leighou, R. B.: *Chemistry of Engineering Materials*, 4th ed., McGraw-Hill, New York, 1942, 684 pp.
1371. Mantell, C. L. (ed.): *Engineering Materials Handbook*, McGraw-Hill, New York, 1958, 1906 pp.
1372. Clapp, W. H., and D. S. Clark: *Engineering Materials and Processes—Metals and Plastics*, 2d ed., International Textbook, Scranton, Pa., 1949, 526 pp.
See also Refs. 131–135, 140–143.

I. USEFUL TABLES

***Table* I.1. Cross-sectional Areas of Round Bars**
(*D* = diameter; in.; *A* = area, sq in.)

D	*A*	*D*	*A*	*D*	*A*	*D*	*A*	*D*	*A*	*D*	*A*	*D*	*A*	*D*	*A*
0.200	0.0314	0.250	0.0491	0.350	0.0962	0.450	0.1590	0.500	0.1964	0.700	0.3848	0.750	0.4418	0.950	0.7088
0.201	0.0317	0.251	0.0495	0.351	0.0968	0.451	0.1598	0.501	0.1971	0.701	0.3859	0.751	0.4430	0.951	0.7103
0.202	0.0320	0.252	0.0499	0.352	0.0973	0.452	0.1605	0.502	0.1979	0.702	0.3870	0.752	0.4441	0.952	0.7118
0.203	0.0324	0.253	0.0503	0.353	0.0979	0.453	0.1612	0.503	0.1987	0.703	0.3881	0.753	0.4453	0.953	0.7133
0.204	0.0327	0.254	0.0507	0.354	0.0984	0.454	0.1619	0.504	0.1995	0.704	0.3893	0.754	0.4465	0.954	0.7148
0.205	0.0330	0.255	0.0511	0.355	0.0990	0.455	0.1626	0.505	0.2003	0.705	0.3904	0.755	0.4477	0.955	0.7163
0.206	0.0333	0.256	0.0515	0.356	0.0995	0.456	0.1633	0.506	0.2011	0.706	0.3915	0.756	0.4489	0.956	0.7178
0.207	0.0337	0.257	0.0519	0.357	0.1001	0.457	0.1640	0.507	0.2019	0.707	0.3926	0.757	0.4501	0.957	0.7193
0.208	0.0340	0.258	0.0523	0.358	0.1007	0.458	0.1647	0.508	0.2027	0.708	0.3937	0.758	0.4513	0.958	0.7208
0.209	0.0343	0.259	0.0527	0.359	0.1012	0.459	0.1655	0.509	0.2035	0.709	0.3948	0.759	0.4525	0.959	0.7223
0.210	0.0346	0.260	0.0531	0.360	0.1018	0.460	0.1662	0.510	0.2043	0.710	0.3959	0.760	0.4536	0.960	0.7238
0.211	0.0350	0.261	0.0535	0.361	0.1024	0.461	0.1669	0.511	0.2051	0.711	0.3970	0.761	0.4548	0.961	0.7253
0.212	0.0353	0.262	0.0539	0.362	0.1029	0.462	0.1676	0.512	0.2059	0.712	0.3981	0.762	0.4560	0.962	0.7268
0.213	0.0356	0.263	0.0543	0.363	0.1035	0.463	0.1684	0.513	0.2067	0.713	0.3993	0.763	0.4572	0.963	0.7284
0.214	0.0360	0.264	0.0547	0.364	0.1041	0.464	0.1691	0.514	0.2075	0.714	0.4004	0.764	0.4584	0.964	0.7299
0.215	0.0363	0.265	0.0552	0.365	0.1046	0.465	0.1698	0.515	0.2083	0.715	0.4015	0.765	0.4596	0.965	0.7314
0.216	0.0366	0.266	0.0556	0.366	0.1052	0.466	0.1706	0.516	0.2091	0.716	0.4026	0.766	0.4608	0.966	0.7329
0.217	0.0370	0.267	0.0560	0.367	0.1058	0.467	0.1713	0.517	0.2099	0.717	0.4038	0.767	0.4620	0.967	0.7344
0.218	0.0373	0.268	0.0564	0.368	0.1064	0.468	0.1720	0.518	0.2107	0.718	0.4049	0.768	0.4632	0.968	0.7359
0.219	0.0377	0.269	0.0568	0.369	0.1069	0.469	0.1728	0.519	0.2116	0.719	0.4060	0.769	0.4645	0.969	0.7375
0.220	0.0380	0.270	0.0573	0.370	0.1075	0.470	0.1735	0.520	0.2124	0.720	0.0472	0.770	0.4657	0.970	0.7390
0.221	0.0384	0.271	0.0577	0.371	0.1081	0.471	0.1742	0.521	0.2132	0.721	0.4083	0.771	0.4669	0.971	0.7405
0.222	0.0387	0.272	0.0581	0.372	0.1087	0.472	0.1750	0.522	0.2140	0.722	0.4094	0.772	0.4681	0.972	0.7420
0.223	0.0391	0.273	0.0585	0.373	0.1093	0.473	0.1757	0.523	0.2148	0.723	0.4106	0.773	0.4693	0.973	0.7436
0.224	0.0394	0.274	0.0590	0.374	0.1099	0.474	0.1765	0.524	0.2157	0.724	0.4117	0.774	0.4705	0.974	0.7451
0.225	0.0398	0.275	0.0594	0.375	0.1104	0.475	0.1772	0.525	0.2165	0.725	0.4128	0.775	0.4717	0.975	0.7466
0.226	0.0401	0.276	0.0598	0.376	0.1110	0.476	0.1780	0.526	0.2173	0.726	0.4140	0.776	0.4729	0.976	0.7482
0.227	0.0405	0.277	0.0603	0.377	0.1116	0.477	0.1787	0.527	0.2181	0.727	0.4151	0.777	0.4742	0.977	0.7497
0.228	0.0408	0.278	0.0607	0.378	0.1122	0.478	0.1795	0.528	0.2190	0.728	0.4162	0.778	0.4754	0.978	0.7512
0.229	0.0412	0.279	0.0611	0.379	0.1128	0.479	0.1802	0.529	0.2198	0.729	0.4174	0.779	0.4766	0.979	0.7528
0.230	0.0415	0.280	0.0616	0.380	0.1134	0.480	0.1810	0.530	0.2206	0.730	0.4185	0.780	0.4778	0.980	0.7543
0.231	0.0419	0.281	0.0620	0.381	0.1140	0.481	0.1817	0.531	0.2215	0.731	0.4197	0.781	0.4791	0.981	0.7558
0.232	0.0423	0.282	0.0625	0.382	0.1146	0.482	0.1825	0.532	0.2223	0.732	0.4208	0.782	0.4803	0.982	0.7574
0.233	0.0426	0.283	0.0629	0.383	0.1152	0.483	0.1832	0.533	0.2231	0.733	0.4220	0.783	0.4815	0.983	0.7589
0.234	0.0430	0.284	0.0633	0.384	0.1158	0.484	0.1840	0.534	0.2240	0.734	0.4231	0.784	0.4828	0.984	0.7605
0.235	0.0434	0.285	0.0638	0.385	0.1164	0.485	0.1847	0.535	0.2248	0.735	0.4243	0.785	0.4840	0.985	0.7620
0.236	0.0437	0.286	0.0642	0.386	0.1170	0.486	0.1855	0.536	0.2256	0.736	0.4254	0.786	0.4852	0.986	0.7636
0.237	0.0441	0.287	0.0647	0.387	0.1176	0.487	0.1863	0.537	0.2265	0.737	0.4266	0.787	0.4865	0.987	0.7651
0.238	0.0445	0.288	0.0651	0.388	0.1182	0.488	0.1870	0.538	0.2273	0.738	0.4278	0.788	0.4877	0.988	0.7667
0.239	0.0449	0.289	0.0656	0.389	0.1188	0.489	0.1878	0.539	0.2282	0.739	0.4289	0.789	0.4889	0.989	0.7682
0.240	0.0452	0.290	0.0661	0.390	0.1195	0.490	0.1886	0.540	0.2290	0.740	0.4301	0.790	0.4902	0.990	0.7698
0.241	0.0456	0.291	0.0665	0.391	0.1201	0.491	0.1893	0.541	0.2299	0.741	0.4312	0.791	0.4914	0.991	0.7713
0.242	0.0460	0.292	0.0670	0.392	0.1207	0.492	0.1901	0.542	0.2307	0.742	0.4324	0.792	0.4927	0.992	0.7729
0.243	0.0464	0.293	0.0674	0.393	0.1213	0.493	0.1909	0.543	0.2316	0.743	0.4336	0.793	0.4939	0.993	0.7744
0.244	0.0468	0.294	0.0679	0.394	0.1219	0.494	0.1917	0.544	0.2324	0.744	0.4347	0.794	0.4951	0.994	0.7760
0.245	0.0471	0.295	0.0683	0.395	0.1225	0.495	0.1924	0.545	0.2333	0.745	0.4359	0.795	0.4964	0.995	0.7776
0.246	0.0475	0.296	0.0688	0.396	0.1232	0.496	0.1932	0.546	0.2341	0.746	0.4371	0.796	0.4976	0.996	0.7791
0.247	0.0479	0.297	0.0693	0.397	0.1238	0.497	0.1940	0.547	0.2350	0.747	0.4383	0.797	0.4989	0.997	0.7807
0.248	0.0483	0.298	0.0697	0.398	0.1244	0.498	0.1948	0.548	0.2359	0.748	0.4394	0.798	0.5001	0.998	0.7823
0.249	0.0487	0.299	0.0702	0.399	0.1250	0.499	0.1956	0.549	0.2367	0.749	0.4406	0.799	0.5014	0.999	0.7838

***Table* I.2. Percentage reduction of area for nominal 0.505-in. bars**

Final diam, in.	Original diameter, in.																				
	.495	.496	.497	.498	.499	.500	.501	.502	.503	.504	.505	.506	.507	.508	.509	.510	.511	.512	.513	.514	.515
0.300	63.2	63.5	63.6	63.8	63.9	64.0	64.1	64.4	64.5	64.5	64.7	64.9	65.1	65.2	65.2	65.3	65.6	65.9	65.9	66.0	66.0
0.302	62.8	63.0	63.1	63.4	63.4	63.5	63.6	64.0	64.0	64.0	64.2	64.5	64.7	64.6	64.8	64.8	65.1	65.5	65.5	65.5	65.5
0.304	62.2	62.4	62.6	62.9	63.0	63.0	63.1	63.5	63.5	63.5	63.8	64.0	64.3	64.1	64.4	64.3	64.6	65.0	65.0	65.0	65.0
0.306	61.7	61.9	62.1	62.4	62.5	62.5	62.6	63.0	63.0	63.1	63.3	63.5	63.8	63.6	64.0	63.8	64.2	64.5	64.5	64.5	64.6
0.308	61.2	61.4	61.6	61.9	62.0	62.0	62.1	62.5	62.5	62.6	62.8	63.0	63.3	63.2	63.5	63.4	63.7	64.0	64.0	64.0	64.1
0.310	60.7	60.9	61.1	61.4	61.5	61.5	61.6	62.0	62.0	62.1	62.3	62.5	62.8	62.8	63.0	63.0	63.2	63.5	63.5	63.6	63.6
0.312	60.2	60.4	60.6	60.9	61.0	61.0	61.1	61.5	61.6	61.6	61.8	62.0	62.3	62.3	62.5	62.5	62.7	63.0	63.0	63.1	63.2
0.314	59.8	60.0	60.1	60.4	60.5	60.5	60.6	61.0	61.1	61.2	61.4	61.5	61.8	61.9	62.0	62.0	62.2	62.5	62.5	62.7	62.8
0.316	59.3	59.4	59.5	59.8	60.0	60.0	60.1	60.4	60.6	60.7	60.9	61.0	61.3	61.4	61.5	61.5	61.8	62.0	62.0	62.3	62.3
0.318	58.8	58.9	59.0	59.3	59.5	59.5	59.6	59.9	60.1	60.2	60.4	60.5	60.8	60.9	61.0	61.0	61.3	61.5	61.6	61.8	61.8
0.320	58.3	58.4	58.5	58.8	59.0	59.0	59.1	59.4	59.6	59.8	60.0	60.0	60.3	60.3	60.5	60.5	60.8	61.0	61.1	61.3	61.3
0.322	57.8	57.9	58.0	58.3	58.5	58.5	58.6	58.9	59.1	59.3	59.4	59.5	59.8	59.8	60.0	60.0	60.3	60.5	60.6	60.8	60.8
0.324	57.3	57.4	57.5	57.8	58.0	58.0	58.1	58.4	58.6	58.8	58.9	59.0	59.3	59.4	59.5	59.5	59.8	60.0	60.1	60.3	60.3
0.326	56.5	56.7	57.0	57.3	57.5	57.5	57.6	57.9	58.1	58.1	58.3	58.5	58.8	58.8	59.0	59.0	59.3	59.5	59.5	59.7	59.8
0.328	56.0	56.1	56.5	56.8	57.0	57.0	57.1	57.4	57.6	57.6	57.8	58.0	58.3	58.3	58.5	58.5	58.8	59.0	59.1	59.3	59.3
0.330	55.5	55.6	56.0	56.2	56.4	56.4	56.6	56.9	57.1	57.2	57.4	57.5	57.8	57.8	58.0	58.0	58.4	58.5	58.6	58.8	58.8
0.332	55.0	55.1	55.5	55.6	55.8	55.9	56.0	56.4	56.5	56.5	56.8	57.0	57.3	57.2	57.5	57.5	57.8	58.0	58.0	58.3	58.3
0.334	54.4	54.6	55.0	55.1	55.2	55.4	55.5	55.8	56.0	56.0	56.2	56.5	56.8	56.7	57.0	57.0	57.4	57.5	57.6	57.8	57.9
0.336	53.8	54.1	54.4	54.5	54.6	54.9	55.0	55.3	55.4	55.5	55.7	55.9	56.2	56.2	56.4	56.5	56.9	57.0	57.1	57.2	57.4
0.338	53.2	53.5	53.8	54.0	54.1	54.3	54.5	54.8	54.9	55.0	55.2	55.5	55.8	55.7	55.9	56.0	56.4	56.5	56.6	56.7	56.9
0.340	52.7	53.0	53.2	53.5	53.5	53.7	54.0	54.2	54.4	54.5	54.8	54.9	55.1	55.2	55.4	55.5	55.9	56.0	56.0	56.2	56.4
0.342	52.2	52.5	52.6	52.9	53.0	53.2	53.5	53.6	53.7	54.0	54.2	54.4	54.6	54.5	54.9	55.0	55.4	55.4	55.5	55.7	55.9
0.344	51.7	52.0	52.1	52.3	52.5	52.7	53.0	53.1	53.2	53.5	53.7	54.0	54.1	54.0	54.4	54.5	54.9	55.0	55.0	55.2	55.4
0.346	51.1	51.4	51.5	51.7	52.0	52.2	52.4	52.5	52.6	53.0	53.2	53.4	53.5	53.5	53.9	53.9	54.3	54.4	54.5	54.7	55.0
0.348	50.6	50.7	51.0	51.1	51.4	51.6	51.8	51.9	52.1	52.4	52.6	52.8	53.0	53.0	53.3	53.4	53.7	53.8	53.9	54.2	54.4
0.350	50.0	50.1	50.4	50.6	50.8	51.0	51.2	51.4	51.6	51.8	52.0	52.1	52.4	52.5	52.8	52.8	53.0	53.3	53.4	53.7	53.9
0.352	49.4	49.6	49.8	50.0	50.2	50.4	50.6	50.9	51.0	51.4	51.5	51.6	51.9	52.0	52.3	52.2	52.5	52.8	52.9	53.1	53.4
0.354	48.8	49.0	49.3	49.5	49.6	49.8	50.0	50.3	50.5	50.7	51.0	51.0	51.4	51.5	51.7	51.7	52.0	52.4	52.5	52.6	52.8
0.356	48.2	48.5	48.7	48.9	49.0	49.2	49.5	49.6	50.0	50.1	50.4	50.5	50.8	50.9	51.1	51.1	51.5	51.8	51.9	52.0	52.2
0.358	47.6	47.8	48.0	48.3	48.4	48.6	48.9	49.1	49.4	49.5	49.8	49.9	50.3	50.3	50.5	50.5	51.0	51.3	51.3	51.5	51.6
0.360	47.0	47.2	47.5	47.7	47.9	48.1	48.3	48.6	48.7	49.0	49.2	49.4	49.7	49.7	50.0	50.0	50.5	50.7	50.7	51.0	51.1
0.362	46.5	46.7	47.0	47.1	47.4	47.5	47.7	48.0	48.1	48.4	48.6	48.9	49.0	49.1	49.5	49.5	50.0	50.1	50.2	50.4	50.6
0.364	45.9	46.1	46.4	46.5	46.8	47.0	47.1	47.4	47.6	47.8	48.0	48.2	48.5	48.5	48.9	48.9	49.4	49.5	49.6	49.9	50.0
0.366	45.2	45.5	45.7	46.0	46.2	46.4	46.5	46.9	47.0	47.2	47.5	47.6	48.0	48.0	48.4	48.3	48.7	49.0	49.1	49.4	49.5
0.368	44.7	44.9	45.1	45.4	45.6	45.7	46.0	46.2	46.4	46.6	46.9	47.0	47.4	47.4	47.7	47.8	48.1	48.4	48.5	48.8	49.0
0.370	44.1	44.3	44.5	44.8	45.0	45.1	45.5	45.6	45.9	46.0	46.3	46.5	46.8	46.9	47.1	47.2	47.6	47.8	48.0	48.2	48.4
0.372	43.1	43.7	44.0	43.2	44.4	44.6	44.8	45.0	45.3	45.5	45.8	46.0	46.2	46.3	46.5	46.6	47.0	47.2	47.4	47.6	47.8
0.374	42.8	43.1	43.4	43.6	43.8	44.0	44.2	44.5	44.7	44.9	45.1	45.4	45.6	45.7	46.0	46.1	46.4	46.6	46.8	47.0	47.2
0.376	42.3	42.5	42.8	43.0	43.2	43.4	43.6	43.9	44.1	44.4	44.5	44.8	45.0	45.2	45.5	45.5	45.9	46.1	46.3	46.5	46.6
0.378	41.6	41.9	42.1	42.4	42.6	42.8	43.0	43.3	43.5	43.8	44.0	44.1	44.5	44.5	44.9	45.0	45.3	45.5	45.7	45.9	46.0
0.380	41.0	41.3	41.5	41.7	42.0	42.2	42.5	42.7	43.0	43.1	43.4	43.6	43.9	44.0	44.3	44.4	44.7	45.0	45.1	45.3	45.5
0.382	40.6	40.7	40.9	41.1	41.4	41.6	41.9	42.0	42.4	42.5	42.8	43.0	43.3	43.4	43.6	43.8	44.1	44.6	44.7	44.8	45.0
0.384	39.8	40.0	40.3	40.5	40.8	41.0	41.2	41.5	41.7	42.0	42.2	42.4	42.7	42.8	43.1	43.2	43.5	43.9	43.9	44.2	44.4
0.386	39.1	39.4	39.7	39.9	40.2	40.4	40.6	40.8	41.1	41.4	41.5	41.8	42.1	42.2	42.5	42.6	43.0	43.2	43.3	43.6	43.8
0.388	38.5	38.8	39.0	39.3	39.6	39.8	40.0	40.2	40.5	40.7	41.0	41.2	41.5	41.6	42.0	42.0	42.4	42.6	42.8	43.0	43.2
0.390	37.9	38.1	38.4	38.6	38.9	39.2	39.4	39.6	39.9	40.0	40.4	40.5	40.8	41.0	41.4	41.4	41.8	42.0	42.2	42.4	42.6
0.392	37.2	37.5	37.8	38.0	38.3	38.5	38.8	39.0	39.3	39.5	39.8	40.0	40.2	40.4	40.7	40.8	41.2	41.4	41.6	41.8	42.0
0.394	36.6	36.8	37.2	37.4	37.7	37.8	38.2	38.4	38.7	38.9	39.2	39.4	39.7	39.8	40.1	40.2	40.6	40.8	41.0	41.2	41.4
0.396	35.9	36.2	36.5	36.8	37.0	37.2	37.5	37.7	38.0	38.2	38.5	38.8	39.0	39.2	39.5	39.6	39.9	40.2	40.4	40.6	40.8
0.398	35.3	35.6	35.9	36.1	36.4	36.6	36.8	37.1	37.4	37.6	37.9	38.2	38.4	38.6	38.9	39.0	39.4	39.6	39.8	40.0	40.2

Table I.2. Percentage reduction of area for nominal 0.505-in. bars (*Continued*)

Final diam, in.	Original diameter, in.																				
	.495	.496	.497	.498	.499	.500	.501	.502	.503	.504	.505	.506	.507	.508	.509	.510	.511	.512	.513	.514	.515
0.400	34.7	34.9	35.2	35.5	35.7	36.0	36.2	36.5	36.8	37.0	37.2	37.5	37.7	37.9	38.3	38.4	38.7	39.0	39.2	39.4	39.6
0.402	34.0	34.3	34.6	34.8	35.1	35.3	35.6	35.9	36.2	36.4	36.6	36.9	37.2	37.4	37.7	37.8	38.1	38.4	38.6	38.8	39.0
0.404	33.3	33.6	33.9	34.2	34.4	34.7	35.0	35.2	35.5	35.7	36.0	36.2	36.5	36.7	37.0	37.2	37.5	37.8	37.9	38.2	38.4
0.406	32.6	33.0	33.2	33.5	33.8	34.0	34.3	34.6	34.8	35.1	35.4	35.6	35.9	36.0	36.4	36.5	36.9	37.1	37.3	37.6	37.8
0.408	32.0	32.3	32.6	32.9	33.2	33.4	33.7	34.0	34.2	34.5	34.8	35.0	35.3	35.5	35.8	35.9	36.3	36.6	36.8	37.0	37.2
0.410	31.4	31.6	32.0	32.3	32.5	32.7	33.0	33.3	33.6	33.8	34.2	34.4	34.6	34.8	35.2	35.2	35.7	35.9	36.1	36.4	36.6
0.412	30.7	31.0	31.3	31.6	31.8	32.1	32.4	32.7	32.9	33.2	33.7	33.8	34.0	34.2	34.5	34.7	35.0	35.3	35.5	35.8	36.0
0.414	30.0	30.4	30.6	30.9	31.2	31.4	31.7	32.0	32.3	32.5	32.9	33.1	33.3	33.6	33.8	34.0	34.4	34.7	34.9	35.2	35.4
0.416	29.4	29.6	30.0	30.2	30.5	30.8	31.0	31.3	31.6	31.9	32.2	32.4	32.7	32.9	33.2	33.4	33.8	34.1	34.2	34.5	34.7
0.418	28.6	29.0	29.3	29.6	29.9	30.1	30.4	30.7	31.0	31.2	31.6	31.8	32.0	32.3	32.6	32.8	33.1	33.4	33.6	33.9	34.1
0.420	28.0	28.3	28.6	28.9	29.2	29.4	29.7	30.0	30.3	30.6	30.8	31.1	31.4	31.6	32.0	32.2	32.5	32.8	33.0	33.2	33.5
0.422	27.3	27.6	27.9	28.2	28.5	28.7	29.0	29.3	29.6	29.9	30.2	30.4	30.7	30.9	31.3	31.4	31.8	32.1	32.3	32.6	32.8
0.424	26.6	26.9	27.2	27.5	27.8	28.1	28.4	28.7	29.0	29.2	29.5	29.8	30.1	30.3	30.6	30.8	31.2	31.5	31.6	32.0	32.2
0.426	25.9	26.2	26.6	26.8	27.2	27.4	27.7	28.0	28.3	28.6	28.9	29.2	29.4	29.6	30.0	30.2	30.6	30.8	31.0	31.3	31.6
0.428	25.2	25.5	25.8	26.1	26.5	26.7	27.0	27.3	27.6	27.8	28.2	28.5	28.8	29.0	29.3	29.5	29.9	30.2	30.4	30.6	30.9
0.430	24.5	24.8	25.2	25.4	25.8	26.0	26.3	26.6	26.9	27.2	27.5	27.8	28.1	28.3	28.7	28.9	29.2	29.5	29.7	30.0	30.3
0.432	23.8	24.1	24.4	24.8	25.1	25.3	25.6	25.9	26.2	26.5	26.8	27.1	27.4	27.6	28.0	28.2	28.6	28.8	29.1	29.4	29.6
0.434	23.1	23.4	23.8	24.1	24.4	24.6	24.9	25.3	25.6	25.9	26.2	26.5	26.8	27.0	27.4	27.5	27.9	28.2	28.4	28.7	29.0
0.436	22.4	22.7	23.0	23.4	23.7	23.9	24.2	24.6	24.9	25.2	25.5	25.8	26.1	26.4	26.7	26.8	27.2	27.6	27.8	28.0	28.3
0.438	21.6	22.0	22.3	22.7	23.0	23.2	23.6	23.8	24.2	24.5	24.8	25.1	25.4	25.6	26.0	26.2	26.5	26.8	27.1	27.4	27.7
0.440	20.9	21.3	21.6	22.0	22.3	22.5	22.8	23.1	23.5	23.8	24.1	24.4	24.7	25.0	25.3	25.5	25.8	26.2	26.4	26.7	27.0
0.442	20.3	20.6	20.9	21.3	21.6	21.8	22.2	22.5	22.8	23.1	23.6	23.8	24.0	24.3	24.6	24.9	25.2	25.5	25.8	26.1	26.3
0.444	19.6	19.9	20.2	20.6	20.9	21.2	21.4	21.8	22.1	22.4	22.7	23.0	23.4	23.6	24.0	24.2	24.5	24.9	25.1	25.4	25.6
0.446	18.8	19.2	19.5	19.8	20.2	20.4	20.8	21.1	21.4	21.7	22.0	22.3	22.7	22.9	23.2	23.6	23.8	24.2	24.4	24.7	25.0
0.448	18.1	18.4	18.8	19.1	19.5	19.7	20.0	20.4	20.7	21.0	21.3	21.6	22.0	22.2	22.6	22.8	23.2	23.5	23.8	24.0	24.3
0.450	17.4	17.7	18.0	18.4	18.7	19.0	19.3	19.7	20.0	20.3	20.6	21.0	21.3	21.6	21.9	22.1	22.5	22.8	23.1	23.4	23.6
0.452	16.6	16.9	17.3	17.6	18.0	18.3	18.6	19.0	19.4	19.6	19.9	20.2	20.5	20.8	21.2	21.4	21.8	22.1	22.4	22.7	23.0
0.454	15.9	16.2	16.6	16.9	17.2	17.6	17.9	18.2	18.6	18.8	19.2	19.5	19.9	20.1	20.4	20.7	21.0	21.4	21.6	22.0	22.2
0.456	15.1	15.5	15.8	16.2	16.5	16.8	17.2	17.5	17.8	18.1	18.4	18.8	19.1	19.4	19.8	20.0	20.4	20.7	21.0	21.3	21.6
0.458	14.4	14.7	15.1	15.4	15.8	16.1	16.4	16.9	17.1	17.4	17.7	18.1	18.4	18.7	19.0	19.3	19.7	20.0	20.3	20.6	20.9
0.460	13.6	14.0	14.3	14.7	15.0	15.3	15.7	16.0	16.4	16.6	17.1	17.4	17.7	18.0	18.3	18.6	19.0	19.3	19.6	19.9	20.2
0.462	12.9	13.2	13.6	13.9	14.3	14.6	15.0	15.3	15.6	16.0	16.3	16.6	17.0	17.3	17.5	17.9	18.3	18.6	18.9	19.2	19.5
0.464	12.1	12.5	12.8	13.2	13.6	13.9	14.2	14.6	14.9	15.2	15.6	15.9	16.3	16.6	16.9	17.2	17.5	17.9	18.2	18.5	18.8
0.466	11.4	11.8	12.1	12.5	12.8	13.1	13.5	13.8	14.2	14.5	14.8	15.2	15.5	15.8	16.2	16.5	16.8	17.2	17.5	17.8	18.1
0.468	10.6	11.0	11.3	11.7	12.0	12.4	12.7	13.1	13.4	13.8	14.1	14.5	14.8	15.1	15.5	15.8	16.1	16.5	16.8	17.1	17.4
0.470	9.9	10.2	10.6	10.9	11.3	11.7	12.0	12.4	12.7	13.1	13.4	13.7	14.1	14.4	14.7	15.1	15.4	15.7	16.0	16.4	16.6
0.472	9.1	9.4	9.8	10.2	10.5	10.9	11.2	11.6	11.9	12.3	12.6	13.0	13.3	13.7	14.0	14.3	14.7	15.0	15.3	15.7	16.0
0.474	8.3	8.7	9.0	9.4	9.8	10.1	10.5	10.9	11.2	11.6	11.9	12.3	12.6	13.0	13.3	13.6	14.0	14.3	14.6	15.0	15.3
0.476	7.5	7.9	8.3	8.6	9.0	9.4	9.7	10.1	10.4	10.8	11.1	11.5	11.9	12.2	12.5	12.9	13.2	13.6	13.9	14.2	14.5
0.478	6.7	7.1	7.5	7.9	8.3	8.6	9.0	9.3	9.7	10.0	10.4	10.8	11.2	11.5	11.8	12.2	12.5	12.9	13.2	13.5	13.8
0.480	6.0	6.3	6.7	7.1	7.5	7.8	8.2	8.6	8.9	9.3	9.7	10.0	10.4	10.7	11.1	11.4	11.8	12.1	12.4	12.8	13.1
0.482	5.2	5.6	5.9	6.3	6.7	7.1	7.4	7.8	8.2	8.5	8.9	9.3	9.6	10.1	10.4	10.7	11.0	11.4	11.7	12.1	12.4
0.484	4.4	4.8	5.2	5.5	5.9	6.3	6.7	7.0	7.4	7.8	8.2	8.5	8.9	9.2	9.6	9.9	10.3	10.7	11.0	11.3	11.7
0.486	3.6	4.0	4.4	4.8	5.1	5.5	5.9	6.3	6.6	7.0	7.4	7.7	8.1	8.5	8.9	9.2	9.5	9.9	10.2	10.6	10.9
0.488	2.8	3.2	3.6	4.0	4.4	4.7	5.1	5.5	5.9	6.2	6.6	7.0	7.4	7.7	8.1	8.4	8.8	9.2	9.5	9.9	10.2
0.490	2.0	2.4	2.8	3.2	3.6	4.0	4.3	4.7	5.1	5.5	5.8	6.2	6.6	7.0	7.4	7.7	8.0	8.4	8.8	9.1	9.4

Table **I.3. Wire and sheet-metal gages**

Name of gage	**U.S. Standard gage***		**The U.S. steel wire gage**	**American or Brown & Sharpe wire gage**	**New Birmingham Standard sheet & hoop gage**	**British Imperial or English legal standard wire gage**
Principal use	**Uncoated steel sheets and light plates**		**Steel wire except music wire**	**Nonferrous sheets and wire**	**Iron and steel sheets and hoops**	**Wire**
Gage No.	**Weight, oz per sq ft**	**Approx. thick-ness, in.**	**Thickness, in.**			
7/0	......	......	0.4900	......	0.6666	0.500
6/0	......	......	0.4615	0.5800	0.6250	0.464
5/0	......	......	0.4305	0.5165	0.5883	0.432
4/0	......	......	0.3938	0.4600	0.5416	0.400
3/0	......	......	0.3625	0.4096	0.5000	0.372
2/0	......	......	0.3310	0.3648	0.4452	0.348
0	......	......	0.3065	0.3249	0.3964	0.324
1	......	......	0.2830	0.2893	0.3532	0.300
2	......	......	0.2625	0.2576	0.3147	0.276
3	160	0.2391	0.2437	0.2294	0.2804	0.252
4	150	0.2242	0.2253	0.2043	0.2500	0.232
5	140	0.2092	0.2070	0.1819	0.2225	0.212
6	130	0.1943	0.1920	0.1620	0.1981	0.192
7	120	0.1793	0.1770	0.1443	0.1764	0.176
8	110	0.1644	0.1620	0.1285	0.1570	0.160
9	100	0.1495	0.1483	0.1144	0.1398	0.144
10	90	0.1345	0.1350	0.1019	0.1250	0.128
11	80	0.1196	0.1205	0.0907	0.1113	0.116
12	70	0.1046	0.1055	0.0808	0.0991	0.104
13	60	0.0897	0.0915	0.0720	0.0882	0.092
14	50	0.0747	0.0800	0.0641	0.0785	0.080
15	45	0.0673	0.0720	0.0571	0.0699	0.072
16	40	0.0598	0.0625	0.0508	0.0625	0.064
17	36	0.0538	0.0540	0.0453	0.0556	0.056
18	32	0.0478	0.0475	0.0403	0.0495	0.048
19	28	0.0418	0.0410	0.0359	0.0440	0.040
20	24	0.0359	0.0348	0.0320	0.0392	0.036
21	22	0.0329	0.0318	0.0285	0.0349	0.032
22	20	0.0299	0.0286	0.0253	0.0313	0.028
23	18	0.0269	0.0258	0.0226	0.0278	0.024
24	16	0.0239	0.0230	0.0201	0.0248	0.022
25	14	0.0209	0.0204	0.0179	0.0220	0.020
26	12	0.0179	0.0181	0.0159	0.0196	0.018
27	11	0.0164	0.0173	0.0142	0.0175	0.0164
28	10	0.0149	0.0162	0.0126	0.0156	0.0148
29	9	0.0135	0.0150	0.0113	0.0139	0.0136
30	8	0.0120	0.0140	0.0100	0.0123	0.0124
31	7	0.0105	0.0132	0.0089	0.0110	0.0116
32	6.5	0.0097	0.0128	0.0080	0.0098	0.0108
33	6	0.0090	0.0118	0.0071	0.0087	0.0100
34	5.5	0.0082	0.0104	0.0063	0.0077	0.0092
35	5	0.0075	0.0095	0.0056	0.0069	0.0084
36	4.5	0.0067	0.0090	0.0050	0.0061	0.0076
37	4.25	0.0064	0.0085	0.0045	0.0054	0.0068
38	4	0.0060	0.0080	0.0040	0.0048	0.0060
39	......	......	0.0075	0.0035	0.0043	0.0052
40	......	......	0.0070	0.0031	0.0039	0.0048

* U.S. Standard gage is officially a weight gage, in ounces per square foot as tabulated.

Table I.4. Summary of useful values

Weights, Measures, Temperatures:

1 in. = 2.540 cm	1 cm = 0.3937 in.
1 lb = 0.4536 kg	1 kg = 2.2046 lb
1°F = 0.5556°C	1°C = 1.80°F

Concrete Technology:

1 sack cement = 1 cu ft = 94 lb = 0.25 bbl
1 bbl cement = 4 cu ft = 376 lb = 4 sacks
1 cu ft water = 62.4 lb = 7.48 gal
1 gal water = 8.33 lb

Comparative water-cement ratios. On various bases:

Volume	Weight	Gal per sack
1.0	0.664	7.48
1.505	1.0	11.25
0.134	0.089	1.0

Example: W/C by volume = 1.505 × W/C by weight

$$\textit{Solid volume} = \frac{\text{weight of given quantity}}{\text{bulk sp gr} \times 62.4}$$

Specific gravity of portland cement. Varies from about 3.05 to 3.20. Assume 3.15 when value is unknown.

Specific gravity of aggregates. For siliceous sands and gravels, use average value = 2.65.

Unit weights of common aggregates.

Material	Moisture condition	Unit weight, pcf	
		Loose	Compact
Sand	Dry Damp	90–100 85–95	95–115
Gravel ¼–1½ in.	Dry or damp	90–97	98–105
Gravel ⅜–¾ in.	Dry or damp	90–94	92–100
Mixed sand and gravel, 1½ in. max	Dry Damp	 100–115	110–125

Free moisture in aggregates. See Table E.2.

Effective absorption of mixing water by air-dry aggregates. For siliceous sands and gravels (excluding sandstones) the effective absorption commonly ranges from about 0.2 to 0.5 percent by weight. See Table E.1 for absorption capacities.

INDEX